Global Energy

Global Energy

Issues, Potentials, and Policy Implications

Edited by
Paul Ekins, Michael Bradshaw,
and Jim Watson

OXFORD
UNIVERSITY PRESS

Great Clarendon Street, Oxford, OX2 6DP,
United Kingdom

Oxford University Press is a department of the University of Oxford.
It furthers the University's objective of excellence in research, scholarship,
and education by publishing worldwide. Oxford is a registered trade mark of
Oxford University Press in the UK and in certain other countries

Published in the United States of America by Oxford University Press
198 Madison Avenue, New York, NY 10016, United States of America

British Library Cataloguing in Publication Data
Data available

Library of Congress Control Number: 2015933880

ISBN 978–0–19–871952–6 (hbk.)
ISBN 978–0–19–871953–3 (pbk.)

Printed and bound by
CPI Group (UK) Ltd, Croydon, CR0 4YY

Links to third party websites are provided by Oxford in good faith and
for information only. Oxford disclaims any responsibility for the materials
contained in any third party website referenced in this work.

■ FOREWORD AND ACKNOWLEDGEMENTS

This book is one of the major outputs of the last five years' work of the UK Energy Research Centre (UKERC). The great majority of the chapters are the result of UKERC projects or special collaborations between UKERC researchers. The book seeks to respond to UKERC's main remit of adopting, exploring, and explaining a 'whole system approach' to the complex issues raised by the supply and demand of energy globally in the twenty-first century.

Our first acknowledgement, therefore, must be to Research Councils UK (RCUK), which funds the interdisciplinary research of UKERC through its Energy Programme, and thereby enabled this book to be produced.

Our second acknowledgement is to our authors and peer reviewers. Each chapter was peer reviewed by two other authors expert in the field, thereby contributing significantly to the quality of this work.

Finally we would like to acknowledge the support staff, Katherine Welch, Alison Parker, Aimee Walker, and Kiran Dhillon at University College London's Institute for Sustainable Resources, and the editorial team at Oxford University Press (OUP), for ensuring that the book came in more or less on time and was produced with OUP's usual efficiency and excellence.

This is not the first book on global energy issues in this century, but it is shorter and we hope, therefore, more accessible than some other notable publications, such as the Global Energy Assessment of 2012 or the annual World Energy Outlook of the International Energy Agency. While it inevitably goes into less detail than these much longer publications, its coverage of the issues is comparable, and it includes one or two new topics, such as the material use of energy systems and the impacts of energy technologies on ecosystems and ecosystem services. We very much hope that this book will help the present and next generation of teachers, students, policy makers, and citizens with an interest in energy issues to develop a clearer understanding of the 'whole system approach' to these issues, so that they may be better able to contribute to the resolution of the urgent, complex and interacting energy system problems which now face humanity.

Paul Ekins, Deputy Director, UKERC, and University College London
Michael Bradshaw, Warwick Business School
Jim Watson, Research Director, UKERC

■ CONTENTS

LIST OF FIGURES   ix
LIST OF TABLES   xv
LIST OF CONTRIBUTORS   xvii

Introduction   1

PART I GLOBAL ENERGY: CONTEXT AND IMPLICATIONS   7

1 **The global energy context**   9
  Jim Skea

2 **Energy systems and innovation**   34
  Jim Watson, Xinxin Wang, and Florian Kern

3 **Deepening globalization: economies, trade, and energy systems**   52
  Gavin Bridge and Michael Bradshaw

4 **The global climate change regime**   73
  Joanna Depledge

5 **The implications of indirect emissions for climate and energy policy**   92
  Katy Roelich, John Barrett, and Anne Owen

6 **Energy production and ecosystem services**   112
  *Robert Holland, Kate Scott, Tina Blaber-Wegg, Nicola Beaumont,
  Eleni Papathanasopoulou, and Pete Smith*

7 **Technical, economic, social, and cultural perspectives on energy demand**   125
  Charlie Wilson, Kathryn B. Janda, Françoise Bartiaux, and Mithra Moezzi

8 **Energy access and development in the twenty-first century**   148
  Xavier Lemaire

PART II GLOBAL ENERGY: OPTIONS AND CHOICES   161

9 **Improving efficiency in buildings: conventional and alternative approaches**   163
  Kathryn B. Janda, Charlie Wilson, Mithra Moezzi, and Françoise Bartiaux

10 **Challenges and options for sustainable travel: mobility, motorization, and
  vehicle technologies**   189
  Hannah Daly, Paul E. Dodds, and Will McDowall

11 **Shipping and aviation**   209
  Antony Evans and Tristan Smith

12 **Carbon capture and storage**   229
Jim Watson and Cameron Jones

13 **Fossil fuels: reserves, cost curves, production, and consumption**   244
Michael Bradshaw, Antony Froggatt, Christophe McGlade, and Jamie Speirs

14 **Unconventional fossil fuels and technological change**   268
Michael Bradshaw, Murtala Chindo, Joseph Dutton, and Kärg Kama

15 **The geopolitical economy of a globalizing gas market**   291
Michael Bradshaw, Joseph Dutton, and Gavin Bridge

16 **Nuclear power after Fukushima: prospects and implications**   306
Markku Lehtonen and Mari Martiskainen

17 **Bioenergy resources**   331
Raphael Slade and Ausilio Bauen

18 **Solar energy: an untapped growing potential?**   354
Chiara Candelise

19 **Water: ocean energy and hydro**   377
Laura Finlay, Henry Jeffrey, Andy MacGillivray, and George Aggidis

20 **Global wind power developments and prospects**   404
Will McDowall and Andrew ZP Smith

21 **Network infrastructure and energy storage for low-carbon energy systems**   426
Paul E. Dodds and Birgit Fais

22 **Metals for the low-carbon energy system**   452
Jamie Speirs and Katy Roelich

23 **Electricity markets and their regulatory systems for a sustainable future**   476
Catherine Mitchell

PART III **GLOBAL ENERGY FUTURES**   497

24 **Global scenarios of greenhouse gas emissions reduction**   499
Christophe McGlade, Olivier Dessens, Gabrial Anandarajah, and Paul Ekins

25 **Energy and ecosystem service impacts**   525
Eleni Papathanasopoulou, Robert Holland, Trudie Dockerty, Kate Scott,
Tina Blaber-Wegg, Nicola Beaumont, Gail Taylor, Gilla Sünnenberg,
Andrew Lovett, Pete Smith, and Melanie Austen

26 **Policies and conclusions**   538
Paul Ekins

AUTHOR INDEX   569
GENERAL INDEX   577

LIST OF FIGURES

1.1	Global primary energy demand by region	12
1.2	Evolution of energy use and GDP per capita 1971–2011	13
1.3	Global primary energy demand by fuel	15
1.4	Energy demand by sector in OECD and non-OECD countries	16
1.5	Proportion of final energy demand met by electricity	17
1.6	Markets for oil products	18
1.7	Supply costs of liquid fuels	20
1.8	Projected US natural gas production	21
1.9	Regional imports of crude oil	22
1.10	Regional imports of natural gas	23
1.11	Regional exports of coal	23
1.12	Oil and coal prices 1971–2012	24
1.13	Regional gas prices 1971–2012	25
1.14	Global energy related CO_2 emissions	26
1.15	Primary energy demand in different global energy scenarios/projections for 2040	29
2.1	Total government spending by IEA member countries (1974–2012)	40
2.2	Government R&D spending by the EU, USA, and Japan (1991–2010)	41
2.3	Government energy R&D spending in the 'BRICS' countries, Mexico, and the USA	42
2.4	Energy technology patent applications filed under the Patent Co-operation Treaty	43
2.5	Global new investment in clean energy by sector 2004–13 ($bn)	45
3.1	Labour productivity and energy use	54
3.2	International convergence in energy intensity	65
5.1	Consumption, production, and territorial greenhouse gas emissions for the UK	95
5.2	Uncertainty associated with UK consumption-based CO_2 Emissions (as calculated using EE-MRIO analysis)	97
5.3	Future GHG emissions showing domestic (production minus exports) and indirect (those associated with the production of goods imported to the UK) emissions for the UK	99
5.4	Lifecycle GHG emissions of electricity generation technologies	100
5.5	Contribution of domestic (UK production minus exports) and indirect (those associated with the production of goods imported to the UK) emissions by sector for the UK 2010	102
5.6	Decomposition of net exports of indirect emissions for selected countries (2009)	104

5.7   Time-series of decomposition of net exports of indirect emissions for the UK   105

6.1   Key findings from case study mapping during phase 1 research; overview of the sugarcane bioethanol supply chain identified as the basis of the case study showing stakeholder groups connected at a high level and at local sites of production and consumption, and likely impacts and equity issues that may be identified during phase 2   119

6.2a Key findings from site of production and processing (impacts affecting local communities and stakeholders in Brazil)   120

6.2b Key findings from site of consumption (impacts mainly felt by consumers in the UK)   121

7.1   Global energy flows (in EJ) from primary to useful energy and end-use sector applications in 2005   126

7.2   UK energy service demand since 1800, measured by final energy inputs   128

7.3   Drivers of UK energy service demand since 1800   129

7.4   Per capita primary energy by service category over time and across different populations   132

9.1   'Micro' space heating Conservation Supply Curve for a hypothetical house   166

9.2   Ladder of citizen participation   179

11.1 Average real round-trip airfares in the domestic United States from 1979 to 2012   213

11.2 Relationship between world GDP (1985–2009) and transport demand for different marine transport commodities   213

11.3 Relationship between global output-side GDP at chained PPPs, against global air passenger and airfreight (including mail) traffic, 1950–2005   214

11.4 Variation of aircraft energy intensity (measured in Mega-Joules per Available Passenger Seat Kilometre) with stage length   216

11.5 Historical and forecast improvements in aircraft energy intensity, 1955–2015   217

11.6 Estimated carbon intensity of newbuild tankers in size categories from build year 1980 to 2010   218

12.1 Project status and geographical spread   231

13.1 Comparison of three reserve and resource classification schemes: the SPE/PRMS, the UN 2009, and the Russian Federation Classification   247

13.2 Long-term gas supply cost curve   248

13.3 Global coal production, 1981–2012 (mtoe)   253

13.4 Global coal consumption, 1965–2012 (mtoe)   254

13.5 Oil production, 1970–2012   258

14.1 The resource triangle   269

14.2 The global distribution of oil sands and bitumen   275

14.3a,b,c Oil sands production processes   277

14.4	Comparing production costs to other energy sources	278
14.5	The shale gas production process	284
15.1	Trends in global gas prices 1995–2012	294
16.1	Number of reactors, operating capacity (1954–2013), and nuclear generation (1990–2012)	309
16.2	Nuclear electricity production in the world (1990–2012)	309
16.3	Average annual nuclear construction times (1954–2013)	315
16.4	Typical cost breakdown of nuclear electricity (O&M: Operating and Maintenance.)	321
17.1	The major bioenergy conversion pathways	333
17.2	Global net electricity generation from biomass and waste	336
17.3	Biogas production in Germany 1992–2013	338
17.4	Global production of bioethanol and biodiesel 2000–11	339
17.5	Estimates for the contribution of energy crops, wastes, and forest biomass to future energy supply	341
17.6	Pre-conditions for increasing levels of biomass production	342
18.1	Technical potential of renewable energy sources global	355
18.2	Shares of energy sources in total global primary energy supply in 2008	355
18.3	Renewable technologies total capacity in operation and produced energy, 2012	357
18.4	Evolution of global PV annual installations, 2000–12 (MW)	363
18.5	PV module price (1968–2000)	365
18.6	PV module retail price index (2003–12, €2012 and $2012)	366
18.7	PV system price across European countries	368
19.1	OES Membership (as of September 2012)	379
19.2	Theoretical [1], Technical [2], and Practical Resource [3] Visualization Diagram	380
19.3	Global potentially significant tidal resource	382
19.4	Distribution of average annual global wave power level (kW per metre)	383
19.5	Likely steps in ocean energy device cost development	384
19.6	Wave energy converter types (Oscillating Water Column [1]; Oscillating Wave Surge Converter [2]; Heaving Buoy [3]; and Attenuator [4])	385
19.7	Tidal Stream Energy Converters—Foundation and Mooring Options (Monopile [1]; Pinned [2]; Gravity [3]; and Buoyant Moored [4])	386
19.8	High level challenges facing the ocean energy sector	387
19.9	Priority topic areas for the areas for the wave and tidal sectors	388
19.10	Levelized cost of energy (€/MWh) for ocean energy, wind, and offshore wind	389
19.11	Cost breakdown	390
19.12	Early array costs	391
19.13	Projected cost reduction in tidal and wave energy	392

20.1   Global cumulative deployment of wind power (both onshore and offshore) 2000–12   405

20.2   Power generated from offshore wind in leading European markets   406

20.3   Offshore wind energy deployment: (a) depths and (b) distances from shore   406

20.4   Offshore wind unit costs per Watt capacity, by (a) depth and (b) by distance from shore   412

22.1   Historical production of ten critical metals from 1971 to 2011   454

22.2   Approaches to estimating future metal demand in the literature and their relative sophistication   455

22.3   Review of different low-carbon vehicle deployment scenarios   457

22.4   A simplified diagram of a generic thin-film layer structure   458

22.5   McKelvey box presenting metal resource classification   461

22.6   The relationship between base metals and their associated by-products found in the same ore deposits, many of which are critical metals   462

22.7   Example of a cumulative availability curve with cumulative resources in tonnes on the x-axis and extraction costs on the y-axis   465

22.8   Relationship between primary metal production and recycling rate   467

22.9   Comparison of historical lithium production, forecast supply, and forecast lithium demand from electric vehicles   469

22.10 Normalized criticality range of eleven materials for low carbon energy technologies found in eleven criticality assessments   471

24.1   The relationship between the rate of annual decrease in emissions for a specific peaking year, level of temperature rise, and the maximum rate of post-peak emissions reduction   504

24.2   Temperature changes in the scenarios from Figure 24.1 and in the three temperature scenarios implemented   506

24.3   Cumulative production of fossil fuels in the three emissions mitigation scenarios generated in this work and comparison with scenarios generated by a variety of other modelling groups   508

24.4   The global primary energy mix in the four scenarios generated in this work   509

24.5   Gas production in the four scenarios (top) and changes in gas and coal production between 2DS and REF over time (bottom)   510

24.6   Contribution of each sector to global emissions in 2DS (top) and 3DS (bottom)   512

24.7   Global electricity generation in the four scenarios (top) and its GHG intensity (bottom)   513

24.8   Per capita GHG emissions in the different economic regions in the four scenarios implemented in this work   514

24.9 CO_2 prices generated in 3DS, 2DS, and 2DS-nobioCCS (top), and ranges of the 2050 CO_2 prices in the 2 °C, 2.5 °C, and 3 °C scenarios presented in Figure 24.1 (bottom)   516

24.10 Percentage increase in total annual energy system cost in 2DS compared with REF in the different economic regions ... 517

24.11 Historical rates of installation of new electricity capacity globally for a selection of technologies ... 518

24.12 Rates of installation of new electricity capacity globally for a selection of technologies in 2DS (top) and 2DS-nobioCCS (bottom) ... 519

25.1 Value of economic activity derived from UK electricity demand for 2007 demonstrating the international reach of demand in the UK ... 527

25.2 Sankey diagram illustrating global activity ($) associated with UK electricity demand ... 528

25.3 Nuclear impacts on global marine ecosystem services ... 531

25.4 Global ecosystem service impacts associated with nuclear, gas, wind, and biomass ... 532

26.1 Actual and projected global per capita electricity consumption (kWh/year) ... 543

26.2 Longitudinal trends in final energy (GJ) versus income (at PPP, in Int 1990 $) per capita for six megacities ... 545

26.3 Atlanta or Barcelona, the range of possible urban futures ... 546

26.4 Additional investment needs in the IEA's 2DS, compared to its 6DS, scenario ... 552

26.5 Total assets by type of institutional investor ... 553

26.6 Carbon intensity of primary energy in mitigation and baseline scenarios, normalized to 1 in 2010 ... 557

26.7 Summary map of existing, emerging, and potential regional, national, and sub-national carbon pricing instruments (ETS and Tax) ... 562

26.8 Domains, policy pillars, and outcomes ... 564

▨ LIST OF TABLES

1.1 Primary energy consumption in selected countries in 2011 (tonnes of oil equivalent per capita) 14

1.2 Fossil fuel reserves, resources, and consumption 19

1.3 Key scenario/projection indicators for 2040 30

1.4 Public Sector Energy RD&D/R&D Spend in IEA Countries 2011 ($bn) 31

4.1 Key milestones in the climate change negotiations 75

4.2 The top 20 aggregate CO_2 emitters from energy consumption in 2011 (compared with their 1990 rankings) and their obligations to 2020 76

5.1 Emissions inventory definitions used in this chapter 94

7.1 Energy conversion globally from resources into useful energy 127

9.1 Factors influencing energy-efficient refurbishment (Denmark, Germany, Netherlands, England) 173

11.1 Transport demand in 2050 214

12.1 Incentives for six CCS demonstrations in Canada, the USA, and UK 240

13.1 Brief descriptions of resource and reserves for oil used in this report 245

13.2 Examples of factors limiting the short-term availability of oil and gas resources presented in a supply cost curve 250

13.3 Categorization of coal 252

13.4 The global oil balance in 2012 258

13.5 The EIA's top ten countries with technically recoverable shale gas resources 261

14.1 'Shale-to-power' and 'shale-to-liquids' discourses of oil shale development 272

14.2 Costs of production using major oil sands recovery methods 278

14.3 Some technologies at various phases of development 280

14.4 Top ten countries with technically recoverable shale gas resources 285

17.1 The Global Bioenergy Partnership's sustainability indicators for bioenergy 349

19.1 Historic wave energy technology examples 385

19.2 Mature tidal energy technology examples 386

20.1 Landmarks of cumulative installed offshore wind capacity 410

22.1 Five critical metals studies and the critical metals that they include 453

22.2 List of ten critical metals and the low-carbon technologies in which they are used 454

22.3 Estimated future demand in 2030 over current production for ten metals included in two critical metals studies 464

22.4 Factors considered in typical metals criticality assessments 470

23.1 Summary of key features in various markets   481

24.1 TIAM-UCL regions, abbreviations, and economic groups to which each
      have been assigned   501

24.2 Changes in the latest year in which global emissions can peak for different
      annual post-peak emissions reductions if overshooting of 2 °C is permitted or not   505

25.1 Lifecycle stages and impacts of nuclear power on global marine ecosystem services   529

26.1 Global mitigation costs in cost-effective scenarios to meet different GHG
      concentrations in 2100   549

26.2 Range of reported one-factor and two-factor learning rates for electric power
      generation technologies   556

■ LIST OF CONTRIBUTORS

George Aggidis is the Director of Lancaster University Renewable Energy Group and Fluid Machinery Group. He has research, design, development, and patent contributions in the field of Fluid Machinery, Energy and Renewable Energy. His main research interests include Energy, Renewable Energy, Fluid Machinery, and Energy Policy.

Gabrial Anandarajah is a lecturer in Energy Systems Modelling. Gabrial also worked for King's College London and Asian Institute of Technology (AIT), Thailand. Gabrial gained his PhD from University College Dublin (2006), Master degree from AIT (2000), and Bachelor degree in Engineering from University of Peradeniya, Sri Lanka (1996).

Melanie Austen leads the Sea and Society area at PML. Her interdisciplinary research includes marine ecosystem services, the social, economic, and health value of their multiple benefits and marine renewable energy. She was Chief Scientific Advisor to the UK Marine Management Organisation (2010–13). She coordinates and leads UK and EU funded marine research.

John Barrett holds a Chair in Sustainability Research at the University of Leeds. His research interests include sustainable consumption and production (SCP) modelling, carbon accounting, and exploring the transition to a low-carbon pathway. John has been an advisor to the UK Government on the development of carbon footprint standards and the future of consumption-based emissions in the UK.

Françoise Bartiaux is a sociologist with a PhD in demography. Her research interests include energy-consuming practices and social interactions using both qualitative and quantitative data, social practice theories, and energy policy directed towards consumers. She has coordinated and participated in several pluridisciplinary research projects, including projects on everyday practices with impacts on the environment, dwelling renovation, and currently energy poverty. She is Professor of environmental sociology and methodology of the social sciences at the Université catholique de Louvain (Belgium), where she also received her PhD.

Ausilio Bauen is a Senior Research Fellow at Imperial College London's Centre for Energy Policy and Technology and a Director of E4tech, a strategic consultancy focused on sustainable energy. His work covers techno-economic, environmental, market, business, and policy aspects of bioenergy systems.

Nicola Beaumont works at the Plymouth Marine Laboratory as an experienced interdisciplinary scientist, specializing in the combination of marine science with environmental economics. Nicola's research is focused around marine ecosystem services, including the quantification and valuation of these services, and translating complex natural science into terms which are meaningful in a social and economic context.

Tina Blaber-Wegg's First Class BSc (Hons) Environmental Sciences degree led to a UKERC-funded PhD project: 'The Social Acceptability of Biofuels: Equity Matters'. This work identifies important relationships between perceptions of injustice and the social acceptability of renewables and advocates systematic inclusion of social and equity issues into energy technologies' sustainability appraisals.

Michael Bradshaw joined Warwick Business as Professor of Global Energy in January 2014, where he teaches a course on their Global Energy MBA entitled Energy in Global Politics. Prior to that he spent thirteen years at the University of Leicester as Professor of Human Geography. He has a PhD in Human Geography from the University of British Columbia, Canada. His research deals with the geopolitical economy of oil and gas, with a particular emphasis on developments in Russia. He has recently completed a project funded by the UK Energy Research Centre (UKERC) that examined the Geopolitical Economy of Global Gas Security and Governance and its implications for the UK. He is also involved in both UK-based and EU-wide research programmes on the social science aspects of shale gas development. In 2014 Polity Press published his book: *Global Energy Dilemmas*. He is currently writing a book on the geopolitics of natural gas.

Gavin Bridge is Professor of Economic Geography at Durham University. He is interested in the political economy and political ecology of extractive industries. His research on mining, oil, and gas has been funded by the US National Science Foundation, National Geographic Society, European Commission, and the UK Energy Research Centre. He is a founder member of the Energy Geographies Working Group of the Royal Geographical Society—Institute of British Geographers.

Chiara Candelise is an experienced energy economist and leading solar energy specialist. Her research interests span from techno-economic assessment of PV technologies to wider economic and policy analysis of energy and climate change issue. She is Research Fellow at Imperial College London and Bocconi University. She has been worked as an economist for several private and public institutions, including the UK Department for Environment, Food and Rural Affairs (Defra).

Murtala Chindo obtained a BSc (Hons) in Geography, an MSc and DIC in Mineral Deposit Evaluation (Mineral Exploration) from Imperial College in 2003, and a PhD in Geography at the University of Leicester in 2012. His PhD thesis explored Nigerian oil sands' potential impact on the environment and nearby communities.

Hannah Daly, a researcher at the UCL Energy Institute, builds, develops, and applies models for issues facing sustainability in the areas of energy and transport. She was an architect of UK TIMES, an energy systems model. Her PhD examined the technological and behavioural drivers of energy and pollution in Irish mobility.

Joanna Depledge is an affiliated lecturer at the Department of Politics and International Studies, Cambridge University (UK). She has published widely on climate change and other environmental issues, including as author of *The Organization of global negotiations: Constructing the climate change regime* and co-author of *The International Climate Change Regime: A Guide to Rules, Institutions and Procedures*. Joanna also worked for several years with the UN Climate Change Secretariat, and as writer/editor for the *Earth Negotiations Bulletin*.

Olivier Dessens is Senior Research Associate at the Energy Institute of University College London with expertise in climate representation within Integrated Assessment Models. Previously he was employed at the Centre for Atmospheric Sciences of the University of Cambridge where he was working on climate change and atmospheric composition.

Trudie Dockerty is an environmental researcher at the University of East Anglia interested in scenarios and stakeholder engagement for evaluating environmental impacts, with work on topics including climate change and rural landscapes, diffuse pollution from agriculture, water quality, expansion of biomass crops, and evaluation of different potential energy futures.

Paul E. Dodds is a Senior Research Associate at the UCL Energy Institute of University College London. He is an energy economist specializing in energy system modelling. Paul's areas of interest include gas networks, the integration of renewables into energy systems, hydrogen systems, and bioenergy systems.

Joseph Dutton is a market reporter for Argus Media, covering the UK gas market and upstream industry. He previously worked on the Global Gas Security Project at the University of Leicester researching the globalization of the gas industry, UK gas supply, and the development of shale gas in the UK and Europe. He also worked as an analyst for upstream oil and gas consultancy Douglas Westwood, and he holds an MA in International Relations and European Studies from the University of Kent.

Paul Ekins has a PhD in economics from the University of London and is Professor of Resources and Environmental Policy and Director of the UCL Institute for Sustainable Resources at University College London. He is also Deputy Director of the UK Energy Research Centre, in charge of its Energy Resources and Vectors theme. Paul Ekins' academic work focuses on the conditions and policies for achieving an environmentally sustainable economy. He is an authority on a number of areas of energy–environment–economy interaction and environmental policy, including energy scenarios, modelling and forecasting, and sustainable energy use. He is the author of numerous papers, book-chapters, and articles in a wide range of journals, and has written or edited twelve books, including *Global Warming and Energy Demand* (Routledge, 1995), *Carbon-Energy Taxation: Lessons from Europe* (Oxford University Press, Oxford, 2009), *Hydrogen Energy: Economic and Social Challenges* (Earthscan, London, 2010), and *Energy 2050: the Transition to a Secure, Low-Carbon Energy System for the UK* (Earthscan, London, 2011).

Antony Evans is a lecturer in Energy and Air Transport at the UCL Energy Institute, and has fifteen years of experience researching air transport. He was previously a research fellow at Cambridge University, MIT and NASA. Antony has two Masters degrees from MIT, and a PhD from Cambridge University.

Birgit Fais is a Research Associate in the Energy Systems Group at the UCL Energy Institute. She holds a PhD from the Institute for Energy Economics and the Rational Use of Energy at the University of Stuttgart focusing on the modelling of policy instruments in energy system models.

Laura Finlay has a background in marine science, and has been working in the field of marine renewable energy for several years. Her research in this field has focused on tidal energy in the past and more recently in policy and innovation of both wave and tidal energy.

Antony Froggatt studied energy and environmental policy at the University of Westminster and the Science Policy Research Unit at Sussex University. He is currently an independent consultant on international energy issues, a Senior Research Fellow at Chatham House and an Associate of the Energy Policy Group at Exeter University.

Robert Holland is an ecologist at the University of Southampton whose work examines the relationship between energy production and ecosystem services. With a particular interest in freshwater systems, Robert's research draws on techniques from the physical and social sciences to examine questions of global policy relevance.

Kathryn B. Janda is an interdisciplinary, problem-based scholar and senior researcher at the Environmental Change Institute at the University of Oxford. Prior to joining the ECI, she worked at Lawrence Berkeley National Laboratory, served as an American Association for the Advancement of Science (AAAS) Environmental Policy Fellow, and taught Environmental Studies at Oberlin College (USA). She holds degrees in energy and resources (PhD and MS), electrical engineering and English literature.

Henry Jeffrey is a specialist in low-carbon roadmaps, action plans, and strategies. He has been instrumental in the development of IEC standards and guidelines for the developing marine renewable industry and has collaborated on the production of marine roadmaps and research strategies for the European commission, Canada, USA, Korea, Taiwan, and Chile.

Cameron Jones works as an analyst for the Ministry of Energy in Alberta, Canada. His primary subject focus is on policy support for electricity distribution, carbon capture and storage (CCS) and micro-generation technology. Cameron holds an MSc in Energy Policy for Sustainability from the University of Sussex, England.

Kärg Kama is a Research and Teaching Fellow in the School of Geography and the Environment and St Anne's College, University of Oxford. She is currently writing a book on the science and politics of oil shale development based on her DPhil research. She has also published on carbon trading and electronic waste management.

Florian Kern is Co-Director of the Sussex Energy Group and Senior Lecturer at SPRU—Science Policy Research Unit at the University of Sussex. His research combines ideas and approaches from innovation and policy studies to investigate innovation processes for low-carbon energy systems, and the governance of sustainability transitions more generally.

Markku Lehtonen is Research Fellow at Science Policy Research Unit (SPRU), Sussex Energy Group, University of Sussex and a visiting researcher at the Groupe de Sociologie Pragmatique et Réflexive (GSPR) at EHESS, Paris. His current research focuses on governance, participation, and public controversies in the area of nuclear energy and radioactive waste management.

Xavier Lemaire is a socio-economist working at the UCL-Energy Institute on clean energy policies in the Global South. His current research projects are on energy transition in African municipalities and on thermal energy access in Africa. Previously, he was working on regulation to promote clean electricity in developing countries.

Andrew Lovett is a Professor of Environmental Sciences at the University of East Anglia, Norwich, UK. His academic specialism is geographical information systems and he is currently involved in a range of research projects concerned with future rural land-use change, ecosystem services, renewable energy systems, and catchment management.

Andy MacGillivray has been actively involved in the field of renewable energy for six years. His research at the University of Edinburgh is focused on energy policy and the risk and uncertainty within the development and deployment of emerging marine renewable energy technologies.

Mari Martiskainen is Research Fellow at the Centre on Innovation and Energy Demand (CIED), based at Science and Policy Research Unit (SPRU), University of Sussex. Her research focuses in the area of transitions to sustainable socio-technical energy systems, including, for example, debates linked to energy technologies such as nuclear power.

Will McDowall is a Senior Research Associate at the UCL Energy Institute and Institute of Sustainable Resources. His research focuses on climate and energy policies, particularly focused on energy innovation policy, and on long-term energy scenarios. He also lectures on energy innovation policy.

Christophe McGlade is an energy systems modeller with extensive experience in using and developing models. He completed his PhD at the UCL Energy Institute and is currently a Research Associate at the UCL Institute for Sustainable Resources. He is lead researcher for the Resources theme of the UK Energy Research Centre.

Catherine Mitchell is Professor of Energy Policy at the University of Exeter. She previously worked for the universities of Sussex, Berkeley, and Warwick. She is currently an established career fellow of the EPSRC and PI on the Innovation and Governance for a Sustainable Economy Project. She has just finished being a lead author in the Policy Chapter of the IPCC's AR5 WG3.

Mithra Moezzi is on the research faculty at Portland State University in Oregon, USA, and is a member of HELIO International, an NGO devoted to energy sustainability. She holds a PhD in Anthropological Folkloristics and a MA in Statistics, both from the University of California Berkeley, and specializes in combining quantitative and qualitative approaches in her research.

Anne Owen is a research fellow at the University of Leeds specializing in the development, construction and application of Multi-regional Input-Output models. In particular, Anne's research aims to develop MRIO and hybrid MRIO construction methodologies and understand the implication that construction assumptions and decisions have on the model outcomes.

Eleni Papathanasopoulou is an economist at the Plymouth Marine Laboratory whose research focuses on how economic activities, including energy technologies, impact the marine environment, as well as how changes in marine ecosystem services impact economic activities. She uses input–output and general equilibrium models to measure whole economic system impacts.

Katy Roelich is a Senior Research Fellow at the University of Leeds. Prior to this she was co-leader of the Rethinking Development theme at the Stockholm Environment Institute and worked in environmental and engineering consulting in the UK and overseas for nine years. Her current research centres on the governance of sustainable transitions.

Kate Scott is a Research Fellow at the University of Leeds whose key areas of research include the theoretical analysis, practical development and application of models to assess the effectiveness of policies aimed at climate change mitigation.

Jim Skea is Research Councils UK Energy Strategy Fellow and Professor of Sustainable Energy at Imperial College. He is a member of the UK Committee on Climate Change and a Vice Chair of the IPCC Working Group III. He was awarded a CBE for services to sustainable energy in 2013.

Raphael Slade is a lecturer in Environmental Sustainability at Imperial College London, where he specializes in resource systems analysis. He has a longstanding interest in renewable energy technology and innovation.

Andrew ZP Smith trained as a mathematician, has worked as a stage manager in a circus, a photovoltaic power-plant designer, and a transport planner; in 2010 he join the UCL Energy Institute, where he is now the Academic Head of the RCUK Centre for Energy Epidemiology.

Pete Smith, FSB, FRSE, is the Professor of Soils and Global Change at the University of Aberdeen (Scotland, UK), Science Director of the Scottish Climate Change Centre of Expertise (Climate XChange), Director of Food Systems for the Scottish Food Security Alliance-Crops, and leads the University's Environment & Food Security Theme.

Tristan Smith is a Lecturer at UCL Energy Institute. He is the Director of the Research Councils UK and industry funded project Shipping in Changing Climates. He leads a research group that maintains a number of techno-economic models including GloTraM (Global Transport Model), which is used by industry to explore shipping's future scenarios and technology evolution.

Jamie Speirs is based in Imperial's Colleges Centre for Energy Policy and Technology (ICEPT), where he conducts research on the social, technical, and economic issues affecting energy policy in the UK, Europe, and globally. His work has included research into global oil depletion, shale gas resources, and the availability of critical metals.

Gilla Sünnenberg (Dipl MSc) is a specialist in geographical information systems (GIS) in the School of Environmental Sciences at the University of East Anglia. Gisela has been involved in a wide variety of research projects concerned with applications of GIS, ranging from renewable energies through to hydrology and ecology.

Gail Taylor is Director of Research for Biological Sciences and co-Chair of the university-wide Energy group at the University of Southampton. She has published over 120 peer-review papers on bioenergy and allied topics on topics that extend from the molecular to landscape scales.

Xinxin Wang is an Insights Analyst at the Energy and Environment division of Haymarket Business Media, responsible for renewable energy market research. Previously, she worked at the UK Energy Research Centre as a Research Associate for 7 years. She received her PhD in Electrical Engineering from Imperial College London in 2008.

Jim Watson is Research Director of the UK Energy Research Centre and Professor of Energy Policy at the University of Sussex. He was Director of the Sussex Energy Group at Sussex from December 2008 to January 2013. He has twenty years of research experience on climate change, energy, and innovation policies. He has advised several UK government departments, and has been a specialist advisor to two House of Commons select committees. He also has extensive international experience, particularly in China. He is a Council Member of the British Institute for Energy Economics, and a member of the advisory boards of several research and policy organisations.

Charlie Wilson is a lecturer in energy and climate change at the Tyndall Centre for Climate Change Research at the University of East Anglia. His research interests lie at the intersection between innovation, behaviour, and policy, in the field of energy and climate change mitigation.

Introduction

Energy, and access to energy, are essential to human life, civilization, and development.

One characteristic of industrial societies, that has both allowed them to evolve into their current form and continues to fuel their activities, is their greatly enhanced use of energy per person, enabled by the discovery of fossil fuels and the development of technologies that enable these fuels to be exploited at an increasing scale and from less accessible locations. Another characteristic has been the greater efficiency with which societies turn energy into economic output as industrialization proceeds.

Fossil fuels are still plentiful in the earth's crust, and they continue to supply the great majority of the world's demand for energy. But they are increasingly associated with problems that are becoming more prominent on the world stage.

The first is emissions. The old industrial societies have already grappled with, and to a considerable extent resolved, the local air pollutants associated with fossil fuel combustion. Fast-growing emerging economies, especially those that burn a lot of coal, are now struggling with the same problems. Less-developed economies without access to modern energy sources also may have indoor air pollution problems from burning biomass for such activities as cooking. The focus of this book is *global* energy issues, so that it does not address directly the issue of local air pollution. But, as has been stressed in both the Global Energy Assessment (GEA, 2012), and the most recent report of the Intergovernmental Panel on Climate Change (Chapter 11 in the WG2 report of IPCC 2014),[1] actions to address the emissions from fossil fuels that have a global impact—primarily emissions of carbon dioxide (CO_2)—can also have a beneficial effect in terms of the reduction of both indoor and outdoor local air pollution.

The world has so far been far less successful in controlling CO_2 emissions than emissions of local air pollutants. This is hardly surprising because of the far more obvious and immediate impacts of these local air pollutants. But, as the science of climate change has firmed up on the reality of anthropogenic climate change, and on the severe risks that unmitigated greenhouse gas emissions would impose on human societies, the world's attention is increasingly turning to the CO_2 emissions from the global energy system that are the major contributor to climate change. How to reduce

---

[1] IPCC, 2014. *Climate Change 2014: Impacts, Adaptation, and Vulnerability. Part A: Global and Sectoral Aspects. Contribution of Working Group II to the Fifth Assessment Report of the Intergovernmental Panel on Climate Change* (Field, C.B., V.R. Barros, D.J. Dokken, K.J. Mach, M.D. Mastrandrea, T.E. Bilir, M. Chatterjee, K.L. Ebi, Y.O. Estrada, R.C. Genova, B. Girma, E.S. Kissel, A.N. Levy, S. MacCracken, P.R. Mastrandrea, and L.L. White (eds)). Cambridge University Press, Cambridge, United Kingdom and New York, NY, USA.

CO_2 emissions, and to a lesser extent emissions of other greenhouse gases such as methane, globally, therefore provides one of the major areas of focus of this book.

If the prominence of the issue of CO_2 emissions from the global energy system is relatively recent, this is not the case with one of the other pressing issues relating to energy use today, namely the desire for energy security. This relates particularly to fossil fuels, of course, historically in respect of oil, and more recently in respect of both oil and gas. As noted above, the issue is not so much one of physical availability in the earth's crust, which seems to have plenty of hydrocarbons for many years of human use. The question is whether the countries that need to import fossil fuels will be able to import them in the quantities and at the time they desire—and to develop and maintain reliable infrastructures that can transport these fuels to final consumers (either in their original form or as electricity). Many factors influence the answer to this question, including the location and mode of occurrence of the fossil fuels, their ease of access, their ownership (state or private company), levels of investment in exploration, supply, and distribution, anticipated levels of demand, and geopolitical considerations.

Another important issue of global concern is that of universal access to modern energy services. At present around 1.4 billion people have no access to electricity, and around 3 billion rely on the traditional, inefficient use of biomass or coal for cooking and heating. The former condition acts as a major constraint on increasing the welfare and economic development of these people, while the latter is associated with major negative health impacts from bad indoor air quality. An essential part of sustainable development is to bring modern energy services to these large numbers of people.

And then there is the issue of the price of energy. While in the light of past experiences it seems foolhardy to seek to predict the price of fossil fuels in the future (in the six months from July 2014 the oil price halved from its then level of around US$100, a level it had more or less maintained since the global recovery from the financial downturn in 2010), there are good reasons for thinking that the oil price at least will not fall back to the levels seen at the end of the last century, when for some months the crude oil price was below US$10 per barrel. Given current and projected levels of demand, and in the absence of either a non-oil technological breakthrough in transport that greatly reduces demand for oil, or a very stringent climate agreement, the marginal future barrel of oil seems most unlikely to maintain for significant periods a price of below US$50, and supply–demand or other market volatilities may push it, and keep it for long periods, well above this. Gas prices, on the other hand, have fallen dramatically in recent years in the USA, with much ongoing speculation as to whether shale gas exploitation will cause the same to happen elsewhere, and whether these or other developments will lead to the formation of a global gas market, with a single price, as opposed to the regional markets, with some indexation to the oil price, seen today. Coal prices, however, at the time of writing, were weak, with substantial new supplies coming on the market, gas displacing coal on price in the USA, and gas being preferred to coal on convenience and environmental grounds elsewhere. In contrast to oil, it is hard to see where upward price pressure on coal is going to come from. One consequence is that coal demand in the power sectors of some European countries has begun to rise again. Whatever the market prices of fossil fuels, carbon pricing, whether through carbon taxes or emissions trading, could serve to favour less carbon-intensive gas over coal and increase the relative competitiveness of low-carbon energy sources compared to all fossil fuels.

The prices of other modern energy sources have tended to be above those of fossil fuels, but these too are changing, with wind and solar photovoltaics in particular experiencing significant price reductions over the past ten years as investment in and deployment of these technologies has ramped up. In some electricity markets, these renewables are now cost-competitive with incumbent fossil fuel generation technologies, but this obviously depends on the quality of the renewable resource, the prices of the fossil fuels with which they are competing, and the nature of the existing electricity infrastructure.

In addressing these and other global energy issues, the purpose of this book is to lay out the broad global energy landscape, exploring how these issues might develop in coming decades and the implications of such developments for energy policy. There are great uncertainties, which will be identified, in respect of some of these issues, but many of the defining characteristics of the landscape are clear, and the energy policies of all countries will need to be broadly consistent with these if they are to be feasible and achieve their objectives.

The book therefore provides information about and analysis of energy and related resources, and the technologies that have been and are being developed to exploit them, that is essential to understanding how the global energy system is developing, and how it might develop in the future. But its main focus is the critical economic, social, political, and cultural issues that will determine how energy systems will develop and which technologies are deployed, why, by whom, and who will benefit from them.

The book has three Parts. Part I sets out the current global context for energy system developments, filling in the brief sketch that has been provided here. Chapter 1 outlines the essential trends of global energy supply and demand, and atmospheric emissions, from the past and going forward, and their driving forces. It sets out the main dimensions of energy policy for all countries, namely energy security, competitiveness and affordability, and environmental considerations. It reviews the synergies and tensions between energy security and climate change, as well as illuminating the macroeconomic and energy challenges engendered by global competition for resources, skills, and technologies. It considers the development of carbon markets and their role, among other things, in technology deployment and transfer. It identifies the most important technologies and institutions that seem likely to shape what energy is delivered, how, when, and to whom, drawing on the global scenarios produced by such bodies as the International Energy Agency (IEA), and drawing attention to the issues, obstacles, and barriers relating to the development and deployment of these technologies. And it discusses public attitudes towards energy and climate change issues, and the public acceptability of different approaches to them. This creates the context for the more detailed discussions of many of these issues and topics in subsequent chapters, which explore them in greater depth.

Chapter 2 looks at how technologies come to be developed and deployed at scale, showing what is sometimes called 'the innovation chain' to be actually an innovation system with multiple simultaneous interactions between the science base, education, engineering and business culture, and political, financial, and commercial structures and institutions. Chapter 3 explores the phenomenon of the deepening globalization of the energy system over time, with the energy system and the globalization of trade and investment being closely intertwined. Greater energy interdependence through markets can be a source of security or vulnerability, at least partly related to energy prices. Another major potential source of vulnerability is climate change. Chapter 4 brings in the dimension of global climate diplomacy and negotiations, with special reference to the UN Framework Convention on Climate Change, and assesses attempts to put in place a global regime that can respond appropriately to the challenge of climate change, both in terms of mitigating future change and adapting to the change that seems already inevitable. It makes clear that the climate issue has already transformed the way energy sources are judged and assessed, but it remains to be seen whether this transformation will be translated into lower-carbon energy sources being deployed on the ground at the kind of scale required to make a real difference to climate outcomes.

Another implication of globalization and trade has been a widening difference for many countries between the carbon emissions from the use of fossil fuels in their territory and those emitted in other countries that derive from the production of the goods that they import. Chapter 5 sets out the essential dimensions of this growing trend and its possible implications for energy policy. Chapter 6 shows that the impact of energy technologies on the environment extends far beyond carbon emissions. All major energy technologies have significant environmental impacts, often in countries distant from where the technologies are actually deployed. These impacts need to be considered when technological choices are being made.

Chapter 7 shifts the focus to energy demand, exploring how this demand is related to cultural practices and aspirations as well as to economic activity. These dimensions of energy demand are complex, and better understanding of them is needed if policies are to persuade people to make use of energy efficiency technologies that are or could be available, or to move towards less energy-intensive lifestyles. Chapter 8 brings this part to a close by considering issues of how to bring access to modern energy services to the millions of people who do not currently have it.

Part II of the book explores the options and choices facing national and international policymakers as they confront the challenges of the global context outlined in Part I. The options and choices cover both the energy demand side, explored in Chapters 9, 10, and 11, and the fuels and technologies through which energy demands will be met: fossil fuels (Chapters 12, 13, 14, 15), nuclear power (Chapter 16), bioenergy (Chapter 17), and the other renewables, deriving from the sun, water, and wind (Chapters 18, 19, 20).

Chapter 9 builds on the theoretical framework set out in Chapter 7 to explore the prospects for far more efficient energy use in buildings, which are responsible for around 32 per cent of total final energy consumption and 40 per cent of primary energy consumption, in most of the countries belonging to the International Energy Agency (IEA).[2] Chapters 10 and 11 look at the transport sector, which is responsible for another 30 per cent of final energy demand. Road vehicles, and the factors that affect their efficiency and how they might develop, are the focus of Chapter 10, while Chapter 11 plots the growth in energy demand and associated emissions from aviation and shipping, along with the trend of globalization and the growth of world trade. The other major demand sector, industry, is not covered in this book, because of the huge heterogeneity of different industrial sectors. This applies to different processes within the same sector, to different sectors within the same country, and the same sector between countries. Doing such heterogeneity justice would have required another book, so suffice it to say here that studies have shown that, as with buildings, there are numerous opportunities for energy efficiency measures in different industrial sectors that could substantially reduce the need for new energy supplies as emerging economies industrialize.[3]

Chapters 12, 13, and 14 all look at different aspects of fossil fuels, which, as already noted, still comprise the great majority of the world's energy supply. Chapter 12 presents the evidence of the existence of ample fossil fuel resources for the future. Chapter 13 plots the rise of the so-called 'unconventional' fossil fuels, including shale gas, and the development of the new technologies that have increasingly permitted their economically viable extraction. The geopolitics of oil has a long history and has been well covered in numerous other publications, so that it is not dealt with in detail here, with Chapter 14 focusing on the far more recent phenomenon of the geopolitical economy of global gas markets. Chapter 15 closes these chapters on fossil fuels by exploring the prospects of a technology, carbon capture and storage (CCS), without which the widespread use of fossil fuels beyond about 2040 is likely to be inconsistent with global aspirations to reduce carbon emissions.

The next chapters look at the energy sources that are intrinsically low carbon, on which the hopes of deep carbon reductions in the coming decade are largely pinned. Chapter 16 explores the prospects of nuclear power after the power station meltdown at Fukushima in Japan. Chapter 17 looks at the potential of the world's bioenergy resources, with the recognition that there are numerous other necessary and desirable uses of biomass, most obviously food, and that not all methods of producing bioenergy are low carbon. Chapters 18, 19, and 20 deal with the truly zero-carbon renewables (at the point of generation at least; all technologies need energy for, and therefore potentially are the source of carbon emissions during, their construction)—solar, water, and wind energy respectively. The rates of technical change in solar and wind energy technologies, and the associated price reductions, have been dramatic in recent years, as described in Chapters 18 and 20, with the prospect of plenty more such change in the future, bringing further price decreases. In respect of water, large dams already provided about 17 per cent

[2] <http://www.iea.org/aboutus/faqs/energyefficiency/>.
[3] See, for example, McKinsey Global Institute 2011 *Resource Revolution*, McKinsey Global Institute, London, <http://www.mckinsey.com/insights/energy_resources_materials/resource_revolution>.

of the world's electricity in 2011, but often with substantial negative environmental and social effects in their construction. The major focus of Chapter 19 is the far newer field of marine energy, and the development of wave, tidal, and tidal current technologies to exploit it. In each case for the chapters in this part, while the resource availability, and the nature of the various technologies is briefly described, the emphasis in each chapter is on the policy, institutional, economic, and social issues that will jointly determine whether the resources and technologies become available at scale, and on the various barriers that will need to be overcome if this is to occur.

The ambient renewables that are currently being deployed at some scale, wind and solar, are both variable and intermittent. These characteristics give a new importance and value to energy storage technologies, as well as raising special issues for the network infrastructures through which the energy will be delivered, and this is the focus of Chapter 21. Chapter 22 explores the materials that will be required for the various technologies and infrastructures described in the preceding chapters, some of which will be standard bulk materials, such as steel, that may be needed in great quantities, and some of which may be relatively rare materials, such as neodymium, adequate supplies of which will need to be assured if these new technologies are to deployed at scale as desired. The final chapter of Part II (Chapter 23) focuses on the possible governance mechanisms for electricity systems, comprising a mix of markets and regulation, which will be needed to ensure timely and efficient delivery of electricity to end-users when and where they want it.

Part III of the book brings together the discussion in Parts I and II with consideration of possible global energy and environmental futures. Chapter 24 describes a global integrated assessment energy system model that integrates all the component fuels, technologies, and infrastructures that have been discussed, and their associated carbon dioxide emissions. This is used to generate three possible scenarios for the development of the global energy system, characterized by very different levels of greenhouse gas emissions (mainly carbon dioxide, CO_2), corresponding broadly to an average global temperature increase, over pre-industrial levels, of $2\,^{\circ}C$ (the current globally agreed desirable limit to avoid dangerous anthropogenic climate change), $3\,^{\circ}C$, and a Reference scenario with no constraint on CO_2 emissions, in which the global average temperature rises to $4.5\,^{\circ}C$ above pre-industrial levels by 2100. For sensitivity purposes a scenario is also reported in which the $2\,^{\circ}C$ limit is attained on the assumption that CCS technology does not become widely available. This chapter also includes comparisons with projections of energy, and especially fossil fuel, use from other global scenarios. As already made clear in Chapter 6, CO_2 emissions are by no means the only impact of energy system technologies on ecosystems and the services they generate. Chapter 25 develops further the analysis of the earlier chapter by investigating more specifically the impacts on ecosystems of the various technology mixes in the global scenarios described in Chapter 24. Chapter 26 brings Part III, and the book as a whole, to a close by discussing the policy and governance implications, at global and national level, of the issues explored in the book, highlighting relevant issues for energy policy making in all countries and drawing conclusions.

Part I
Global Energy: Context And Implications

1 The global energy context

Jim Skea

1.1 Setting the scene

Energy underpins almost every human activity. In 2009, energy production, transformation, and distribution accounted for almost 4 per cent of economic activity in OECD countries reporting this data (OECD, 2011). Liquid petroleum products enable transport and therefore trade and commerce. Electricity powers lighting systems, office machinery, domestic appliances, and electronic goods, as well as enhancing comfort levels in hotter climates. A shifting balance of fossil fuels and electricity maintains comfort for people living in colder climates. Manufacturing industry depends on the supply of energy.

Primary energy resources are unevenly distributed round the globe and their exploitation, especially of globally traded fossil fuels such as oil, is intimately linked to economic development. Those countries that are not well endowed with energy resources are often sensitive to their exposure to imports, and their potential vulnerability to supply interruption.

Given the unique role that energy plays, policymakers have neither wanted, nor have they been able, to play a detached role in energy markets. For a variety of reasons, they have intervened to incentivize or discourage specific sources of energy, promote energy efficiency and conservation, regulate natural monopolies and market power where it is deemed to be excessive, regulate environmental impacts, set the rules for spatial planning, and stimulate and direct technological innovation. Internationally, energy is the subject of diplomacy both within and between producer and consumer nations.

The energy policy challenge is often framed as a 'trilemma' (e.g. World Energy Council, 2013)—a balancing of three main policy drivers that are in tension as often as they reinforce each other. Although the term 'trilemma' is recent, the basic concept of a triangle of forces shaping policy trade-offs goes back decades (McGowan, 1989).

The first policy driver concerns the cost of energy to consumers and its impact on a country's competitiveness, now frequently captured in the short-hand term 'affordability'. Major shifts in the price of globally traded forms of energy can have significant macro-economic consequences for both consumers and producers. The 1970s oil crises still cast a long shadow over energy policymaking.

The second driver is 'energy security'. This is perhaps the most nebulous of the three drivers. The term can be used to refer to access to, and the price of, primary energy resources (e.g. oil, natural gas) as well as to the availability of plants (e.g. power stations) that convert energy into a form suitable for consumption (e.g. electricity). Recent work has made considerable progress in structuring thinking about energy security (Mitchell et al., 2014). Energy security can be threatened by natural disasters, economic disturbances, politically motivated supply interruptions (whether inside a country or internationally), or simply inadequate planning. Memories of the 1970s oil crises mean that an association is often made between the reduction of import dependency and the promotion of energy security. Though the evidence for this link can be tenuous (Stern, 2004), the notion helps to legitimize the quest for energy independence as a political goal.

The third driver concerns the management of the environmental impacts of energy. The energy sector makes a disproportionately large contribution to environmental problems. For example, it accounts for two thirds of the radiative forcing from human activities leading to climate change. Climate change is a dominant concern both nationally and internationally in current discussions of the energy sector. Energy activities still contribute disproportionately to regional and local air pollution problems, such as acid rain and urban smog, in low-income countries and emerging economies, though technological solutions have addressed these problems in most developed countries. There are rising concerns about the interaction of energy activity with water and land, especially if the use of biomass for energy develops.

The purpose of this introductory chapter is to set the scene for a fuller exposition of these issues in the chapters that follow. The major contention is that, after a period starting in 1990 and ending with the 2008 economic crisis in which trends in global energy evolved fairly smoothly, the global energy system is now deeply unsettled and there are major uncertainties about its development. The global economic crisis has played a role, but longer term drivers include rapid growth in energy demand in emerging economies, new possibilities for exploiting unconventional oil and gas, and the challenge of dealing with climate change.

The chapter is structured as follows. Section 1.2 looks at global energy trends focusing on three main themes: regional trends in energy and economic development; energy supply; and energy demand. Section 1.3 opens up the energy security issue by looking first at resource availability and then at trends in global energy trade and markets. Section 1.4 looks, fairly briefly, at the climate change challenge in relation to energy, putting it in the context of resource availability. Section 1.5 reviews energy projections and scenarios produced by some major businesses and public sector organizations, concluding that energy futures are not only uncertain but contested. A short concluding section draws the chapter together (Box 1.1).

BOX 1.1 MEASURING ENERGY

Various terminologies and conventions are associated with the quantification of energy supply and demand. *Primary energy* measures energy supply at the point at which resources are harvested, in the case of renewables, or extracted, in the case of non-renewable energy. The International Energy Agency (IEA), whose data is used extensively in this chapter, has adopted the principle that the primary energy form should be the first energy form downstream in the production process for which multiple energy uses are practical. This still leaves some choice as to the method for measuring non-combustible forms of energy. The IEA, like the UN and the EU, uses the *physical energy content* of the primary energy source as the primary energy equivalent. For combustible fuels (fossil and bioenergy), this is straightforward. For nuclear, geothermal, and solar thermal energy, heat input (e.g. from the nuclear reactor pile) is used. For renewables such as hydro, wind, and solar PV, the energy content of the electricity generated is used. The *partial substitution method* is an alternative to the physical energy content method. This attributes to renewables such as hydro the amount of primary energy that would have been required to generate the same amount of electricity in a fossil plant. This increases the apparent contribution of renewables to primary energy supply.

Some primary energy is transformed into a different form before being sold to final consumers, that is households, businesses, and the public sector. Energy is transformed in power stations, petroleum refineries, and in the production of manufactured fuels. *Final energy* measures the amount of energy in the products sold to final consumers. If this is in the form of electricity, it includes the actual energy content of the electricity, not the primary energy used in its production. Final energy is always less than primary energy because of transformation losses and energy industry own use.

There is also a choice as to how to measure the heat content of fuels. The *gross calorific value* of a fuel includes the heat that can be recovered by condensing water vapour in flue gases. The *net calorific value* excludes this energy. Gross calorific values are 5 per cent higher than net calorific values for coal and oil and 9–10 per cent higher for gas. The IEA, the UN, and the EU all use net calorific values in published data. The UK uses gross. The international data reported in this chapter are based on the *physical energy content* method using *net calorific values* unless otherwise stated.

1.2 **Trends in the global energy system**

1.2.1 ENERGY AND ECONOMIC DEVELOPMENT

Global demand for energy has more than doubled in the last forty years (Figure 1.1), with an annual average growth rate of 2.2 per cent. At 1.4 per cent per annum, demand grew more slowly in the 1990s, but growth has accelerated to 2.6 per cent per annum since 2000. The economic crisis of 2008 had a pronounced downward impact on demand, but the previous high rate of growth has resumed.

Growth has been unevenly distributed between developed and developing countries. Countries with developed market economies belonging to the Organisation for Economic Co-operation and Development (OECD) accounted for only 40 per cent of global energy demand in 2011 compared to 61 per cent in 1971. In the 2000s, energy demand in

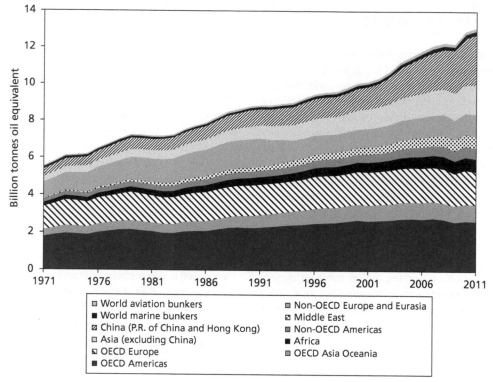

Figure 1.1 Global primary energy demand by region

Source: International Energy Agency, 2013a.

these developed countries levelled off and it has been in decline since 2007. The evidence suggests that mature developed economies can expand without increasing their use of energy.

This stands in sharp contrast with the situation in China, where energy demand has grown at an annual average rate of 3.8 per cent per annum 1971–2001, accelerating to 8.6 per cent since 2001. Growth in other parts of Asia has also been rapid, though not nearly as fast as in China. Demand in the Middle East, though less important in absolute terms, has grown fastest in the longer term, at an annual average rate of 7.0 per cent between 1971 and 2011. Demand in Africa has been growing at a relatively modest 3.0 per cent per annum and the continent still accounts for only 5 per cent of global energy demand.

The complex relationships between energy use and levels of economic development are shown in Figure 1.2, which traces energy use per capita in relation to GDP per capita in nine world regions over the period 1971 to 2011. This shows the wide economic divide between developed countries and the rest. GDP per capita in OECD countries averages around $30,000 while the wealthiest non-OECD regions, the non-OECD Americas and the Middle East, are currently at around $5,000 per capita.

Among developed countries, per capita energy use appears to have stabilized in North America and Europe at around 6 tonnes of oil equivalent (toe) and 3 toe per capita, respectively. However, per capita energy use continues to rise with income in OECD Asia

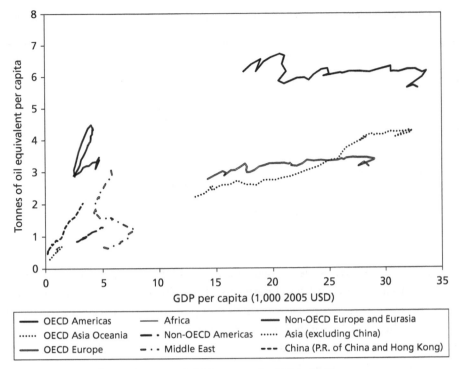

Figure 1.2 Evolution of energy use and GDP per capita 1971–2011

Source: International Energy Agency, 2013a.

and Oceania (Japan, Australia, and New Zealand), where it has reached just over 4 toe. This compares with a global average of 1.9 toe per capita.

The picture in the developing world is more mixed. There are strong and consistent relationships between economic activity and energy use in China, other parts of Asia, and the non-OECD Americas. Chinese energy use has now reached 2 toe per capita and the growth path is more energy intensive than in other regions.

Energy use in the Middle East and non-OECD Europe and Eurasia (consisting largely of former members of the Soviet Union) has been more erratic. GDP per capita in the Middle East is now little higher—at around $6,000—than it was in 1971, having fallen post 1979 before rising again from the late 1990s onwards. However, energy use per capita has risen by a factor of five over the same period to roughly the European level. In non-OECD Europe and Eurasia, GDP and energy use both fell signficantly after the collapse of the Soviet Union in 1991, before starting to recover since the year 2000. Energy use, if not GDP per capita, is now at Western European levels.

Africa remains the least advantaged region in terms of both GDP and energy. GDP per capita, at $1,200, is only 20 per cent higher in 2011 than it was in 1971, while energy use, at 0.7 toe, is 30 per cent higher. Progress has been so slow that the development path in the bottom left-hand corner of Figure 1.2 is hard to pick out.

The range in terms of energy use is even wider at the level of individual countries. Table 1.1 compares per capita energy use in four groups of countries: a) exceptionally

Table 1.1 Primary energy consumption in selected countries in 2011 (tonnes of oil equivalent per capita)

High consuming countries		Major developed economies		Emerging economies		Lower-income countries	
Iceland	17.9	United States	7.0	South Africa	2.8	DR Congo	0.4
Qatar	17.8	Australia	5.4	PR China	2.0	Tajikistan	0.3
Trinidad and Tobago	15.5	Korea	5.2	Argentina	2.0	Nepal	0.3
Kuwait	11.5	Russian Federation	5.2	Thailand	1.7	Cameroon	0.3
Netherlands Antilles	10.9	Netherlands	4.6	Mexico	1.7	Haiti	0.3
Brunei Darussalam	9.3	France	3.9	Turkey	1.5	Yemen	0.3
Oman	8.9	Germany	3.8	Brazil	1.4	Myanmar	0.3
United Arab Emirates	8.4	Japan	3.6	Indonesia	0.9	Senegal	0.3
Luxembourg	8.0	United Kingdom	3.0	Nigeria	0.7	Bangladesh	0.2
Canada	7.3	Italy	2.8	India	0.6	Eritrea	0.1

Source: International Energy Agency, 2013a.

high energy users; b) major developed countries; c) emerging economies; and d) a selection of low-income countries. Energy use in major developed countries is in the range 3–7 toe. However, energy use in a small group of countries exceeds 7 toe per capita, greatly above the global average. These are generally countries with major energy resources, such as those in the Middle East. However, some developed countries with exceptional characteristics—such as Canada, Luxembourg, and Iceland—are also major energy consumers. Iceland's per capita energy consumption, for example, is the world's highest as the result of access to cheap geothermal energy and hydro-electric power and the consequent location of energy intensive industry in that country.

Per capita energy use in emerging economies varies widely, from 0.6 toe per capita in India through to 2.0 toe in China and 2.8 toe in South Africa, similar to the level in developed countries. The low level of energy consumption in India reflects the fact that many people in that country still live in poverty. Energy use in low-income countries is covered fully in Chapter 8. Higher levels of energy consumption in some emerging economies are the result of higher levels of industrialization and emerging middle classes whose patterns of consumption are as energy intensive as those in developed economies. In countries at the lower end of the range, although there is an emerging middle class, the lifestyles of a significant proportion of the population are similar to those in the least-developed countries. Column 4 of Table 1.1 shows the ten countries with the lowest per capita energy consumption. Eritrea, for example, has a per capita energy use that is less than a tenth of the global average. The per capita consumption of every country in the column is less than a fifth of the global average.

There are three main messages from this review of energy use and economic development: a) energy use appears to have saturated in many developed economies; b) energy use is growing strongly in emerging economies, closely linked to expanding economic activity; and c) low incomes are associated with very low levels of per capita energy use in many low-income countries. Improved access to energy and greatly expanded energy use will be a prerequisite for economic development.

1.2.2 ENERGY SUPPLIES

Figure 1.3 looks at global energy from a different perspective, focusing on the sources of energy used to meet demand. The global energy system is still very much based on fossil fuels. They met 82 per cent of primary energy demand in 2011, down only slightly from 86 per cent in 1971. However, the fossil fuel mix has changed over this period. Oil still has the largest market share (32 per cent in 2011) but this is down from 44 per cent in 1971. The most rapid decline in oil's share came in the 1980s as it was increasingly substituted in uses other than transport following the 1973 and 1979 oil crises. Coal's market share (29 per cent in 2011) is now approaching that of oil, with a very rapid growth in use having taken place from 2002 onwards. This largely reflects the rapid growth of the Chinese economy which has relied on the expanding use of coal. Natural gas currently has the lowest market share of all the fossil fuels (21 per cent) but has experienced the strongest and most consistent growth, 2.9 per cent per annum on average, over the period 1971–2011.

Despite policy efforts, non-fossil sources of energy still make only a modest contribution to the global energy mix. The largest contribution comes from biomass and waste (10 per cent in 2011). Much of this is traditional biomass used for heating and cooking in low-income countries rather than 'modern' bioenergy. Nuclear's contribution peaked at 6.8 per cent in 2001 having started from a level of 0.5 per cent in 1971. Its share is down to 5.1 per cent as a result of the closure of older plants and little replacement or new investment in most developed countries. However, nuclear capacity has expanded in

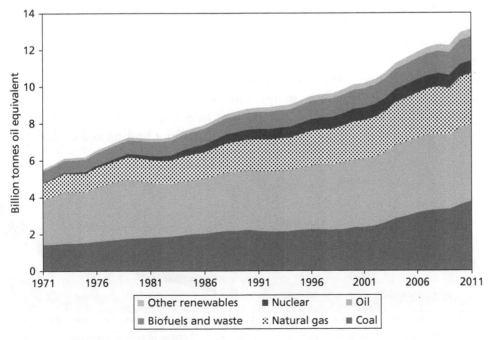

Figure 1.3 Global primary energy demand by fuel

Source: International Energy Agency, 2013a.

Asia. Renewables other than bioenergy contribute only 3.1 per cent of the global energy mix in spite of aspirations and considerable policy effort. Hydro-electricity accounted for 70 per cent of these other renewables in 2011, but there has been extremely rapid growth in the use of geothermal energy (now 15 per cent) and wind and solar (together now accounting for 15 per cent) in recent years. Between 2001 and 2011, wind and solar output grew at 21 per cent per annum. If this rate of growth continues, their contribution will be much greater in the future. In 2012, investment in renewable energy, including large hydro, was US$260bn, roughly on a par with the US$262bn investment in fossil fuels (Frankfurt School-UNEP Centre, 2013). However, some of the fossil fuel investment was in replacement capacity and net new investment in fossil fuels was lower at US$148bn.

1.2.3 ENERGY DEMAND

Figure 1.4 compares the evolving patterns of energy demand by sector in OECD and non-OECD countries. In OECD countries, transport (22 per cent) and energy transformation (31 per cent) are the most important sectors. Energy transformation comprises mainly electricity generation but also includes oil refining and manufactured fuels, for example in the iron and steel sector. Remaining energy demand is spread fairly evenly round industry, the residential sector, and other consumers in the service sector, the public sector, and agriculture. Since overall energy demand levelled off in the last decade, sectoral shares have remained fairly stable.

The picture looks very different in non-OECD countries. Much of the surge in energy demand since the year 2000 has come in the energy transformation sector and in industry. These now account for 34 and 23 per cent respectively of primary energy demand. The rate of growth in these two sectors in the decade 2001–11 has been very rapid, 6.1 per cent per annum in energy transformation and 5.5 per cent in industry. The residential sector (18.4 per cent) is the third largest area of demand in non-OECD countries. This includes the use of traditional biomass as a heating and cooking fuel in

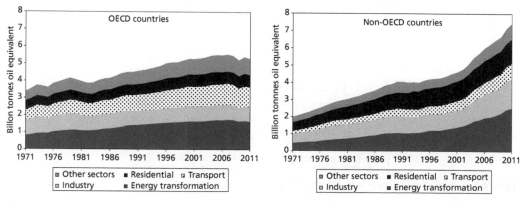

Figure 1.4 Energy demand by sector in OECD and non-OECD countries

Source: International Energy Agency, 2013a.

lower income countries. The transport and 'other' sectors account for a smaller proportion of demand than in OECD countries.

There are two main underlying explanations for the divergent patterns of demand in OECD and non-OECD countries. The first lies in the growth of consumption in emerging economies. The second relates to comparative trade advantages in sectors such as heavy manufacturing and textiles, which has led to a shift in the location of such industries away from the OECD to emerging economies. Large amounts of carbon-based energy are used in emerging economies to manufacture goods that are then consumed in developed countries (see Chapter 5). For example, a considerable amount of electricity is used in manufacturing textiles and clothing in China which are then consumed overseas (Barrett et al., 2013). This offshoring of manufacturing is a second reason for the divergent energy use patterns.

There are two further notable trends in energy demand relating to the roles of electricity and liquid petroleum products. Figure 1.5 shows that the proportion of final energy demand met by electricity has been rising steadily in both OECD and non-OECD countries. Electricity now accounts for 22 per cent of OECD final energy demand and 16 per cent of non-OECD demand. The rate at which electricity has increased its share of non-OECD demand has accelerated since the mid 1990s. This partially explains the increasing amount of energy consumed in the energy transformation sector, as shown in Figure 1.4.

The main reason for this trend is that electricity is either uniquely placed, or offers considerable advantages, in meeting demand for lighting, air conditioning, appliances, and IT equipment. Growing demand for these applications has been driving increases in electricity consumption. In the future, a key uncertainty is whether electricity will be increasingly adopted outside its core applications, for example in heating or transportation,

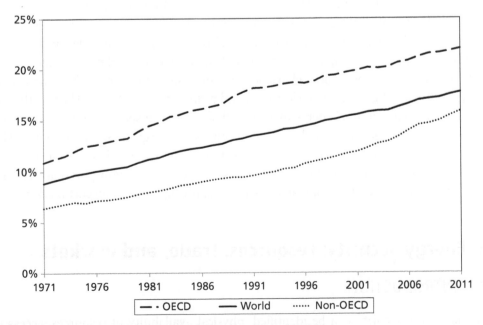

Figure 1.5 Proportion of final energy demand met by electricity

Source: International Energy Agency, 2013a.

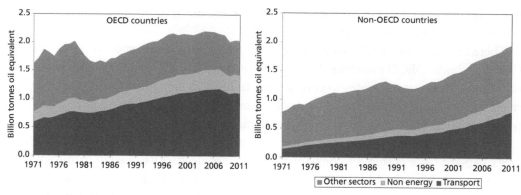

Figure 1.6 Markets for oil products

Source: International Energy Agency, 2013a.

in competition with the direct use of fossil fuels. An increasing role for electricity has a wider system-level significance because there is flexibility in generating electricity from both fossil and non-fossil sources, such as nuclear or renewables.

The second trend concerns the use of oil products. Figure 1.6 tracks the use of oil products in OECD and non-OECD countries in three main markets: transport; non-energy uses (e.g. chemical feedstocks, lubricants); and other uses (e.g. electricity generation, heating). The use of oil is becoming increasingly concentrated in its core markets for transport and non-energy applications. In 1971, non-OECD countries used oil predominantly for electricity generation and heat. Demand for these applications has increased by 45 per cent since then, while transport demand has increased by a factor of more than five and non-energy uses by a factor of eight. OECD countries substituted out of oil for electricity generation and heat in the late 1970s and early 1980s, and once again in the post-2000 period. These applications now account for less than 30 per cent of demand. This changing pattern of demand has been driven by a combination of high oil prices and policies for reducing oil dependence.

Oil products meet 93 per cent of global demand for transport energy, with the remainder coming from natural gas, biofuels, electricity, and a tiny amount of coal. There is thus an increasing bifurcation in final energy markets with a strong transport–oil nexus on the one hand, and a more contested market for non-transport energy uses on the other. This bifurcation is most pronounced downstream in the energy supply chain; upstream there are exceptions. For example, the production of gas destined for non-transport markets tends to be undertaken by the same companies that are supplying oil for transport markets.

1.3 **Energy security: resources, trade, and markets**

1.3.1 INTRODUCTION

Three aspects of security can be identified: physical availability of resources; access to and the cost of utilizing these resources; and the adequacy and reliability of the infrastructure needed to move energy from suppliers to consumers. Infrastructure reliability is largely a national or even local issue and is not discussed further here. However, trade and regional inter-connection can enhance reliability by diversifying sources of supply and building redundancy into networks.

1.3.2 RESOURCE AVAILABILITY

Although issues such as 'peak oil' have been cited as a potential problem (Sorrell et al., 2009; Leggett, 2006), the general consensus is that the world is physically well endowed with fossil fuel resources, uranium and other fissile material, and renewable energy. Table 1.2, for example, compares estimated fossil fuel reserves and resources with global consumption levels in 2011. Resources refer to the energy content of fuels physically in place, while reserves refer to known resources that can be recovered using current technology and under current economic conditions.[1] New discoveries, technological change, and market conditions can all result in resources being turned into reserves.

The reserves-to-production ratio for conventional oil is in the range of 30–40 years and there is some consensus that conventional oil production is either close to, or even past, its peak (Sorrell et al., 2009). However, moving to 'unconventional' oil sources such as tar/oil sands, deepwater/Arctic production, tight oil, coal-to-liquids (CTL) conversion, or gas-to liquids (GTL) conversion could greatly extend the resource base. Figure 1.7 shows technically recoverable resources and the range of potential production costs for conventional and unconventional oil. The remaining resource of 6,800 billion barrels of oil compares to current annual consumption of 32 billion barrels. Most of this oil could be produced at costs at or below $100/barrel. Actual market prices could be higher due to the exercise of market power or stickiness in investment processes.

Gas and coal resources are even more ample. The reserve-to-production ratio for conventional gas is 40–60 years. Reserves of unconventional tight and shale gas are an order of magnitude higher and the overall resource base is higher still. The production of

Table 1.2 Fossil fuel reserves, resources, and consumption

	Reserves		Resources		Production in 2011	
	EJ	b t CO_2	EJ	b t CO_2	EJ	b t CO_2
Conventional oil	4,900–7,610	98–152	4,170–6,150	83–123	173.0	11.1
Unconventional oil	3,750–5,600	75–112	11,280–21,000	226–420		
Conventional gas	5,000–7,100	76–108	7,200–10,650	110–163	117.5	6.3
Unconventional gas	20,100–67,100	307–1,026	40,200–121,900	614–1,863		
Coal	17,300–21,000	446–542	291,000–435,000	7,510–11,230	161.2	13.8
Total	51,050–108,410	1,102–1,940	353,850–594,700	8,543–13,799	551.7	31.2

Source: Rogner et al., 2012; IEA, 2013b.

---

[1] Terminology varies across fuels. See Sorrell et al. (2009) and Chapter 13 for a fuller explanation.

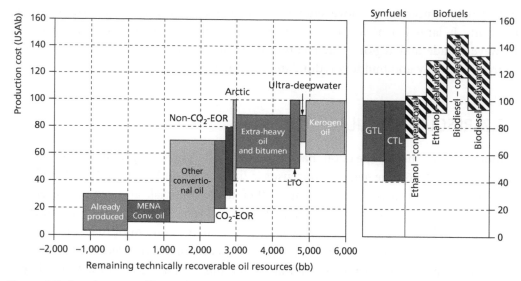

Figure 1.7 Supply costs of liquid fuels

Source: International Energy Agency, 2013b.

shale gas using technological developments associated with hydraulic fracturing ('fracking') and horizontal drilling has transformed gas supply in the USA (see Chapter 14). Figure 1.8 shows that unconventional shale gas already accounts for a third of US natural gas output. This is expected to rise to more than half of a much higher overall output level by 2030. The USA expects to be a natural gas exporter by 2016. There is the prospect of shale gas production in other countries but the geology and the economics remain uncertain. However, production in the USA alone has had, and will continue to have, an impact on global energy markets.

The coal resources that could ultimately be extracted exceed even those of gas. The reserves-to-production ratio for coal is in the range of 100–130 years.

1.3.3 ENERGY TRADE AND INTERDEPENDENCY

Energy products are extensively traded on global markets. This acts as a spur to economic activity in countries that are well endowed with resources and enables industrial activities in countries without their own energy resources. However, this interdependency also engenders tensions and suspicions. Some major oil producing countries coordinate production through the Organization of Petroleum Exporting Countries (OPEC)[2] and, as a result, the price of oil is much higher than it would be in a free market. Major consuming countries formed the International Energy Agency

[2] OPEC members are: Algeria, Angola, Ecuador, Iran, Iraq, Kuwait, Libya, Nigeria, Qatar, Saudi Arabia, United Arab Emirates, and Venezuela.

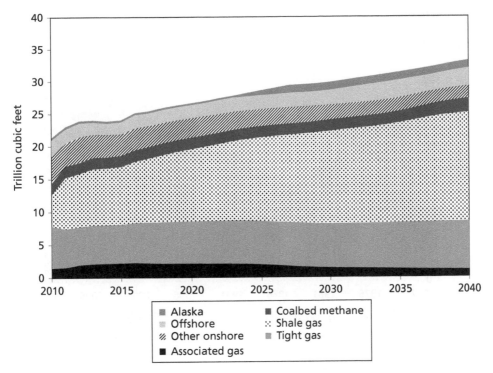

Figure 1.8 Projected US natural gas production

Source: Energy Information Administration, 2014.

(IEA) in 1974 in response to oil embargoes by OPEC members with the aim of coordinating emergency response mechanisms and facilitating diversification away from oil. Although energy trade, if smoothly conducted, brings clear economic benefits, lack of supply diversity, the consequences of supply interruptions, and the macro-economic implications of high prices create strong political incentives for consuming countries to reduce import dependency or even aspire to energy independence.

Oil is the most widely traded form of energy and raises the greatest sensitivities because of the lack of short-term alternatives, especially in the transport sector which is critical to the functioning of modern economies. In the 1970s, 40–45 per cent of world oil was traded between the major world regions (Figure 1.9). Inter-regional trade fell by almost half by the mid 1980s with an even more dramatic fall in North American imports. This was mirrored by a significant fall in Middle East exports. The rapid decline in trade was partly a market response driven by high prices and partly a policy response. The response included energy conservation, the substitution of oil by other energy sources, and investment in new, higher cost oil production outside the OPEC region. When oil prices fell in the mid 1980s, market and policy pressures receded. Inter-regional trade is now higher than in it was in the 1970s and represents 40 per cent of global production. Middle East exports are now back to levels similar to those in the

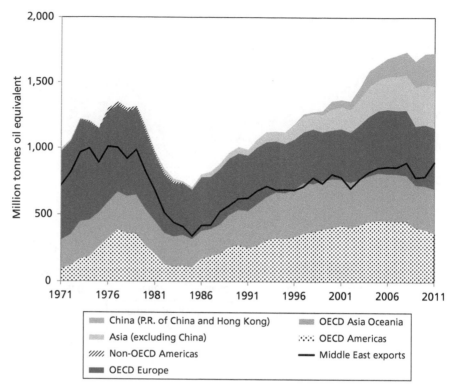

Figure 1.9 Regional imports of crude oil

Source: International Energy Agency, 2013a.

1970s. The big difference is the important role of China and other Asian countries, which now import about one third of the oil traded in global markets.

International trade in natural gas is far less developed (Figure 1.10) but has expanded rapidly in the last decades: 13 per cent of global gas production is now traded between regions. Imports are concentrated in two regions—OECD Europe and OECD Asia/Oceania—though China has recently emerged as a significant actor in global markets. European consumers have become heavily reliant on pipeline gas from Russia and the Central Asian republics. Russia's key role as a major gas exporter is influencing the energy policies of the EU and its member states. International trade in liquefied natural gas (LNG) is expanding rapidly, with the Middle East and Africa being the most important sources of supply.

Only 8 per cent of coal enters inter-regional trade (Figure 1.11), but growth has been rapid in the last ten years and the aggregate trade volume is similar to that of natural gas. However, coal gives rise to less policy anxiety than other fossil fuels as there are diverse sources of supply. There have been notable changes since 2006. China has switched from being an exporter to an importer of coal in a short space of time and has driven the overall rise in trade volume. North American exports, having been in decline since 1990, are rising again. This is attributable to the availability of shale gas in the USA, described

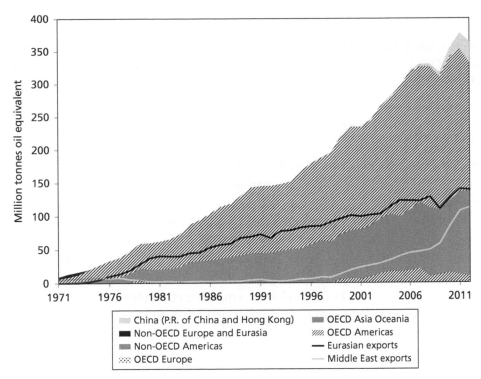

Figure 1.10 Regional imports of natural gas

Source: International Energy Agency, 2013a.

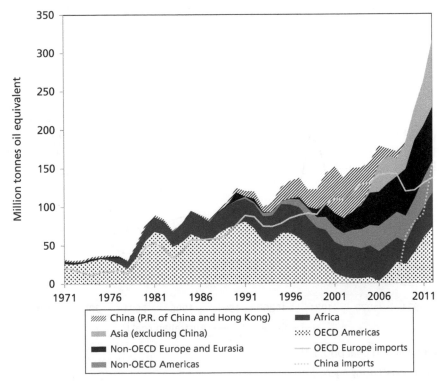

Figure 1.11 Regional exports of coal

Source: International Energy Agency, 2013a.

above, so that gas has displaced coal from power stations and made it available on export markets. OECD Europe has until recently been the main destination for internationally traded coal, but Chinese imports surpassed European imports in 2012.

1.3.4 ENERGY PRICES

Coal and oil are globally traded commodities and it is easy to identify global marker prices; there is more 'stickiness' in gas markets due to the need to invest in pipelines and LNG facilities. Figure 1.12 shows average prices for crude oil and Australian coal since the 1970s.The crude oil curve shows clearly the impact of the 1970s oil crises, the subsequent decline in prices that followed the market and policy responses, and the subsequent rise in prices, and instability, since the early 2000s. The latter rises are not due to the physical scarcity of oil, as demonstrated above. A number of factors are at play, including the surge in demand from emerging economies, the depletion of conventional oil reserves developed some time ago in response to the 1970s crises, the higher costs of alternative sources and the time it takes to develop them, and management of production by OPEC countries so as to maintain the prices and revenue streams needed to diversify their economies. There is a reasonable degree of correlation between coal and oil prices that reflects competition in markets such as power generation and industry and a coincidence of demand and investment cycles.

It is more meaningful to look at gas prices in regional markets. Figure 1.13 shows that Japanese, European, and US gas prices were reasonably well correlated until

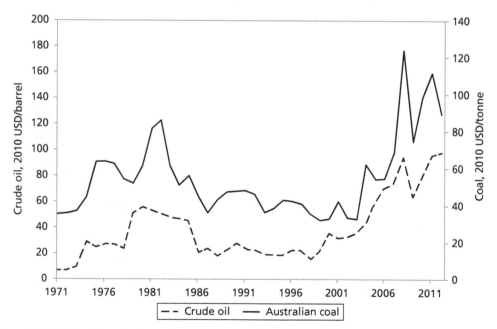

Figure 1.12 Oil and coal prices 1971–2012

Source: International Energy Agency, 2013a.

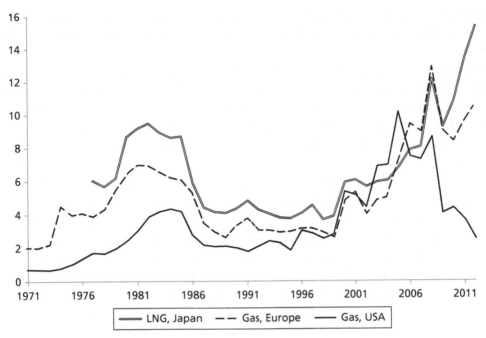

Figure 1.13 Regional gas prices 1971–2012

Source: International Energy Agency, 2013a.

about 2005, prices having converged in the mid 1980s. The traditional practice of indexing gas prices to those of oil in long-term contracts helps to explain the strong correlation with oil prices. The rationale for oil-indexation is disappearing as gas trade expands, spot markets develop, and the degree of direct competition between gas and oil in final markets declines. Regional markets have diverged significantly since 2005, with Asian prices rising and those in North America falling. The latter is due to the development of shale gas, which has yet to find its way on to international markets due to the absence, so far, of LNG export facilities. The rise in Asian prices can be attributed to emerging demand in China plus the Fukushima accident in Japan, which led to nuclear generation being replaced with natural gas. The recent instability in prices and markets suggests a profound structural change in the way that gas markets operate.

1.4 Energy and climate change

Climate change is the greatest environmental challenge the world has faced. There are significant costs associated both with its consequences and with reducing the rate at which human activities are modifying the climate system. This chapter is not the place to assess adaptation to or mitigation of climate change. However, climate change is an important driver of energy policy in many countries and understanding the energy sector's contribution is essential. Climate change and energy is discussed more fully in Chapter 4.

Members of the UN Framework Convention on Climate Change (UNFCCC) set the aim in 2010 of making deep cuts in global greenhouse (GHG) gas emissions with the aim of holding the increase in global average temperature below 2°C above preindustrial levels. The aim is to conclude a comprehensive international agreement at the Paris meeting of the UNFCCC Conference of the Parties (CoP) in late 2015. Global GHG emissions were estimated to be 49.5 billion tonnes of CO_2 equivalent (CO2e) in 2010 (IPCC, 2014) of which 37.6 billion tonnes (76 per cent) were associated with CO_2 and, of these, 32.3 billion tonnes (86 per cent) originated in the energy sector.

The Intergovernmental Panel on Climate Change (IPCC) Fifth Assessment Report concluded that global GHG emissions would need to be 40–70 per cent lower in 2050 than in 2010 to make it likely that global temperature increases could be held to below 2°C above preindustrial levels (IPCC, 2014). This would represent a radical departure from current trends in energy related CO_2 emissions as shown in Figure 1.14. Given the strong link between fossil fuel use and CO_2 emissions, the patterns

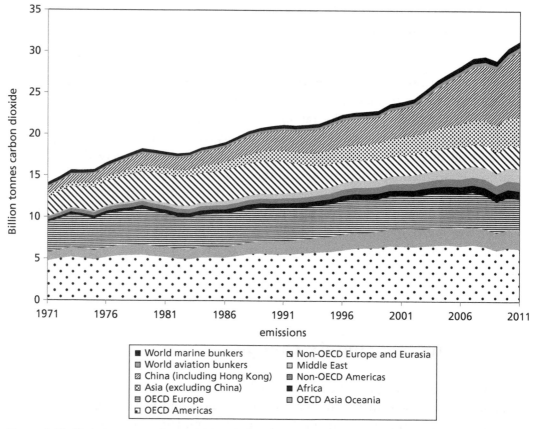

Figure 1.14 Global energy related CO_2 emissions

Source: International Energy Agency, 2013a.

in Figure 1.14 are very similar to those in Figure 1.1 showing primary energy demand. However, the role of China since 2001 is even more pronounced since much of its increased energy use has been fuelled by coal, the most carbon intensive of the fossil fuels. Growth in global CO_2 emissions accelerated to 2.7 per cent per annum, with China's emissions growing at 8.8 per cent.

Securing a global agreement that achieves radical cuts in GHG emissions is an enormous challenge. Fundamental issues that remain unresolved include: the distribution of emissions reductions between countries, especially between the developed and developing; whether or how 'historic responsibility' for emissions that have accumulated in the atmosphere should be taken into account; the financing of investments in developing countries; and different modes of technology cooperation.

At the 2010 CoP, a number of countries and regions confirmed (non-binding) emission reduction pledges (known as the 'Cancún pledges'). IPCC has concluded that these pledges, by themselves, far from guarantee meeting the UNFCCC 2 °C goal. The EU and individual countries such as the UK and the USA have indicated a long-term aim of reducing GHG emissions by at least 80 per cent by 2050. In the UK's case this is binding under legislation. Even though the Cancún pledges may fall short of the 2 °C goal, the steps required to secure their achievement would still require substantial changes to patterns of energy demand and supply.

Table 1.2 includes an estimate of the CO_2 that would be released if fossil fuel reserves were to be used unabated, that is without carbon capture and storage technology. IPCC (2013) has estimated that cumulative emissions of CO_2 starting from a 2011 baseline would need to be held below 3,010bn tonnes to give a 50 per cent chance of keeping the global temperature increase below the UNFCCC objective of 2 °C above preindustrial levels. This is higher than the CO_2 associated with current fossil fuel reserves, but well below the CO_2 associated with total resources. From this, two conclusions emerge: a) resource scarcity cannot be relied on to resolve the climate challenge; and b) if the UNFCCC 2 °C objective is to be pursued, policies need to set a regulatory and/or market framework that incentivizes the energy industry to leave fossil fuel resources in the ground.

1.5 **Energy futures: uncertain and contested**

The evidence in the earlier part of this chapter points to a global energy system that is undergoing profound change, the smooth curves shown in Figures 1.1 and 1.3 notwithstanding. A variety of public and private organizations have put their thinking about longer-term energy futures into the public domain. This section reviews the recent outputs of six organizations—three supported by governments (the US Energy Information Administration (EIA), the IEA, and OPEC)—and three in the private sector (BP, ExxonMobil, and Shell). The central message from the review is that the future of global energy is not only uncertain but also *contested*, that is the futures exercises reflect different normative views about how the energy system *ought* to develop. The response to the climate change challenge is the key differentiator.

Different philosophies are used in undertaking energy futures scenarios. Conscious of inherent uncertainties, no organization currently uses terms such as 'prediction' or 'forecast' to describe their forward thinking. However, the majority produce 'outlooks' (EIA, OPEC, BP, ExxonMobil), the currently favoured term for projecting forward current trends combined with foreseeable changes in the economic, policy, and technology background. These have been described as business-as-usual (BAU) scenarios. The time horizon for these outlooks ranges from 2035 (BP, OPEC) through to 2040 (EIA, ExxonMobil). These outlooks effectively constitute a best estimate of the future for planning purposes. The EIA also publishes four 'projections' that explore sensitivities round a central 'reference case'. The sensitivities cover 'high' and 'low' variants on economic growth and energy prices.

Those organizations looking further into the future (IEA to 2050 and Shell to 2060) investigate deeper uncertainties. Shell has a well-known practice of exploring diverse futures in which political, social, and economic trends play out in different ways (Shell, 2013). The scenarios start with qualitative 'storylines' that are then quantified and formally modelled. The Shell scenarios are intended to challenge the organization to develop strategies that are robust against a range of different futures. They are designed to avoid any individual scenario becoming identified with BAU. However, it is worth noting that Shell's two scenarios start to diverge significantly only in the longer term (post 2035–40).

IEA's recent long-term futures work has emphasized the role of technology in response to the climate change challenge (IEA, 2012). Two of their three scenarios could reasonably be described as 'outlooks', the six degrees (6DS) scenario assuming that current trends and policies continue, and their four degrees (4DS) scenario assuming that the Cancún pledges described earlier are implemented. The titles of the scenarios refer to the global temperature rise that could be expected to follow from these assumptions. Their third scenario—two degrees (2DS)—is of a different character. It is a 'normative' scenario that starts with the assumption that the UNFCCC goal of holding the increase in global average temperature below 2°C above preindustrial levels is reached, and then works backwards to identify which combination of technologies would meet the goal at least cost. A large academic literature based on this philosophy, covering 900 scenarios, has been assessed by the IPCC (IPCC, 2014).

Figure 1.15 compares the primary energy mix in 2040 in projections produced by each of the six organizations. The 2011 baseline is included for reference. The BP projections have been extrapolated to 2040, while the OPEC projections for 2035 are reported for reference. The projections are organized from lowest to highest primary energy demand in 2040 going from left to right. The three IEA scenarios embrace a wide range of climate change responses. Under more ambitious climate change policies: a) energy efficiency measures play a larger role in reducing demand; b) coal use is squeezed and declines

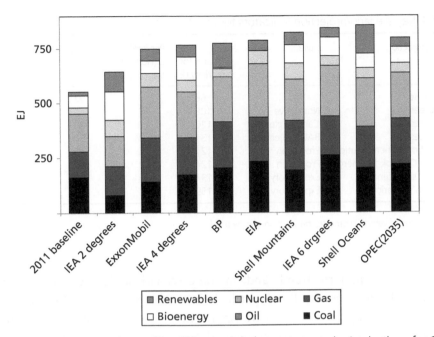

Figure 1.15 Primary energy demand in different global energy scenarios/projections for 2040

Source: BP, 2014; EIA, 2014; ExxonMobil 2014; IEA, 2012; OPEC, 2013; Shell 2013.

Note: OPEC is to 2035; BP is extrapolated to 2040; BP and EIA include liquid biofuels in oil and include other bioenergy in renewables; BP accounts for renewables using the partial substitution method; EIA generates four 'sensitivities' reflecting different assumptions about energy prices and economic growth—only the central reference case is presented here.

substantially in the 2DS scenario; c) oil use is also squeezed; d) the use of all forms of renewable energy, including bioenergy, expands; and e) nuclear power expands. While it is not apparent from Figure 1.15, the use of carbon capture and storage (CCS) also expands.

The projections produced by the other organizations are broadly similar to the IEA 4DS/6DS scenarios, though there are differences that reflect different views as to how uncertainties might be resolved. Key features worth noting are:

- There is widespread agreement that the use of natural gas will expand. This expansion is more muted in the IEA and Shell Oceans scenarios.
- Most scenarios foresee some expansion in the use of coal, except in the IEA 2DS scenario and also the ExxonMobil outlook.
- Oil use also expands except in the IEA 2DS scenario. Oil use expands to a lesser extent in the Shell Mountains scenario where biofuels play a greater role.
- All scenarios foresee a greater role for non-combustible renewables (mainly wind and solar). The expansion is largest in the Shell Oceans scenario, which posits a breakthrough in solar PV.
- Bioenergy plays a much bigger role in all of the IEA scenarios and the Shell Mountains scenario. However, there is little expansion in the ExxonMobil, Shell Oceans, and OPEC projections.

Table 1.3 looks more closely at some of the assumptions underlying the projections. All scenarios envisage substantially rising electricity demand. This links partly to the

Table 1.3 Key scenario/projection indicators for 2040

Scenario/projection	Style	Final demand for electricity (EJ)	Per cent of non-oil transport	Energy-related CO_2 emissions (bn t)	Per cent fossil fuel generation CCS
2011 baseline		66	7	31	–
IEA 2 degrees	Normative	108	33	20	44.7
IEA 4 degrees	Cancún pledges	120	10	40	3.2
IEA 6 degrees	BAU projection	131	8	52	0.1
EIA International Energy Outlook	Projection	127	6	45	
ExxonMobil Outlook for Energy	Projection	124	12	36	
Shell Mountains scenario	Exploratory	131	35[1]	37	major post-2040
Shell Oceans scenario	Exploratory	159		41	

Notes: [1] In terms of vehicle kilometres rather than energy used.

continuation of long-term trends and, in some scenarios, electricity developing new markets, notably in transport. This is particularly notable in the IEA 2DS scenario, where oil accounts for only two thirds of transport energy needs by 2040, and in the Shell Mountains scenario, where electricity and hydrogen both displace liquid transport fuels. Oil is also displaced from transport in the ExxonMobil outlook, but this is mainly in the form of natural gas.

All scenarios, apart from IEA 2DS, envisage considerable growth in energy related CO_2 emissions. In the 2DS scenario, a substantial proportion of the CO_2 reductions come from installing CCS at coal- and gas-fired power stations. There is also substantial investment in CCS in the Shell Mountains scenario, which helps to keep emissions low relative to most other scenarios. Other than IEA 2DS, the lowest CO_2 emissions are in the ExxonMobil outlook. This is essentially due to more ambitious assumptions regarding energy efficiency that lead to a relatively low primary energy demand.

Views of the future affect energy strategy in the present. Introducing a theme covered more thoroughly in Chapter 2, Table 1.4 shows the allocation of budgets for energy research, development, and demonstration (RD&D) in the public sector in IEA countries and in the private sector globally. The information for the public and private sectors comes from different sources and is different in scope. However, the pictures are so contrasting that even a broad-brush comparison is meaningful. The scope of the private sector data is narrower because it excludes technology demonstration and excludes the R&D activities of diversified engineering companies and equipment suppliers whose R&D portfolios may include substantial energy related components.

The striking feature is the huge divergence in the level of effort—an order of magnitude—expended on R&D relating to fossil fuels and more specifically oil and gas. In oil and gas, private sector producers in emerging economies, notably China and Brazil, are starting to play a major role. Private sector R&D has underpinned the development of unconventional oil and gas, notably shale gas, in recent years. Much of

Table 1.4 Public Sector Energy RD&D/R&D Spend in IEA Countries 2011 ($bn)

Public sector RD&D spend IEA countries		Private sector R&D spend in selected sectors			
Area	Spend	Sector	OECD	Non-OECD	TOTAL
Energy efficiency	3.1	Oil & gas producers	6.5	6.1	12.6
Fossil fuels	1.8	Oil services & distribution	3.0	0.3	3.3
Renewable energy	3.7	Electricity	3.3	0.0	3.3
Nuclear	4.4	Other utilities	1.3	0.1	1.4
Hydrogen and fuel cells	0.7	Alternative energy	1.0	0.0	1.0
Power and storage	0.9				
Cross-cutting	2.6				
TOTAL	17.2	TOTAL	15.1	6.5	21.6

Source: IEA 2014; Hernández et al., 2014.

public sector RD&D in the fossil fuel area currently focuses on CCS. The public/private balance tips in the other direction for non-fossil R&D. Nuclear research (including fusion) still takes the largest share of public sector R&D, but renewables and efficiency also account for significant shares.

The evidence is that patterns of R&D investment mirror expectations about the future of the energy system embodied in scenarios and projections. Fossil fuel producers see a world in which fossil fuels continue to dominate, and invest R&D resources accordingly. Many governments envisage a re-shaping of the energy system largely to address climate change. Their R&D investments focus on efficiency and non-fossil alternatives. The energy future can therefore be said to be contested rather than simply uncertain.

1.6 **Conclusions**

This chapter has shown that the global energy system is undergoing profound change. The key aggregate indicator, global primary energy demand, is on a smooth upward trend following a small dip after the 2008 economic crisis. However, a number of factors are unsettling the system, principally rapid energy demand growth in emerging economies, new possibilities for exploiting unconventional oil and gas, and the challenge of dealing with climate change. At the same time there is a continuing need for improved access to energy in lower-income countries, without which economic development will not take place.

The future is not only uncertain but contested. The key differentiator between different views on the future of the energy system is whether and to what extent the challenge of climate change is addressed. Using all, or even a substantial part, of the available physical resource of fossil energy, is not compatible with reaching the climate goal formally agreed by the UNFCCC. The following chapters explore many of these issues, and the complex relations between them, in more detail.

■ REFERENCES

Barrett, J., Peters, G., Wiedmann, T., Scott, K., Lenzen, M., Roelich, K., Le Quere, C. (2013) 'Consumption-based GHG emission accounting: a UK case study', *Climate Policy*, vol. 13 no. 4, pp. 451–70, DOI:10.1080/14693062.2013.788858.

BP (2014) *BP Energy Outlook 2035*. London. <http://bp.com/energyoutlook>. (Accessed April 2015).

Energy Information Administration (2013) *Annual Energy Outlook 2013: Early Release Overview*. Washington DC. <http://www.eia.gov/forecasts/aeo/er/pdf/0383er%282013%29.pdf>. (Accessed April 2015).

Energy Information Administration (2014) *Annual Energy Outlook 2014 with Projections to 2040*. Washington D.C.: US Department of Energy.

ExxonMobil (2014) *The Outlook for Energy: A View to 2040*. Irving, Texas. <http://corporate.exxonmobil.com/en/energy/energy-outlook>. (Accessed April 2015).

Frankfurt School-UNEP Centre (2013) *Global Trends in Renewable Energy Investment 2013*. Frankfurt am Main. <http://www.fs-unep-centre.org>. (Accessed April 2015).

Hernández, H., Tübke, A., Hervás, F., Vezzani, A., Dosso, M., Amoroso, S., and Grassano, R. (2014) *The 2014 EU Industrial R&D Investment Scoreboard*, European Commission, Brussels. http://iri.jrc.ec.europa.eu/documents/10180/354280/EU%20R%26D%20Scoreboard%202014 (Accessed May 2015).

Intergovernmental Panel on Climate Change (2013) *Climate Change 2013: The Physical Science Basis- Summary for Policymakers*. Geneva.

Intergovernmental Panel on Climate Change (2014) *Climate Change 2014: Mitigation of Climate Change—Summary for Policymakers*. Geneva.

International Energy Agency (2012) *Energy Technology Perspectives 2012: Pathways to a Clean Energy System*. Paris. <http://www.iea.org/etp/etp2012/>. (Accessed April 2015).

International Energy Agency (2013a) *World Energy Balances* (Edition: 2013). University of Manchester: Mimas,. <http://dx.doi.org/10.5257/iea/web/2013>.

International Energy Agency (2013b) *Resources to Reserves: Oil, Gas and Coal Technologies for the Energy Markets of the Future*. Paris. <https://www.iea.org/w/bookshop/add.aspx?id=447>. (Accessed April 2015).

International Energy Agency (2014) *Detailed Country RD&D Budgets*. Paris: IEA. <http://wds.iea.org/WDS/TableViewer/dimView.aspx?ReportId=1399>. (Accessed April 2015).

Leggett, J. (2006) *Half Gone: Oil, Gas, Hot Air and the Global Energy Crisis*. London: Portobello Books.

McGowan, F. (1989) 'The single energy market and energy policy: conflicting agendas?', *Energy Policy*, vol. 17 no. 6, pp. 547–53.

Mitchell, C., Watson, J., and Whiting, J. (eds). (2014) *New Challenges in Energy Security: The UK in a Multipolar World*. Basingstoke: Palgrave MacMillan.

OECD (2011) *STAN Database for Structural Analysis (ISIC Rev. 3)*. OECD. Paris. <http://www.oecd.org/industry/ind/stanstructuralanalysisdatabase.htm>. (Accessed April, 2015).

OPEC (2013) *World Oil Outlook*, Vienna. <http://www.opec.org/opec_web/static_files_project/media/downloads/publications/WOO_2013.pdf>. (Accessed April 2015).

Rogner, H.-H., Aguilera R. F., Archer C., Bertani R., Bhattacharya S. C., Dusseault M. B., Gagnon L., Haberl H., Hoogwijk M., Johnson A., Rogner M.L., Wagner H., and Yakushev V., (2012)

Chapter 7—Energy Resources and Potentials. In Johansson, T.B., Nakicenovic, N., Patwardham, A., and Gomez-Echeverri, L. (eds) *Global Energy Assessment—Toward a Sustainable Future*. Cambridge, UK and New York, NY, USA: Cambridge University Press and Laxenburg, Austria: the International Institute for Applied Systems Analysis, pp. 423–512.

Shell (2013) *New Lens Scenarios: A Shift in Perspective for a World in Transition*. The Hague. <http://www.shell.com/global/future-energy/scenarios/new-lens-scenarios.html>. (Accessed April 2015).

Sorrell, S., Speirs, J., Bentley, R., Brandt, A., and Miller R. (2009) *Global Oil Depletion: An Assessment of the Evidence for a Near-term Peak in Global Oil Production*. London: UK Energy Research Centre.

Stern, J. (2004) 'UK gas security: Time to get serious', *Energy Policy*, vol. 32 no. 17, pp. 1967–79.

The World Bank (2014) *Commodity Markets: Prices (Pink Sheet)*. Washington DC, 2014. <http://econ.worldbank.org/WBSITE/EXTERNAL/EXTDEC/EXTDECPROSPECTS/0,,content MDK:21574907~menuPK:7859231~pagePK:64165401~piPK:64165026~theSitePK:476883,00. html>. (Accessed April 2015).

World Energy Council (2013) *World Energy Trilemma 2013: Time to Get Real—The Case for Sustainable Energy Investment*. London: World Energy Council.

2 Energy systems and innovation

Jim Watson, Xinxin Wang, and Florian Kern

2.1 Introduction

The transition to sustainable, low-carbon energy systems will require the development and deployment of a range of new and existing energy technologies. These technologies will be required to deliver substantial reductions in greenhouse gas emissions, for example by lowering the carbon intensity of energy use or increasing energy efficiency. They could also have wider economic benefits through the establishment of new firms, industries, skills, and jobs. They include centralized supply-side options such as carbon capture and storage, infrastructure technologies such as decentralized energy networks, and technologies adopted by consumers such as LED lighting, cleaner vehicles, and micro-generation.

This chapter will provide some context to subsequent chapters in Part II of the book that focus on particular energy technologies or fuels. The main aim of the chapter is to discuss different approaches to the analysis of innovation for more sustainable energy systems, to examine some recent global trends in energy technology innovation, and to discuss some implications for government policies and for the subsequent chapters that focus on specific energy technologies.

The definition of innovation used for this chapter and other contributions to this book is a broad one. It encompasses the different stages of innovation (from research and development, R&D, to commercial deployment); different types of innovation (including technical, social, and organizational); and different scopes of innovation (from incremental to radical; from product to system). It also encompasses innovation on the supply side (e.g. renewable energy technologies providing heat or electricity) and the demand side (e.g. innovations improving energy efficiency or reducing energy demand). While it is widely accepted that energy efficiency and energy demand reduction have a major role to play in meeting energy policy goals (DECC, 2012; IEA, 2013a), much of the existing analytical work on energy innovation focuses on the supply side. This partly reflects the greater level of public support for innovation in energy supply technologies (Grubler et al., 2012).

Of particular importance for this book is a recognition that innovation is shaped by a combination of national and global factors. Many of the technologies that are expected to be part of future energy systems are being developed, manufactured, and deployed by multinational firms. Therefore national innovation policies need to be analysed and understood with this context in mind.

In the three sections that follow, Section 2.2 reviews innovation systems concepts and frameworks, and what insights they provide for the analysis of innovation in energy systems—including the role of government innovation policies. Section 2.3 examines global trends in energy innovation using some of the available indicators. The section also includes a discussion of the usefulness of these and other indicators of innovation. Finally, Section 2.4 draws conclusions and discusses some implications for subsequent chapters of the book.

2.2 **Innovation systems**

Innovation includes several distinctive but related stages—from research and development (R&D) to prototyping, demonstration, commercialization, and deployment. Early conceptions of innovation characterized the process of moving through these stages from R&D to deployment as a linear one. However, this 'linear model' was soon abandoned as too simplistic by many of those engaged in innovation, as well as some of those trying to understand and support it. Models of innovation have since become more sophisticated to reflect real processes of innovation observed by empirical studies (Rothwell, 1994). A key feature of these more sophisticated models is that they incorporate feedback between innovation stages—a process that is sometimes referred to as 'learning by doing'. Lessons from prototyping, demonstration, and the commercial deployment of new technologies are used to underpin further innovation. This might yield further improvements or solve problems that become apparent when technologies are incorporated into commercial products. These models also incorporate the increasingly networked character of innovation—including parallel activities by different functional departments within innovating firms; closer relationships between technology suppliers and customers; and a focus on speed and flexibility of product development to respond to changing needs—especially in consumer goods and services markets.

This increasingly sophisticated understanding of innovation is further enhanced by a recognition that the scale and scope of innovation varies widely. Chris Freeman (Freeman, 1992) drew attention to the contrast between incremental innovations that lead to improvements in existing products, and radical innovations that yield new inventions and/or methods of production. He also showed how a series of radical innovations in different parts of the economy can lead to changes in technological systems, for example through the adoption of a series of low-carbon technologies. Going further, changes of techno-economic paradigms can occur when a set of innovations has a pervasive effect on the whole economy. An example of this is the widespread uptake of information technology (IT).

Many studies of the innovation process emphasize economics as a key driver for technical change. Indeed, one of the most important rationales for public policy support for energy innovation is to reduce the costs of more sustainable, low-carbon technologies (Gallagher et al., 2006). However, this does not mean that the relationship between relative costs and the success of new innovations is a simple one. Freeman and Louçã note that wide-ranging shifts in techno-economic paradigm are driven by the prospect of 'super profits' for innovators. Such super profits help to offset the risks of investing in radical new innovations (Freeman and Louçã, 2001). In the early stages of new innovations, however, incumbent technologies can have a price advantage. For example, when electric lighting was first introduced in the 1880s, it was four times more expensive than gas lighting (Pearson and Fouquet, 2006). Parity in cost was only achieved in the 1920s. While the diffusion of electric lighting was driven by the potential for cost reductions, it also occurred due to other non-economic benefits it offered to users, such as convenience and novelty.

Empirical analysis also showed that patterns of innovation are also very much shaped by national institutions. Therefore attention has focused on national innovation systems, including the main actors, their linkages (e.g. the extent of collaboration between universities and firms), and the institutional context within which this occurs (Freeman, 1987). The most well-cited definition of national innovation systems states that a national innovation system comprises 'the elements and relationships which interact in the production, diffusion and use of new, and economically useful, knowledge...and are either located within or rooted inside the borders of a nation state' (Lundvall, 1992).

In addition to national characteristics, one of the key insights from the innovation literature is that innovation processes vary significantly between sectors (Pavitt, 1984). Innovation and technological change on the supply side of the energy sector are characterized by being particularly capital intensive, having long lead times, needing to fit in with existing infrastructures, and so on. All of these features mean that energy innovation is particularly difficult for private firms and therefore often involves substantial state involvement in the development of key technologies (e.g. nuclear reactors, electricity networks, or modern gas turbines). On the demand side, technologies vary widely—and include similar, capital-intensive investments (such as buildings or industrial process equipment) alongside mass-produced consumer goods (such as vehicles and domestic appliances).

These and other insights have led to a number of standard rationales for government intervention and financial support for innovation. Most of these rationales focus on the existence of one or more market failures (Scott and Steyn, 2001; Jaffe et al., 2005). In the field of sustainable energy, two market failures are most commonly cited. First, that the social costs of carbon emissions from the energy system are not fully internalized. This means that technologies that emit less carbon are at a disadvantage. Second, that there is a tendency of the private sector to under-invest in R&D because individual firms cannot fully capture the returns from their investments. Further market failures are sometimes added to these two—for example, the tendency of markets to under-invest in other relevant public goods such as energy security. The natural response to the first two market failures is to create a policy framework that emphasizes market mechanisms (such as emissions trading), that prices carbon emissions, and that provides government funding for R&D. Government technology policies often need to attend to other stages of the innovation process. There has been an increasing focus on the 'valley of death' that faces developers as they try to move technologies from demonstration or prototype phase to incorporation in commercial products (Department of Trade and Industry, 2004; Gallagher et al., 2006). To cross this 'valley of death', developers often need to commit increasing financial resources to technology development at a stage where the risks of failure remain high (Trezona, 2009).

There are further rationales for intervention that go beyond market failures. An innovation systems perspective also focuses on wider system failures. Such system failures are particularly important for low-carbon and sustainable technologies (Foxon, 2003; Stern, 2006). The adoption of some of these technologies often requires both technological change and institutional change. For example, the diffusion of smart metering technology is not just a simple technical challenge but also implies a new approach to information provision to energy consumers and new information technology infrastructure. Others require new links between established but hitherto separate actors within the innovation system. For example, carbon capture and storage (CCS) technologies require new collaborations between utilities, oil and gas companies, and power equipment companies and can also require amendments to existing regulations (e.g. those that govern marine pollution or issues around liability). Electric vehicles require infrastructure changes and cooperation between vehicle manufacturers and electricity companies. The diffusion of technologies such as micro-renewables or business models like energy service contracting might also require changes to the ways in which consumers engage with their energy suppliers.

One of the most important system failures for sustainable technologies is 'lock-in' (Unruh, 2000). The term lock-in was originally coined by Brian Arthur to explain how some technologies become widely adopted due to increasing returns to scale, even though they may not, objectively speaking, be the best technologies for a particular application (Arthur, 1989). Examples include the QWERTY computer keyboard and the VHS video format. This lock-in concept was subsequently scaled up to describe the pervasiveness of fossil fuels, and hence high carbon emissions, within modern industrialized economies (Unruh, 2000, 2002). This means that if markets are left to themselves, energy systems tend to change slowly. Transitions such as the historical shifts in the UK from wood fuel to coal, and from coal to other fossil fuels, have taken many decades (e.g. Pearson and Fouquet, 2006).

The key insight from the concept of lock-in is that it is not simply a case of making low-carbon technologies more attractive and cost-effective. This is because many parts of our high-carbon energy system consist of long-lived capital assets including electricity grids, gas pipelines, and buildings. Furthermore, these assets are supported by interacting systems of rules, regulations, and institutions that coordinate energy flows, market relationships, and investment decisions. Technologies and institutions co-evolve and are closely integrated (Geels, 2004; Weber and Hemmelskamp, 2005), and many of those that currently exist were designed for a fossil fuel-based energy system.

To analyse sustainable technologies from this wider systems perspective, specific frameworks have been developed and applied. For example, the Technological Innovation Systems (TIS) framework (e.g. Jacobsson and Bergek, 2004) has been developed to analyse the innovation system for a particular technology or group of technologies. The TIS literature not only focuses on the structure and the key actors involved in these innovation systems. It also has a normative aim, which is to analyse how well such innovation systems perform in supporting sustainable technologies. The performance of a TIS is assessed with respect to a set of functions that include creating new knowledge, stimulating entrepreneurial activity, creating legitimacy, and forming markets (Hekkert and Negro, 2009; Dewald and Truffer, 2011). For example, a recent analysis of offshore wind innovation concluded that the innovation system functions reasonably well on a European level (Luo et al., 2012), and that system functions that are lacking in one country (e.g. lack of market creation in the Netherlands) are 'compensated' by actions in others (e.g. the UK).

The emerging literature on sustainability transitions takes a broader approach—and places technological innovation in a wider context of system change. It focuses on so-called socio-technical systems that are defined on the basis of the societal needs they fulfil (e.g. the provision of electricity or mobility services). Socio-technical systems are defined as encapsulating not only the technologies but also the physical infrastructures, market arrangements, industry structures, knowledge, consumer preferences, policy, and institutional frameworks which all shape the way in which a certain societal need is fulfilled (Geels, 2004). The argument is that all of these elements are well aligned towards the optimization of current energy systems (e.g. the improvement of the efficiency of current technologies) but that transitions towards low-carbon energy systems require radical changes in all of these elements.

For example, new and existing infrastructures played a major role in the transition from a transportation system based on horse-drawn carriages to automobiles. This included major infrastructure investments in new roads, fuelling stations, and car repair garages. It also helped the gasoline car to conquer early niche markets (such as touring and long-distance racing), because 'of technical characteristics of the internal combustion engine and because it could build on an existing petrol infrastructure, and an existing maintenance network and repair skills (crucial aspects that were hindering the use of electric vehicles outside cities)' (Geels, 2005c: 469). New infrastructures will also play a key role in future energy transitions. For example, a possible transition to hydrogen as a dominant transport fuel would necessitate a roll-out of filling stations within urban settings and on highways, but analysis shows that the costs of this would be very small compared to subsidies for vehicles and fuel (Köhler et al., 2010). The socio-technical transitions literature also sheds light on the complex ways in which such systems change (Geels, 2002) and what role governments can play in steering the direction and influencing the speed of such processes (Kern, 2010; Verbong and

Loorbach, 2012). Within this literature there is an ongoing discussion about whether long-lived network infrastructures such as electricity networks, gas distribution pipe-lines, road and rail networks reinforce lock-in and system inertia or whether they can also facilitate change (Frantzeskaki and Loorbach, 2010; Loorbach et al., 2010).

2.3 Global trends in energy innovation

This section examines some global trends in energy technology innovation, drawing on published data sources. There are a number of proxy indicators that can be used to analyse patterns of energy innovation. Given the systemic, long-term, and uncertain nature of innovation processes, such indicators can only provide a partial picture of these patterns. Furthermore, changes in the structure of many industrialized economies over time, and the increasing importance of globalization, mean that some indicators may become less useful over time. Despite well-documented drawbacks (Freeman and Soete, 2009), efforts to analyse innovation often focus on input measures, particularly levels of research, development, and demonstration (RD&D) funding devoted to technology development by the public and private sectors. These indicators are sometimes comple-mented by indicators of innovation outputs, particularly patents. In addition, innovation outcomes can also be represented using indicators such as rates of deployment of new technologies or trends in technology costs.

Taken together, these indicators can provide a useful starting point for understanding broad trends in energy technology innovation patterns on a national, sectoral, or global basis. The remainder of this section presents some of the available data in the three categories identified (inputs, outputs, and outcomes) and critically examines the extent to which such data are useful representations of global innovation patterns.

2.3.1 INNOVATION INPUTS: R&D SPENDING

The total spending on energy Research, Development, and Deployment (RD&D) by governments in International Energy Agency (IEA) member countries is shown in Figure 2.1. Since the first oil price shock in 1973/74, overall energy RD&D spending by these countries has undergone both rapid increases and significant declines. As shown in Figure 2.1, these trends have broadly followed movements in global oil prices. Between 1974 and 1980, RD&D spending increased by 83 per cent in the wake of the first and second oil price shocks. This was followed by a sharp decline in the early 1980s as oil prices dropped and programmes to develop substitutes (particularly the US synfuels programme) were scaled back or cancelled.

Compared to the spending levels in 1974, spending in 2012 was over 50 per cent higher in real terms. This is partly due to the renewed high profile of global energy issues due to higher energy prices and strengthening evidence about climate change. During

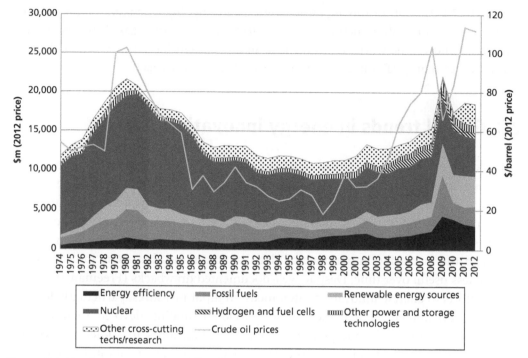

Figure 2.1 Total government spending by IEA member countries (1974–2012)

Source: (BP, 2013; IEA, 2013a).

the last few years, spending has also been boosted by government stimulus packages in response to the global financial crisis of 2008. This has had a particular impact in the United States (IEA, 2013a), but it has also led to spending increases in other countries.

Spending on renewable energy has dramatically increased by more than 10 times and energy efficiency spending has increased five-fold. At the same time, spending on nuclear power—which used to dominate public RD&D budgets in many countries, has declined by 45 per cent as the popularity of this technology has waned. These technology trends are also reflected in patterns of spending by some of the largest OECD countries. Figure 2.2 shows government RD&D spending by the EU, USA, and Japan. In general, RD&D spending of EU, USA, and Japan for the period 2001–10 has increased compared to the previous decade. All three countries or regions increased their R&D investment in renewable energy and energy efficiency. Similarly, spending on nuclear power has declined in the USA and EU, with the steepest drop occurring in the EU. It has stayed fairly constant in Japan, though post-Fukushima a different trend may emerge.

One important weakness of IEA data is that it does not include spending by non-member governments, particularly emerging economies such as China, India, and Brazil. Spending data for these countries is more difficult to find. Figure 2.3 shows a recent time series for government RD&D spending by the 'BRICS' countries, Mexico, and the United States from 2000 to 2008 (Kempener et al., 2010). It shows that Chinese government spending is growing the most rapidly. Spending has quadrupled since 2000, and

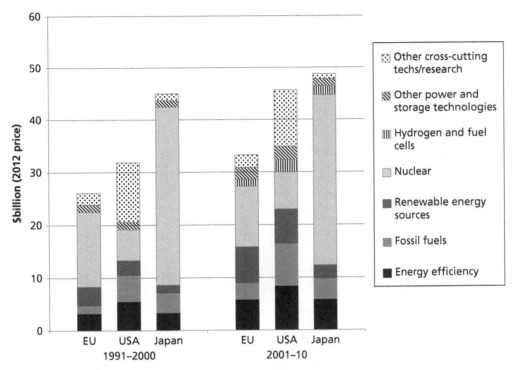

Figure 2.2 Government R&D spending by the EU, USA, and Japan (1991–2010)

Source: (IEA, 2013a).

overtook spending in the United States in the mid 2000s. It also shows that 20–50 per cent of China's spending is focused on fossil fuel RD&D. It is important to note that some of these figures for emerging economies may not be directly comparable to figures for OECD countries. In many cases, particularly China, the figures include significant spending by state-owned enterprises as well as spending directly by governments.

There are other limitations of the RD&D spending figures shown in Figures 2.1–2.3. First, some of these data only include direct spending on energy technologies. In some cases, spending on important but related areas of science and technology, such as hybrid electric vehicle drivetrains, information and communication technologies, or advanced materials that have energy sector applications, may be excluded. Second, the data does not fully capture all stages of the innovation process. The figures tend to focus on research and development. Some demonstration spending is included, but spending on the deployment of early stage technologies is likely to be excluded. Third, the data do not include significant spending by some public agencies (e.g. the devolved administrations in the UK) or the private sector.

It is difficult to find comprehensive data on private sector energy RD&D spending. Private sector spending tends to track government spending. This is not surprising given that both public and private spending are subject to some of the same drivers—and that private funding is sometimes required to 'match' public funding. Figures from the OECD

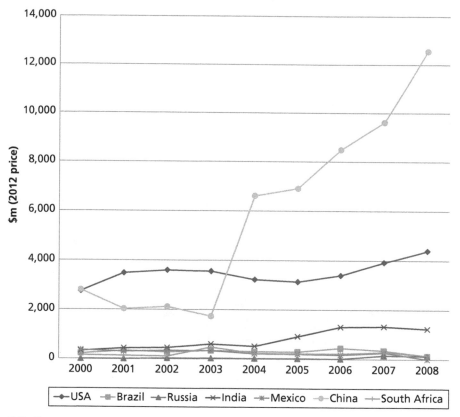

Figure 2.3 Government energy R&D spending in the 'BRICS' countries, Mexico, and the USA
Source: (Kempener et al., 2010).

show that private sector R&D in OECD countries has been dominated by spending on nuclear power and oil and gas technologies (Doornbosch and Upton, 2006). Overall private sector spending on energy R&D fell faster than public R&D after the peak in the early 1980s: from $8.5bn in the late 1980s to $4.5bn in 2003 (in 2004 dollars).

Some more detailed data are available for private sector spending in the United States. Data from the period 1997–2008 show a broadly similar overall trend to government spending on RD&D in that country (Dooley, 2010). These data are also similar to the international figures, and show that the majority of private sector spending by 'major energy producers' in recent years has focused on fossil fuels. This is confirmed by analysis by Dan Kammen and Greg Nemet which shows that private sector R&D spending on nuclear power fell during the 1990s as electricity markets were deregulated and the outlook for nuclear power investment became less positive (Kammen and Nemet, 2007). By 2004, 90 per cent of all nuclear power R&D in the United States was funded by the Federal government.

More recent data from the European Commission Joint Research Centre based on company reporting confirms these overall patterns of spending. It shows that in 2011, private sector spending was dominated by the RD&D budgets of large oil and gas companies. In his analysis of these data, Jim Skea in Chapter 1 shows that companies in the oil and gas industry had combined spending in that year that was comparable to spending by all OECD governments on all energy technologies.

2.3.2 INNOVATION OUTPUTS: PATENTS

Figure 2.4 shows global trends in patent applications for six energy technology areas between 1976 and 2010. The data only show applications filed under the international Patent Co-operation Treaty (PCT). These data show a rapid rise in applications since the early 1990s, particularly for non-fossil energy generation technologies and technologies for energy efficiency. As with energy R&D spending, the main drivers of this rapid increase in applications are likely to be increasing energy prices (from the mid 2000s) and increased emphasis on climate change mitigation. More specifically, they also reflect the increasing importance of specific policy incentives in many countries that are designed to accelerate the deployment of renewable energy and to improve energy efficiency. In general, trends in patenting often track trends in investment and deployment (Lee et al., 2009).

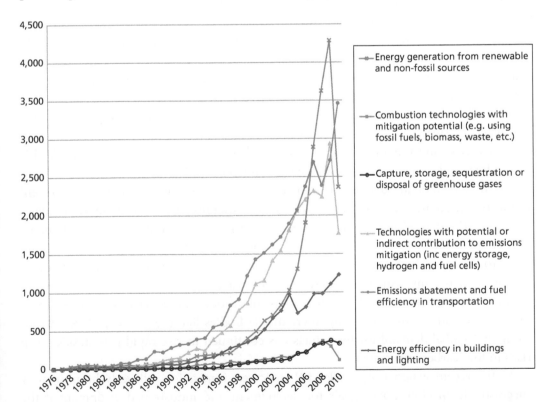

Figure 2.4 Energy technology patent applications filed under the Patent Co-operation Treaty

Source: (OECD, 2013).

The geographical spread of patenting is also changing over time. While developed OECD countries are still the origin of most patents for cleaner energy technologies, patents in China have become more important—especially in wind energy and other renewable technologies such as solar PV and concentrated solar. There are a number of reasons for this including the growth in the Chinese market for these technologies, the location of production in China by multinational firms, and the increasing importance of domestic Chinese firms (Lee et al., 2009). For similar reasons, there has been hardly any patenting of CCS technologies under the PCT in China.

2.3.3 INNOVATION OUTCOMES

As discussed earlier, a primary aim of energy innovation policies in many countries, and of some private sector investments in RD&D, is to accelerate the deployment of cleaner energy technologies. Another important aim for many countries is support for a domestic industry to develop and manufacture, install and maintain these technologies. Later chapters of this book discuss the extent to which these aims have been met for many of the most important cleaner energy technologies.

Figure 2.5 shows the global new investment in clean energy technologies between 2004 and 2012. Note that the data focus on renewable energy and smart-grid technologies. In addition to investment in the deployment of these technologies, these data include public and private investment in RD&D. In 2013, RD&D spending accounted for 10 per cent of total investment. Given the drivers of RD&D spending and patent trends presented earlier, it is not surprising that overall investment in these cleaner energy technologies has also risen steeply since 2004. There was some levelling off immediately after the 2008 financial crisis, and a significant drop in 2012 and 2013 due to falls in the cost of solar PV technology and changes in policy frameworks for renewable energy.

It is harder to obtain data for investment in energy efficiency technologies. A recent report by the environmental NGO WWF suggests that investment in these technologies was approximately 25 per cent of the total investment in renewable energy and energy efficiency technologies in 2011 (van der Slot and van den Berg, 2012). Such figures are open to criticism, however, given that judgements about what should be included as energy efficiency investment are likely to be very subjective.

A further important caveat for this data is that it does not cover trends in fossil fuel investments. Successive reports by international bodies like the IEA (e.g. IEA, 2013b) show that global demand for fossil fuels has continued to rise rapidly alongside steep rises in low-carbon investment.

As the recent experience of solar PV indicates, innovation will only be ultimately successful in meeting policy goals for deployment and industrial development if the costs of cleaner energy technologies fall over time. The costs of many cleaner energy technologies remain higher than those of incumbent fossil fuels and their associated

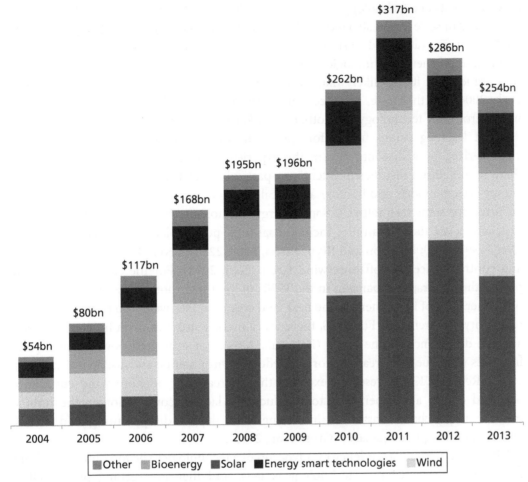

Figure 2.5 Global new investment in clean energy by sector 2004–13 ($bn)

Source: Bloomberg New Energy Finance.

electricity generation technologies, though in some cases (notably solar PV and onshore wind), those costs have fallen significantly in recent years (Gross et al., 2013).

Innovation processes are often incorporated into models of the energy system using learning curves to represent the expectation that costs will fall as cumulative investment or deployment increases. There is an extensive literature on learning curves (van der Zwaan and Rabl, 2004; Klaassen et al., 2005; Sagar and van der Zwaan, 2006) and experience curves (Neij, 1997; Rubin et al., 2004; Junginger et al., 2005; Neij, 2008; Ferioli et al., 2009). The basic assumption is that increasing deployment of a technology over time will lead to 'learning by doing' and a reduction in unit costs. These curves produce an understanding of the long-term patterns of cost development by using extrapolations of past learning (Neij, 1997; Ferioli et al., 2009). By calculating the experience curve one can try to predict how much investment is necessary to bring down the costs of a technology to a competitive level. If a technology is widely deployed, economies of scale might also reduce the cost of the technology further (Junginger et al., 2005). Learning rates tend to slow down over time and are generally lower for conventional, mature energy technologies compared to new(er) renewable energy technologies. It is argued that it is unlikely that learning rates can be sustained indefinitely (Rubin et al., 2004). Furthermore, the mechanisms through which learning takes place differ widely from one technology to another (Winskel et al., 2014).

What learning curves fail to foresee are technological breakthroughs, which are characterized by a discontinuity in the experience curve (Neij, 1997). Furthermore, there is sometimes insufficient attention—particularly in the use of learning curves in energy system models and in other analytical tools—to negative learning rates that have occurred for some technologies. Some energy technologies have actually become more expensive over time. Examples include coal-fired power plants and nuclear reactors (Neij, 1997: 1100; Hansson and Bryngelsson, 2009: 2280), LNG plants (Rai et al., 2009), and contract prices for offshore wind farms (Neij, 2008). For the case of US nuclear power, the very rapid expansion in the 1970s did not incorporate lesson-learning from one generation of investment to the next, and cost increases were also driven by stricter regulatory requirements. For LNG, there was a market structure with little competition and the dominance of one buyer (Rai et al., 2009). For coal-fired power plants, price increases were due to increased efforts to reduce environmental and accident risks (Neij, 1997). Recent UKERC research explains that increases in offshore wind costs have occurred due to a number of factors including a lack of competition in the turbine market, supply chain constraints, increasing water depths, disappointing load factors, and commodity price increases (Gross et al., 2013).

There is sometimes a tendency by advocates of a technology to be too optimistic about costs—particularly where a technology is new, and they are seeking public policy support. This so-called 'appraisal optimism' is a generic phenomenon when new energy technologies are being developed (Hansson and Bryngelsson, 2009: 2275; Scrase and Watson, 2009: 4539). There is a temptation for industry to produce low-cost estimates to secure regulatory support and to reveal 'real' costs once that support is committed. Recent research by UKERC suggests that such appraisal optimism may have abated in recent years, at least in the UK (Gross et al., 2013).

2.4 Conclusions and Implications

This chapter has discussed some of the different approaches to the analysis of innovation within energy systems and critically analysed recent global trends in energy innovation. It has shown that, in common with innovation in general, innovation in energy technologies is a complex process that should be analysed from a systems' perspective. Innovation is not only concerned with the development of new technologies through RD&D—it also concerns the deployment of these technologies in markets. There is a close inter-relationship between technological change within energy systems and other dimensions of change in these systems. The direction and intensity of energy innovation processes are heavily influenced by consumer preferences and needs, market structures and rules, government policy frameworks and incentives, and by the priorities and strategies of incumbent energy firms.

The chapter has shown that government policies have a particularly important influence over the direction of innovation—both nationally and internationally. Such policies are critical if energy innovation is to continue to be steered in a more sustainable direction—and global climate change targets are to be met. However, it is important to recognize that while national governments and international organizations can implement policy frameworks including funding and incentives for low-carbon energy technologies, innovation is often carried out by firms, universities, and research organizations rather than governments themselves.

The data presented in Section 2.3 confirms that, particularly during the last decade, the scale and intensity of innovation activities worldwide have been growing significantly. In addition to this, innovation within emerging economies (particularly China) has become more globally significant. Public RD&D spending has been growing, though arguably it needs to grow further. While it is not possible to determine how much larger public RD&D budgets should be, because such spending does not guarantee the desired outcomes from the innovation process, the IEA has suggested an increase to 3–6 times current levels is required (Skea et al., 2013). From a global perspective, efforts to decarbonize energy systems (including through the deployment of low-carbon technologies) are still proceeding too slowly.

Given that many countries are resource-constrained, a situation that has been exacerbated by the financial crisis of 2008, public funding for R&D, technology demonstration, and market creation is subject to prioritization and specialization. Few countries can afford to support the full range of technologies and the associated organizational, institutional changes to the same degree. Furthermore, support for innovation in one country (e.g. through the feed-in-tariff policy in Germany) can benefit firms in another country (e.g. solar PV manufacturers in China). Therefore, national innovation policies can only go so far. Where it is possible, international collaboration (e.g. within the EU or between developed and developing countries) will be important. But competition between countries and trading blocs will limit the scope for such collaboration.

This does not only mean that particular countries may wish to prioritize particular low-carbon technologies. It is also important to focus on the right 'stage' of the innovation process. Some technologies, such as offshore wind and carbon capture and storage, have not yet navigated the notorious 'valley of death' between early stage development and commercial deployment. Furthermore, there is a need to recognize that the risky nature of innovation means that there will be failures as well as successes, and that it is important that lessons are learned from both.

Finally, the understanding of innovation in energy systems—and policies to steer innovation processes—needs to take account of the history and momentum of these systems. The systemic innovation required to shift energy systems in a more low-carbon and sustainable direction is likely to include challenges to existing, unsustainable technologies and their associated actors and interests. This means that efforts by governments to support low-carbon technologies—and of firms seeking to develop and deploy them—will have a strong political dimension. The history of innovation shows that incumbent technologies and their associated interests tend to fight back as new technologies emerge.

■ REFERENCES

ARTHUR, B. 1989. Competing technologies, increasing returns and lock-in by historical events. *Economic Journal*, 99, 116–31.

BP 2013. *BP Statistical Review of World Energy 2013*. London: BP.

DECC 2012. *Energy Efficiency Strategy*. London: Department of Energy and Climate Change.

DEPARTMENT OF TRADE AND INDUSTRY 2004. *The Renewables Innovation Review*. London: DTI.

DEWALD, U. & TRUFFER, B. 2011. Market formation in technological innovation systems—diffusion of photovoltaic applications in Germany. *Industry and Innovation*, 18, 285–300.

DOOLEY, J. J. 2010. The Rise and Decline of U.S. Private Sector Investments in Energy R&D since the Arab Oil Embargo of 1973. Pacific Northwest National Laboratory.

DOORNBOSCH, R. & UPTON, S. 2006. Do we have the right R&D priorities and programmes to support energy technologies of the future. *Background Paper, 18th Round Table on Sustainable Development*. Paris: OECD.

FERIOLI, F., SCHOOTS, K. & VAN DER ZWAAN, B. C. C. 2009. Use and limitations of learning curves for energy technology policy: A component-learning hypothesis. *Energy Policy*, 37, 2525–35.

FOXON, T. 2003. Inducing innovation for a low-carbon future: drivers, barriers and policies. *Report for the Carbon Trust*. London: Carbon Trust.

FRANTZESKAKI, N. & LOORBACH, D. 2010. Towards governing infrasystem transitions: Reinforcing lock-in or facilitating change? *Technological Forecasting and Social Change*, 77, 1292–301.

FREEMAN, C. 1992. *The Economics of Hope*. London, New York: Pinter.

FREEMAN, C. 1987. *Technology and Economic Performance: Lessons from Japan*. London: Pinter.

FREEMAN, C. & LOUCA, F. 2001. *As Time Goes By: From the Industrial Revolutions to the Information Revolution*. Oxford: Oxford University Press.

FREEMAN, C. & SOETE, L. 2009. Developing science, technology and innovation indicators: What we can learn from the past. *Research Policy*, 38, 583–9.

GALLAGHER, K. S., HOLDREN, J. P., & SAGAR, A. D. 2006. Energy-technology innovation. *Annual Review of Environmental Resources*, 31, 193–237.

GEELS, F. 2005. The dynamics of transitions in socio-technical systems: A multi-level analysis of the transition pathway from horse-drawn carriages to automobiles (1860–1930). *Technology Analysis & Strategic Management*, 17, 445–76.

GEELS, F. W. 2002. Technological transitions as evolutionary reconfiguration processes: a multi-level perspective and a case-study. *Research Policy*, 31, 1257–74.

GEELS, F. W. 2004. From sectoral systems of innovation to socio-technical systems: insights about dynamics and change from sociology and institutional theory. *Research Policy*, 33, 897–920.

GROSS, R., HEPTONSTALL, P., GREENACRE, P., CANDELISE, C., JONES, F., & CASTILLO, A. C. 2013. Presenting the future: An assessment of future costs estimation methodologies in the electricity generation sector. London: UK Energy Research Centre.

GRUBLER, A., AGUAYO, F., GALLAGHER, K., HEKKERT, M., JIANG, K., MYTELKA, L., NEIJ, L., NEMET, G., & WILSON, C. 2012. Chapter 24—Policies for the Energy Technology Innovation System (ETIS). In: *Global Energy Assessment—Toward a Sustainable Future*. Cambridge University Press, Cambridge, UK and New York, NY, USA and the International Institute for Applied Systems Analysis, Laxenburg, Austria.

HANSSON, A. & BRYNGELSSON, M. 2009. Expert opinions on carbon dioxide capture and storage—A framing of uncertainties and possibilities. *Energy Policy*, 37, 2273–82.

HEKKERT, M. P. & NEGRO, S. O. 2009. Functions of innovation systems as a framework to understand sustainable technological change: Empirical evidence for earlier claims. *Technological Forecasting and Social Change*, 76, 584–94.

IEA 2013a. IEA Energy Technology Research and Development Database 2013 edition. Paris: International Energy Agency.

IEA 2013b. *World Energy Outlook 2013*. Paris: OECD/International Energy Agency.

JACOBSSON, S. & BERGEK, A. 2004. Transforming the energy sector: the evolution of technological systems in renewable energy technology. *Industrial and Corporate Change*, 13, 815–49.

JAFFE, A. B., NEWELL, R. G., & STAVINS, R. N. 2005. A tale of two market failures: Technology and environmental policy. *Ecological Economics*, 54, 164–74.

JUNGINGER, M., FAAIJ, A., & TURKENBURG, W. C. 2005. Global experience curves for wind farms. *Energy Policy*, 33, 133–50.

KAMMEN, D. M. & NEMET, G. 2007. U.S. energy research and development: Declining investment, increasing need, and the feasibility of expansion. *Energy Policy*, 35, 746–55.

KEMPENER, R., ANADON, L. D., & CONDOR, J. 2010. Database of Energy RD&D Investments in Brazil, Russia, India, Mexico, China, and South Africa. *Database of Energy RD&D Investments in Brazil, Russia, India, Mexico, China, and South Africa*. Cambridge, MA, USA: Kennedy School of Government, Harvard University.

KERN, F. 2010. *The politics of governing 'system innovations' towards sustainable electricity systems*. Doctoral thesis, University of Sussex.

KLAASSEN, G., MIKETA, A., LARSEN, K., & SUNDQVIST, T. 2005. The impact of R&D on innovation for wind energy in Denmark, Germany and the United Kingdom. *Ecological Economics*, 54, 227–40.

KÖHLER, J., WIETSCHEL, M., WHITMARSH, L., KELES, D., & SCHADE, W. 2010. Infrastructure investment for a transition to hydrogen automobiles. *Technological Forecasting and Social Change*, 77, 1237–48.

LEE, B., ILIEV, I., & PRESTON, F. 2009. *Who Owns our Low Carbon Future?* London: Chatham House.

LOORBACH, D., FRANTZESKAKI, N., & THISSEN, W. 2010. Introduction to the special section: Infrastructures and transitions. *Technological Forecasting and Social Change*, 77, 1195–202.

LUNDVALL, B.-A. 1992. *National Systems of Innovation: Towards a Theory of Innovation and Interactive Learning.* London: Pinter.

LUO, L., LACAL-ARANTEGUI, R., WIECZOREK, A. J., NEGRO, S. O., HARMSEN, R., HEIMERIKS, G. J., & HEKKERT, M. P. 2012. A Systemic Assessment of the European Offshore Wind Innovation. Insights from the Netherlands, Denmark, Germany and the United Kingdom. Petten, The Netherlands: European Commission Joint Research Centre Institute for Energy and Transport.

NEIJ, L. 1997. Use of experience curves to analyse the prospects for diffusion and adoption of renewable energy technology. *Energy Policy*, 25, 1099–107.

NEIJ, L. 2008. Cost development of future technologies for power generation—A study based on experience curves and complementary bottom-up assessments. *Energy Policy*, 36, 2200–11.

OECD 2013. OECD StatExtracts. Paris: OECD.

PAVITT, K. 1984. Sectoral patterns of technical change: towards a taxonomy and a theory. *Research Policy*, 13, 343.

PEARSON, P. & FOUQUET, R. 2006. Seven centuries of energy services: the price and use of light in the United Kingdom (1300–2000). *The Energy Journal*, 27, 139–77.

RAI, V., VICTOR, D. G., & THURBER, M. C. 2009. Carbon Capture and Storage at Scale: Lessons from the Growth of Analagous Energy Technologies. *Working Paper #81.* Freeman Sprogli Institute for International Studies.

ROTHWELL, R. 1994. Towards fifth-generation process innovation. *International Marketing Review*, 11, 7–31.

RUBIN, E. S., TAYLOR, M. R., YEH, S., & HOUNSHELL, D. A. 2004. Learning curves for environmental technology and their importance for climate policy analysis. *Energy*, 29, 1551–9.

SAGAR, A. D. & VAN DER ZWAAN, B. 2006. Technological innovation in the energy sector: R&D, deployment, and learning-by-doing. *Energy Policy*, 34, 2601–8.

SCOTT, A. & STEYN, G. 2001. The Economic Returns to Basic Research and the Benefits of University-Industry Relationships: A Literature Review and Update of Findings. *Report to the Office of Science and Technology.* Brighton: SPRU.

SCRASE, J. I. & WATSON, J. 2009. Strategies for the deployment of CCS technologies in the UK: a critical review. *Energy Procedia*, 1, 4535–42.

SKEA, J., HANNON, M., & RHODES, A. 2013. Investing in a Brighter Energy Future: Energy Research and Training Prospectus. London: Imperial College/Research Councils Energy Programme.

STERN, N. 2006. *Stern Review: The Economics of Climate Change.* London: HM Treasury.

TREZONA, R. 2009. *Bridging the Valley of Death in Low Carbon Innovation.* London: Carbon Trust.

UNRUH, G. C. 2000. Understanding carbon lock-in. *Energy Policy*, 28, 817–30.

UNRUH, G. C. 2002. Escaping carbon lock-in. *Energy Policy*, 30, 317–25.

VAN DER SLOT, A. & VAN DEN BERG, W. 2012. Clean Economy, Living Planet. *Report commissioned by WWF.* Amsterdam, Netherlands: Roland Berger Strategy Consultants.

VAN DER ZWAAN, B. & RABL, A. 2004. The learning potential of photovoltaics: implications for energy policy. *Energy Policy*, 32, 1545–54.

VERBONG, G. & LOORBACH, D. (eds.) 2012. *Governing the Energy Transition. Reality, Illusion or Necessity?* New York and London: Routledge.

WEBER, M. & HEMMELSKAMP, J. (eds.) 2005. *Towards Environmental Innovation Systems.* Berlin and Heidelberg: Springer.

WINSKEL, M., MARKUSSON, N., JEFFREY, H., CANDELISE, C., DUTTON, G., HOWARTH, P., JABLONSKI, S., KALYVAS, C., & WARD, D. 2014. Learning pathways for energy supply technologies: bridging between innovation studies and learning rates. *Technological Forecasting and Social Change*, 81, 96–114.

3 Deepening globalization: economies, trade, and energy systems

Gavin Bridge and Michael Bradshaw

This chapter considers the relationship between energy systems and the spatial integration of economies through trade and investment. It characterizes this relationship as one of 'deepening globalization' over time, associated with increases in both the spatial extent and intensity of international economic interdependence via trade and investment (Held et al., 1999; Yeung, 2009). Energy systems and the globalization of trade and investment are closely intertwined. Energy is a significant 'factor of production': energy costs vary spatially, influencing the profitability of producing goods in particular places and transporting them to market (Strange, 1988). Historically energy systems have had a significant influence on the location of production and the extent to which production and consumption may be separated geographically. The 'global' manufacturing systems that characterize the contemporary world economy—and which grew to prominence in the second half of the twentieth century—were shaped in significant ways by a five-fold expansion in the availability of primary energy (and oil in particular) during this period (Dicken, 2011; Grubler, 2012). In the last couple of decades, 'national' energy systems have themselves become more spatially integrated and interdependent: policies to liberalize trade and investment have driven a re-scaling of energy systems, with energy companies, infrastructures, and markets expanding beyond national borders. This growing international interdependence of critical energy systems—experienced as changes to their territorial form, ownership, and control—poses a major challenge for national energy policy.

The two-way relationship between energy systems and globalization is at the heart of this chapter. Section 3.1 focuses on energy systems as a driver of globalization and examines the influence of energy technologies and energy costs on how economic activity is organized geographically. It highlights how energy is harnessed to overcome 'the friction of distance' and achieve what economic geographer's have come to call 'time-space compression'—the simultaneous speeding up of economic activity and erosion of spatial barriers that defines the experience of globalization (Harvey, 1989; Dicken, 2011); and it shows how declines in the significance of transport and other energy costs have historically created new territorial forms of economic activity, such as the emergence of global trade in bulk goods in the nineteenth century and the globalization of manufacturing supply chains in the second half of the twentieth century. Section 3.2 focuses on the liberalization of energy trade and investment and the increasing spatial integration of energy markets, technologies, and infrastructures. It highlights how markets for coal, oil, gas, and electricity are spatially integrated ('globalized') in different ways, and how geographies of energy trade and investment have changed over time. A short Section 3.3 considers the implications of transition to a low-carbon economy for a further deepening of globalization, and the chapter then concludes.

3.1 Energy systems as a driver of globalization: making, moving, and consuming

Three general processes underpin the way energy influences economic integration and the geographies of economic activity: energy as a factor of *production*, as an enabler of *transportation* and trade, and as a vital ingredient of *consumption*. Each of these processes has a geographical component (reflecting variability across space) and an historical component (associated with change over time). This section explores each in turn.

A primary way in which energy systems influence the widening and deepening of globalization concerns the cost at which they deliver energy as an input to economic activity: that is, energy as a factor of production.[1] Over time there has been a long-run decline in the cost of energy relative to other factors of production (Sorrell, 2009). These declines are clearest when measured in terms of the energy services delivered, such as tons of freight moved or lumens of light delivered (Smil, 2010b): by 2000, for example, the cost of artificial illumination in the UK had fallen to one three-thousandth of its value in 1800 (Fouquet and Pearson, 2006). The decline in relative energy costs is linked to improvements in the productivity of production processes, whereby a greater amount of output can be achieved for a given energy input. Such improvements are typically attributed to technological innovation, although ecological economists point out how they are also a result of fuel-shifting and the substitution of higher quality (lower entropy) energy carriers such as oil, gas, or electricity for coal or wood.[2] Ayres and Warr (2009: xxi), for example, argue that economic growth 'has been driven primarily not by "technological progress" in some general or undefined sense, but specifically by the availability of ever-cheaper energy ... from coal, petroleum (and gas).' The shift to higher quality fuels is credited with driving a six-fold increase in energy availability in

---

[1] Susan Strange (1988) refers to energy as a fifth factor of production: while the classical economists noted only three (land, labour, and capital), Strange adds technology and energy. Ayres and Warr (2009) refine this further in the context of their critique of neo-classical models of economic growth, adding energy (specifically 'useful work') as a necessary complement to capital and labour as factors of production.

[2] Biophysical and ecological economists note that 'technology' is a poor explanatory device as 'technology is simply the specific methods by which energy is applied to upgrade and transform natural resources' (Hall et al., 1986: 42).

OECD economies on a per capita basis over the last 200 years, to approximately 8 tons of oil equivalent per person (Ruhl et al., 2012). The expansion of energy availability over the last century—and particularly since the Second World War—has underpinned high (although geographically very uneven) rates of productivity growth, economic development, and wealth creation.

Over time, a growing supply of higher-quality energy inputs and falling prices for energy services have enabled manufacturers to expand energy use, substituting capital equipment (electric motors, internal combustion engines, steam turbines) for the more diffuse energies provided by water, wind, and human and animal muscle. This process has occurred at different speeds in different sectors and locations: fossil fuels and mechanical power were introduced to cotton manufacturing in the UK nearly 100 years before similar transitions in agriculture, for example. Overall, however, a growing availability of energy— and the progressive substitution of increasingly high-quality (i.e. concentrated) forms of energy in production—has underpinned the remarkable increases in labour productivity that characterize twentieth century economic growth (Figure 3.1).[3] Economic historians

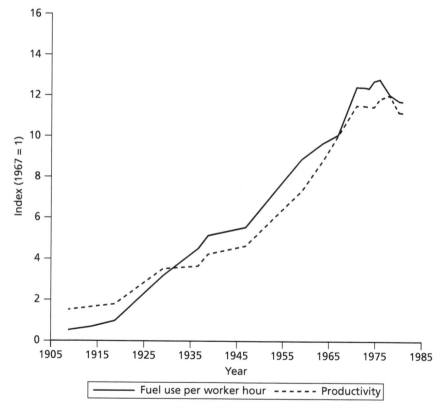

Figure 3.1 Labour productivity and energy use (based on Hall et al., 1986)

---

[3] Hall et al. (1986: 43) use an instructive example from coal mining in the United States: between 1954 and 1967, extraction techniques shifted away from a dominance of labour power towards a reliance on

highlight how substituting electric motors for steam and waterpower drove significant gains in industrial productivity in the UK, USA, and Sweden in the early twentieth century (David, 1990; Enflo et al., 2009). Similarly the mid-twentieth-century dominance of the USA in global manufacturing has been attributed to its 'unsurpassed access to affordable energy in general and to inexpensive electricity in particular' (Smil, 2013: 67).

These examples illustrate how it is not only the availability of energy that determines the significance of energy as a factor of production, but also how that energy is harnessed to deliver economic output. The relationship between the long-run expansion in energy availability and economic development is complex, as it is influenced by the range of energy sources, technologies of energy conversion, and the structure of the economy. Historic trends in the energy intensity of economic output—the amount of energy required to produce a unit of GDP—are a matter of some debate (Gales et al., 2007; Ruhl et al., 2012). The energy intensity of many economies has declined over time: over the past 200 years energy intensity has fallen in Western economies by at least a factor of three, and possibly much more (Ruhl et al., 2012). Analyses that focus only on commercially traded fuels suggest an inverted-U pattern: energy intensity rises from a low base during the early phases of economic development (reflecting the growing role of markets in the allocation of energy sources as well as a shift in economic structure from agriculture to industry) to a peak and then subsequently declines (due to improvement in energy efficiency and, in some contexts, further evolution in economic structure towards services).[4] However, other studies that include traditional fuels, fodder, and food in their analyses challenge the assumption that energy intensity was low in the pre-industrial period (Gales et al., 2007): they indicate not a 'peaking' in energy intensity but a longer-run decline driven primarily by technological change and the substitution of higher-quality energy sources. Overall, the global economy has been shaped in significant ways by the growing availability of high-quality energy sources and by a general decline in the energy intensity of economic activity. However, these changes have not occurred everywhere equally and, as a consequence, the significance of energy as a factor of production is geographically uneven. This geographical diversity has important implications: it underpins patterns of economic integration (spatial differences in energy availability and energy intensity influence international energy trade, the distribution of certain forms of manufacturing, and territorial carbon emissions, for example); and it explains why the 'global energy dilemma' takes distinctive forms in different parts of the world (Bradshaw, 2014).

The effect of transitions towards higher-quality energy sources—what Smil (2010b) terms the 'grand fuel sequence'—has been a general loosening over time of those

---

continuous mining machines. As a consequence, total employment fell by 55 per cent, the amount of fuel used per miner rose nearly 100 per cent, and labour productivity rose 102 per cent.

[4] Ruhl et al. (2012), for example, indicate energy intensity peaking in the UK in the 1890s and declining by a factor of four over the following century, with the USA, China, and India peaking in the 1920s, 1980s, and 1990s respectively.

locational ties associated with energy supply. The electrification of industrial power, for example, introduced a new degree of spatial flexibility by enabling manufacturing to spread beyond traditional sites based on water power and/or proximity to coal, generating new geographies of industry at urban, regional, and national scales. In a similar way, the superior mobility of oil over coal 'reinforced a net increase in the mobility of the major factors of production' in the second half of the twentieth century (Strange, 1988: 190). The net effect has been that other factors—notably the cost and productivity of labour—have emerged as key determinants of industrial location. The 'global shift' in manufacturing over the last 40 years—associated with the de-industrialization of traditional manufacturing heartlands in Europe and North America and the rise of new 'workshops of the world' within Asia (Dicken, 2011)—may be traced, in part, to this decline in the relative significance of energy as a factor of production and a much greater sensitivity to the cost of labour.

Although the relative cost of energy services has declined, there has been only limited spatial convergence in the price of fuels. There remain significant geographical price spreads for oil, gas, coal, and electricity that reflect the persistence of regional market structures (in the case of gas and electricity, for example), as well as policies on energy taxation and energy subsidy that create wide national variances from international prices (IEA, 2010, 2011). In the case of oil, although crude is internationally priced, consumer prices for petrol worldwide vary by a factor of 100 (from $0.09 gallon in Venezuela to $9.00 gallon in Turkey and Norway) with subsidy and taxation the most significant source of this variation (Davis, 2013). Consumer subsidies on fossil fuels (and related under-pricing of electricity) account for the lion's share of an estimated $700 billion in producer and consumer energy subsidies per year (approximately 1 per cent of world GDP), with the highest rates of subsidy among oil and gas exporters in the Middle East, North Africa, and Central Asia (IEA, 2010, 2011).

Energy prices, then, retain a 'distinct geographic structure ... (that) combines both international markets and intranational policies and supply structures' (Mulhall and Bryson, 2014). This spatially uneven character of energy prices affects firms and sectors in different ways and is particularly significant for industries that are energy intensive. For cement manufacturers and primary aluminum producers, for example, energy can account for over a third of total costs: as a consequence, the geography of aluminium production is associated with locations with large power surpluses (typically derived from hydro-electricity) where favourable energy supply contracts can be obtained. By contrast, in engineering, light manufacturing, and services, energy typically represents a much lower proportion of input costs (although issues around reliability of supply can be significant, for example in the context of data servers that form critical hubs of the information economy). Geographic variations in the price of energy can have implications for national competitiveness, particularly in sectors like chemicals, steel, and metals processing that are relatively energy intensive. In the United States, shale gas production—and relatively limited energy efficiency and greenhouse gas requirements at the national level[5]—have enabled industrial power costs to fall, raising concerns in the EU that its higher energy price regime may place it at a disadvantage. There is also some limited evidence that lower energy prices in the United States may be reversing the trend for industries serving the US market to 'offshore' production to regions with lower energy costs (Boston Consulting Group, 2012; PWC, 2012). Shale gas has been popularly credited with leading this 're-shoring' of energy-intensive production (via its effect on energy costs and the enhanced availability of associated natural gas liquids as a petrochemical feedstock), although talk of a wider 'manufacturing renaissance' in the USA appears premature (to the extent it is occurring, it may be attributed more to a convergence in the cost of labour between the USA and China, rising transportation costs, and a weak US dollar).

A second process by which energy systems are implicated in the progressive deepening of globalization concerns the role of trade in accelerating economic integration. A defining characteristic of economic globalization is the way international trade has grown faster than economic output: while the production of goods increased six-fold between 1960 and 2010, international trade in these goods increased twenty-fold (Dicken, 2011). At the world scale, the value of trade now accounts for over 51 per cent of GDP, compared to 24.5 per cent in 1960 (there is, however, wide variation at the level of individual countries).[6] Behind such figures is the mundane business of moving goods around: the economic history of globalization—of accelerating economic integration via trade—is, at least in part, a history of the application of energy in water, rail, road, and air transportation. More significantly, the contemporary global economy—a pattern of economic integration predicated on extensive international trade—has taken form over the last half-century around a particular suite of transportation technologies (diesel engines and gas turbines in particular) whose function depends on inexpensive and plentiful supplies of liquid fossil fuels: these engines are, in a physical sense, 'fundamentally... more important to the global economy than are any particular corporate modalities or international trade agreements' (Smil, 2010a: 18).

Energy historians trace a series of innovations in fuels and energy technologies ('prime movers') that, over time, have been associated with distinctive waves of economic integration. Each of the 'Kondratieff waves' that characterizes the emergence of industrial economies over the last 250 years is based on a distinctive assemblage of energy technologies and infrastructures, from water power and roads associated with early mechanization in the 1770s; through steam power, railways, and shipping in the 1830s and electricity and steel in the 1890s; to automobiles, highways, aircraft, and airports in the 1950s (Dicken, 2011). Smil (1999, 2010b), for example, comments on developments in construction, design, and rigging of sailing ships from the seventeenth century onwards

---

[5] For a cross-national analysis of the impact of energy and climate change policies on energy intensive industries (including the United States), see ICF (2012).

[6] For China, the figures are 9.3 per cent (1960) and 67.8 per cent (2007); for India; 12.5 per cent and 30.8 per cent respectively.

which enabled more effective conversion of the wind's kinetic energy to forward motion, enabling an increase in the speed of movement and expanding the scope of long-distance ocean trade. It was developments in steam propulsion, however, that enabled the development of inter-continental bulk cargo shipping and commodity market integration in the first half of the nineteenth century. While the expanding power of steamships reduced the costs of ocean freight, railways connected the continental interiors of the Americas, Eurasia, and Africa to worldwide trade (Chisholm, 1990). The greater speeds afforded by steam power and the capacity to transport large tonnages at lower cost drove a process of time–space compression and (uneven) spatial integration. In the United States, for example, expanding rail networks served to integrate regional commodity markets, such as those for wheat, meat, and forest products, leading to a convergence of prices: for example, the wheat price spread between New York City and Iowa fell from 69 per cent to 19 per cent between 1870 and 1910; similarly, the difference between the price of wheat in Liverpool and Chicago fell from 57.6 per cent in 1870 to 15.6 per cent in 1913 (O'Rourke and Williamson, 1999). The application of steam power to transportation increased both the intensity and extensity of economic integration, drove further the geographical separation of production and consumption to the continental scale, and expanded significantly the 'supply zones' of the global economy: for example, the average distance travelled by agricultural imports to the UK increased from 2,929 km in the period 1831–35 to 9,463 km in the period 1909–13 (Chisholm, 1990). The expanding capacity to move greater tonnages across space at higher speeds and/or lower costs rested on a critical transition in energy supply: the shift from 'organic' to 'mineral' resources (Wrigley, 1988).

The twentieth century saw further waves of 'time–space compression' associated with the development and application of new transportation technologies and fuels, consolidating the decisive nineteenth-century shift in economic integration.[7] A significant transition occurred in the early twentieth century from coal to oil as the primary fuel of transportation, reflecting the superior energy services that oil could provide as a result of its liquid character and higher energy density (twice that of coal by weight and around 50 per cent more than gas by volume). Oil's higher energy density changed the economies of scale required for crossing space, allowing the size of vehicle units to fall—from the train and tram to the automobile—and an increase in the power output for a given size or weight of engine. Oil's energy density enabled the evolution of the internal combustion engine (where oxidation/combustion on a small scale releases a sufficient amount of energy to enable the direct movement of a piston), as opposed to the much larger, external combustion engines associated with steam power (Smil, 1999). Oil was not the first fossil fuel to have significantly shrunk distance: the introduction of coal-fired steamships in the second half of the nineteenth century drove down shipping costs and further facilitated long-distance trade in bulk commodities like wheat and wool. Yet oil consolidated this process and drove it further (Bridge and Le Billon, 2013). In cars and

---

[7] This paragraph draws on Bridge and Le Billon (2013).

planes and on buses and trains, the number of passenger-kilometres travelled increased ten-fold between 1950 and 2000, rising from an average of 1,400 km per person to 5,500 km per person. Such a startling increase in the global average hides significant regional disparities (see Schäfer, 2006): the process of 'deepening globalization' creates new patterns of uneven development rather than flattening spatial difference.

The tonnage of seaborne international freight has continued to increase much faster than both world population and GDP: whereas international trade was around 100 kg per person in 1900, it had risen to 1,240 kg per person in 2012.[8] Smil (2010a: 18) argues that the contemporary global economy of trade rests, in particular, on the diesel engines and gas turbines that have become the 'indispensable driving forces of the global economy'. Marine diesels have facilitated a 'true planetary economy' (Smil, 2010a) via the inexpensive transportation of bulk raw materials and finished goods. The price of bulk shipping declined steadily in the second half of the twentieth century, so that by 2004 it was half as much per ton as in 1960 (Hummels, 2007: 142): seaborne cargoes now account for around 90 per cent of world merchandise trade by volume, and 70 per cent by value (WTO, 2009).[9] Where an earlier wave of globalization expanded the geographical reach of economic interaction, the second half of the twentieth century has witnessed a deepening of globalization as the intensity of international economic interactions has increased. Scale economies in transportation have accelerated economic integration by decreasing 'frictional costs': the speed of freight movements via land and sea has not changed substantially on a kilometre per hour basis, but significant time–space compression is apparent over the last fifty years when measured in terms of tons per kilometre per hour (crude oil, for example, continues to move around the world—in ships and pipelines—at about the same speed as forty years ago—around 20 kilometres per hour—but the volume of global oil trade has risen by about 50 per cent). Achieving economies of scale in marine transportation has rested on improved ship design, the introduction of containerization and the scaling up of tankers and bulk carriers, increases in the size and efficiency of marine engines, and the continuing availability of inexpensive bunker fuels (prices for which are related to crude oil and currently carry no 'carbon tax'). The net result is that, for many goods, shipping costs are now a relatively small proportion of final product costs. Transportation costs declined from 'an average of 8% of total import costs in 1970 to about 3% in 2002' (Dean and Sebastia-Barriel, 2004: 314, cited in Dicken, 2011: 83). Furthermore, above a certain threshold shipping costs do not increase substantially with distance, making it possible to geographically separate different phases in the production of a given product over very large distances to take advantage of spatial differences in input costs. The growth of global production chains—in which a series of discontinuous sites spread across the globe are functionally integrated in the production of a single product—reflects this reduction in the significance of transportation costs.

[8] The figure for 1900 is from Chisholm (1990); the figure for 2012 is calculated based on world seaborne trade of 8.7 billion tons (UNCTAD 2013a: 5) divided by a world population of 7 billion people.
[9] This figure excludes trade in the EU.

Like seaborne trade, a similar process of time–space compression can be seen with global air freight and air passenger travel. While average cruising speeds of aircraft have plateaued since the optimization of the jet engine, ton-kilometre and passenger-kilometres have expanded greatly: average air shipping costs ($ per kg) declined dramatically following the introduction of jet engines, so that global air freight—which in 1965 stood at around 1.8 billion ton-miles—had by 2004 increased to 79.2 billion ton-miles, an annual increase of over 11 per cent and significantly higher than total trade (Hummels, 2007: 133). In the second half of the twentieth century, air freight assumed an increasingly significant role in driving economic integration via trade: in reference to the United States, for example, it went from being an 'insignificant share of trade to a third of U.S. imports by value and half of U.S. exports outside of North America' (Hummels, 2007: 152). Falling air shipping costs have extended the transportable range for perishable, high-value agricultural products—such as fresh fruits, vegetables, and flowers—enabling retailers (in the Global North) and producers (in the global South) to take advantage of 'counter-seasonality'. The effect has been to collapse both space and time as it relates to global food provisioning: distant agricultural landscapes in the global South become tightly linked to the purchasing desires of consumers in the Global North, while consumers' experience of seasonality is transformed (via the 'endless summer' of fresh fruits and vegetables available through supermarkets) (Cook, 2004; Freidberg, 2004). The quest for time–space compression via higher speeds continues to be a feature in ground transportation, particularly around the construction of highways and high-speed rail. Proponents of both cite the economic value of increased connectivity and a greater intensity of economic interaction at regional, national, or international scales. Higher speeds, however, require higher energy inputs.

A third way in which energy systems are associated with globalization is via consumption and, more generally, the profound transformation to the experience of time and space associated with modernity (this is treated briefly here: see Chapter 5 for a more extensive discussion of this issue related to energy demand). The ability to call on progressively larger quantities of energy—and to access energy in the high-quality forms of liquid fuels and electricity—has historically underpinned significant transformations in household practices and routines of social reproduction (e.g. food provisioning, personal mobility, hygiene, and sanitation). Access to modern energy services has long been a measure of development. Indeed, strategies of economic development have frequently *built in* expectations and dependencies around the availability of modern energy services, such that their involuntary withdrawal constitutes an 'energy crisis'—from electricity blackouts to interruption of oil and gas supplies (Bridge, 2011).

Increasing access to energy services is now an explicit development goal, as shown by initiatives like the UN's *Sustainable Energy for All* and recognition that a post-2015 successor to the Millennium Development Goals needs to include reference to energy. This acknowledgement of the importance of energy access reflects the way standards of living and consumption norms are closely linked to the availability of energy services. In industrial economies (and particularly in North America), cheap energy in the post-war period enabled households to significantly expand consumption norms—around personal mobility, car ownership, and average house size, for example—and played an important role in driving unprecedented increases in standards of living (Huber, 2013). The embedding of mass consumption norms and expanding access to energy services rested on the ability of industrial economies to draw on energy surpluses that were distant in space (e.g. the capture, transformation and spatial transfer of current solar energies via agriculture) and/or time (e.g. the extraction and combustion of fossil fuels). In this way the dramatic expansion of energy services at the household level—a characteristic of modern norms of mass consumption—has played an important role in driving economic integration via trade. Consumption-based accounting of energy use and carbon emissions reveals the significance of trade and economic structure in determining a country's final energy demand (Minx et al., 2011). The globalization of manufacturing supply chains makes possible a geographic displacement of energy use and emissions, enabling consuming countries to effectively 'outsource' the demand for energy and emissions associated with all stages of product manufacture other than final consumption (Peters et al., 2011). The scale of energy and emission transfers via trade can be significant (Machado et al., 2001; Liu et al., 2010). Recent work on the UK, for example, reveals how changing patterns of production and trade (and growing demand) have caused emissions outside the UK but associated with UK final consumer demand to rise, while the UK's territorial emissions of greenhouse gases (as monitored under the United Nations Framework Convention on Climate Change) have fallen (Baiocchi and Minx, 2010). Overall the large net transfers of energy to industrial economies associated with economic integration through trade have ensured an expanding availability of cheap energy in these economies, enabling the gains in industrial productivity and improved standards of living described above. Ecological economists have introduced the notion of 'ecological debt' to highlight the social and spatial transfers of energy and raw materials associated with deepening globalization in the post-war period (Martinez-Alier, 2002).

In summary, this first section has identified three general processes by which energy systems are drivers of globalization. In practice these processes are not separate: to the extent that falling transport costs and growing international trade realize the principle of comparative advantage, they also drive overall gains in productivity within the tradable goods sector, for example. The primary point, however, is that the historical evolution of these 'economic' processes has been underpinned by significant changes in the amount and quality of available energy and in the technologies ('prime movers') by which energy is converted into useful work. In short, the contemporary global economy—characterized by a high degree of (geographically uneven) economic integration—is predicated on the availability of cheap fossil fuels (and, for transportation, liquid fuels in particular). While it has not been possible to estimate the contribution of energy to the rate or pattern of economic growth (but see Ayres and Warr, 2009), it has surely played a role in respect of both these variables, as well as being a significant conditioning influence on the degree of economic integration. The lower energy density of low-carbon, renewable energy sources, and the relatively strong linkage between global economic integration and the enhanced spatial mobility of goods and people via fossil fuels, are major challenges for a low-carbon energy transition.

3.2 **Energy market integration: liberalization, investment, and technology**

Like trade, cross-border investment is a key metric of economic integration: growth in international investment relative to the size of the global economy is a defining characteristic of contemporary globalization. Foreign direct investment (FDI)—investment by one firm in another across a national border for the purposes of gaining a degree of control—has not only grown faster than the world economy but, since the 1980s, has also grown faster than trade. This suggests that, for the global economy as a whole, 'the primary mechanism for interconnectedness . . . has shifted from trade to FDI' (Dicken, 2011: 20). Trends in cross-border investment in the energy sector exemplify the spatial and scalar shifts characteristic of globalization: many national energy markets have been liberalized, international investments in the energy and power sectors now account for some of the largest FDI flows worldwide, and a number of once 'national' firms are now significant agents of transnational investment. This section focuses on the ways in which globalization (as a set of policies designed to encourage cross-border investment) is transforming the predominantly 'national' energy systems developed in the post-war period. It highlights the national and international policy regimes that have encouraged the liberalization and/or geographic integration of energy markets, and provides examples of transnational organizations emerging as key actors in shaping patterns of investment. It also comments on the geographically and sectorally variegated character of energy market integration, and the dynamic and unsettled balance between states and markets. Although Europe has pursued energy market liberalization since the 1990s it is not a universal experience, and state ownership of upstream and/or downstream parts of the production chain remains common. And in several of the countries that embraced liberalization—involving a rolling back of the state's ownership and control of energy assets and associated with the growth of cross-border energy trade and investment–concerns are now being expressed about import dependency and external control (Helm, 2005). One of the legacies of energy market liberalization, however, is that governments that embraced liberalization now lack the capacity for a coordinated response to public concerns around energy security, climate change, and the affordability of energy.

A number of national economies have pursued policies of energy market liberalization since the 1980s. The UK was a leader in comprehensive electricity sector privatization: in England and Wales, the major incumbent and vertically integrated state monopoly—the Central Electricity Generating Board—was 'unbundled', with generation being separated from transmission and distribution, and encouragement of competition in energy supply (Joskow, 2008). The liberalization of gas and electricity markets has been a central EU policy objective for nearly two decades: liberalization directives were first issued for electricity in 1996 and gas in 1998, and the more ambitious 'Third Package' of legislative proposals was adopted in 2007. Liberalization policies deepen the globalization of energy markets by dismantling the 'natural monopoly' position of energy in national markets, opening up production, generation, and distribution of assets to private ownership and international investment. The transformation of ownership and introduction of commercial logic also change the function and scale of energy trade: work on the Nordic Power Pool, for example, shows how electricity market liberalization in member countries led to a dramatic increase in the scale and frequency with which market players used interconnectors (i.e. international trade) (IEA, 2005).

While liberalization policies have created a degree of convergence in market structures for energy, geographic variation persists because of the way liberalization policies articulate with existing national political economies. In some cases (such as Argentina), market reforms have been rolled back while, in the UK, state intervention is an increasingly ubiquitous feature in the electricity market in the name of securing public goods such as energy security or decarbonization: examples include 'strike prices' significantly above market rates for nuclear power and renewables (providing the generator with government-backed guaranteed prices) to encourage investment in low-carbon electricity generation, and 'capacity payments' to ensure sufficient generating capacity to meet peak loads. The extent of market integration also varies widely by fuel type, so that markets for coal, oil, gas, and electricity are spatially integrated ('globalized') in different ways: around 14 per cent of coal and 25 per cent of gas production are traded internationally, compared to over 50 per cent of oil (IEA, 2012). Similarly, international electricity trade has grown but traded quantities remain low as a proportion of total generation (around 1 per cent of world electricity production is traded across a national border). As a consequence there is variation among fuel types in the relative significance of trade and investment as a mode of global economic integration: in the electricity sector, for example, it is cross-border investment—rather than trade—that is the dominant mode of global integration (see Kirkegaard et al., 2009 on the value chain in the wind industry).

There are several distinctive consequences of market integration for energy systems. Among the most significant is a growth in transnational investment in, and ownership of, the energy sector. Between 1990 and 2011 the worldwide stock of FDI in the upstream (extractive) part of the energy sector increased eight-fold to 1.4 trillion dollars; annual flows of FDI increased by a similar amount. Downstream investments in the energy sector (in electricity and gas services) increased by a much larger amount: worldwide FDI stock in electricity, gas, and water services rose over fifty-fold to 516 billion dollars (UNCTAD, 2013b).[10] Energy companies have long been among the most significant

---

[10] Disaggregated figures that encompass the energy sector are not available. The figures quoted here are aggregates for the 'mining, quarrying and petroleum' sector and 'electricity, gas and water services' sectors respectively, published annually by UNCTAD (2013b).

transnational corporate actors. The vertically integrated 'oil majors' continue to be a feature of the global energy economy: Shell, BP, Total, Exxon, and Chevron account for five of the top ten non-financial companies when ranked by foreign assets (UNCTAD, 2013b). However, a key feature of the contemporary global energy economy is the emergence of new transnational corporate actors from outside the traditional 'triad' economies (North America, Europe, and Japan), associated with a shift in the centre of gravity of the world economy towards Asia and the emergence of a multipolar world system (de Graaf, 2011; Bridge and Le Billon, 2013). While not limited to fossil fuels, this process is exemplified by the growing transnational activities of *state-owned* oil and gas companies such as PetroChina, ONGC-Videsh, Statoil, Rosneft, Petronas, and Petrobras. However, at least as significant as their new global reach is the way transnationalization of these state-owned firms simultaneously involves an increasing integration with private companies (de Graaf, 2011). The deepening globalization of international energy markets, then, relates not only to the growing extensity and intensity of trade and investment but also to significant changes in its organizational structures. Overall, the net effect of this process of globalization is that once-national systems of provision are increasingly porous, with 'component parts of the world economy . . . increasingly interconnected in qualitatively different ways from the past' (Dicken, 2011: 52). The implication is that energy systems are no longer 'nationally' contained but increasingly are shaped by decisions and interactions at multiple scales.

A second consequence of energy market integration is a degree of convergence in technology and market dynamics across space. This can be seen at both the supply (e.g. the selection of CCGTs as a technology of choice for delivering new electricity generating capacity) and demand ends of the equation (e.g. the global expansion of air conditioning technology and, more generally, norms of thermal comfort). However, perhaps the clearest indication can be seen in national trends in energy intensity (the quantity of energy needed to produce a unit of economic output), which illustrate 'a massive and accelerating convergence since about 1990, toward lower and lower levels of global energy intensity' (Ruhl et al., 2012: 114; see Figure 3.2. The causes of this significant trend can be traced to the liberalization of trade and investment which have expanded the geographical scope of trade in fuels, 'greatly accelerate(d) the allocation of fuels to their most efficient use' and driven the geographical diffusion of energy conversion technologies (including those associated with consumption).

A third consequence is the fundamentally uneven character of globalization. The 'deepening' of globalization may lead to some forms of spatial convergence (as above), but it also reproduces significant patterns of spatial difference (or produces them in alternative forms). The call for 'universal energy access' is a recognition that patterns of energy trade and investment have been remarkably 'sticky' over time: flows of energy— and flows of investment into energy infrastructure—remain highly concentrated, reproducing large differences in per capita energy consumption at the global scale. Geographical differences within countries can also be significant, particularly in parts of the global South where access to energy services often reflects a broader urban/rural divide. While

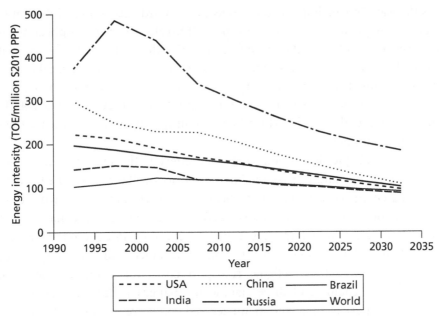

Figure 3.2 International convergence in energy intensity (based on Ruhl et al., 2012: 115)

patterns of international trade in key commercial fuels like oil and gas have changed with geographical shifts in the centre of gravity of the world economy, the overall level of geographical concentration has not altered significantly. The international trade in coal, oil, and gas remains dominated by a handful of major exporters and importers (the five most significant account for 60 per cent of all oil imports; and 48 per cent of exports). A tightening oil market has encouraged a growth in bilateral trade deals that effectively remove oil from the market. Such deals lock out as much as they bind together, illustrating how patterns of energy trade and investment produce geographies of both connection *and* disconnection. These new geographies are politically and economically significant: by strengthening some flows of trade and investment at the expense of others, for example, they have the potential to effectively 'reposition' a country like the UK within international political economy. They highlight, therefore, the importance of thinking about 'national' energy policy as transcending the domestic sphere and being fundamentally linked to foreign policy and international trade and development.

3.3 Implications of low-carbon transition for deepening globalization

The previous sections characterize globalization as an 'integrating set of tendencies that ... intensify connections and flows across territorial borders and regions' but which at the same time are geographically very uneven (Yeung, 2002: 288). This section considers the ways in which the historic trajectory towards deepening globalization may change as a consequence of carbon constraints. It breaks down the implications of low-carbon transition into three different effects: a scale effect relating to the intensity and extensity of global integration; a composition effect describing the potential to substitute for high-quality fossil fuels in the energy mix; and a technique effect associated with reductions in the carbon intensity of economic output (Bridge, 2002; WTO, 2009). These represent different 'pathways' by which low-carbon transition may influence the spatial integration of economies.

The fundamental question around the scale effect is the extent to which steps to address climate change and/or sustained high oil prices will attenuate the historic trend towards falling energy costs so that 'the wheels of globalization get thrown into reverse' (Rubin, 2009). Because the dynamics of increasing availability and falling cost have underpinned the stretching and deepening of economic integration over time, any significant change that reasserts the friction of distance will have implications for patterns of economic activity. Higher transport costs, for example, may encourage shorter production chains and drive significant changes in how global provisioning systems for raw materials, food, and manufactured goods are organized. Because current patterns of economic production have been laid down at a time when the carbon consequences of energy use have been largely external to the decision-making of firms and states, policies that significantly internalize these costs are likely to drive a re-evaluation of locational decisions. Thus carbon pricing, together with the 'end of cheap oil', 'have the potential to re-draw the map of economic activity in some interest-ing ways—just as previous revolutions in the cost and availability of energy to society produced distinctive geographical forms' (Bridge et al., 2013: 335; see also Jiusto, 2009).

However, a distinctive characteristic of low-carbon energy transition is that, unlike previous energy transitions, it is not propelled by the 'economic advantages of higher-quality fuels' and requires extensive and sustained political intervention (Cleveland, 2008). There are, then, considerable uncertainties over whether scientific concern about climate change will translate into policies that effectively internalize the carbon-costs of fossil fuels: the normalization of cheap travel and global provisioning systems, expanding demand for modern energy services among the world's growing middle class, and continued growth in reserves of oil and gas are sources of significant political inertia. That carbon pricing has not yet been extended to transportation in a comprehensive way illustrates some of the uncertainties over whether efforts to address climate change will constrain deepening globalization. Without the requisite political will, a more likely outcome is a continuation of historical trends in which unabated fossil fuel use remains dominant: in such a scenario technological change may further reduce both carbon and energy intensity of production in relative terms, but absolute emissions will continue to rise.

Behind both the composition and technique effects is the question of how far the historic trajectory towards deepening globalization may be separated from its carbon consequences. The composition effect concerns the mix of energy sources (i.e. the capacity to replace the use of fossil fuels with alternative sources), while the technique effect relates to the extent to which fossil fuels may be used sustainably by decoupling their superior capacity for time–space compression from carbon emissions (Jaccard, 2006). Fossil fuels may be substituted in some applications, while potential emissions may be captured in others (e.g. via carbon capture and storage (CCS) in power generation). The UK's policy of reducing carbon emissions by 80 per cent by 2050, for example, centres on decarbonizing electricity generation by increasing the penetration of renewables and nuclear power and the application of CCS, replacing fossil fuels with electricity in heating and transportation, and improving energy efficiency (DECC, 2011). Effecting this shift in a commercial context, however, is constrained by the comparatively favourable economics of conventional thermal power plants (from an investor perspective): commercial application of CCS in power generation, for example, is severely hampered by the low price of carbon in the EU ETS, while investment in non-fossil electricity generation has to compete with the relatively quick construction and low capital and operating costs of gas-fired power plants. There are also technical constraints to the substitution of non-fossil sources. The relatively low power densities of renewables imply significant spatial trade-offs in securing equivalent energy flows from renewable sources (Howard et al., 2009). The dominance of liquid fuels in road, rail, air, and marine transportation present a particular challenge: there are few alternatives that provide the energy density and flexibility of liquid fuels, these diffuse applications are not amenable to carbon capture, and replacing the current volume of fossil-derived transport fuels with biofuels would require dedicating very large areas of land to fuel production.

Because the historic tendency to capture/consume a greater quantity of energy from fossil fuels has not occurred everywhere equally, the composition of national energy systems varies widely. The energy intensity of production, carbon intensity, and efficiency of energy conversion technologies all demonstrate wide geographic variation. This diversity constitutes a series of 'regional laboratories' in which distinctive approaches and technologies for energy capture and conversion have been developed. It also provides a series of socio-political 'experiments' in how energy technologies may be owned, managed, and governed. The globalization of energy trade and investment creates potential pathways for the spread of these technologies and practices, as well as for developing political alliances that may promote energy transition. Significant emergent structures associated with low-carbon transition include global supply chains for nuclear power and renewable electricity generation (particularly hydropower and wind), the growing international trade in biofuels, production networks for rare-earths and other materials associated with electrification (such as lithium), and global governance networks that integrate a series of urban-scale experiments with low-carbon energy (Bulkeley and Castán Broto, 2013). Low-carbon transition generates not only new patterns and scales of spatial integration, but also geo-economic and geo-political consequences: the emergence of a global supply chain for wind turbines and photovoltaics, for example, has sidelined a number of national producers (while the UK has seen extensive investments in wind power generation over the last decade it has simultaneously lost domestic turbine manufacturing capacity).

3.4 **Conclusion**

This chapter has presented some general characteristics of the relationship between global economic integration and energy systems. It has outlined the influence of energy costs, technologies, and fuels on the unfolding of a global economy and its progressive deepening over time; it has suggested how energy systems are themselves becoming increasingly 'globalized' through the opening up of national energy markets, and the stretching of energy infrastructure and diffusion of technologies across space; and it has considered the ways in which a low-carbon transition may modify the historic trajectory towards deepening globalization. At the core of the chapter is the complex relationship between energy use, economic growth, and geographical integration. While there is a good deal of work in economics on the causes of economic growth, there has been only limited research (e.g. Ayres and Warr, 2009) on the critical linkage between energy use and economic expansion, and there is less still on the relationship between growing energy use and the spatial integration of economic activity. This chapter has drawn widely, therefore, on work in economic geography, ecological economics, and environmental history to understand the role that energy systems have played in the deepening of globalization that has occurred over the past 200 years.

Although economic integration via trade and investment is an ongoing process and its origins do not neatly coincide with any singular shift in energy regime, historians of energy and environment highlight the distinctiveness of the last couple of centuries in energetic terms: the marked disjuncture associated with transitions from 'biomass to fossil fuels and from animate to inanimate prime movers' (Smil, 2010b: 25; see also Wrigley, 1988; McNeill, 2000). A world economy characterized by long-distance trade in high-value goods preceded the widespread use of coal, but fossil energies—and particularly cheap fossil energies—played a decisive role in the emergence of a *highly integrated* global economy predicated on the large-scale movement of people, goods, and information. The critical point here is not so much the fossil character of such energies as the abundance of supply, and the higher-quality energy services that such resources could provide. Over time innovation in energy services (around mobility, heating, and lighting, for example) and the progressive refinement of consumption efficiencies has driven down the cost of using energy (as a proportion of total costs). This in turn has enabled wider applications of energy leading to an expansion in energy consumption overall, in the manner of Jevon's classic paradox. The striking expansion in energy availability (in absolute and per capita terms) that is so peculiar to the last 200 years enhanced the productivity of labour and reduced transportation costs, thereby surely contributing to economic growth, and transformed the experience of time and space for millions of people around the world. The unprecedented economic expansion and geographical integration that followed has, however, been highly uneven so that globalization is best understood as a simultaneous process of spatial equalization and differentiation. This can be seen with regard to figures on energy use: while there is evidence for convergence in the energy intensity of GDP, there is also very wide variation in energy access per capita as a result of a highly uneven pattern of investment in energy infrastructure. The same principle of simultaneous equalization and differentiation can also been seen in regard to the liberalization of energy markets, trade, and investment.

Deepening globalization is tied historically to an increasing availability of cheap energy from fossil fuels. Key features of the contemporary world economy developed in a period when cheap and abundant energy was the norm: these include high levels of integration through trade, a competitive international state system, expectations around rates of economic growth, and the availability of the atmosphere as a carbon sink. Cheap energy, then, was a foundational circumstance for the contemporary economy but 'to regard these circumstances as enduring and normal, and to depend on their continuation, is an interesting gamble' (McNeill, 2000: xxiii). If sustained over time, significant changes in the cost and availability of energy that breach historic norms are likely to have an effect on the form and scale of economic integration. Similarly, a high and sustained carbon price would increase the relative prices of carbon-intensive energy sources. While this could drive fuel switching in some sectors, in parts of the transportation sector where alternatives are limited a high carbon price could significantly increase the costs of traversing space. There are some intriguing indications that a combination of tightening oil markets, falling domestic gas prices (in the USA), and concerns over carbon (e.g. freight air miles) may—in some very specific instances—be associated with cases of 'relocalization' (or de-globalization) of production. This chapter has outlined how energy systems, economic growth, and geographical integration were linked during the evolution of a global economy, and the specific conditions that produced a deepening of globalization over time. To the extent that future energy transitions depart from the historic pattern of progressively higher quality energy services and/or lower energy costs for the end user, there is every reason to expect that the 'friction of distance' will become more important and over time could significantly re-work the cost-surfaces of the global economy.

■ REFERENCES

Ayres, Robert and Warr, Benjamin B. 2009. *The Economic Growth Engine: How Energy and Work Drive Material Prosperity*. Cheltenham: Edward Elgar.

Baiocchi, Giovanni and Minx, Jan C. 2010. Understanding changes in the UK's CO_2 emissions: A global perspective. *Environmental Science & Technology*, 44(4), 1177–84.

BCG. 2012. *Made in America, Again: why manufacturing will return to the US*. Boston Consulting Group.

Bradshaw, Michael. 2014. *Global Energy Dilemmas*. Cambridge: Polity Press.

Bridge, Gavin. 2002. Grounding globalisation: the prospects and perils of linking economic processes of globalization to environmental outcomes. *Economic Geography* 78(3): pp. 361–86.

Bridge, Gavin. 2011. 'Past Peak Oil: political economy of energy crises.' In *Global Political Ecology*, edited by Richard Peet, Paul Robbins, and Michael Watts, pp. 307–24. London: Routledge.

Bridge, Gavin and Le Billon, Phillipe. 2013. *Oil.* Cambridge: Polity Press.

Bridge, Gavin, Bouzarovski, Stephan, Bradshaw, Michael, and Nick Eyre. 2013. Geographies of energy transition: space, place and the low-carbon economy. *Energy Policy* 53: pp. 331–40.

Bulkeley, Harriet and Castán Broto, Vanesa. 2013. Government by experiment? Global cities and the governing of climate change. *Transactions of the Institute of British Geographers* 38: pp. 361–75.

Chisholm, Michael. 1990. 'The increasing separation of production and consumption.' In *The Earth as Transformed by Human Action: Global and Regional Changes in the Biosphere over the Past 300 Years*, edited by Turner, B. L., Clark, William, Kates, Richard, Richards, John, Mathews, Jessica and Meyer, William, pp. 87–102. Cambridge: Cambridge University Press.

Cleveland, Cutler. 2008. Energy transitions past and future. *Encyclopaedia of Earth*. Available at <http://www.eoearth.org/view/article/152562/>. (Accessed April 2015).

Cook, Ian. 2004. 'Follow the Thing: papaya'. *Antipode* 36(4): pp. 642–64.

David, Paul. 1990. The dynamo and the computer: an historical perspective on the modern productivity paradox. *The American Economic Review* 80 (2): pp. 355–61.

Davis, Lucas. 2013. *The Economic Cost of Global Fuel Subsidies*. Energy Institute at Haas Working Paper Series, Haas School of Business, University of California-Berkeley. <http://www.nber.org/papers/w19736>. (Accessed April 2015).

Dean, Mark and Sebastia-Barriel, Maria. 2004. Why has world trade grown faster than world output? *Bank of England Quarterly Bulletin* 44(3): pp. 310–20.

deGraaf, Nana. 2011. A global energy network? The expansion and integration of non-triad national oil companies. *Global Networks* 11: pp. 262–83.

Department of Energy and Climate Change (DECC) 2011. *The Carbon Plan: Delivering our Low Carbon Future*. London: HM Government.

Dicken, Peter. 2011. *Global Shift: Mapping the Changing Contours of the World economy*. Sage, 6th Edition.

Enflo, Kerstin, Kander, Astrid, and Schön, Lennart. 2009. Electrification and energy productivity. *Ecological Economics* 68(11): pp. 2808–17.

Fouquet, Roger and Pearson, Peter. 2006. Seven centuries of energy services: the price and use of light in the United Kingdom (1300–2000). *Energy Journal* 27(1): pp. 139–77.

Freidberg, Susanne. 2004. *French Beans and Food Scares: Culture and Commerce in an Anxious Age*. Oxford: Oxford University Press.

Gales, Ben, Kander, Astrid, Malanima, Paolo, and Rubio, Mar. 2007. North versus south: Energy transition and energy intensity in Europe over 200 years. *European Review of Economic History* 11: pp. 219–53.

Grubler, Andreas. 2012. Energy transitions research: insights and cautionary tales. *Energy Policy* 50: pp. 8–16.

Hall, Charles, Cleveland, Cutler, and Kaufmann, Robert. 1986. *The Ecology of the Economic Process: Energy and Resource Quality*. New York: John Wiley.

Harvey, David. 1989. *The Condition of Postmodernity: An Enquiry into the Origins of Cultural Change*. Oxford: Blackwell Press.

Held, David, McGrew, Anthony, Goldblatt, David, and Perraton, Jonathan. 1999. *Global Trans-formations: Politics, Economics and Culture*. Cambridge: Polity Press.

Helm, Dieter. 2005. The assessment: the new energy paradigm. *Oxford Review of Economic Policy* 21(1): pp. 1–18.

Howard, David, Wadsworth, Richard Whitaker, Jeanette, Hughes, Nick, and Bunce, Robert. 2009. The impact of sustainable energy production on land use in Britain through to 2050. *Land Use Policy* 26, SUPPL. 1, pp. S284—92.

Huber, Matthew. 2013. *Oil, Freedom and the Forces of Capital*. University of Minnesota Press.

Hummels, David. 2007. Transportation costs and international trade in the second era of globalization. *Journal of Economic Perspectives* 21(3): 131–54.

ICF. 2012. An international comparison of energy and climate change policies impacting energy intensive industries in selected countries. Report to Department of Business, Innovation and Skills, UK Government.

IEA. 2005. *Lessons from Liberalised Electricity Markets*. Paris: IEA/OECD.

IEA. 2010. Analysis of the Scope of Energy Subsidies and Suggestions for the G20 Initiative. Paris: IEA, OPEC, OECD, World Bank Joint Report. Available at <http://www.oecd.org/env/45575666.pdf#page=33&zoom=auto,83,754> (Accessed April 2015).

IEA. 2011. Joint report by IEA, OPEC, OECD and World Bank on fossil fuel and other energy subsidies: An update of the G20 Pittsburgh and Toronto Commitments. Paris: IEA. Available at <http://www.oecd.org/site/tadffss/49006998.pdf> (Accessed April 2015).

IEA. 2012. *Key World Energy Statistics*. Paris: IEA.

Jaccard, Mark. 2006. *Sustainable Fossil Fuels: The Unusual Suspect in the Quest for Clean and Enduring Energy*. Cambridge: Cambridge University Press.

Jiusto, Scott. 2009. Energy transformations and geographic research. In *A Companion to Environmental Geography*, edited by Noel Castree, David Demeritt, Diana Liverman, and Bruce Rhoads, Blackwells.

Joskow, Paul. 2008: lessons learned from electricity market liberalisation. *The Energy Journal*: Special Issue on the Future of Electricity, pp. 9–42 http://economics.mit.edu/files/2093.

Kirkegaard, Jacob,Weischer, Lutz, and Hanemann, Thilo. 2009. 'It Should Be a Breeze: Harnessing the Potential of Open Trade and Investment Flows in the Wind Energy Industry'. Peterson Institute for International Economics Working Paper No. 09-14. Available at SSRN: <http://ssrn.com/abstract=1521651 or http://dx.doi.org/10.2139/ssrn.1521651>.

Liu, Hongtao, Xi, Youmin, Guo, Ju'e, Li, Xia. 2010. Energy embodied in the international trade of China: an energy input–output analysis. *Energy Policy* 38(8): 3957–64.

Machado, Giovani, Schaeffer, Roberto, and Worrell, Ernst. 2001. Energy and carbon embodied in the international trade of Brazil: an input–output approach. *Ecological Economics* 39(3): 409–24.

Martinez-Alier, Joan. 2002. Ecological debt and property rights on carbon sinks and reservoirs. *Capitalism Nature Socialism* 13(1): pp. 115–19.

McNeill, John. 2000. *Something New Under the Sun: An Environmental History of the Twentieth-century World*. New York: Norton.

Minx, Jan, Baiocchi, Giovanni, Peters, Glen, Weber, Christopher, Guan, Dabo, and Hubacek, Klaus. 2011. A 'Carbonizing Dragon': China's fast growing CO_2 emissions revisited. *Environmental Science and Technology* 45: 9144–53.

Mulhall, Rachel and Bryson, John. 2014. The energy hot potato and governance of value chains: power, risk and organizational adjustment in intermediate manufacturing firms. *Economic Geography* 89(4): pp. 395–419.

O'Rourke, Kevin and Williamson, Jeffery. 1999. *Globalization and History: the Evolution of a Nineteenth-century Atlantic Economy*. Cambridge, MA: MIT Press.

Peters, Glen, Minx, Jan, Weber, Christopher, and Edenhofer, Ottmar. 2011. Growth in emission transfers via international trade from 1990 to 2008. *Proceedings of the National Academy of Sciences* 108(21): 8903–8.

PWC 2012. A homecoming for US manufacturing? Why a resurgence in US manufacturing may be the next big bet. PriceWaterhouseCoopers. Available online at http://www.pwc.com/us/en/industrial-products/publications/us-manufacturing-resurgence.jhtml. Accessed June 4 2015.

Rubin, Jeff. 2009. *Why Your World is About to Get a Whole Lot Smaller: Oil and the End of Globalisation*. London: Virgin Books.

Ruhl, Christof, Appleby, Paul, Fennema, Julia, Naumov, Alexander, and Schaffe, Mark. 2012. Economic development and the demand for energy: a historical perspective on the next 20 years. *Energy Policy* 50: pp. 109–16.

Schäfer, Andreas. 2006. Long-term trends in global passenger mobility. *The Bridge* 36(4): pp. 24–32.

Smil, Vaclav. 1999. *Energies: An Illustrated Guide to the Biosphere and Civilization*. Cambridge, MA: MIT Press.

Smil, Vaclav. 2010a. *Prime Movers of Globalization: The History and Impact of Diesel Engines and Gas Turbines*. Cambridge, MA: MIT Press.

Smil, Vaclav. 2010b. *Energy Transitions: History, Requirements, Prospects*. Santa Barbara, CA: Praeger.

Smil, Vaclav 2013. *Made in the USA: the rise and retreat of American Manufacturing*. MIT Press.

Sorrell, Steven. 2009. Jevons' Paradox revisited: The evidence for backfire from improved energy efficiency. *Energy Policy* 37: pp. 1456–69.

Strange, Susan. 1988. *States and Markets: An Introduction to International Political Economy*. London: Pinter Publishers.

UNCTAD. 2013a. *Review of Maritime Transportation*. Geneva and New York: United Nations.

UNCTAD. 2013b. *World Investment Report 2013—Global Value Chains: Investment and Trade for Development*. Annex Tables 24 and 26 (estimated world inward FDI stock by sector and industry, 1990 and 2011; estimated world inward FDI flow by sector and industry, 1990 and 2011). Geneva and New York: United Nations.

Wrigley, E. 1988. *Continuity, Chance and Change: The Character of the Industrial Revolution in England*. New York, Cambridge University Press.

WTO. 2009. *Trade and Climate Change: WTO-UNEP Report*. Switzerland: World Trade Organisation.

Yeung, Henry. 2002. The limits to globalization theory: a geographic perspective on global economic change. *Economic Geography* 78(3): pp. 285–305.

Yeung, Henry. 2009. Globalisation. *International Encyclopedia of Human Geography*. Elsevier.

4 The global climate change regime

Joanna Depledge

4.1 Introduction

If the world is to respond effectively to human-induced climate change, then energy systems must change radically. The scientific consensus, as reflected in the peer-reviewed academic literature, is clear. Energy-related carbon dioxide (CO_2) emissions are responsible for over half of humanity's contribution to climate change. To stand a likely chance of limiting temperature rise to 2 °C above preindustrial levels—which constitutes the current political (if not universal) definition of a 'safe' warming ceiling—global emissions must peak within the next two decades, and fall steeply thereafter, to about half their 2010 levels by 2050 (Olivier et al., 2012; IPCC, 2014). By 2100, CO_2 equivalent emissions must have fallen to zero, or below[1] (IPCC, 2014). This is against a backdrop of global emissions growth that averaged 2.7 per cent a year over the decade to 2011 (Olivier et al., 2012: 6). Although recent data suggests that global emissions growth may have now peaked, such a dramatic turnaround cannot happen without major transformations in our energy consumption and production patterns towards lower-carbon options. Climate change, and the need to avert it, thus casts an inescapable shadow on the global energy landscape, even if, so far, this most intractable of problems has not yet had the defining influence on energy trends that the science demands.

This chapter considers how the international response to climate change shapes the context in which global energy trends are evolving. It begins with a brief description of the global climate change regime and the ongoing negotiations, focusing on some of the regime's defining features that are particularly relevant to the energy sector. The chapter then turns to the central issue of emission reduction targets: both the collective goal of the regime and national commitments, drawing attention to the disjuncture between the two. The discussion moves on to take stock of the climate regime's impact, and look ahead to the future.

---

[1] This raises the concept of negative emissions: where absorption by sinks, such as the oceans, vegetation, or other forms of sequestration, would outweigh the emissions from sources.

4.2 **The international response to climate change**

The global climate change regime is founded upon two legally binding treaties—the 1992 United Nations Framework Convention on Climate Change (UNFCCC, or 'the Convention') and its 1997 Kyoto Protocol—along with hundreds of decisions, guidelines, and procedures adopted by the parties to those treaties (Yamin and Depledge, 2004). The Convention and Kyoto Protocol both enjoy almost universal membership. The Conference of the Parties (COP) is the main political decision-making body of the regime, with smaller technical bodies focusing on specific issues, such as adaptation, technology transfer, capacity building in developing countries, and compliance.[2] The Clean Development Mechanism (CDM), joint implementation,[3] and emissions trading among the Kyoto Protocol parties are also managed within the regime. The result is a complex and sophisticated set of institutions addressing the concrete practicalities of mitigating and adapting to climate change across the globe and in countries with very different circumstances, going far beyond the contested politics that tend to capture media attention. The regime is powered by continuous intergovernmental negotiations, held at least twice a year, where these contested politics come to the fore. At the time of writing, negotiations were focused on defining the regime's next stage, to take over in 2020, after the expiry of the Kyoto Protocol's second commitment period (2013–20) and transitional voluntary arrangements that were made under the 2009 Copenhagen Accords and 2010 Cancun Agreements. Key milestones in the climate change negotiations are presented in the table below (Table 4.1).

4.2.1 THE GREAT DIVIDE: ANNEX I AND NON-ANNEX I PARTIES

Decisions taken by the Convention's drafters over twenty years ago have had profound repercussions on the international response to climate change, and with that probably also on global energy trends. Perhaps the most momentous decision, although one that seemed rather obvious in the early 1990s, was to divide the regime's parties between what were, at the time, considered to be developed and developing countries. The developed countries—members of the Organisation for Economic Cooperation and Development (OECD) and the more advanced countries of the former Soviet bloc—were listed by name in the Convention's Annex I. These 41 countries are therefore known as 'Annex I parties', with the remaining 150 or so countries known, by default, as 'non-Annex I parties'. The Convention established that the Annex I parties should take 'the lead in

---

[2] For more details, see the UNFCCC website at <http://www.unfccc.int>.
[3] Both mechanisms allow low-carbon projects in one country to generate emission credits to count against a target in another. Joint implementation operates among the Annex I (developed country) parties, while the CDM operates between Annex I and non-Annex I (developing country) parties.

Table 4.1 Key milestones in the climate change negotiations

Treaty/COP decision[4]	Date	Key features
UNFCCC	1992 (in force 1994)	Classification of countries into Annex I/non-Annex I. Leadership role of Annex I parties.
		Annex I parties to 'aim' to return emissions to 1990 levels by 2000.
		General commitments plus reporting obligations for all parties.
Kyoto Protocol, first commitment period	1997 (in force 2005)	Individual, legally binding targets for Annex I parties, amounting to a collective 5% cut from 1990.
		Target period 2008–12.
		Market mechanisms, incl. CDM and emissions trading.
Kyoto Protocol second commitment period	2012 (not yet in force)	Individual, legally binding targets for Annex I parties, amounting to a collective 18% cut from 1990 among participating parties. These include the EU-28, plus up to nine others. Target period 2013–20.
Bali Action Plan (COP decision)	2007	Launched negotiations on new commitments, including for developing countries.
Copenhagen Accords / Cancun Agreements (COP decisions)	2009/2010	2 °C maximum temperature goal.
		Annex I parties to implement 'emissions targets' up to 2020; non-Annex I parties to implement 'nationally appropriate mitigation actions' deviating from business as usual by 2020. Both are voluntary, and nationally determined.
		All Annex I parties, plus around 10 large non-Annex I emitters, have made quantitative pledges. Some 35 more have made qualitative undertakings.[5]
Durban Platform (COP decision)	2011	Launched negotiations on a post-2020 regime for all Parties that will 'raise the level of ambition'.
COP 21 to be held in Paris	2015	Deadline for a deal under the Durban Platform negotiations.

modifying longer-term trends in [GHG] emissions' (Article 4.2(a)), and it is on the basis of this leadership principle that the regime has operated ever since. In concrete terms, this means that only Annex I parties have been subject to specific emission commitments under the Convention, and then the 1997 Kyoto Protocol. Until the negotiations launched at the Bali COP in 2007 (see Table 4.1), most non-Annex I parties, including the large emerging economies with rapidly rising emissions, refused to even consider taking on stronger, quantified commitments, drawing attention to the lacklustre leadership of the Annex I parties, most notably the rejection of the Kyoto Protocol in 2001 by the USA, then the world's largest emitter (and still the second).

The dual structure dividing developed (Annex I) and developing (non-Annex I) countries has been much criticized (Prins and Rayner, 2007; Depledge, 2009), and it is

[4] Unlike treaties, which need to be ratified by individual countries before they enter into force, COP decisions apply immediately. However, they have less force in international law.

[5] More information on the pledges can be found at <http://www.unfccc.int> and at <http://www.ecofys.com/en/project/the-climate-action-tracker/>.

Table 4.2 The top 20 aggregate CO_2 emitters from energy consumption in 2011[6] (compared with their 1990 rankings) and their obligations to 2020

Ranking (in 1990)	Country	% share of global emissions (in 1990)	Obligation to 2020
1 (3)	*China*	26.75 (10.12)	CA/intensity
2 (1)	USA	16.85 (23.42)	CA/absolute
3 (2)	Russian Federation	5.49 (17.75[7])	CA/absolute
4 (7)	*India*	5.3 (2.69)	CA/intensity
5 (2)	Japan	3.62 (4.86)	CA/absolute
6 (5)	Germany[8]	2.3 (4.6)	KP2
7 (21)	*Iran*	1.92 (0.94)	-
8 (16)	*South Korea*	1.88 (1.12)	CA/BAU
9 (8)	Canada	1.79 (2.19)	CA/absolute
10 (20)	*Saudi Arabia*	1.58 (0.97)	-
11 (6)	UK	1.52 (2.8)	KP2
12 (17)	*Brazil*	1.46 (1.1)	CA/BAU
13 (12)	*Mexico*	1.42 (1.4)	CA/BAU
14 (13)	*South Africa*	1.41 (1.38)	CA/BAU
15 (24)	*Indonesia*	1.31 (0.72)	CA/BAU
16 (9)	Italy	1.23 (1.93)	KP2
17 (15)	Australia	1.2 (1.24)	KP2
18 (10)	France	1.15 (1.71)	KP2
19 (18)	Spain	0.98 (1.04)	KP2
20 (11)	Poland	0.95 (1.55)	KP2

Italics—non-Annex I countries; KP2—country expected to ratify the second commitment period of the Kyoto Protocol, with a legally-binding reduction target for absolute emissions; CA—Copenhagen Accords/Cancun Agreements; absolute—pledge to cut absolute emissions; intensity—pledge to cut emissions intensity, that is, emissions per unit of GDP; BAU—pledge to cut emissions from an unspecified business-as-usual baseline.

easy to see why. For a start, the large group of non-Annex I parties unhelpfully bundles together the tiniest and poorest emitters, as well as the carbon giants, and many in between. Since 1990, energy consumption and emissions in large parts of the developing world, especially Asia, have risen (see Chapter 1); in China, emissions have more than quadrupled. While the largest non-Annex I emitters (those in the global top 20, see Table 4.2) contributed to 18.8 per cent of global emissions in 1990, they accounted for over 43 per cent of the 2011 total, which itself had climbed by over 50 per cent over that period. As shown in Table 4.2, this mostly reflects a steep rise in the Chinese (and, to a lesser extent, the Indian) share, but also the dramatic decline in emissions in the former Soviet Union and its European allies, following the economic and political collapse of that bloc.

---

[6] Data from the Energy Information Administration (EIA) at <http://www.eia.gov/cfapps/ipdbproject/iedindex3.cfm?tid=90&pid=45&aid=8&cid=all,&syid=1980&eyid=2010&unit=MTCDPP>.

[7] The 1990 figure is for the Soviet Union, and so includes many high emitting and other states that are now independent, e.g. Ukraine.

[8] Figures for the former East and West Germany have been aggregated for the 1990 figure.

It made perfect sense in 1990 to single out the more industrialized nations for immediate emission curbs; these accounted for the vast majority of global emissions, and even more so from a historical perspective, with far higher emissions and economic wealth per capita. The real mistake of the early negotiators—if it can be called that—was in failing to define, upfront in the Convention, a clear 'graduation mechanism' establishing how and when developing countries would be expected to follow the lead of the Annex I parties.[9] The classification of countries under the climate change regime, and in particular the absence of a graduation mechanism, has therefore led to a situation whereby, through the 1990s and the early 2000s, a small[10] group of countries were subject to some kind of binding, treaty-based emission target, whereas the remainder of the world did not even face a clear prospect of future constraints. Although the Kyoto Protocol's emission targets have been criticized for their leniency, they at least imposed the discipline and expectation of a binding, carbon-constrained world upon most of the Annex I parties. It would be difficult to imagine that this uneven pattern of expectations has not impacted on patterns of energy production and consumption over the globe. Indeed, concerns have been raised about so-called 'carbon leakage', whereby industries have migrated to countries without carbon constraints in order to enjoy cheaper or less regulated energy supplies (e.g. *Climate Strategies*: Dröge, S. et al. (2009)).

4.2.2 THE CLEAN DEVELOPMENT MECHANISM

Although without quantified commitments, the non-Annex I parties have still been profoundly engaged in the regime, notably through national reporting commitments and, later, the CDM. Under the CDM, Annex I parties to the Kyoto Protocol can fund emission curbing projects in developing countries, and claim the resulting 'certified emission credits' (CERs)[11] against their own emission targets. At the same time, the CDM was supposed to assist developing countries in 'achieving sustainable development' and, crucially, 'in contributing to the ultimate objective of the Convention' (KP Article 12); in essence, although this was not explicitly stated, to start to prepare for their eventual assumption of more formal emission targets. The merits and flaws of the CDM are well-rehearsed in the literature (e.g. Paulsson, 2009; Newell, 2012). Nonetheless, the CDM had issued over 1.5 billion CERs by end 2014 from projects in developing countries. By 2020, existing registered CDM projects should have generated over 7 billion CERs. These are not trivial figures. Moreover, around 75 per cent of CDM

---

[9] By contrast, the 1987 Montreal Protocol on Substances that Deplete the Ozone Layer set threshold levels of production/consumption of ozone-depleting substances (ODS), beyond which countries 'graduate' into emission targets. It also set a limited grace period for developing countries, after which they assumed quantified reduction goals. This ensured that all countries either had targets, or knew they soon would.

[10] Made smaller when the USA refused to ratify the Kyoto Protocol in 2001, and Canada withdrew in 2011.

[11] A CER is equivalent to 1 tonne of CO_2.

projects are currently in the energy sector, the bulk of these in low-carbon energy production.[12] However, CDM projects are entirely voluntary in nature, and are clearly not on the scale needed to make a noticeable dent in overall developing country emissions. The 7 billion CERs that the CDM hopes to have saved up to 2020, for example, will amount to less than China's emissions for the year 2011.

4.2.3 METHODOLOGICAL DECISIONS

Working with the IPCC, the climate change regime has been at the centre of developing methodologies for estimating and reporting greenhouse gas emissions. Indeed, boosting the capacity of all countries, Annex I and non-Annex I, to calculate their national emissions and produce comparable and reliable data has been one of the great unsung successes of the regime (Depledge and Yamin, 2010). In the process of establishing reporting rules, however, decisions have been taken—or not taken—that have had far-reaching ramifications, including for the effectiveness of the international response to climate change.

4.2.3.1 Aviation and Marine Transport

One of these decisions was to exclude emissions from international aviation and marine transport from national totals, and thereby from emission targets (although countries must still report on these; see UNFCCC, 2006; IPCC, 2006). The rationale was that allocating responsibility for such emissions across countries was complex: Should emissions for a journey be assigned to the registered country of the carrier? The country of departure, or of arrival? Or perhaps shared between these? While admittedly complex, the question is not unfathomable, and was debated in the subsidiary body for scientific and technological advice (SBSTA)—one of the two main subsidiary bodies of the COP—in the 1990s. Politics intervened, however, and high-emitting, oil-exporting countries such as Saudi Arabia, and, though less stridently, the USA, blocked any progress. The decision taken during the Kyoto Protocol negotiations to assign responsibility for these transport emissions to the International Civil Aviation Organisation (ICAO) and the International Maritime Organisation (IMO) (KP Article 2.2) effectively bypassed the regime, although the issue remains on the SBSTA's agenda. As discussed further in Chapter 11, it is only very recently that potentially significant action has been taken within ICAO and IMO to curb international transport emissions; in the case of aviation, largely prompted by unilateral action by the EU. The result of excluding aviation and marine transport emissions from national totals under the climate change regime, without a concrete, time-bound plan for agreeing allocation, has inevitably been to

[12] Figures from UNFCCC-CDM website, at <http://cdm.unfccc.int/Statistics/Public/CDMinsights/index.html>.

weaken expectations of strong carbon constraints in these sectors, therefore leading, in all likelihood, to higher emissions from these sources.

4.2.3.2 Production vs Consumption Emissions

A second element of the climate change regime's reporting guidelines, again based on IPCC methodologies, is the assumption that emissions associated with manufacturing are always allocated to the country producing those manufactured goods, rather than the country consuming them. This constitutes a default position within the IPCC and the regime, and there is no documentary evidence to suggest that an alternative means of allocating emissions—for example, shared between the producing and consuming country—was ever seriously proposed. Nonetheless, it is a default assumption with important consequences, which have recently come to the fore as the globalization of manufacturing has accelerated. It is of specific relevance to China, where about a third of its emissions are related to exported goods. This issue, which has triggered a lively debate over 'indirect' emissions, also known as emissions 'embodied' in manufactured goods, is the focus of Chapter 5. The point here is that the regime's assumption that emissions are only counted at the point of production, within a structure whereby only a minority of (mostly importing and consuming) countries have emission targets, will inevitably impact on global energy patterns, while also threatening the effectiveness of the climate regime itself, as emissions are, in effect, exported, rather than reduced.

4.2.4 ENERGY AND TECHNOLOGY IN THE CLIMATE CHANGE REGIME

Given that energy and climate change are inextricably bound, one might expect that the climate change regime would have a lot to say about the energy sector. This is, however, far from the case; instead, the climate change regime keeps energy very much at arm's length. To a large extent, this reflects the general permissiveness of the regime in giving countries absolute flexibility in choosing how to curb their GHG emissions. Nowhere do the regime's rules constrain or prescribe the energy and technology choices of parties, and they are remarkably light-touch in even encouraging specific low-carbon options, or discouraging high-carbon ones. This approach is partly a function of the extreme reluctance of key players—notably China and the USA—to cede any regulatory authority to a supranational body. The EU's attempt, for example, to include some mandatory and coordinated policies (e.g. removal of fossil fuel subsidies) under the Kyoto Protocol was quickly shot down (UNFCCC, 2000). The absence of any policy prescription or proscription is accompanied by a remarkable sidelining of energy in the regime. This reflects the geopolitics of climate change, whereby energy exporters—especially OPEC—have worked hard to downplay the contribution of fossil fuels to climate change, and the role of clean energy sources in addressing the problem (Depledge, 2008). References to 'energy' in the regime's legal texts are thus few and far between, while limited attention has been paid specifically to energy in the negotiations and supporting activities. This contrasts, for example, with the forestry sector, which features centrally in the process, not only in the negotiations, but also in informal debates, workshops, technical papers, and other initiatives. At the very least this 'air-brushing out' of energy from the regime has discouraged a potentially much stronger response from the energy sector to the challenges posed by climate change.

4.2.4.1 Carbon Capture and Storage and Nuclear

Where the regime has engaged more forcefully with specific technologies and energy sources is under the CDM, which has had to decide whether to allow emission reduction projects involving two controversial technologies, carbon capture and storage (CCS) and nuclear. In the case of CCS, rules allowing CDM projects involving CCS in geological formations were eventually adopted in 2011 (although many issues are still pending before such projects can actually start) (decision 10/CMP.7, in UNFCCC 2011b). This decision, however, came in the face of political opposition, on the grounds that allowing CCS would perpetuate fossil-fuel use and therefore not promote sustainable development, as required by the CDM's rules. There are also continuing practical fears over environmental integrity and liability issues, related in particular to possible seepage. It is significant, however, that in the end negotiators were able to agree to a framework that would allow investment in CCS projects through the CDM, in effect establishing pioneering legal and technical rules. The de facto endorsement of CCS technology by a UN-sponsored process, as well as the existence of internationally accepted rules, is now likely to help promote the intro-duction of CCS across the globe.

The outcome on nuclear energy, however, has been different. Nuclear energy was effectively excluded from the CDM from the very start,[13] although countries with large nuclear programmes, notably China and India, continue to lobby in its favour. The CERs generated under the CDM would, in effect, have provided a subsidy that would make foreign investment in these countries' nuclear programmes more attractive, and might have acted as a decisive game-changer. However, strong opposition against nuclear in the CDM, especially among European countries phasing out their own nuclear capacity, makes it unlikely that the CDM would ever move in this direction. In any case, the uncertainty now hanging over the CDM, as a result of the shift away from legally binding targets in the negotiations (see Section 4.3), along with the low price of credits, has effectively put paid to this idea, while also lessening the impact of the positive decision on CCS.

---

[13] Annex I parties are to 'refrain from using certified emission reductions generated from nuclear facilities to meet their [emission] commitments under Article 3' (preambular text, decision 17/CP.7).

4.3 **Emission reduction targets**

4.3.1 COLLECTIVE AMBITION

The central purpose of the climate change regime is to set the international community's collective ambition in addressing climate change. The Convention first did so through its ultimate objective, namely, to '...stabilize atmospheric concentrations of greenhouse gases at a level that would avoid dangerous anthropogenic interference with the climate system...' (Article 2). The possibility of quantifying this ambition, and in particular giving meaning to the term 'dangerous', remained largely dormant in the negotiations for many years. By defining 'dangerous', the international community would, in effect, be both placing an implicit cap on global emissions (necessarily implying a limit on developing country emissions, which was taboo for many years), and accepting a certain degree of climate damage as 'acceptable', with all the profound political and economic consequences this would entail. The regime was not ready to deal with this, especially given that developing countries had not yet agreed to curb their emissions. Work was carried out, however, in the IPCC, which, while steering clear of making policy recommendations, sought to frame the risks that might be involved at different levels of temperature rise.[14]

Formal attempts to pin down the regime's overall ambition more precisely began as part of the 2007 Bali Action Plan, which launched negotiations on the future of the climate regime, including possible new commitments for developing countries. The debate, which took place under the rubric of a 'shared vision for long-term cooperative action' (decision 1/CP.13, UNFCCC, 2007), eventually culminated in agreement, in the 2009 Copenhagen Accords/2010 Cancun Agreements, '...to hold the increase in global average temperature below 2 °C above pre-industrial levels...' (decision 1/CP.16, para. 4, UNFCCC, 2010). Agreement on such a figure was both supremely momentous, and rather predictable. Momentous, because putting a figure on 'dangerous' had been such a politically sensitive issue; predictable, because 2 °C was a clear 'focal point' for agreement. The figure of 2 °C had already featured in climate debates back in the 1990s, as the temperature rise that was thought to correspond to a doubling of CO_2 concentrations in the atmosphere from preindustrial levels (about 550 ppm). Since then, improved climate sensitivity models have shown that lower concentrations, more akin to 400–450 ppm, are needed to ensure a good chance of limiting temperature rise to 2 °C. The 2 °C target—with a 550 ppm concentration level—was proposed by the EU as part of the negotiations on the Kyoto Protocol, and supported, at that time, by the Alliance of Small Island States (AOSIS) (UNFCCC, 2000). Prior to the 2009 Copenhagen COP, G-8 leaders also

---

[14] For an early attempt in this regard, see IPCC (1997). An important step forward was the 'reasons for concern' or 'burning embers' diagram in the Third Assessment Report (IPCC, 2001(a), figure TS-12; IPCC, 2001(b), figure SPM 2), which was then updated and elaborated by Smith et al. (2009).

recognized the 'broad scientific view' coalescing around 2 °C in their 2009 Summit communiqué (G-8, 2009).[15] The IPCC's 'burning embers' diagram (see footnote 14), and subsequent scenario work, clearly flagged anything above 2 °C as implying a step-change in potential negative consequences. Agreeing on anything above 2 °C would therefore be very difficult to justify.[16]

As shown in Chapter 24, achieving a 2 °C world would imply major transformations to current patterns of energy production and consumption, and indeed all areas of economic activity. The 2 °C target, however, remains little more than a rhetorical aspiration, and as such the political signal it sends out to energy decision-makers is rather weak. It is not actually written down as a target, but rather as a 'long-term goal', and even then in typically convoluted language that includes no conventionally-accepted legally-binding language. This is not surprising, and indeed is a more honest approach. Even if it were written in legal language, such a goal could never be enforced, because it applies collectively to all countries.[17] Moreover, from a scientific point of view, a goal articulated in terms of temperature rise is an extremely blunt instrument; it would be difficult to identify when temperature had stabilized, and even if this were possible, then very long time lags would be involved.

The 2 °C goal would therefore be much more meaningful if it were accompanied by shorter-term, more measurable markers, such as setting a maximum atmospheric concentration level for GHGs, or a peaking year for global emissions, both of which are being considered in the current negotiations. Another powerful way of upholding at least the 2 °C target would be to declare a long-term goal of achieving a 'zero carbon' or 'carbon/climate neutral'[18] world by the end of this century, consistent with the IPCC's findings that peaking at 2 °C requires nothing less. This notion of a 'zero carbon' global goal is gaining traction in the climate debate. Christiana Figueres, the Executive Secretary of the Climate Change Secretariat, has called for 'a carbon neutral world by the second half of the century' (UNFCCC, 2014), while the 2014 UN General Assembly Summit on Climate Change similarly saw 'many leaders' advocating this same goal (UN, 2014).

A new approach used in the IPCC's fifth assessment report (IPCC, 2013 and 2014) is to consider climatic limits in terms of cumulative emissions, that is, the amount of carbon that can be released into the atmosphere over time to be consistent with a particular temperature goal. Using this approach, the IPCC (2014: 25) finds that, to have a 50 per cent chance of limiting warming to 2 °C, cumulative emissions (since 1861–1880) cannot exceed some 1210 GtC; about 515 GtC have already been emitted, which means that humanity has already used up just over 40 per cent of its all-time carbon 'budget'.[19] The cumulative approach underlines the fact that higher emissions in

[15] 'Responsible leadership for a sustainable future', G8 Leader's Declaration, L'Aquila, Italy, 2009.

[16] For more on the 2 °C target, see Jordan et al. (2013).

[17] Although arguably the largest emitters could be held to account for failing to curb emissions to such an extent as to make the 2 °C target unattainable for everyone else.

[18] The terms are subtly different, but are often used interchangeably.

[19] The upper limits are reduced if non-CO_2 gases are taken into account.

the short term—using up more of the budget—will require steeper cuts in the medium to long term to stay within the limits. At the same time, barring some major technological breakthrough, the costs of overall mitigation would be higher and choices available would be constrained, due to the lock-in of existing carbon-intensive technologies (UNEP, 2013; IPCC, 2014). The point is that, although the 2 °C target can still technically be met even if countries continue to procrastinate in their response, it becomes increasingly more difficult and more expensive to do so. Indeed, UNEP (2013: xiii) notes that the steeper cuts required under a 'later action' scenario would be 'without historic precedent'.

Meanwhile, in May 2013, the concentration of CO_2 in the atmosphere, as measured at the Earth System Research Laboratory in Mauna Loa, Hawaii, temporarily topped 400 ppm for the first time.[20] Although average values remain slightly below this landmark figure[21]—397 ppm in January 2014—concentrations are rising at about 2 ppm annually, meaning that recordings below 400 ppm will soon be a thing of the past. The long-term stabilization of concentrations at 400 or 450 ppm, therefore, would almost certainly involve a temporary 'overshoot' to higher concentrations, before a steep cutback.

While the window of opportunity for achieving 2 °C becomes ever narrower, the evidence is accumulating that even a 2 °C ceiling may expose large swathes of humanity to dangerous climatic risks (Hansen et al., 2013; Schaeffer et al., 2012; Smith et al., 2009). AOSIS now argues strongly for a 1.5 °C ceiling (with a 350 ppm concentration target), supported by the Least Developed Countries and many environmental groups. Thanks to insistence by these countries, the Copenhagen Accords/Cancun Agreements do provide for the possible 'strengthening' of the 2 °C target to 1.5 °C, through a review on the adequacy of the long-term global goal (decision 1/CP.16, paras 4 and 138, in UNFCCC, 2010). The review is due to conclude in 2015, at the same time as the negotiations on the next stage of the climate change regime. The prospects of cranking up the level of collective ambition to 1.5 °C, however, are very slim. A 1.5 °C goal might be far more consistent with the scientific evidence of what is needed to make quite sure that dangerous climate change is averted, but science is only one of many factors competing for attention in climate politics, not to mention the energy sector. As discussed below, emission cuts pledged so far do not even allow for a 2 °C world. A 1.5 °C target would translate into a far more stringent cap on global and national emissions and, crucially, on China, which has strongly objected to even mentioning 1.5 °C (Lynas, 2009). The international community has set its collective ambition and, for all its weaknesses, if the 2 °C goal were taken seriously and the measures needed to achieve it actually implemented, then this would go a very long way to mitigating the worst effects of climate change.

---

[20] Data available at <http://www.esrl.noaa.gov/gmd/ccgg/trends/weekly.html>.
[21] CO_2 concentrations fluctuate throughout the year, notably in line with the autumn/winter dieback and spring growth of vegetation in the northern hemisphere.

4.3.2 THE AMBITION GAP: NATIONAL TARGETS

Herein lies the problem. The most cursory glance at energy market trends indicates no seismic shift in 2009/2010, when the 2 °C goal was adopted. Clearly, key players in the energy markets do not really believe that governments are about to start legislating for a 2 °C world. To send a serious political signal that energy markets would take notice of, the 2 °C goal would need to be backed up by concrete national emission targets, accompanied by strong legislation. At present, this is not the case. The current emission commitments up to 2020 (pledged under the Copenhagen Accords/Cancun Agreements and the second period of the Kyoto Protocol) account for less than half of the cuts needed to have a good chance of limiting temperature rise to 2 °C. UNEP (2013) estimates that, under a business as usual scenario, global GHG emissions in 2020 would reach 59 $GtCO_2e$ per year. However, a pathway towards 2 °C would require emissions of 44 $GtCO_2e$.[22] If implemented fully, pledges and commitments would result in a reduction of just 3–7 $GtCO_2e$ per year, leaving surplus emissions of 8–12 $GtCO_2e$. A similar analysis (Climate Action Tracker, 2013) suggests that, far from 2 °C, the world is on track for warming of 3.7 °C by 2100. The IPCC (2014) agrees that the current set of commitments under the Cancun Agreements are only likely to keep the temperature rise to 3 °C. This discrepancy between national targets on the one hand, and collective ambition on the other (which is itself weak, compared to scientific warnings), is widely known as the 'ambition gap' or 'emissions gap' (UNEP, 2013; decision 1/CP.19, para. 1, in UNFCCC, 2013; Climate Action Tracker, 2013).

What are the prospects for closing this ambition gap? At present, the climate change regime is in a transition period towards a more comprehensive regime post-2020, which is scheduled to be agreed at the Paris COP in 2015. A core group of Annex I countries, mostly the EU and allies, are still operating under the Kyoto Protocol, having adopted legally-binding targets for a second commitment period (2013–20). Most other parties have taken on *voluntary* commitments under the Copenhagen Accords/Cancun Agreements up to 2020; these, too, are due to be succeeded by the new post-2020 regime. The Copenhagen Accords/Cancun Agreements marked a turning point for the climate change regime, with the major developing country emitters, including most of those in the top 20, pledging concrete emission reductions for the first time (see Tables 4.1 and 4.2). Unlike under the Kyoto Protocol, however, these pledges are not based on common metrics. That is, they are mostly based on emission intensity or deviation from (often unspecified) business-as-usual trajectories, rather than absolute emission reductions. They also vary in terms of base years, gases covered, treatment of the land-use sector, and other factors. Nonetheless, the pledges are hugely significant in suggesting that these major emerging economies are now starting to operate on the assumption of a

---

[22] Median level of the range 41–7 $GtCO_2e$. For more details on the data, see UNEP (2013).

carbon-constrained future. Moreover, they are just the starting point. Under the 2011 Durban Platform, all countries signed up to a negotiating mandate for the new 2020 regime that will define commitments for beyond 2020, and specifically aims to 'raise the level of ambition' (decision 1/CP.17, in UNFCCC, 2011a).

The question is whether the international negotiations can 'raise the level of ambition' high enough to get on track for the collective 2 °C goal (which becomes harder to do as action is delayed). To date, targets under the regime have always been adopted through a de-facto 'bottom-up' process. Under the 1997 Kyoto Protocol, countries simply proposed their own targets, which were inscribed into the agreement with minimal negotiation. The individual targets were then totted up, and the resulting number, 5 per cent, was inserted into the Kyoto Protocol (Article 3). Now, however, the regime has a 2 °C 'top down' collective goal, to which national emission commitments will be expected to correspond. In an international system resolutely based on sovereignty and competitiveness, it is unlikely that the 2 °C 'pie' will be shared out into emission targets in any objective or rational way, despite the elegance of proposals put forward to this effect (e.g. Höhne et al., 2006). Nonetheless, the existence of the 2 °C goal will act as a yardstick against which proposals can be independently judged as adequate or not. This provides some grounds for cautious optimism.

However, there is also a strong case for profound pessimism. The international relations literature has tended to assume that regimes start off weak, but, as scientific evidence hardens and political will increases, parties agree to ratchet up their commitments and the regime strengthens; this was clearly the assumption of the early climate change negotiators (Bodansky, 1993; Susskind, 1994). It is difficult to say, however, whether the climate change regime is now getting stronger or weaker. On the one hand, the regime's coverage is expanding and deepening among the non-Annex I parties. As noted above, the Copenhagen Accords/Cancun Agreements have engaged a much wider group of countries, including all major emitters, into national target-setting, while the Durban Platform mandate implies that all countries, not just Annex I, are expected to raise their ambition in the new post-2020 regime. On the other hand, the engagement of Annex I parties is weakening compared to the 1990s. There are now only a maximum of 37 Annex I parties that will be subject to legally-binding, comparable targets under the second phase of the Kyoto Protocol,[23] with major emitters such as Canada, Japan, the Russian Federation, and, of course, the USA, now operating only under the Copenhagen Accords/Cancun Agreements, whose targets are voluntary and not subject to common metrics.

The voluntary, permissive model of the Copenhagen Accords/Cancun Agreements is absolutely appropriate for smaller emitters and poorer countries; indeed, a 'hybrid', or 'mosaic' approach (Flannery, 2014) that would assign obligations according to capability and emission levels would seem logical (and has always been the intent of the regime).

---

[23] It is still unclear exactly who will ratify the Kyoto Protocol's second commitment period.

However, if the voluntary, permissive model becomes generally applicable post-2020 to *all* parties—including the wealthy countries of Annex I and the large developing country emitters—it would represent a huge step backwards for the regime. The idea under the regime was for developing (indeed, all) countries to gradually assume stronger commitments, and not for the Annex I parties to backtrack to weaker ones.

The problem is not so much the voluntary nature of pledges. It is a fair assumption that the big emitters will, in any case, only sign up to what they are prepared to do domestically, and any country can always withdraw from an international treaty (as Canada did from the Kyoto Protocol). What is critical, however, is for the targets of the giant emitters to be based on *absolute reductions* (not intensity targets) and *common metrics* (same base/target year, same gases covered, same assumptions about the land-use sector), while being subject to *strong monitoring and verification procedures*, all of which applied under the Kyoto Protocol, but are now in question. At the time of writing, the mixed signals continued. On the one hand, the EU—still hanging on to its leadership role—had proposed a relatively strong binding cut of 40 per cent on absolute emissions from 1990 levels by 2030, accompanied by a renewable energy target of at least 27 per cent (EU, 2014). On the other hand, other recent policy announcements, including from Australia and Japan, indicated a backtracking of ambition. Climate Action Tracker (2013) has characterized the current situation as 'a race to the bottom'. It is not surprising if decision-makers in the energy sector have failed to take the 2 °C goal very seriously.

4.4 **Taking stock and looking ahead**

The climate change regime, and the wider climate change problem, is a defining feature on the global energy landscape; a large mountain, perhaps, but one that is shrouded in mist, and therefore too easy to ignore. If the science is correct—and all the world's governments have endorsed the IPCC's findings—then curbing GHG emissions must surely be the major determinant of energy trends; the alternative means risking a climatically dangerous world. The problem, as demonstrated in this chapter, is that the political messages emanating from the climate change regime are still deeply contradictory: an ambitious (although probably still inadequate) 2 °C global goal, yet national emission commitments that fall far short of that goal. Moreover, it is doubtful whether the regime is actually getting any stronger, as the framework of legally-binding, comparable, absolute reduction targets adopted in Kyoto is (almost certainly) abandoned in favour of voluntary, nationally-defined goals. Most serious is the absence of political leadership. The collective nature of climate change has allowed the largest emitters to adopt a 'you go first' attitude, with the predictable result that no one goes very far. The EU has taken the lead as far as it can, 'saving' the Kyoto Protocol in 2000 after the US repudiation of the treaty, then agreeing to a second commitment period in the face of withdrawal from other large Annex I parties. The mathematics of global emissions, along with geopolitical realities, now requires strong leadership by China and the USA. Both these countries, however, have traditionally resisted any kind of internationally-imposed constraints on their emissions (or indeed in any other field). The prospect of them leading the way to a strong, binding global agreement that adds up to at most 2 °C seems absurdly optimistic.

And yet it is important to look beyond the international negotiations. The climate change regime—and wider concern over climate change—has already started to nudge the world towards a lower-carbon energy future. Without climate change, debates over energy futures and energy choices would be very different. There would be no need for CCS, for example, and, while renewable energy options would still be pursued, this would likely be on a much more relaxed time-scale. National regulation to address climate change, especially in the Kyoto Protocol parties who are subject to legally binding targets, has already transformed the incentive structures facing energy players, as detailed throughout this book, even if such legislation remains controversial and subject to roll-back.

Following the Copenhagen Accords/Cancun Agreements, all but two of the top 20 GHG emitters now have emission pledges on record (see Table 4.2). Many of the emerging economies are developing innovative regulatory mechanisms, including emissions trading schemes, to curb the growth in their emissions, notably China, which launched several regional schemes in 2013 (Jotzo, 2013; Yeo, 2013). It is also striking that the focus for renewable energy investment is now shifting to these emerging economies. In 2012, the developing world as a whole saw investment in renewable energy rise by nearly 20 per cent, compared with a drop of nearly a third in the OECD countries (Frankfurt School-UNEP Centre, 2013). At the same time, the CDM has engaged a wide group of countries for which the notion that saving carbon could generate value was previously alien (Grubb, 2014).

Moreover, responding to climate change no longer involves (if it ever did) just a top-down process of negotiating country targets at the international level, and then translating these into national legislation, with which private sector and other sub-national entities must then comply. Economic, social, and political actors are implementing initiatives to curb their own emissions, often in the absence of any imperative to comply with legislation. A 'voluntary' low-carbon transformation is underway, in addition to the 'compliance'-induced transformation triggered by internationally-agreed national targets. Nearly half of US states have emission targets,[24] for example, while aspiration to certified carbon neutrality is becoming fashionable among household business names.[25] Over 1,000 cities worldwide are involved in programmes to cut local emissions,[26] while large funding bodies are seeking out opportunities to leverage widespread change

---

[24] See data from the US-based Center for Climate and Energy Solutions, at <http://www.c2es.org/>.
[25] For example, see data from the Carbon Neutral Company, at <http://www.carbonneutral.com/>.
[26] See the Local Governments for Sustainability, at <http://www.iclei.org/>.

towards a less carbon-intensive world.[27] Why businesses, local governments, philanthropic organizations, community groups, individuals, and others should champion low-carbon alternatives often in the absence of legislation forcing this, is an interesting question with many possible answers: anticipation of future legislation; reaping of first-mover advantages; aspirations to a 'greener' image; awareness of co-benefits; genuine concern. Whatever the motivation, it is an important and hopeful development, even if there remains a massive gap between the pace of action on the ground, and the scale and urgency of change required.

Overall, the climate change regime, and broader concern over climate change, has started to set in motion a profound transformation in the way in which we evaluate and judge energy choices. The question, of course, is whether, and how quickly, that transformation—still too slow and too timid—can translate into the dramatic reversal of emission trends that is needed to ensure a safe climate. At present, diplomatic wrangling, inertia, vested interests, sheer denial (Harris, 2013) and other 'dragons of inaction' (Gifford, 2011) still have the upper hand. Despite the threat that climate change poses to human development, it remains but one of many factors influencing decision-makers in the energy sector, and rarely the most important one. Concern about climate change has started to influence energy choices, but, at present, prevailing patterns of energy production and consumption are still driving the world towards a dangerous climatic future.

■ REFERENCES

Bodansky, D. (1993). The United Nations Framework Convention on Climate Change: A commentary. *The Yale Journal of International Law*, Vol. 18, No. 2, pp. 451–558.

Climate Action Tracker (2013). *Warsaw Unpacked: A Race to the Bottom?* Policy brief, available at <http://climateactiontracker.org/publications/briefing/152/Warsaw-unpacked-A-race-to-the-bottom.html>. (Accessed April 2015).

Climate Strategies: Dröge, S. (2009). Tackling Leakage in a World of Unequal Carbon Prices, Cambridge, UK, report available from: <http://www.climatestrategies.org>. (Accessed April 2015).

Depledge, J. (2008). Striving for no: Saudi Arabia in the climate change regime (2008). *Global Environmental Politics*, Vol. 8, No.4, pp. 9–35.

Depledge, J. (2009). The road less travelled: difficulties in moving between annexes in the climate change regime. *Climate Policy*, Vol. 9, No. 3, pp. 273–87.

Depledge, J. and Yamin, F. (2010). The global climate change regime: A defence (2009). With Farhana Yamin. In Helm, D. and Hepburn, C. (eds) *The Economics and Politics of Climate Change*. Oxford: OUP, pp. 433–53.

---

[27] For example, see the Children's Investment Fund Foundation (CIFF), at <http://ciff.org/priority-impact-areas/climate/>.

EU (2014). Press release. <http://europa.eu/rapid/press-release_IP-14-54_en.htm>. (Accessed April 2015).

Flannery, B. (2014). Negotiating a Post-2020 Climate Agreement in a Mosaic World. *Resources*, No. 185, pp. 26–31. Available at <http://www.rff.org>. (Accessed April 2015).

Frankfurt School-UNEP Centre (2013). *Global Trends in Renewable Energy Investment 2013*, available at <http://www.fs-unep-centre.org>. (Accessed April 2015).

G-8 (2009). 'Responsible leadership for a sustainable future', G8 Leader's Declaration, L'Aquila, Italy. Available at <http://www.wcoomd.org/en/topics/key-issues/~/media/21E3FE6008454F9D8EC0B73A8BB7F5E2.ashx>. (Accessed April 2015).

Gifford, R. (2011). The dragons of inaction: Psychological barriers that limit climate change mitigation and adaptation. *American Psychologist*, Vol. 66, No. 4, pp. 290–302.

Grubb, M. (2014). *Planetary Economics: Energy, Climate Change and the Three Pillars of Sustainable Development*. London: Routledge.

Hansen, J, Kharecha, P., Sato, M., Masson-Delmotte, V., Ackerman, F., Beerling, D. J., Hearty, P. J., Hoegh-Guldberg, O., Hsu, S.-L., Parmesan, C., Rockstrom, J., Rohling, E.J., Sachs, J., Smith, P., Steffen, K., Van Susteren, L., Schuckmann, K., and Zachos, J.C. (2013). Assessing 'dangerous climate change': required reduction of carbon emissions to protect young people, future generations and nature, *Public Library of Science (PLoS one)*, Vol. 8, No. 12, e:18648. Available online at <http://www.ncbi.nlm.nih.gov/pmc/articles/PMC3849278/#!po=0.675676>. (Accessed April 2015).

Harris, P. G. (2013). *What's Wrong with Climate Politics and How to Fix It*. Cambridge: Polity Press.

Höhne, N., Den Elzen, M., and Weiss, M. (2006). Common but differentiated convergence (CDC): a new conceptual approach to long-term climate policy. *Climate Policy*, Vol. 6, No. 2, pp. 181–99.

IPCC (1997). *Technical Paper III, Stabilization of Atmospheric Greenhouse Gases: Physical, Biological and Socio-economic Implications*. Geneva: Intergovernmental Panel on Climate Change (IPCC).

IPCC (2001a). *Technical Summary, Contribution of Working Group II to the Third Assessment Report of the Intergovernmental Panel on Climate Change*. Cambridge, UK: Cambridge University Press.

IPCC (2001b). *Summary for Policy Makers, Contribution of Working Group II to the Third Assessment Report of the Intergovernmental Panel on Climate Change*. Cambridge, UK: Cambridge University Press.

IPCC (2006). *2006 IPCC Guidelines for National Greenhouse Gas Inventories, Prepared by the National Greenhouse Gas Inventories Programme* (Eggleston H.S., Buendia L., Miwa K., Ngara T., and Tanabe K. (eds)). Hayama, Japan: Institute for Global Environmental Studies (IGES).

IPCC (2013). Summary for Policymakers. In: *Climate Change 2013: The Physical Science Basis. Contribution of Working Group I to the Fifth Assessment Report of the Intergovernmental Panel on Climate Change* (Stocker, T.F., D. Qin, G.-K. Plattner, M. Tignor, S.K. Allen, J. Boschung, A. Nauels, Y. Xia, V. Bex and P.M. Midgley (eds)). Cambridge, United Kingdom and New York, NY, USA: Cambridge University Press.

IPCC (2014). Summary for Policymakers. In: *Climate Change 2014, Mitigation of Climate Change. Contribution of Working Group III to the Fifth Assessment Report of the Intergovernmental Panel on Climate Change* (Edenhofer, O., R. Pichs-Madruga, Y. Sokona, E. Farahani,

S. Kadner, K. Seyboth, A. Adler, I. Baum, S. Brunner, P. Eickemeier, B.Kriemann, J. Savolainen, S. Schlomer, C. von Stechow, T. Zwickel and J.C. Minx (eds)).Cambridge, United Kingdom and New York, NY, USA: Cambridge University Press.

Jordan, A., Rayner, T., Schroeder, H., Adger, N., Anderson, K., Bows, A., Le Quere, C., Joshi, M., Mander, S., Vaughan, N., and Whitmarsh, L. (2013). Going beyond two degrees: The risks and opportunities of alternative options. *Climate Policy*, Vol. 13, No. 6, pp. 751–69.

Jotzo, F. (2013). *Emissions trading in China: Principles, design options and lessons from international practice*, CCEP Working Papers 1303, Centre for Climate Economics & Policy, Crawford School of Public Policy, The Australian National University.

Lynas, M. (2009). How do I know China wrecked the Copenhagen deal? I was in the room. *The Guardian*, 22 December. Retrieved from <http://www.guardian.co.uk/environment/2009/dec/22/copenhagen-climate-change-mark-lynas>. (Accessed April 2015).

Olivier, J.G.J., Janssens-Maenhout, G., and Peters, J.A.H.W. (2012). *Trends in global CO2 emissions; 2012 Report*. The Hague: PBL Netherlands Environmental Assessment Agency; Ispra: Joint Research Centre.

Newell, P. (2012). The political economy of carbon markets: The CDM and other stories. *Climate Policy* Vol. 12, No 1, pp. 135–9.

Paulsson, E. (2009). A review of the CDM literature: From fine-tuning to critical scrutiny? *International Environmental Agreements*, Vol. 9, pp. 63–80.

Prins, G. and Rayner, S. (2007). *The wrong trousers: Radically rethinking climate policy*. Joint discussion paper of the James Martin Institute for Science and Civilization, University of Oxford, and the MacKinder Centre for the Study of Long-Wave events, London School of Economies and Political Science.

Schaeffer, M., Hare, W., Rahmstorf, S., Vermeer, M. (2012). Long-term sea-level rise implied by 1.5 °C and 2 °C warming levels. *Nature Climate Change*, Vol. 2, pp. 867–70, available online at <http://www.nature.com/nclimate/journal/v2/n12/full/nclimate1584.html>. (Accessed April 2015).

Smith, Joel B., Schneider, S.H., Oppenheimer, M., Yohe, G.W., Hare, W., Mastrandrea, M.D., Patwardhan, A., Burton, I., Corfee-Morlot, J., Magadza, C.H.D., Füssel, H.M., Pittock, A.B., Rahman, A., Suarez, A. and van Ypersele, J-P (2009). Assessing dangerous climate change through an update of the Intergovernmental Panel on Climate Change (IPCC) "reasons for concern". *Proceedings of the National Academy of Sciences* Vol. 106, No. 11, pp. 4133–7.

Susskind, L.E. (1994). *Environmental Diplomacy: Negotiating More Effective International Agreements*. Oxford: Oxford University Press.

UN (2014). UN Climate Summit 2014: Chair's Summary. Available at <http://www.un.org/climatechange/summit/2014/09/2014-climate-change-summary-chairs-summary/>. (Accessed April 2015).

UNEP (2013). *The Emissions Gap Report 2013: A UNEP Synthesis Report*. Available at <http://www.unep.org/pdf/UNEPEmissionsGapReport2013.pdf>. (Accessed April 2015).

UNFCCC (2000). *Tracing the origins of the Kyoto Protocol: An article by article textual history*. FCCC/TP/2000/2. Geneva: United Nations.

UNFCCC (2006). *Updated UNFCCC reporting guidelines on annual inventories following incorporation of the provisions of decision 14/CP.11*. FCCC/SBSTA/2006/9. Geneva: United Nations.

UNFCCC (2007). *Report of the Conference of the Parties on its thirteenth session, held in Bali from 3 to 15 December 2007*. FCCC/CP/2007/6/Add.1.

UNFCCC (2010). *Report of the Conference of the Parties on its sixteenth session, held in Cancun from 29 November to 10 December 2010.* FCCC/CP/2010/7/Add.1. Geneva: United Nations.

UNFCCC (2011a). *Report of the Conference of the Parties on its seventeenth session, held in Durban from 28 November to 11 December 2011.* FCCC/CP/2011/9/Add.1. Geneva: United Nations.

UNFCCC (2011b). *Report of the Conference of the Parties serving as the meeting of the Parties to the Kyoto Protocol on its seventh session, held in Durban from 28 November to 11 December 2011.* FCCC/KP/CMP/2011/10/Add.2. Geneva: United Nations.

UNFCCC (2013). Report of the Conference of the Parties at its nineteenth session, held in Warsaw, from 11 to 23 November 2013. Addendum. Part two: Action taken by the Conference of the Parties at its nineteenth session. Geneva: UN.

UNFCCC (2014). Press release. UN Climate Chief: IPCC Science Underlines Urgency to Act Towards a Carbon Neutral World. 13 April. Bonn, Germany: UN Climate Change Secretariat.

Yamin, F. and Depledge, J. (2004). *The International Climate Change Regime: A Guide to Rules, Institutions and Procedures.* Cambridge: Cambridge University Press.

Yeo, S. (2013). *Shanghai and Beijing latest Chinese cities to launch carbon trading schemes*, Report from Responding To Climate Change (RTCC) at: <http://www.rtcc.org/2013/11/26/shanghai-and-beijing-latest-chinese-cities-to-launch-carbon-trading-schemes/#sthash.0SeaP911.dpuf>. (Accessed April 2015).

5 The implications of indirect emissions for climate and energy policy

Katy Roelich, John Barrett, and Anne Owen

5.1 Introduction

Rapid growth in international trade and a shift of industrialized country economic activity towards services has been shown to facilitate Greenhouse Gas (GHG) emission reductions for countries listed in Annex I[1]of the Kyoto Protocol (Peters and Hertwich 2008; Le Quéré et al., 2009; Davis and Caldeira, 2010; Peters et al., 2011). However, this reduction in Annex I countries has been offset by corresponding increases in emission in non-Annex I countries, which have expanded production of goods that are traded with Annex I countries. The scale of such so-called indirect emissions, which have been produced in one country to serve consumption of goods and services in another, is significant and increasing; growth in indirect emissions from all international trade is five times higher than territorial emission reductions achieved by the Annex I countries (Peters et al., 2011). Globally, it is estimated that 20–25 per cent of global GHG emissions are from the production of traded goods (Peters and Hertwich, 2008; Davis and Caldeira, 2010).

The increasing significance of indirect emissions has implications for energy and climate policy:

- GHG emission reductions within a particular territory could be offset by emission increases in countries where imported goods are produced, offsetting the net reduction in global emissions and generally discouraging emission reductions efforts; so that
- sub-global climate policy (i.e. national or regional policy) becomes less effective in terms of both emissions reductions and cost efficiency in the country concerned.

To understand the full environmental, social, and economic effects of consumption and to successfully promote climate change mitigation policies, there is a need to capture

---

[1] Not all countries listed in Annex I have ratified the Kyoto Protocol. For example, neither the USA nor Canada have ratified and, as such, are not subject to binding targets. See Chapter 4 for further details.

the whole lifecycle impacts of products and services across international supply chains (Wiedmann et al., 2011). We cannot consider the UK energy system in isolation from the global energy system, upon which we rely not only for energy fuels and technologies but also for energy consumed and emissions created during production processes that occur outside the UK territory. It is important that we understand the scale and implications of this reliance and identify action to mitigate the additional risks.

In this chapter we discuss the quantification of indirect emissions, analyse the driving forces of these emissions, and present some additional policy responses that might more effectively manage indirect emissions. We use the UK to illustrate key elements of this discussion but the analysis is applicable to other countries.

5.2 Accounting for carbon emissions

5.2.1 EMISSIONS INVENTORIES

The recognized approach to accounting for carbon emissions is to use territorial-based emission inventories. The United Nations Framework Convention on Climate Change (UNFCCC) requires Annex I countries to submit annual National Emission Inventories in accordance with the Intergovernmental Panel on Climate Change's (IPCC) methodology. These inventories include 'emissions and removals taking place within national (including administered) territories and offshore areas over which the country has jurisdiction' (IPCC, 1996: 5). Territorial inventories do include GHG emissions emitted in international territory, as well as through international aviation and shipping, but these emissions are listed separately, and figures are not internationally reconciled. Importantly, these emissions are excluded from national inventories for compliance purposes (see also Chapters 4 and 11).

In parallel to territorial emissions, some countries quantify production-based emission inventories, which use the same system boundary as the System of National Accounts (SNA). This has the benefit of being the same system boundary as that used for reporting Gross Domestic Product (GDP). This includes emissions from international aviation and shipping, which are allocated to the country where the transport operator is based. This boundary is set to make the emission statistics consistent with economic data used in economic modelling. These inventories are often called 'National Accounting Matrices including Environmental Accounts (NAMEAs)' (for example Office for National Statistics (2013)). In the EU, NAMEAs are reported to Eurostat; however, most other developed countries develop NAMEAs but do not report them internationally.

Neither of the above standard approaches, however, account for the indirect emissions that occur because some goods and services required to support national consumption are produced outside the reporting territory. Recognizing this deficiency, additional consumption-based accounting methods have recently been developed. Consumption-based emission accounting has been developed to allocate indirect emissions to the country

Table 5.1 Emissions inventory definitions used in this chapter

Term	Definition
Territorial-based emissions	Emissions produced from industrial, commercial, and household activity within borders of territory.
Production-based emissions	Emissions from industrial, commercial, and household activity within borders of territory plus emissions from aviation and shipping (allocated to country where operator is based).
Consumption-based emissions	Emissions produced to satisfy consumption of residents within borders of territory (production emissions minus exports plus indirect emissions).
Indirect emissions[2]	Emissions associated with the production of goods imported to country of interest.
Domestic emissions	Emissions produced within borders of territory to support consumption *in that territory* (production emissions minus from the production of exports).

that consumes the final products, usually based on final consumption as in the SNA but also as trade-adjusted emissions (Peters, 2008). Conceptually, consumption-based inventories can be thought of as production-based emissions minus the emissions from the production of exports plus the emissions from the production of imports (Barrett et al., 2013).

The three emissions inventories described above are summarized for clarity in Table 5.1 along with the terms *indirect emissions* and *domestic emissions*, which are used subsequently in this chapter.

The inventories have different system boundaries and provide complementary information about the drivers of emissions resulting from a particular nation. This places emphasis on different approaches to emissions reduction which are largely complementary. The inventories should be considered together to identify the most effective mitigation strategies for global emissions reductions; consumption-based emissions inventories are not an alternative to territorial inventories.

5.2.2 CALCULATING CONSUMPTION-BASED INVENTORIES

A European Scientific Knowledge for Environmental Protection project identified Environmentally Extended Multi-Region Input–Output (EE-MRIO) Analysis as an ideal basis for the assessment of environmental impacts of trade (Wiedmann et al., 2009). Input–output analysis uses an economic model of sectorally disaggregated accounts to quantify inputs to each industrial sector and the subsequent uses of the outputs of those sectors. This allows the user to analyse the interdependency between sectors in an economy and to track the flows of resources between them (Leontief, 1970). The basic model can be extended to include trade flows between all countries or regions (Wiedmann et al., 2011b), allowing users to track the impacts of supply chains spanning multiple sectors in multiple countries. Environmental and other extensions can be added to the MRIO framework in the form of inputs or 'factors' of production, for example as amounts of

---

[2] Referred to in much of the literature as emission transfers.

raw materials extracted or pollutants emitted by industrial sectors. Most current MRIOs use IEA energy consumption data, combined with emissions factors for energy vectors for the benefits of consistency (Andrew and Peters, 2013; Lenzen et al., 2013; Tukker et al., 2013).

A number of statistical offices and other government agencies have started to calculate consumption-based emissions, including Australia, Canada, Denmark, France, Germany, Netherlands, Sweden, and the UK. However, these are rarely seen as 'official statistics' meaning that there is no clear commitment to annually update the indicator and provide an official statistical release of the data (Edens et al., 2011). Only two countries update and release consumption emissions account data; Australia (Australian Bureau of Statistics, 2013) and the UK (Department for Environment Food and Rural Affairs, 2012). We use the UK as a case study to demonstrate whether attribution of emissions to the consumer generates robust results that are useful in the formulation of climate policy beyond what can be offered by territorial accounting.

5.2.3 COMPARING EMISSIONS ACCOUNTS

The UK has adopted consumption-based emissions as an official government indicator, complementary to but not replacing the current territorial accounting system. Figure 5.1 illustrates the latest time-series results published by the UK government (Department for Environment Food and Rural Affairs, 2012) for the three main approaches (i.e. territorial-, production-, and consumption-based). Consumption-based GHG emissions

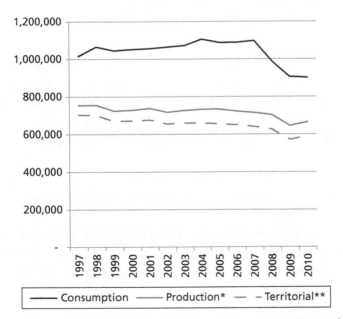

Figure 5.1 Consumption, production, and territorial greenhouse gas emissions for the UK * ONS figures, ** UNFCCC reported (DECC 2013)

grew by 8 per cent between 1997[3] and 2007, followed in 2007–10 by an 18 per cent reduction, predominately due to the global financial crisis (Wiedmann et al., 2010). Consumption emissions showed an overall decrease of 11 per cent between 1997 and 2010 but were still 53 per cent higher than territorial emissions reported to the UNFCCC in 2010.

The UK GHG emissions reported to the UNFCCC (i.e. their 'territorial emissions') in relation to commitments under the Kyoto Protocol show a 16 per cent reduction in territorial GHG emissions between 1997 and 2010, which represents an annual decline of around 1.25 per cent per annum. GHG emissions were 113 million tonnes lower in 2010 than in 1997, and the UK government achieved its target established under the Kyoto Protocol. Production-based GHG emissions reduced by only 12 per cent during the same time period. There has thus been a greater reduction in emissions as accounted for under the Kyoto Protocol than in those that are not. In fact, production-based GHG emissions not originally accounted for under the Kyoto Protocol (i.e. the emissions from aviation and shipping) increased by 46 per cent between 1997 and 2010, from 50 million tonnes to 74 million tonnes of CO_2e emissions.

From a consumption perspective, the UK's GHG emissions rose at a rate of almost 0.7 per cent per annum between 1997 and 2007 (before the recession-related reduction). These figures stand in stark contrast to the 1.25 per cent decrease each year in territorial GHG emissions. The gap between consumption-based and territorial emissions has continued to grow year on year with the exception of 2009, when a comparatively large reduction was recorded as a result of the recession.

5.2.4 UNCERTAINTY IN CONSUMPTION-BASED EMISSIONS INVENTORIES

All emission inventories have some uncertainty, including territorial emissions; however, consumption-based estimates will have a larger uncertainty due to the incorporation of more input data. The additional uncertainties stem from data calibration, balancing and harmonization, use of different time periods, different currencies, different country classifications, levels of disaggregation, inflation, and raw data errors (Lenzen et al., 2010). Many of the uncertainties arise as a result of inconsistent reporting practices in different countries and regions, and a process of harmonization in data collection methodologies could greatly reduce the necessary manipulations and, hence, uncertainties (Peters and Solli, 2010).

It is possible to account for such uncertainties by applying error propagation methods to determine their influence on the analytical results of carbon footprint studies, for example by employing Monte-Carlo simulation techniques. A detailed uncertainty

---

[3] Note that 1997 is used as a baseline year, rather than 1990, because this is the earliest year for which consumption-based accounts were available at the time of writing.

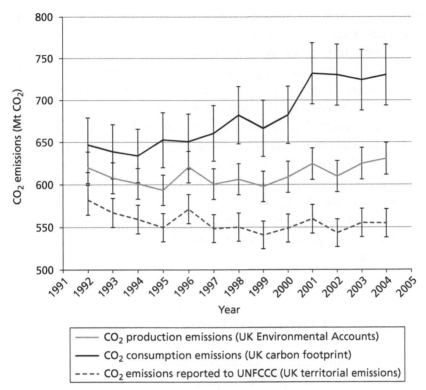

Figure 5.2 Uncertainty associated with UK consumption-based CO_2 Emissions (as calculated using EE-MRIO analysis)

Source: Lenzen et al., 2010.

analysis of the UK consumption-based emissions using EE-MRIO modelling was undertaken using these techniques for Defra (Lenzen et al., 2010). Figure 5.2 provides the results of this uncertainty analysis and demonstrates that there is an additional uncertainty in the region of 3 per cent between consumption- and production-based accounting.

Note that these data relate to CO_2 emissions, not total GHG, and analysis was undertaken of a limited dataset. Single data items, such as sector level emissions, may be associated with a high degree of uncertainty; however, aggregate measures, such as total consumption-based emissions, are usually known with much more certainty. The analysis demonstrated that the trends in UK consumption-based emissions, despite uncertainties in the absolute levels, are robust, as is the difference from UK territorial-based emissions, despite the greater uncertainty in the consumption-based estimates.

5.3 **Future consumption-based carbon emissions**

A number of countries and regions, including the UK, have policy packages designed to reduce GHG emissions to levels that avoid dangerous climate change. These packages include binding (territorial) emissions reduction targets, renewable energy targets, and incentives to encourage decarbonization of domestic energy systems. If successful, these policies will significantly reduce the UK's territorial GHG emissions and may give the impression of progress towards the global objective of emissions reduction, to increase chances of avoiding dangerous climate change (IPCC, 2007). However, considering emission reductions from a solely territorial approach has two limitations. First, the historical analysis presented in Section 5.2.3 has shown that in some situations, even if domestic decarbonization is occurring, increasing reliance on indirect emissions is outweighing these decreases. There is no guarantee that there will be a corresponding decarbonization of the energy systems of our trading partners; therefore global emissions may not reduce significantly. Secondly, the decarbonization of energy systems relies on the rapid roll out of low-carbon technologies, which can have high levels of emissions produced during their lifecycle. We discuss these two issues in more detail below.

5.3.1 SCENARIOS OF FUTURE EMISSIONS

The UK's Climate Change Act 2008 introduced a series of carbon budgets to help the UK to achieve its target of reducing territorial greenhouse gas emissions by at least 80 per cent by 2050 based on 1990 levels (HM Government, 2008). These targets are in line with the Copenhagen Accord, which aims to limit global warming to less than 2 °C. The UK targets, and indeed the targets of most other countries, have taken global reduction requirements and translated them into national targets, without appreciating the significance of indirect emissions. To illustrate the limitations of this approach we have constructed a simple scenario from a consumption-based perspective. In the scenario we assume that:

- the UK achieves its territorial emissions reduction target of 80 per cent by 2050 based on 1990 levels; and
- emissions associated with imported products continue to grow in line with historical growth rates for the past twenty years.

Figure 5.3 shows the results of this scenario, which demonstrates the potential scale of indirect emissions without either a radical change in global production efficiency or specific policies to address consumption. Domestic emissions (emissions produced in the UK for UK consumption, i.e. excluding emissions associated with exported goods and services) comprise less than 25 per cent of total UK emissions by 2050 and total emissions only reduce by 47 per cent from 1990 levels of territorial emissions. Analyses such as this show that without due attention to indirect emissions, it is likely that the scope of existing territorial emissions reduction strategies will be significantly undermined over time.

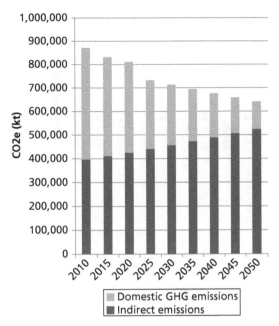

Figure 5.3 Future GHG emissions showing domestic (production minus exports) and indirect (those associated with the production of goods imported to the UK) emissions for the UK

(*Source*: own calculations).

5.3.2 EMISSIONS RELEASED OVER THE LIFECYCLE OF LOW-CARBON TECHNOLOGIES

The process of energy system decarbonization will require an extensive roll out of low-carbon technologies, such as wind turbines, solar panels, carbon capture and storage, and efficient end-use technologies as well as low-carbon heat and transport technologies (HM Government, 2011). GHG emissions are caused at each stage of the life of these technologies, from the extraction of raw materials and manufacturing right through to use and final re-use, recycling or disposal (often termed lifecycle emissions). If low-carbon technologies available now had significant lifecycle emissions, whether in the UK or indirect emissions, this might suggest a need to develop alternative technologies with a lower lifecycle footprint, as well as a need for additional abatement in the UK or abroad (Committee on Climate Change, 2013a). A recent assessment by the IPCC generated estimates of the lifecycle emissions of low-carbon electricity technologies (IPCC, 2012). High-level results are presented in Figure 5.4. These indicate that while renewable electricity generation and carbon capture and storage technologies have some emissions associated with their production, they are still much less carbon-intensive over their lifecycle than the use of unabated fossil fuels.

In assessing the need for additional abatement, it is important to understand whether lifecycle emissions of low-carbon technologies are addressed by current emission reduction schemes as well as assessing the scale of the emissions from the technologies

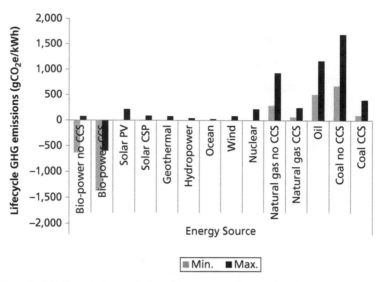

Figure 5.4 Lifecycle GHG emissions of electricity generation technologies
Source: IPCC, 2012.

themselves. The UK Committee on Climate Change (CCC) estimated that total lifecycle emissions from the deployment of key low-carbon electricity technologies are around 260 $MtCO_2e$ over the period to 2030, of which around 40 per cent occur outside the UK (Committee on Climate Change, 2013b: 68). The report concluded that the lifecycle impacts of low-carbon electricity technologies were not substantial enough to warrant a change in strategy and that a sufficient proportion of emissions was captured by the UK emissions target; therefore no additional abatement was required.

However, both the IPCC analysis and the CCC report are based on process lifecycle analysis alone. Wiedmann et al. (2011a), estimated lifecycle emissions for wind turbines using a combination of input–output-based methods and process lifecycle analysis (LCA). The hybrid methods yielded estimates of more than twice the LCA result. This difference was predominantly a result of the avoidance of truncation error that occurs when using LCA in isolation (Wiedmann et al., 2011a). This could affect the CCC's conclusion about the small scale of lifecycle emissions negating the need for further action.

Lifecycle emissions are not currently accounted for in some energy system models, particularly national models, which are used to inform national policy. These models minimize the energy system costs in scenario projections of future energy systems under different assumptions. A cost is assigned to carbon emitted by technologies during operation that is combined with technology cost; however, no cost is assigned to the emissions generated during the manufacturing or decommissioning stages of the technology lifecycle. Therefore, cost optimization in these models is only based on emissions reductions in the country of interest, not globally, and technology selection may not be optimal when global emissions reductions are taken into account (Arvesen et al., 2011). Obviously such problems do not arise in global models which optimize the global energy system and take all carbon emissions into account, but these models to not allow emissions to be associated with particular technologies.

5.4 **Drivers of indirect emissions**

Indirect emissions are created when goods are traded between non-Annex I and Annex I countries and may be thought of as a kind of carbon leakage, whereby emissions that would have been covered and therefore regulated by targets under the Kyoto Protocol[4] are instead produced in another country without emissions targets. Carbon leakage reveals the extent to which emissions are just shifted between countries, rather than abated (Peters and Hertwich, 2008). Peters and Hertwich describe two mechanisms for carbon leakage (Peters and Hertwich, 2008):

- Strong carbon leakage occurs when there is an increase in indirect emissions specifically due to climate policy (e.g. a UK climate change policy).
- Weak carbon leakage occurs when there is an increase in indirect emissions due to increased consumption (rather than one due to a specific government policy).

There is little evidence that strong carbon leakage is significant (Carbon Trust, 2010) and the Committee on Climate Change concluded that 'costs and competitiveness risks associated with measures to reduce direct emissions to 2030 are manageable' (Committee on Climate Change, 2013b). Conversely, weak carbon leakage, where a country has increased its production (and associated emissions) to support increased consumption in another country, is considered to be large and growing (Peters et al., 2011).

Indirect emissions are not consistent across all economic sectors and are not directly proportional to the volume of trade. Therefore understanding past trends of indirect emissions requires detailed analysis of the sources of emissions and the drivers of indirect emission growth. In the following sections we discuss the sectors that contribute most to indirect emissions and the determinants of those emissions.

5.4.1 INDIRECT EMISSIONS BY SECTOR

The proportion of domestically produced[5] and indirect emissions that serve UK consumption is not the same across different sectors in the UK; therefore, action required to address consumption emissions will differ significantly between sectors. Figure 5.5 presents the proportions of domestic and indirect emissions for aggregated sectors of the UK economy to illustrate this difference.

[4] Although not all countries in Annex I have ratified the Kyoto Protocol.
[5] This excludes the emissions from the production of exports from the UK, since they do not service UK consumption.

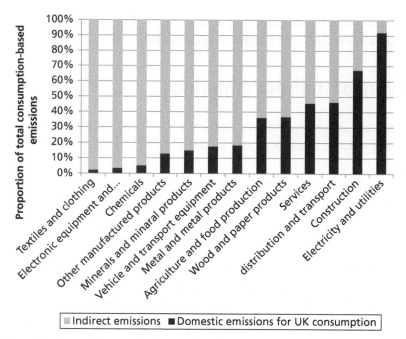

Figure 5.5 Contribution of domestic (UK production minus exports) and indirect (those associated with the production of goods imported to the UK) emissions by sector for the UK 2010

Emissions from electricity production and utility delivery are predominantly produced in the UK (with only 6 per cent imported) and decarbonization of the UK energy system will go a long way toward mitigating these emissions. The only other sector where more emissions are produced domestically than imported is construction (with 33 per cent imported). However, the remaining sectors are dominated by imported emissions, for example, 94 per cent of emissions associated with consumption of chemicals in the UK are generated outside UK territory. Energy-intensive sectors (such as chemicals, metal, and metal products) fall under the EU emissions trading scheme and, therefore, emissions imported from the EU will be captured by this policy. However, only 25 per cent of emissions associated with imports to the UK in these sectors occur inside the EU and only 17 per cent of the total indirect emissions are captured under the EU ETS, meaning that a total of 83 per cent of emissions released through the production of goods for trade are not[6] (Barrett et al., 2013).

5.4.2 DETERMINANTS OF INDIRECT EMISSIONS

Carbon accounting alone does not provide sufficient information to guide the design of effective and fair policies aimed at reducing GHGs (Jakob and Marschinski, 2013).

---

[6] 2009 figures.

We cannot assume a causal relationship between increasing trade and increasing global emissions. Some countries might have higher consumption-based emissions in the absence of trade if their domestic production systems were less efficient than those of their trading partners. We need to understand not just carbon flows but determinants of these flows to assess how indirect emissions would be affected by new policies.

To investigate this phenomenon we have undertaken a decomposition analysis of indirect emissions following a method described by Jakob and Marschinski (2013).[7] The results of this analysis are shown in Figure 5.6 for a series of countries with contrasting emission profiles. Results are shown in terms of the net export of emissions, which is the difference between the emissions associated with goods exported from the country and those imported to the country. An overall positive 'net export' of emissions indicates that emissions associated with exported goods are higher than those associated with imported goods. A negative 'net export', such as is the case with the UK, means that the indirect emissions associated with goods imported to the country are higher than the exported goods. The decomposition analysis has allowed us to quantify the different determinants of this difference including:

- Specialization of a country in the export of carbon-intensive goods (measured by comparing the carbon intensity of exports with that of total production in the country of interest).
- Imbalance of trade—the value of exports and imports are not equal so there is a trade surplus or deficit.
- The difference in economy-wide energy intensity between the trading country and the world average (measured in energy per unit of GDP).
- The difference in carbon intensity of the energy system between the trading country and the world average (measured in tonnes of CO_2 per unit of energy).

We continue our analysis of the UK as an illustration of the value of this decomposition, with the corresponding results for other countries also shown in Figure 5.6. The UK has higher emissions associated with its imports than its exports, so net emission exports show as negative. The principal driver of indirect emissions in the UK is the difference between domestic energy intensity and the world average; the UK trades with countries that have higher energy intensity than itself. It is important to note that this is an economy-wide measure that accounts for the energy intensity of the whole economy. This includes not only the intensity of individual sectors but also the structure of the economy. If the economy is comprised of predominantly less energy-intensive sectors, such as services, then the economy-wide energy intensity would compare favourably with countries whose economy was dominated by energy-intensive sectors, such as steel production. It is not necessarily a sign that production in the UK is more energy-efficient, rather that the UK economy has shifted towards production of less

---

[7] Jakob and Marschinski (2013) used the balance of emissions embodied in trade (BEET) method, rather than a full MRIO, which we replicate in this analysis. We use data from 1995 to 2009.

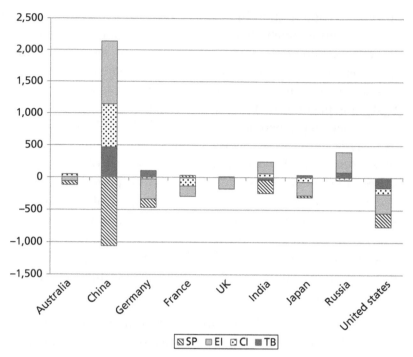

Figure 5.6 Decomposition of net exports of indirect emissions for selected countries (2009). SP= specialization, EI = energy intensity of economy, CI = carbon intensity of energy, TB = trade imbalance

energy-intensive, light manufacturing and services and generally imports energy-intensive products.

There are two potential responses to this; to move production of energy intensive goods back to the UK or to support more efficient production in trading partners. It would be costly and difficult to reinstate the manufacturing infrastructure required to re-start producing energy intensive goods in the UK. There is also no guarantee that this would result in a reduction of global emissions, since the UK energy system currently has a relatively high carbon intensity (as demonstrated by the export of a small quantity of emissions associated with the carbon intensity of the UK energy system in 2009). This would also increase the UK's territorial emissions, which would be politically unfavourable. A more cost and carbon effective approach might be to invest in knowledge and technology transfer to support efficiency improvements in its trading partners, or to have a wider system of emissions commitments so that trading partners were subject to emissions constraints too. This is discussed in more detail in Section 5.5.4.

In contrast to the UK, China has higher emissions associated with its exports as a result of the high energy intensity of its economy and the carbon intensity of its energy system. In 2009, exports had a lower carbon intensity than the average intensity of production in China (shown by the negative result for specialization) indicating that China is exporting less energy-intensive, higher-value products. These indirect emissions

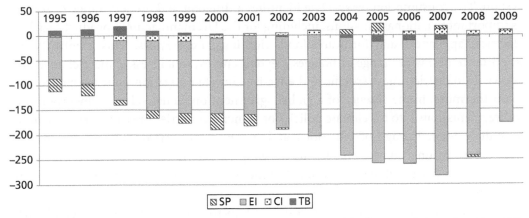

Figure 5.7 Time-series of decomposition of net exports of indirect emissions for the UK. SP = specialization, EI = energy intensity of economy, CI = carbon intensity of energy, TB = trade imbalance

(and hence global emissions) would be reduced if the whole of the Chinese economy became more energy and carbon-efficient.

The determinants of indirect emissions are not static and there is considerable change in the scale of net emissions and the balance between the determinants. Figure 5.7 illustrates a time-series of net exported emissions for the UK. The emissions are dominated by the difference in energy intensity, but there has been more than a two-fold increase from 1995 peaking in 2007. Before 2001, there was a contribution from the UK specializing in export of less carbon-intensive goods, but this specialization is no longer a contributory factor, indicating that UK exports now have a similar carbon intensity to the economy as a whole. There is very little contribution from the carbon intensity of UK energy, indicating that the UK's energy system is roughly equal to the world average. As it decarbonizes its energy system, imported emissions will become an increasingly important source of UK consumption emissions and energy and climate policy will need to address these emissions in order to effectively reduce global emissions.

5.5 Potential national policy responses

A territorial or production perspective on emissions identifies energy production, energy-intensive industries, household energy consumption, and transportation as the dominant sources of emissions. In response, national and regional climate policy frameworks have evolved to focus on energy system decarbonization. It is important that this momentum on energy system decarbonization is maintained, but there must be a parallel debate on approaches to tackle consumption emissions, which includes traded goods within the system boundary for mitigation potential, rather than just the energy system. The consumption perspective identifies a different and broader range of sectors as important to national emissions, including for the UK manufactured products, such as electrical appliances and furniture, food, clothing, and services. This highlights the importance of a wider range of policies, beyond energy, to address climate change and allows us to develop more specific and effective climate mitigation policy responses.

Furthermore, in the absence of a global agreement on climate change, incorporating indirect emissions into the climate policies of Annex I countries presents an opportunity to reduce weak leakage of emissions and increase the environmental efficiency of national or regional climate policies. Identifying the determinants of indirect emissions (as in Section 5.4.2) will help us to identify the most effective strategy to address these emissions.

In this section we discuss potential policy options for incorporating indirect emissions into policymaking. These include: reducing the impact of domestic consumption, border tax adjustments, and offsetting indirect emissions via investments in trading partners.

5.5.1 DOMESTIC CONSUMPTION POLICY

Changing our demand for materials and products could contribute significantly to energy efficiency and carbon emissions mitigation. A number of industrial processes are rapidly approaching the technical limits of their energy efficiency improvement potential; therefore, energy and resource efficiency (delivering the same product with less energy or material input) alone will not achieve the scale of emission reductions required to meet climate change mitigation targets. Consequently, consumption-related policies promoting resource *sufficiency* (delivering the same welfare with less new material or fewer products) offer greater potential to reduce energy demand or carbon emissions than resource efficiency.

This has been clearly demonstrated in the UK, where a study for the Waste and Resources Action Programme (WRAP) demonstrated that a wider scope of emissions were influenced by consumption-based strategies such as extending the lifetime of products, reducing food waste, dietary changes, and product durability (Barrett and Scott, 2012). Savings were significantly higher than those from strategies to improve the efficiency of production. Energy efficiency strategies or product roadmaps should not confine themselves to production-based measures but also include measures that alter or reduce consumption of goods and services.

The promotion of resource sufficiency strategies is by no means a recent phenomenon; work undertaken for the Dutch National Research Programme demonstrated in 2001 that increasing product longevity and moving from products to services can contribute to emission reductions (Nonhebel and Moll, 2001). Despite this history of evidence, consumption policies have remained separate to policies for climate mitigation and their potential is yet to be realized in full.

Many of the policy options that change the composition of consumption could have indirect rebound effects where the saved revenue is spent on another good or service. Furthermore, the mitigation costs and benefits of consumption-based measures have not been quantified in detail. It is essential that these rebound effects and the cost efficiency of consumption-based measures are considered to avoid an overly optimistic picture of the scale of emissions reduction possible.

5.5.2 TRADE POLICY

Border Tax Adjustment (BTA) has been suggested as a way of reducing carbon leakage and indirect emissions (Fischer and Fox, 2012). In the climate change context BTA would involve a charge levied on carbon associated with goods imported from countries without comparable carbon reduction policies, with an exemption from or remissions of taxes on products when they are exported. The charge would be determined in proportion to the carbon content of imports and the price of carbon in the importing country. Other factors might also be invoked, such as whether the exporting country has equivalent policies in place to curb carbon emissions, which might reduce the BTA levied. A similar system is currently in force for VAT, which is politically accepted, so a case could be made that carbon BTA is compliant with World Trade Organisation rules (Ismer and Neuhoff, 2004). Nonetheless, using BTA in such a way in the climate change context remains highly politically contentious.

BTA is intended to address competitiveness issues of domestic or regional decarbonization policy (such as the EU Emissions Trading Scheme) in the absence of a global carbon price. It would do this by internalizing the costs of carbon emissions for producers in countries that are not subject to carbon constraints. To this end it would encourage producers in these countries to reduce their emissions by increasing energy efficiency or reducing the carbon intensity of energy generation. It may also encourage consumers to reduce their demand if prices increase (Ismer and Neuhoff, 2004). It has been argued that the inclusion of trade in climate agreements, via a BTA, may even force a global agreement (Helm, 2012).

BTA has the potential to reduce indirect emissions (Springmann, 2014) but would only reduce global emissions if the exporting country does not increase production of goods in other sectors to mitigate losses in revenue associated with the tax. Furthermore, concerns have been expressed about unequal distribution of economic burden, which is a particular concern when exporting countries are developing countries. The approach also has methodological challenges; it is not likely to ever be possible to measure accurately, robustly, and cost-effectively the emissions associated with the whole lifecycle of individual products and to identify where these emissions occurred. Any practical scheme of BTA would need to be approximate and would be open to challenge.

5.5.3 OFFSETTING INDIRECT EMISSIONS

A more equitable approach to distribution of the costs of abatement could be for importing countries to offset their indirect emissions by financing emission reduction projects in the exporting countries, notably if these are non-Annex I countries that have taken on post-Kyoto commitments. This is similar to the Clean Development Mechanism of the Kyoto Protocol and is based on the premise that emission reductions can be achieved more cost-effectively in developing countries than in developed ones due to the availability of low-cost abatement options.

Offsetting indirect emissions has been shown to have little effect on the scale of emission transfers between countries but could lead to a significant reduction in global emissions through sponsored emissions reductions in non-Annex I countries (Springmann, 2014). This type of mechanism could mitigate against reductions in GDP in non-Annex I countries (in fact it could actually increase GDP in these countries), although there could be a slight reduction in Annex I countries.

However, this type of mechanism is not without challenges. Decisions based entirely on the basis of indirect emissions could direct an even greater proportion of financing to trade-active countries, such as China, while least-developed countries with low volumes of trade would receive very little. The distribution of funds would need to be disbursed by a neutral organization on the basis of abatement potential but take into account development need.

Furthermore, the application of offsetting policies, such as the Clean Development Mechanism, has been subject to criticism. Concerns have been raised about the governance of offsetting schemes, the quality of approved projects, and the contribution to sustainable development. As a result there is ongoing debate about the ability of clean development and offsetting to deliver significant emission reductions (Gillenwater and Seres, 2011). Further work would be required to determine specific detail of this type of policy, which could be politically contentious.

5.6 **Conclusions**

This chapter is concerned with effective climate policy in a fragmented governance environment, without a global agreement on emissions reduction or a global carbon price. In the current situation, emissions are being leaked from those Annex I countries with binding emissions reductions targets, to non-Annex I countries currently without binding targets, through trade. These indirect emissions are increasing and offset territorial emission reductions in Annex I countries. In the absence of a global agreement, therefore, incorporating indirect emissions into the climate policies of industrialized countries presents an opportunity to reduce weak leakage of emissions and increase the environmental and cost-efficiency of global climate policies.

Indirect emissions can be estimated using consumption-based accounting and such accounting reveals that indirect emissions are significant and increasing. Decarbonization of Annex I energy systems alone will not reduce global emissions on the scale and at the speed necessary to avoid dangerous climate change. Furthermore, the decarbonization of energy systems relies on the rapid roll out of low-carbon technologies, which can have high levels of emissions generated during their lifecycle. Current national approaches to decarbonization do not include these emissions in analysis, which might result in overly-optimistic assessments of abatement potential on a global scale.

Measures to incorporate indirect emissions into policymaking should take into account the diverse determinants of these emissions; there is not a direct relationship between increased trade and increased indirect emissions. A more sophisticated understanding of these drivers and the effects of policy on the distribution of economic burden should be used to identify the most cost- and environmentally-efficient policy. This might include domestic consumption policy, border tax adjustment, and offsetting as well as promoting the expansion of emission targets and stronger climate policies in exporting countries.

▒ REFERENCES

Andrew, R.M. & Peters, G.P., 2013. A Multi-Region Input-Output Table based on the Global Trade Analysis Project database (GTAP-MRIO). *Economic Systems Research*, 25(1), pp. 99–121.

Arvesen, A., Bright, R.M., & Hertwich, E.G., 2011. Considering only first-order effects? How simplifications lead to unrealistic technology optimism in climate change mitigation. *Energy Policy*, 39(11), pp. 7448–54.

Australian Bureau of Statistics, 2013. *Towards the Australian Environmental-Economic Accounts.* Australian Bureau of Statistics, Canberra, Australia.

Barrett, J. & Scott, K., 2012. Link between climate change mitigation and resource efficiency: A UK case study. *Global Environmental Change*, 22, pp. 229–307.

Barrett, J. et al., 2013. Consumption-based GHG emission accounting: a UK case study. *Climate Policy*, 13(4), pp. 451–70.

Carbon Trust, 2010. *Tackling carbon leakage: Sector-specific solutions for a world of unequal carbon prices.* London.

Committee on Climate Change, 2013a. *Reducing the UK's carbon footprint: Technical Report.*

Committee on Climate Change, 2013b. *Reducing the UK's carbon footprint and managing competitiveness risks.*

Davis, S.J. & Caldeira, K., 2010. Consumption-based accounting of CO2 emissions. *Proceedings of the National Academy of Sciences*, 107 (12), pp. 5687–92.

Department for Environment Food and Rural Affairs, 2012. *UK's Carbon Footprint 1993–2010.* Latest data available from <https://www.gov.uk/government/statistics/uks-carbon-footprint>. (Accessed April 2015).

Edens, B. et al., 2011. Analysis of changes in Dutch emission trade balance (s) between 1996 and 2007. *Ecological Economics*, 70(12), pp. 2334–40.

Fischer, C. & Fox, A.K., 2012. Comparing policies to combat emissions leakage : Border carbon adjustments versus rebates. *Journal of Environmental Economics and Management*, 64(2), pp. 199–216.

Gillenwater, M. and Seres, S., 2011. *The Clean Development Mechanism: A Review of the First International Offset Program*. Arlington, VA: Pew Center on Global Climate Change.

Helm, D., 2012. *The Carbon Crunch: How We're Getting Climate Change Wrong—And How to Fix It*. New Haven: Yale University Press.

HM Government, 2008. *Climate Change Act—Chapter 27.*

HM Government, 2011. *The Carbon Plan: Delivering our low carbon future.* Department of Energy and Climate Change, London.

IPCC, 1996. *Revised 1996 IPPC Guidelines for National Greenhouse Gas Inventories. Reporting Instructions (Volume 1).*

IPCC, 2007. *Climate Change 2007: The Physical Science Basis. Contribution of Working Group 1 to the Fourth Assessment Report of the Intergovernmental Panel on Climate Change*, Cambridge: Cambridge University Press.

IPCC, 2012. *Renewable Energy Sources and Climate Change Mitigation Special Report of the Intergovernmental Panel on Climate Change.* Cambridge: Cambridge University Press.

Ismer, R. & Neuhoff, K., 2007. Border Tax Adjustments: A feasible way to support stringent emission trading. *European Journal of Law and Economics*, 24(2), pp. 137–64.

Jakob, M. & Marschinski, R., 2013. Interpreting trade-related CO_2 emission transfers. *Nature Climate Change*, 3, pp. 19–23.

Lenzen, M., Moran, D., Kanemoto, K., & Geschke, A., 2013. Building Eora: A global Multi-Region Input-Output database at high country and sector resolution. *Economic Systems Research*, 25(1), pp. 20–49.

Lenzen, M., Wood, R., & Wiedmann, T., 2010. Uncertainty analysis for Multi-Region Input-Output Models—A case study of the UK's Carbon Footprint. *Economic Systems Research*, 22(1), pp. 43–63.

Leontief, W., 1970. Environmental Repercussions and the Economic Structure: an Input-Output Approach. *The Review of Economics and Statistics*, 52(3), pp. 262–71.

Nonhebel, S. & Moll, H.C., 2001. *Evaluation of Options for Reduction of Greenhouse Gas Emissions by Changes in Household Consumption Patterns*, The Hague: PBL Netherlands Environmental Assessment Agency.

Office for National Statistics, 2013. *Statistical Bulletin UK Environmental Accounts, 2013.*

Peters, G.P., 2008. From production-based to consumption-based national emission inventories. *Ecological Economics*, 65(1), pp. 13–23.

Peters, G.P. & Hertwich, E.G., 2008. CO 2 Embodied in International Trade with Implications for Global Climate Policy. *Environmental Science & Technology*, 42(5), pp. 1401–7.

Peters, G.P., Minx, J., Weber, C.L., and Edenhofer, O., 2011. Growth in emission transfers via international trade from 1990 to 2008. *Proceedings of the National Academy of Sciences of the United States of America*, 108(21), pp. 8903–8.

Peters, G.P., & Solli, C., 2010. *Global Carbon Footprints. Methods and Import/Export Corrected Results from the Nordic Countries in Global Carbon Footprint Studies*, Nordic Council of Ministers, Copenhagen, Denmark.

Le Quéré, C., Raupach, M.R., Canadell, J., Marland, G., Bopp, L., Ciais, P., Conway, T.J., Doney, S.C., Feely, R.A., Foster, P., 2009. Trends in the sources and sinks of carbon dioxide. *Nature Geoscience*, 2(12), pp. 831–6.

Springmann, M., 2014. Impacts of integrating emission transfers into policy-making. *Nature Climate Change*, 4, pp. 177–81.

Tukker, A. et al., 2013. EXIOPOL—Development and illustrative analyses of a detailed MR EE SUT/IOT. *Economic Systems Research*, 25(1), pp. 50–70.

Wiedmann, T., Suh, S., Feng, K., Lenzen, M., Acquaye, A., Scott, K. & Barrett, J.R., 2011a. Application of hybrid life cycle approaches to emerging energy technologies-the case of wind power in the UK. *Environmental Science & Technology*, 45(13), pp. 5900–7.

Wiedmann, T., Wilting, H.C., Lenzen, M., Lutter, S., & Palm, V., 2011b. Quo Vadis MRIO? Methodological, data and institutional requirements for multi-region input–output analysis. *Ecological Economics*, 70, pp. 1937–45.

Wiedmann, T. Wood, R., Minx, J.C., Lenzen, M., Guan, D., & Harris, R., 2010. A carbon footprint time series of the UK—Results from a Multi-Region Input-Output Model. *Economic Systems Research*, 22(1), pp. 19–42.

Wiedmann, T., Wilting, H., Lutter, S., Palm, V., Giljum, S., Wadeskog, A., & Nijdam., 2009. *Development of a Methodology for the Assessment of Global Environmental Impacts of Traded Goods and Services*, Report from the ERA-NET SKEP Project EIPOT.

6 Energy production and ecosystem services

Robert Holland, Kate Scott, Tina Blaber-Wegg, Nicola Beaumont,
Eleni Papathanasopoulou, and Pete Smith

6.1 Introduction

The complex network of interactions between organisms and the environment, and among organisms themselves, forms the world's ecosystems. These interactions produce and provide energy, cycle nutrients, store and release carbon, and through such processes provide ecosystem services which ultimately contribute to the goods valued by people (Millennium Ecosystem Assessment, 2005; Mace et al., 2012). Ecosystem services are commonly grouped into four categories; provisioning services such as food, fuel, and freshwater; regulating services such as climate regulation and disease and pest control; cultural services, for example relating to spiritual, aesthetic, and recreational value; and supporting services, such as nutrient cycling, soil formation, and primary production that contribute to the function of all other services (Millennium Ecosystem Assessment, 2005). Ecosystem services flow from the world's natural 'capital', the living and non-living components of nature. We can think of ecosystem services as the 'dividends' that the natural 'capital' provides (Costanza et al., 1997; Sukhdev, 2010).

In 2012 the WWF Living Planet Report estimated that, based on current rates, it would take 1.5 years to renew the resources that humans consume in a single year. Returning to the idea that the environment provides us with both natural 'capital' and 'dividends' in the form of ecosystem services, this suggests that rather than the world's 7 billion people living off the dividends that nature provides us, we are eroding our natural capital base at a rate that is unsustainable (Juniper, 2013).

Meeting the world's energy demands over the coming century represents a fundamental challenge to society. This challenge is not only in providing enough supply to meet growing demand, but doing so while reducing the environmental impacts associated with its production and use (Naik et al., 2010), thus protecting our natural capital base and ensuring that future generations continue to benefit from the ecosystem services that flow from it. In this chapter we examine the routes through which energy production can impact the delivery of ecosystem services and consider the trade-offs between ecosystem services, climate change mitigation, and energy production.

As ecosystem service assessment draws on techniques from environmental, economic, and social sciences it provides a framework that considers the full range of impacts of energy production (Gasparatos et al., 2011) enabling us to consider the full implications of different energy strategies.

We begin by briefly examining the history of ecosystem services and how incorporating this idea into policy may influence decisions about future energy production. We then consider how global environmental footprints can reveal the complex international impact of energy production on ecosystem services driven by the flow of trade around the world. Our attention then turns to the implications of this globalized impact on society. We will see that, together, approaches such as those highlighted here can reveal the implications of different energy strategies across a range of final ecosystem services including provisioning, regulating, and cultural services that exist both within and outside traditional markets.

6.2 The value of ecosystem services and the world's natural capital

The term 'ecosystem services' originated in the late 1970s when conservation groups used the concept as a way of drawing attention to the impact that environmental degradation could have on human well-being (Gomez-Baggethun and Ruiz-Perez, 2011). However, it was in the 1990s, with the publication of a number of influential works (Costanza et al., 1997; Daily, 1997), that the idea began to receive widespread attention. In 2005, the Millennium Ecosystem Assessment (MA) (Millennium Ecosystem Assessment, 2005) provided a global synthesis of the state of the world's ecosystem services. Subsequently a number of national assessments have been undertaken, such as the UK National Ecosystem Assessment (UK NEA, 2011) that built on the MA methodology to provide the first overview of the state of ecosystem services in the UK. Today the idea that ecosystem services are essential for human well-being is increasingly embedded in policy at local, national, and global scales (Daily and Matson, 2008; Gomez-Baggethun and Ruiz-Perez, 2011).

Globally, the MA identified that around two thirds of the earth's ecosystem services are in decline or threatened due to overexploitation (Millennium Ecosystem Assessment, 2005). The MA highlights global impacts on freshwater supply, fisheries, regulation of hazards such as flooding and erosion, and regulation of climate as being particularly severe. The UK NEA reports similar trends in the delivery of services over the last few decades (UK NEA, 2011). For a few services there is an opposite trend, with crops, livestock, and timber all seeing a dramatic increase in production driven by intensive farming and the adoption of modern production techniques (Millennium Ecosystem Assessment, 2005; UK NEA, 2011). This increased production has come at a cost, with agriculture now considered one of the principal drivers of environmental degradation and the loss of biodiversity globally (Vié et al., 2009), which in turn influences the delivery of many ecosystem services (Balvanera et al., 2006; Flynn et al., 2009). This illustrates the often significant trade-offs between the delivery of different ecosystem services (Raudsepp-Hearne et al., 2010; Power, 2010), and the fact that understanding these trade-offs and accounting for them in the decision-making process is a key requirement for sustainability.

From the late 1990s the economic notion of capital was extended to goods and services relating to the natural environment and attempts were made to place an economic value on the world's ecosystem services (Costanza et al., 1997; Hamilton, 2006; Sukhdev, 2010). While such monetary valuation presents significant methodological challenges (Bateman et al., 2011; Gomez-Baggethun and Ruiz-Perez, 2011), and is associated with a high degree of uncertainty (Sukhdev, 2010), ecosystem service valuations acknowledge the substantial contribution that natural processes make to both the economy and social well-being.

Economic techniques, however, capture only one aspect of an ecosystem service's value. Other non-monetary dimensions include health values, which are the benefits gained by individuals through their interaction with the natural environment. Research has shown that increased time spent in the natural environment can increase the physical, mental, emotional, and spiritual well-being of people (Maas et al., 2006; Pretty et al., 2006; DEFRA, 2011). Additionally, shared (social) values have been identified as an area of ecosystem service valuation (UK NEA, 2011). These capture the collective and cultural attitudes of society towards the natural environment, drawing on shared ethical and aesthetic considerations. From an energy perspective, decisions about the roadmap for the future must be based on this broad understanding of ecosystem services and their values (Sukhdev, 2010; Gasparatos et al., 2011; van der Horst and Vermeylen, 2011). This approach ensures a more holistic overview of the costs and benefits associated with energy production, allowing societies to make more informed choices about their futures, including how their energy is sourced, generated and delivered.

6.3 Thinking local and global

We live in an increasingly globalized economy, with commodities transferred around the world through dispersed processing chains before being bought by end consumers (Lenzen et al., 2012). This is true for energy, where areas of extraction, production, and consumption will often be distant spatially. As the implications of energy production and use will be contingent on context, understanding such disaggregated impacts is essential to properly understand ecosystem service impacts. In doing so, the wider social and environmental costs can be considered in a policy designed to address climate change, and informed choices made about whether to forego certain options that may deliver benefits for emission reductions due to high associated ecosystem service costs.

For example, only approximately 22 per cent of biofuels consumed in the UK for transport are derived from domestic feedstocks (Department for Transport, 2013), with the remainder sourced from overseas. Understanding the implications of overseas biofuel production is therefore key to the design of effective policy to minimize negative consequences for environment and society (van der Horst and Vermeylen, 2011).

Conventional production-based accounting techniques measure the ecological impacts directly used or displaced by the production of goods and services within the borders of a sovereign state, ultimately only capturing the domestic impacts of a supply chain. By contrast, consumption-based accounting (or footprinting[1] as it is alternatively known) assesses the total cross-border ecological impact along the whole of a product's supply chain and in so doing captures the domestic as well as global impacts. Different consumption-based accounting approaches and methods are described in Chapter 5, and although the focus there is on greenhouse gas emissions, Ewing et al. (2012) show how approaches can be extended to other ecological indicators. Such analysis can be based on bottom-up or top-down methods, both with their strengths and weaknesses.

In Section 6.4 we discuss examples based on multi-regional input–output analysis (MRIOA), a top-down consumption approach (again, already introduced in Chapter 5), which can calculate the embodied ecological impacts of international trade (Davis and Caldeira, 2010; Lenzen et al., 2012; Peters et al., 2011; Wiedmann, 2009). The idea that impacts may occur along the full supply chain and so extend far beyond national borders can also be applied to examine the social implications of different energy systems. In Section 6.5 we consider how social lifecycle assessment, a bottom-up approach, can be used to examine the societal implications of energy technologies. Together, such techniques provide a powerful tool with which to understand the implications of different energy technologies and so inform the design of policy.

6.4 The export of ecosystem service impacts

MRIOA maps the global production and trade structures of national economies, connecting immediate emitters and resource extractors to the demand for materials and products they generate, based on national economic accounts and international trade data. By extending the economic model by allocating ecological indicators to economic production sectors, ecological multipliers can be calculated showing the total ecological supply chain impact from the production of goods and services for final consumption.

The scope for ecosystem-wide analysis has increased over the last decade, with recent studies going beyond carbon, water, and ecological footprints. For example Yu et al. (2013) integrate a multi-regional trade model with national land-use data for 129

---

[1] Within the literature carbon footprinting is well established, with ecological, land, material, and water footprinting now forming a 'footprinting family' that is beginning to gain traction with decision-makers (Ewing et al., 2012; Galli et al., 2012).

regions, and consider how demand for fifty-seven classes of products in one region drives land-use change in other regions of the world. Similarly, Lenzen et al. (2012) link 25,000 Animalia species threat records from the International Union for Conservation of Nature Red List to more than 15,000 commodities[2] produced in 187 countries. In these examples the authors identify the embodied impact of a product representing both the direct and indirect (i.e. upstream) land and species used or threatened by production respectively. In turn this embodied impact determines how much, and for what purpose, one country has been using land and threatening species in other countries' territories to support the consumption and lifestyles of its own population.

The picture that emerges from these and other (Wyckoff and Roop, 1994; Machado et al., 2001; Li and Hewitt, 2008) studies is of an imbalance between developed and developing countries in their role as drivers of global impacts on ecosystem services. Developed countries consume more goods and services than developing countries, from both domestic and international markets, not only putting pressure on their own domestic resources but also draining other countries' resources. For example, a third of land required for consumption purposes in the USA is displaced abroad, rising to more than 50 per cent in the EU and over 90 per cent in Japan. Conversely approximately half of Brazilian and nearly 90 per cent of Argentinean cropland is used for consumption outside their territories. This land displacement not only occurs due to demand for crops, but more than 50 per cent of land-displacement from developed countries stems from demand for non-agricultural products, such as services, clothing, and household appliances. The tendency for developed countries to be 'net importers' of impacts is likely related to stricter national environmental policies that effectively protect remaining ecosystems and so force impacting industries to locate elsewhere.

Within the literature there are few examples that consider the impact of specific energy technologies on ecosystem services beyond the impacts on carbon and agricultural production. Further work needs to consider the ecosystem service impacts of different energy technologies, and by doing so examine the implications of different energy pathways. Chapter 25 attempts to make headway in this respect, describing the global ecosystem service impacts of biomass, wind, gas, and nuclear energy systems. The consequences for both onshore as well offshore ecosystems are investigated under different energy scenarios and a discussion of the alternative energy futures is presented.

Demonstrating the global implications of economic activity, studies such as those highlighted here make a compelling case for the need to assess impacts of energy production in a way that considers the global context. Failure to do so may lead to a significant underestimate of the true implications of our choices of energy technologies. Assessments that have begun to emerge in recent years suggest that developing countries may be most significantly affected, yet it is within these very countries that the greatest reliance on ecosystem services

---

[2] On average eighty commodities per country, however the MRIO model used (EORA) has different aggregation levels of commodities for individual countries depending on national data availability.

may exist (Millennium Ecosystem Assessment, 2005). This raises important questions relating to social impacts and equity in the use of different natural resources.

6.5 Local to global social impacts of products

Once the degree of ecological impact and its global location have been identified and assessed, there is a natural progression to consider what this means for the well-being of society (UK NEA, 2011). One approach used to capture these social impacts is social lifecycle assessment (S-LCA), an emerging tool that seeks to include the social and economic impacts of a product's lifecycle into sustainability assessments via high levels of stakeholder engagement (United Nations Environment Programme, 2011, 2009). It is a bottom-up approach that can be both local as well as global in its assessment depending on where the system boundaries are drawn. It is in its infancy and lacks application and testing in the energy domain; however, the results of a recently completed energy case study are presented here to highlight its usefulness in identifying particular social issues such as equity.

Lifecycle assessment (LCA) and lifecycle thinking, which seek to identify impacts of different kinds across all stages of production from extraction of raw materials to final consumption, are established, widely-used methods and concepts for evaluating a product's or technology's sustainability. Consideration of a few provisioning and regulating services dominate much of this work, as these aspects are most amenable to assessment, mapping, and economic valuation. In contrast, impacts on cultural services and aspects of human well-being that are not easy to quantify rarely feature in such assessments. Perceptions of the value of services and their benefits are likely to differ among individuals and communities, particularly for cultural services that include aesthetic, spiritual and educational benefits, knowledge and belief systems, and patterns of social, economic, and political organization that fall outside traditional markets (Millennium Ecosystem Assessment, 2005). The growing realization of the importance of these benefits opens up interesting questions about the impacts of energy production on society, both in terms of the ecosystem service impacts and more broadly relating to issues of environment and social justice.

Socio-economic impacts, including matters of social justice, equity, and social inequality, are generally excluded from sustainability appraisals of energy technologies—despite being core concepts of sustainable development ideals (Fast, 2009; Walker and Bulkeley, 2006). The sustainable development agenda seeks to promote human development that improves qualities of life, reduces poverty and social inequalities, while avoiding the degradation or depletion of natural resources faster than they can be naturally replenished (UN WCED, 1987). These aims can be seen to span environmental, economic, and social dimensions. Such thinking is clearly in line with policies designed to protect natural capital and delivery of ecosystem services as a method for supporting human well-being. The ways social and economic impacts are experienced across different communities remain largely unknown, despite the fact that the implications for human well-being can be quite profound. For example, social inequalities have been shown to be correlated with a range of social issues and costs including health, crime rates, and population rises, which threaten the socio-economic stability, or sustainability, of a region (Wilkinson and Pickett, 2010).

Biofuels have been subject to considerable controversy and debate with matters of social or environmental justice featuring as an underlying driver of these concerns. Opposition has been mainly driven by fears that some people may bear more negative impacts, or enjoy more benefits, associated with their production and consumption than others (Nuffield Council on Bioethics, 2011). Using a case study of a bioethanol product's lifecycle, equity issues are explored by firstly identifying social costs and benefits experienced by actors involved or affected and then considering the distribution of these impacts across the chain.

6.5.1 CASE-STUDY: IMPLICATIONS OF BIOETHANOL PRODUCTION AND USE: FROM BRAZIL TO THE UK

Usina Sao Joao (USJ) is a family-owned bioethanol producer based in Sao Paulo, Brazil. It has been in operation for over seventy years producing ethanol and sugar from sugarcane feedstocks and employs around 2,000 staff (Grupo USJ, 2004). Sugarcane is mainly grown on the USJ's own plantations, therefore the production and processing stages of the supply chain (in lifecycle terms) have been integrated into a single 'production' stage of the S-LCA process. USJ processes meet higher levels of biofuel sustainability standards than stipulated by EU law (Grupo USJ, 2004). Greenergy (an energy supplier) purchases and blends the bioethanol with unleaded petrol (as part of mandatory blends) ready for fuel sales at the pumps of Sainsbury's North Walsham outlet in the UK. The site of consumption was chosen purely for the purpose of 'grounding' the study in a specific location (in reality, the same fuel is distributed across the supermarket's other retail outlets), ease of data collection, and accessibility.

Figure 6.1 depicts the supply chain identified and mapped for this case study as part of the first phase of the research. It includes perceptions of the likely socio-economic impacts and equity issues, including their distribution, which may be found during phase 2 research, identified from semi-structured interviews with people working in the biofuels sector. Figures 6.2a and 6.2b summarize findings from the second phase of research and show positive and negative socio-economic issues that emerged as a result of visits to sites of production and consumption and semi-structured interviews with people across different stakeholder groups. It is not possible to explore the findings in detail here but by comparing the two figures, it can be seen that common assumptions made about the socio-economic impacts of biofuels are challenged by this case study,

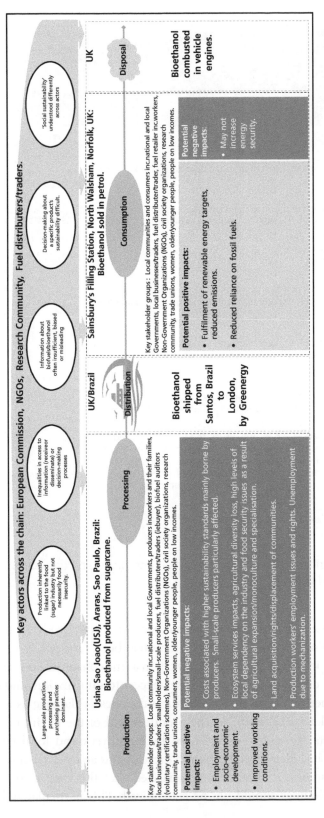

Figure 6.1 Key findings from case study mapping during phase 1 research; overview of the sugarcane bioethanol supply chain identified as the basis of the case study showing stakeholder groups connected at a high level and at local sites of production and consumption, and likely impacts and equity issues that may be identified during phase 2

Positive :

- Improved employment, professional, and economic development opportunities.

- Consistent markets for producers and small-scale farmers (if out-growers) –product diversification (sugar and ethanol).

- Improved working conditions (better than across other parts of agricultural sector).

- Less environmental impacts and improved health effects as a result of higher sustainability standards/certification (incl. reduced field burning).

- Increased social stability, integration and educational attainment as a result of mechanisation (reduced migration of seasonal workers to cut the sugarcane).

- High level of corporate social responsibility (CSR) and investment in local community by the mill (education, health, environment, roads).

- No directly associated food security issues; food crops displaced to surrounding states where environmental/climatic conditions more conducive to cultivation so better productivity between states and high % of sugarcane feedstock used within the food industry for sugar production (if ethanol not produced, sugar production from this crop/lands would continue).

Negative :

- Professional development opportunities less accessible for the least educated.

- Costs of higher sustainability standards mainly borne by producers. Small-scale producers hardest hit/less able to compete; reduced access to information/support/investment.

- Local reliance on industry; specialization.

- Socio-economic benefits may be restricted to sites with large-scale production and high levels of CSR.

- Some environmental impacts (dust from trucks, field burning).

- Exports may jeopardize energy security.

Figure 6.2a Key findings from site of production and processing (impacts affecting local communities and stakeholders in Brazil)

which found more positive socio-economic benefits in the producer region, and more negative implications for UK consumers, than had been foreseen.

Key issues and inequalities in the supply chain at the consumer end relate to consumer ethics, 'exclusive' decision-making processes, access to information, and potential future costs for consumers relating to infrastructural changes or damage as a result of higher mandated fuel blends (i.e. E10). At the producer end of the chain, positive social and economic impacts are widely experienced by employees and local communities, but negative impacts included specialization in the area and an over-reliance on the industry for local communities. It should be emphasized that this case study supply chain is unique and the findings should not be generalized. This particular mill's high level of

Positive:	Negative:
• Compliance with the EU Renewable energy Directive (RED)/ability to meet renewable energy targets. • High level of acceptability for some biofuels under certain condition or for particular uses.	• Unlikely to increase UK energy security (reliance on overseas production, high demand for bioethanol globally, vulnerability to agricultural commodity markets, price volatility and weather conditions/harvests/climatic changes). • Lack of consumer information/choice. Consumer ethics compromised. Inability to 'opt out' of bioethanol consumption without major costs/lifestyle changes (particularly problematic for rural residents). • Inability to influence government spending on alternative sustainable transport initiatives, fuels, consumption-reduction initiatives or different modes or production. • Anomalies in the food and agricultural system; sustainable agriculture in general needed; high levels of wastes and inequalities in food distribution bigger issues than some biofuels production; sugar production from land and feedstocks deemed more acceptable than bioethanol despite major health-related costs and issues associated with high levels of sugar consumption in diets. • Impacts on car engines and infrastructures with higher blends/E10. Likely to hit those on lower incomes, people in rural areas, and smaller fuel retailers hardest. • Lack of government investment in UK domestic production; waste/by-products/district-level schemes.

Figure 6.2b Key findings from site of consumption (impacts mainly felt by consumers in the UK)

corporate social responsibility (CSR), evident through their investment in education, local services, and infrastructures, combined with strong Brazilian employment laws and sustainability standards, are the main reasons that so many positive socio-economic benefits to the local communities are apparent.

This research could be further extended to also consider the impact on society due to the ecosystem changes caused by the production of sugarcane. Environmental assessment could describe changes in the biophysical environment and examine the link between such changes and the requirement of society for the provision of services such as disease and pest regulation, or flood prevention. Combined with the increased understanding of impacts of sugarcane production on communities' identity and culture that the S-LCA technique provides, such an assessment would provide a more complete understanding across the environmental and cultural domains.

6.6 **Conclusion**

We began this chapter by considering natural capital and ecosystem services and their importance for human well-being. We argue that although impacts on climate regulation remain a key driver shaping global energy policy, it is essential to consider the full range of ecosystem service consequences in order to make informed choices about future strategies. Using examples of environmental and social assessment, we have discussed how impacts of energy policy can often be far removed from the point of energy use, and that it is essential to incorporate this global thinking into the decision-making process. As many of the chapters in this book highlight, energy technologies are incredibly diverse. This diversity of technologies interacts with regional variations in climatic, social, political, and economic conditions to affect the sustainability issues experienced. The techniques that we highlight are amongst a suite of emerging tools that enable us to examine costs and benefits associated with different energy pathways and so produce more informed policy guidance that can consider these issues and, in turn, design optimal strategies to meet future energy requirements while reducing the social and environmental costs.

Due to the diverse and global nature of ecosystem services, no one method can inform us of all the impacts. A monetary evaluation excludes benefits derived that are outside the economic realm, such as well-being and social values. Trade models can help us understand wide-reaching global implications of UK energy consumption, but lack the detail a bottom-up case study approach can provide. While time consuming, case-study information can overcome the generalization of high-level trends extracted from top-down methods. There needs to be a balance restricted by resource, data, and time constraints between gaining a comprehensive account of energy impacts and having an in-depth understanding of less measurable impacts on society. Establishing where this balance lies remains a significant challenge for the global community.

▪ REFERENCES

Balvanera, P., Pfisterer, A.B., Buchmann, N., He, J.-S., Nakashizuka, T., Raffaelli, D., Schmid, B., 2006. Quantifying the evidence for biodiversity effects on ecosystem functioning and services. *Ecology Letters* 9, 1146–56.

Bateman, I.J., Mace, G.M., Fezzi, C., Atkinson, G., Turner, K., 2011. Economic analysis for ecosystem service assessments. *Environmental and Resource Economics* 48, 177–218.

Costanza, R., d' Arge, R., Groot, R. de, Farber, S., Grasso, M., Hannon, B., Limburg, K., Naeem, S., O'Neil, R.V., Paruelo, J., Raskin, R.G., Sutton, P., Belt, M. van den, 1997. The value of the world's ecosystem services and natural capital. *Nature* 387, 253–60.

Daily, G.C., 1997. *Nature's Services: Societal Dependence on Natural Ecosystems.* Island Press. Washington, DC.

Daily, G.C., Matson, P.A., 2008. Ecosystem services: from theory to implementation. *Proceedings of the National Academy of Sciences* 105, 9455–6.

Davis, S.J., Caldeira, K., 2010. Consumption-based accounting of CO2 emissions. *Proceedings of the National Academy of Sciences* 107, 5687–92.

DEFRA, 2011. The natural choice: securing the value of nature. Natural Environment White Paper. The Stationery Office Limited.

Department for Transport, 2013 Renewable Transport Fuel Obligation statistics: obligation period 5, 2012/13, report 4. Department for Transport, UK.

Ewing, B.R., Hawkins, T.R., Wiedmann, T.O., Galli, A., Ertug Ercin, A., Weinzettel, J., Steen-Olsen, K., 2012. Integrating ecological and water footprint accounting in a multi-regional input–output framework. *Ecological Indicators* 23, 1–8.

Fast, S., 2009. The biofuels debate: Searching for the role of environmental justice in environmental discourse. *Environments: A Journal of Interdisciplinary Studies* 37 (1), 83–100.

Flynn, D.F., Gogol-Prokurat, M., Nogeire, T., Molinari, N., Richers, B.T., Lin, B.B., Simpson, N., Mayfield, M.M., DeClerck, F., 2009. Loss of functional diversity under land use intensification across multiple taxa. *Ecology Letters* 12, 22–33.

Galli, A., Wiedmann, T., Ercin, E., Knoblauch, D., Ewing, B., Giljum, S., 2012. Integrating ecological, carbon and water footprint into a 'Footprint Family' of indicators: Definition and role in tracking human pressure on the planet. *Ecological Indicators* 16, 100–12.

Gasparatos, A., Stromberg, P., Takeuchi, K., 2011. Biofuels, ecosystem services and human wellbeing: Putting biofuels in the ecosystem services narrative. *Agriculture, Ecosystems & Environment* 142, 111–28.

Gomez-Baggethun, E., Ruiz-Perez, M., 2011. Economic valuation and the commodification of ecosystem services. *Progress in Physical Geography* 35, 613–28.

Grupo USJ, 2004. Usina Sao Joao: 60 Doces Anos. Grupo USJ, Brazil.

Hamilton, K., 2006. Where is the Wealth of Nations?: Measuring Capital for the 21st Century. World Bank Publications. Washington, DC. USA.

Juniper, T., 2013. *What Has Nature Ever Done for Us?: How Money Really Does Grow on Trees.* Profile Books. London.

Lenzen, M., Moran, D., Kanemoto, K., Foran, B., Lobefaro, L., Geschke, A., 2012. International trade drives biodiversity threats in developing nations. *Nature* 486, 109–12.

Li, Y., Hewitt, C.N., 2008. The effect of trade between China and the UK on national and global carbon dioxide emissions. *Energy Policy* 36, 1907–14.

Maas, J., Verheij, R.A., Groenewegen, P.P., Vries, S. de, Spreeuwenberg, P., 2006. Green space, urbanity, and health: how strong is the relation? *Journal of Epidemiology and Community Health* 60, 587–92.

Mace, G.M., Norris, K., Fitter, A.H., 2012. Biodiversity and ecosystem services: a multilayered relationship. *Trends in Ecology & Evolution* 27, 19–26.

Machado, G., Schaeffer, R., Worrell, E., 2001. Energy and carbon embodied in the international trade of Brazil: an input–output approach. *Ecological Economics* 39, 409–24.

Millennium Ecosystem Assessment, 2005. *Ecosystems and Human Well-being: Synthesis.* Island, Washington, DC, USA.

Naik, S.N., Goud, V.V., Rout, P.K., Dalai, A.K., 2010. Production of first and second generation biofuels: A comprehensive review. *Renewable and Sustainable Energy Reviews* 14, 578–97.

Nuffield Council on Bioethics, 2011. Biofuels: ethical issues. Nuffield Council on Bioethics. London.

Peters, G.P., Minx, J.C., Weber, C.L., Edenhofer, O., 2011. Growth in emission transfers via international trade from 1990 to 2008. *Proceedings of the National Academy of Sciences* 108, 8903–8.

Power, A.G., 2010. Ecosystem services and agriculture: tradeoffs and synergies. *Philosophical Transactions of the Royal Society B: Biological Sciences* 365, 2959–71.

Pretty, J., Hine, R., Peacock, J., 2006. Green exercise: The benefits of activities in green places- Little has been said about the potential emotional or health benefits of the natural environment in arguments about conservation. *Biologist-London* 53, 143–8.

Raudsepp-Hearne, C., Peterson, G.D., Bennett, E.M., 2010. Ecosystem service bundles for analyzing tradeoffs in diverse landscapes. *Proceedings of the National Academy of Sciences* 107, 5242–7.

Sukhdev, P., 2010. The economics of ecosystems and biodiversity mainstreaming the economics of nature: a synthesis of the approach, conclusions and recommendations of TEEB.

UK NEA, 2011. The UK National Ecosystem Assessment. Synthesis of the Key Findings. UNEP-WCMC. Cambridge, UK.

UN WCED (World Commission on Environment and Development), 1987. Our common future: The world commission on environment and development. Oxford University Press. Oxford.

United Nations Environment Programme, 2009. Guidelines for social life cycle assessment of products. UNEP.

United Nations Environment Programme, 2011. Towards a life cycle sustainability assessment: Making informed choices on products. UNEP.

Van der Horst, D., Vermeylen, S., 2011. Spatial scale and social impacts of biofuel production. *Biomass and Bioenergy* 35, 2435–43.

Vié, J.-C., Hilton-Taylor, C., Stuart, S.N., 2009. *Wildlife in a Changing World: An Analysis of the 2008 IUCN Red List of Threatened Species^{TM}*. IUCN. Gland, Switzerland.

Walker, G.P., Bulkeley, H., 2006. Geographies of environmental justice. *Geoforum* 37, 655–9.

Wiedmann, T., 2009. A review of recent multi-region input–output models used for consumption-based emission and resource accounting. *Ecological Economics* 69, 211–22.

Wilkinson, R., Pickett, K., 2010. *The Spirit Level: Why Equality is Better for Everyone.* Penguin. London.

Wyckoff, A.W., Roop, J.M., 1994. The embodiment of carbon in imports of manufactured products: Implications for international agreements on greenhouse gas emissions. *Energy Policy* 22, 187–94.

Yu, Y., Feng, K., Hubacek, K., 2013. Tele-connecting local consumption to global land use. *Global Environmental Change* 23, 1178–86.

7 Technical, economic, social, and cultural perspectives on energy demand

Charlie Wilson, Kathryn B. Janda, Françoise Bartiaux, and Mithra Moezzi

7.1 Introduction

This chapter provides a critical overview of the technical, economic, social, and cultural aspects of energy demand. Comparative data from different countries show upward trends in both the efficiency and consumption of energy. These trends are dynamic and heterogeneous, both across and within nations, suggesting that the role energy plays in society is complex and multifaceted. The literature offers several explanations for differences in energy use, as well as the drivers of demand. This chapter briefly explores explanations for energy demand from four different viewpoints, including a physical, technical, and economic model (PTEM), an energy services approach, social practice theories, and socio-technical transitions theory. Each approach provides a window onto a complex landscape, but none provides a complete explanation for historical trends, nor a comprehensive basis for predicting the future. Together, however, the evidence shows that understanding energy demand is a socio-technical problem rather than one that is either social or technical. At the end of the chapter, the authors use the concept of 'energy needs' to discuss the difficulties that these varying (and sometimes competing) explanations pose for policymakers attempting to curb the rising tide of consumption. This chapter focuses largely on problem definitions. For possible solutions, readers should look to Chapters 9, 10, and 11, which address the international implications of selected strategies for improving energy efficiency and reducing consumption in buildings and transportation.

7.2 Global trends

The Energy Primer chapter of the recent Global Energy Assessment (GEA) provides a recent overview of global energy supply and demand (Grubler et al., 2012). To set the stage for our chapter, we refer to selected elements of the GEA work and recommend this source to readers interested in further details. This section begins with a recent snapshot of global energy demand. Next, it introduces several different levels of heterogeneity within energy demand: between end uses, over time, between countries, and within countries. Then it discusses some potential drivers of demand, noting some ways in which our quantitative evidence base struggles to explain differences in energy cultures and associated differences in energy use.

7.2.1. GLOBAL ENERGY DEMAND: A SNAPSHOT

Globally, the demand for energy can be summarized in the energy flow diagram shown in Figure 7.1. Different primary energy sources (coal, crude oil, natural gas, etc.) are converted into fuels and ultimately used in one of several end-use sectors, which include feedstocks, transportation, industry, and residential and commercial buildings. In terms of useful energy, industry is the largest end-use sector at roughly 62 EJ; residential and commercial buildings are close behind at 60 EJ; transportation consumes roughly

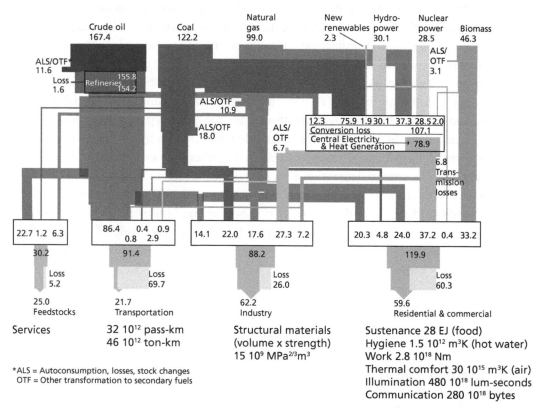

Figure 7.1 Global energy flows (in EJ) from primary to useful energy and end-use sector applications in 2005

Source: GEA Figure 1.5 (Grubler et al., 2012) based on data from IEA (2007a, 2007b), Cullen and Allwood (2010a). Reproduced with permission from the International Institute for Applied Systems Analysis (IIASA).

Table 7.1 Energy conversion globally from resources into useful energy

Energy Resources	GJ	Energy Carriers (Fuels)	GJ	End-Use Devices	GJ	Useful Energy	GJ
		Direct Fuel Use					
		Oil	129				
		Biomass	47				
		Gas	48	Engines	109	Motion	30
Oil	152	Coal	28	Elec. Motors	16	Heat	23
Biomass	54	*Elec. Generation*		Burners	143	Cooling/Light/Sound	2
Gas	97	Oil	13	Heaters	17		
Coal	127	Biomass	4	Lights	6		
Nuclear	30	Gas	40	Electronics	5		
Renewables	15	Coal	87	Other	13		
		Nuclear	30				
		Renewables	15				
Subtotal	475	Subtotal	441	Subtotal	309	Subtotal	55
		Losses (direct)	34	*Losses (direct)*	34	*Losses (direct)*	34
				Losses (generation)[1]	132	*Combustion*[2]	128
						Heat transfer[3]	172
						Other[4]	86
Total	475	Total	475	Total	475	Total	475

Source: Cullen & Allwood, 2010b.

[1] Electricity generation losses (including transmission and distribution).

[2] 'Combustion' includes internal heat exchange, oxidation, mixing (including some generation losses).

[3] 'Heat transfer' includes heat exchange, exhaust, heat loss (including some generation losses).

[4] 'Other' includes electrical resistance, friction, fission, fuel loss (including some generation losses).

22 EJ; and feedstocks are about 25 EJ. Huge losses are evident at each stage of conversion. These losses are shown more clearly in Table 7.1. Using data from Figure 3 in Cullen and Allwood (2010b), Table 7.1 shows, from left to right, the conversion of global energy resources into energy carriers (fuels, electricity) that are then converted by end-use devices into useful energy for the provision of services. Energy at each conversion stage that is lost or wasted and so not available to provide useful services is summarized at the bottom of each column. The reader is recommended to the original Cullen and Allwood (2010b) article for a visual representation of these energy flows, including full details of the data and methodology. Table 7.1 shows that of the 475 EJ of primary energy resources used annually, only about 55 EJ (11.6 per cent) ultimately provide useful energy services (motion, heat, cooking, and light). The remaining 420 EJ (88.4 per cent) are lost in conversion. This very clearly reveals the potential for improving the overall conversion efficiency of our energy system at a number of different stages (Cullen and Allwood, 2010b).

These global snapshots give some sense of the scale of energy use on the planet. However, there is great variation in energy systems across different regions as a result of differences in the nature of economic development, structure of energy demand, and distribution of resource availability, among others. The next sections describe some of these causes of variation.

7.2.2 HETEROGENEITY OVER TIME: UK CASE STUDY

As one of the first countries to industrialize, the UK offers the longest observed record of energy transitions in the modern era (see Figure 7.2). Although distinctive in terms of specific resources, technologies, and economic activities, the broad characteristics of observed energy transitions are informative for other countries. An analysis of energy transitions since before the industrial revolution in the UK explains the dynamics of long-run change as a positive economic and welfare feedback loop (Fouquet, 2010). Demand for energy services grows as: (1) population, GDP and income per capita rise, increasing the aggregate purchasing power for energy services; (2) the efficiency of energy service provision improves, often dramatically so when seen over long time periods; and (3) the effective prices of energy services fall (per GJ service demand or activity level), again often dramatically so, and in a sustained way since the industrial revolution except during the World Wars (Fouquet, 2010). Falling energy service prices stimulate further demand growth and technical innovation in the efficiency of service provision, and so on (see Figure 7.3). This long-run view of energy transitions is consistent with work on the economy-wide rebound effects from improving energy productivity (Sorrell, 2009; Saunders, 2013).

More recent work by Fouquet describes the price and income elasticities of three major energy services (lighting, domestic heating, and passenger transport) over the past 200 years in the UK (Fouquet, 2014). In this context, price elasticities describe the

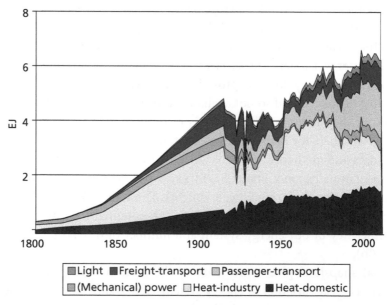

Figure 7.2 UK energy service demand since 1800, measured by final energy inputs

Source: GEA Figure 1.7 (Grubler et al., 2012) based on data from Fouquet (2008). Reproduced with permission from the International Institute for Applied Systems Analysis (IIASA) and Roger Fouquet.

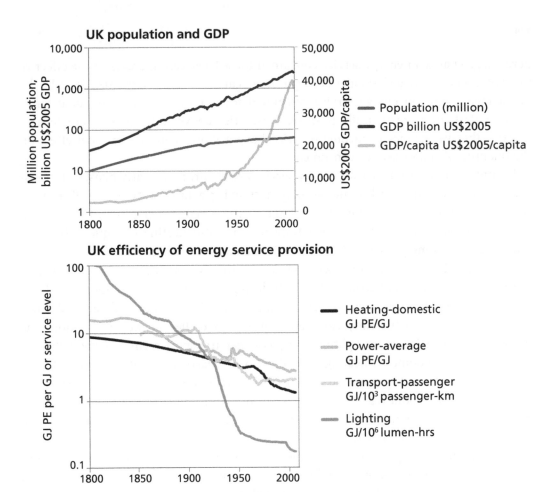

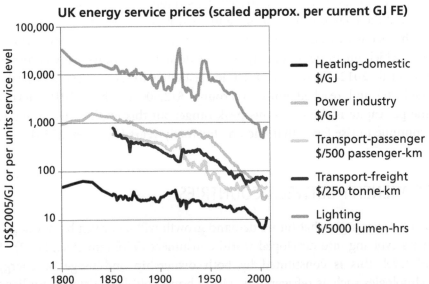

Figure 7.3 Drivers of UK energy service demand since 1800

Source: GEA Figure 1.8 (Grubler et al., 2012) based on data from Fouquet (2008). Reproduced with permission from the International Institute for Applied Systems Analysis (IIASA) and Roger Fouquet.

Note: 'PE' is 'primary energy', which describes energy resources in their natural form (e.g. mined coal, collected fuelwood, drilled oil or gas). 'FE' is 'final energy', which describes energy forms and fuels used by consumers (e.g. electricity at the socket, fuel in the tank, heat in the blast furnace).

percentage change in energy service consumed for a 1 per cent increase in the effective price of a service. Price elasticities are typically negative. Income elasticities describe the percentage change in energy service consumed for a 1 per cent increase in income. Income elasticities are typically positive. Elasticities above unity (+/−1) are termed 'elastic', that is, changes in income or price have a more than proportional effect on the quantity of energy service consumed.

Positive income elasticities for energy services rose during the early phases of industrialization. They then peaked and steadily declined, reaching unity (+1) well into the twentieth century. Even at the current high average income per capita, they remain above zero, particularly for passenger transport but also heating (Fouquet, 2014). In other words, income growth still increases the demand for energy services.

Negative price elasticities for energy services similarly rose (became more negative) during the early phases of economic growth during the industrial revolution (implying greater rebound effects as energy efficiency improved), then peaked and declined. They passed unity (−1) in the early twentieth century, or earlier in the case of domestic heating. Price elasticities for lighting and passenger transport seem to have stabilized around −0.5 with domestic heating generally more inelastic (currently around −0.2).

In other words, each percentage decline in the price of energy services has seen less and less of a corresponding percentage rise in demand, indicating declining marginal utilities of consumption: saturation effects but not saturation, and consequently small but significant rebound effects.

From an emissions perspective, it is dispiriting to apply these historical trends from the UK to developing economies at far lower per capita incomes and at different places on the 'energy services ladder', from traditional fuels (biomass) up to solid and liquid fuels (coal, oil), and then gaseous fuels, grids, and networks (gas, electricity) (Sovacool, 2011). With both price-elastic and income-elastic demand for energy services, it is hard to see a future in which quantities of energy services do not rise dramatically across the world (Wolfram et al., 2012). In the UK, income elasticities for energy services peaked around $4,000–7,000 and reached unity at around $12,000 (both in 2010$ terms). China's income per capita is similar to this peak range; but this implies energy service demand will continue to rise faster than income for many years (Fouquet, 2014).

7.2.3 HETEROGENEITY BETWEEN COUNTRIES

Wolfram and colleagues argue that future demand growth will be driven by developing world consumers evolving into developed world consumers (Wolfram et al., 2012). At the household level, this is constituted by both ownership and usage of energy-conversion technologies such as refrigerators. Using household-level data on appliance ownership in major developing economies, they show that as households rise out of poverty, they become first-time purchasers of the novel energy services provided by refrigerators, cars, air conditioning (AC) units, and so on. Energy consumption therefore grows more rapidly in the lower-income groups as they become first-time asset owners. Growth in demand for energy services will be most rapid in these income groups and countries. These account for a large share of the world's population.

Appliance and car ownership is contingent on the extent of the electricity network and road networks respectively. Around 1.4 billion people do not have access to electricity (Johansson et al., 2012). Electrification rates are correlated with GDP per capita, but not with governance, natural resource endowments, nor urbanization (Wolfram et al., 2012). Electrification is negatively correlated with the amount of money required to lift household incomes up to the poverty line: the presence of many households below the poverty line is associated with low rates of electrification. Causality may run both ways: lower income reduces demand for electricity; lack of access to electricity reduces incomes (Wolfram et al., 2012).

Ma and colleagues (2012) estimate energy flows into useful services for China, and compare their results with estimated global energy flows (Cullen and Allwood, 2010a). There are some striking differences, marking out China's position as a rapidly urbanizing economy with a high proportion of economic activity for industrial production. Factories absorb 60 per cent of China's energy flows, compared to only 32 per cent globally. Conversely, the proportions of China's energy flows directed to buildings and vehicles are lower than the global average: 29 per cent compared to 45 per cent in the case of buildings, and 11 per cent compared to 22 per cent in the case of vehicles (Ma et al., 2012: 184). In terms of energy services provided, 45 per cent of China's energy flows into 'structure' defined as 'materials used to provide structural support' in both buildings and capital goods. This compares with only 14 per cent globally. For all other energy services apart from illumination, China's energy flows are proportionally below the global average. This is most marked for passenger mobility (5 per cent compared to a global 14 per cent) and hygiene, which includes a wide range of domestic services for 'living' (3 per cent compared to a global 12 per cent). The implication is that potential latent demand for domestic energy services in China is vast.

7.2.4 HETEROGENEITY WITHIN COUNTRIES

According to Grubler et al. (2012: 135), 'often the variance in energy use within nations can be of the same or greater order of magnitude as that across nations'. These variations are observable between socio-economic groups, rural and urban populations, and/or across geographical regions of the same country. The following text reproduced from the GEA (Grubler et al., 2012) describes these variations.[1]

---

[1] Excerpt from GEA (Grubler et al., 2012: 135). Reproduced with permission from the International Institute for Applied Systems Analysis (IIASA).

Within nations, substantial differences in energy use exist across geographical regions, rural versus urban residents, and among other socio-economic and demographic sub-groups of the population. Thus, for instance, as shown in Figure 7.4, the poorest 20 per cent of the rural population in India have per capita energy use levels comparable to those estimated for the pre-agrarian European population some 10,000 years ago. Even the richest 20 per cent of the rural population in India uses only about half as much energy per capita as the richest 20 per cent of the urban population, with their energy use levels comparable to the estimates for China in 100 BC. Some of this difference in the quantity of energy used can be explained by disparities in income levels across rural and urban regions. However, large disparities in the structure of energy use are also evident, both in terms of uses of energy and the types of energy used. The starkest disparities in energy use within (and between) nations are those between rich and poor people. Thus, as Figure 7.4 illustrates, the richest decile of the Dutch population uses almost four times as much energy per capita as the poorest decile, which is about the same order of difference as between the richest and poorest urban Indian quintiles. The richest 20 per cent of urban Indians use only a third as much of the energy used by the poorest 10 per cent of the Dutch, albeit the richest 20 per cent in India will include many examples of very wealthy individuals whose energy use vastly surpasses that of the average Dutch top 10 per cent income class. As such, these illustrative numbers reflect the wide disparities in incomes and development levels across and within nations. The richest Dutch also use almost three times as much energy for food on average as their poorest compatriots. This, of course, does not imply that rich people eat three times as much as poor people in the Netherlands. However, the food habits and types of provisions consumed do differ. For instance, the rich Dutch eat more exotic fruits and vegetables (e.g. Kiwi fruit flown in from New Zealand) than the poor which explains their much larger food-related (embodied) energy use. The biggest differences in the structure of energy use between rich and poor people, both within and across nations, is the substantially larger share of energy used for transport and for the consumption of products and services. Poor people, by contrast, use the largest proportion of energy for basic necessities such as food and household fuels (cooking and hygiene). These differences illustrate the substantial variations in lifestyles and growing consumerism evident with rising incomes and retail market sophistication.

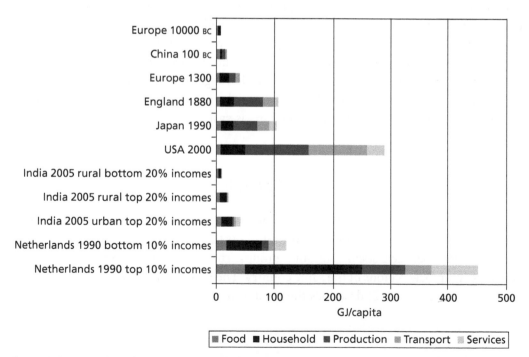

Figure 7.4 Per capita primary energy by service category over time and across different populations

Source: GEA Figure 1.24 (Grubler et al., 2012); see original figure for underlying data. Reproduced with permission from the International Institute for Applied Systems Analysis (IIASA).

7.2.5 DRIVERS OF DEMAND FOR ENERGY SERVICES

At the household level, energy demands are often depicted from an economic perspective as a function of available income, the effective price of energy services, the capital costs of end-use conversion devices providing the services, and the utility derived from consuming the services (Haas et al., 2008). Short-term and long-term drivers of energy service demand differ. Short-term drivers relate to consumer behaviour, and include room temperature settings, kilometres driven in leisure time, and hours watched and stand-by operation of TV sets. Long-term drivers of service demand take into account parameters like area of dwellings, size of cars, number of light bulbs installed in the living room, and so on (Haas et al., 2008). This broadly corresponds to the economic distinction between short-run and long-run change, in which long-run change includes capital stock modifications and turnover.

Ürge-Vorsatz and colleagues (2012) provide a more expansive list of drivers of energy service demands which explicitly references changes in low-income countries: population; urbanization; shifts from traditional fuels to commercial fuels and electricity; income; economic and technological development (which influences the availability and affordability of energy conversion technologies, carriers, and infrastructures); individual behaviour; and cultural practices. They go on to note the wide heterogeneity of energy service demands as a function of geography, culture, lifestyle, climate, as well as the use, ownership, age, and location of buildings (Ürge-Vorsatz et al., 2012: 660). Energy services consumed in similar buildings by similar households with similar socio-economic characteristics still vary widely due to lifestyle, cultural, and behavioural differences (Socolow, 1978; Lutzenhiser, 1992; Lenzen et al., 2006; Lutzenhiser and Lutzenhiser, 2006; Intertek, 2012; Zhang et al., 2012).

Rosa and Dietz (2013) review the drivers of greenhouse gas emissions rather than energy demand. They find that key drivers have been discussed for decades or more. These include population, affluence and consumption, choice of technologies, institutional arrangements, and culture.

Ultimately, most releases of greenhouse gases are driven by consumption of goods and services by individuals, households and organizations, and the manufacturing, transport and waste disposal that underpins that consumption. The rate of emission depends on the scale of consumption, its composition and the techniques used to produce and transport goods and services (Rosa and Dietz, 2013: 582).

As the household is the key domain for this consumption, number of households is a more useful explanatory variable for energy use and emissions than total population. Household size—the number of occupants per household—also helps explain vehicle and appliance usage. Institutional arrangements such as governance, democracy, or environmental movements are conjectured to affect energy use and emissions, but empirical evidence is weak. In terms of culture, Rosa and Dietz (2013: 582) note: 'it is commonplace to expect human values, beliefs, norms, trust and worldviews to be key drivers of environmental change'. However, they stress that the actual influence such factors have over cross-national differences in environmental stress and greenhouse gas emissions 'remains undisciplined by a supporting body of research'.

7.3 Different perspectives on energy demand

The preceding section briefly summarizes the many and varied characteristics, drivers, and trends of energy demand. How do researchers understand energy demand, and using which methods? What kinds of recommendations can they make to policymakers about supporting change? There is no unifying answer to these questions, nor a single theoretical or analytical approach that can be used. Many researchers argue there are generalizable and thus universal quantitative answers to energy demand questions; others find that qualitative research exploring and interpreting contextual factors is the appropriate way forward (Schweber and Leiringer, 2012). Chapter 9 explores the interface between generalizability and specific contexts in greater detail. In this section, we discuss four different perspectives on energy demand, each of which have their adherents among researchers. The four perspectives are: (1) a physical, technical, and economic model (PTEM); (2) an energy services approach; (3) social practice theories; and (4) socio-technical transitions.

7.3.1 PHYSICAL, TECHNICAL, AND ECONOMIC MODEL (PTEM)

As its name indicates, the Physical, Technical, and Economic Model (PTEM) emphasizes technological and economic explanations of energy use and efficiency. For decades, the concept of technical potential of energy efficiency has been a fundamental tool for the energy efficiency industry, although this notion has been criticized (Shove, 1998). Technical potential denotes a best-case energy efficiency scenario. It is based on engineering and economic calculations that are performed 'without concern for the probability of successful implementation' (Rosenfeld et al., 1993: 50), and assumes that the energy efficiency technologies under consideration are appropriate for all building configurations, and are infinitely available at or below the cost considered. At this level, there are no economic, social, or psychological obstacles that would dissuade consumers or organizations from adopting them. Economic potential refers to the subset of technical potential that remains after applying a cost-effectiveness cut-off criterion for saved energy at the current price of delivered energy. To fulfil this techno-economic potential, humans enter only implicitly, as economic agents, as generators of energy service needs, and invisibly as part of the calibration to estimate achievable potentials (Moezzi et al., 2009).

These assumptions about technical and economic potentials provide the backbone of the PTEM, which dominates the energy efficiency field (Lutzenhiser, 1993, 2014). From this perspective, energy demand is a given as people with higher incomes require higher levels of goods and services. Economic and cultural globalization lead to a globalization of energy-consuming activities (e.g. daily showering, car commuting, electronic communications). Energy demand can thus be explained as a function of variables like income (and other socio-demographic characteristics), physical circumstances (e.g. distance from workplace, size of home), and technological availability (e.g. cost and efficiency of end-use appliances).

This perspective on energy demand gives rise to a generalized approach to energy efficiency solutions represented by conservation supply curves. These were first introduced as an analytical tool by US researchers in the early 1980s (Meier et al., 1983; Blumstein and Stoft, 1995), and are now widely used both in modelling and other energy analyses, and for visually summarizing and communicating the potential for energy productivity gains. A more recent variant of this approach is the marginal abatement cost curves (MACCs) for CO_2 or greenhouse gases. MACCs similarly describe a supply curve ordered from low to high marginal cost, but for emission reductions rather than energy savings. McKinsey's MACCs have become particularly widely known and referenced (Enkvist et al., 2007; McKinsey, 2009). Both conservation supply curves and MACCs are discussed further and critiqued in Chapter 9.

Techno-economic potential is one particular construction of how energy use is understood, and of the options available for influence and change. It focuses on 'average' behaviour, leaving little room to account for human variability. This approach can create distortions in an attempt to correct or even remove the variations that real people introduce in real buildings. Recent work on the calculations used to rate housing in Germany, the Netherlands, Belgium, and the UK shows that these models consistently over-predict the amount of energy that dwellings actually use (Sunikka-Blank and Galvin, 2012). Similarly, work on low-energy design commercial buildings shows that assumptions about how people will use 'special' buildings can be very optimistic (e.g. Lenoir et al., 2011). The idealizations embedded in technical potential scenarios are familiar, but they are not necessarily true, and the ability to compel people to act 'properly' is limited (Moezzi and Janda, 2014).

7.3.2 ENERGY SERVICES

Lacking from the PTEM's predominantly technical view of energy demand are insights into why energy is demanded in the first place. An energy services perspective makes salient the 'doings' of daily life: moving around, sheltering from the weather and thermally conditioning indoor environments, cooking and communicating, eating and entertaining. The end of the journey for energy—the final link in the energy conversion chain represented in Figure 7.1—is no longer the final energy 'demanded' at the point of use; it is the useful energy that provides end users with a service.

Final energy is converted in end-use devices such as appliances, machines, and vehicles into useful energy such as heat, light, or motive power (see Figure 7.2). The application of this useful energy is in the provision of energy services such as a moving vehicle (mobility), a warm room (thermal comfort), process heat (for materials manufacturing), or light (illumination) (Grubler et al., 2012: 103).

Understanding demand in terms of energy services makes clearer why—from the end-user's perspective—the energy resources used to supply that demand are out of sight and out of mind. Energy resources are inputs into energy conversion chains of which only the final step—final energy to useful energy to useful service—directly provides for users' needs.

Although energy services explain the ultimate purpose of the energy system, they are not often used in energy assessments as their units are not commensurate. A common unit of measurement of mobility is the two dimensional 'passenger-kilometre', which describes the distances travelled by vehicle or transit users. In contrast, a common unit for measuring lighting as an energy service is 'lumen-hours'. Required energy inputs are therefore used as common proxies for the energy that provides useful energy services. This is highly distortionary as energy services provided at the lowest efficiencies require proportionally more energy inputs and so are inflated in energy accounts (Grubler et al., 2012: 106).

Alongside the potential for improving service efficiency (as well as energy conversion efficiency), the energy services perspective also makes explicit the option of reducing quantities of service demanded: less mobility, less thermal conditioning, less illumination. In PTEM-type analyses, such options are often conspicuously absent, or summarized under the generic and descriptively unhelpful term 'behavioural and lifestyle change' subject to 'socio-cultural barriers' (Levine et al., 2007; Cullen and Allwood, 2010a; Riahi et al., 2012; Ürge-Vorsatz et al., 2012). From the PTEM perspective, reducing levels of service demand implies welfare losses. Options for changing social preferences lie outside the system boundaries of research and policy action.

Self-evidently, the energy services approach is still centred on energy. The drivers of demand and the resulting strategies for addressing inexorable demand growth are cast more widely than in conventional PTEM-type analysis, but energy conversion by end-use technologies in particular contexts remains both important and salient. Other perspectives challenge this premise. Do people really *demand* energy like electricity and gas, or energy services like mobility and cooking and long-distance communication?

7.3.3 SOCIAL PRACTICES AND PRACTICE THEORIES

A social practices approach starts with the idea that practices in general, and thus energy-consuming practices, are socially constructed rather than individually or technologically determined. Practices are always social because their performance and components are socially normalized. There are several different theories used to explain social practices. These often draw upon Giddens's structuration theory (Giddens, 1984) and Bourdieu's outline of a theory of practice (Bourdieu, 1977, 1990). Schatzki (1996) defines a practice as either a performance or a nexus of doings and sayings. These doings and sayings are linked by three components that sustain the practice, according to Schatzki (1996): values and norms, institutionalized rules, and know-how and routines. Reckwitz (2002) adds a fourth component: materials, products, and technologies. He also underlines that routines are either mental or embodied. These components sustain and organize the practices that are 'performed' and so reproduced over daily, weekly, and seasonal cycles and in particular contexts or physical spaces. Examples of energy-consuming practices studied within this theoretical framework are car driving (Warde, 2005), culinary practices (Truninger, 2011), and practices regulating thermal comfort and energy retrofits at home (Gram-Hanssen, 2010, 2011; Hargreaves et al., 2013; Bartiaux et al., 2014).

As observable phenomena, practices may seem equivalent to energy services. However, the social practice lens on the energy system is concerned with the organizing components of practices, and how these evolve and recombine over time and space (Shove et al., 2012). From this perspective, for example, 'mobility' cannot be isolated from road and fuelling networks nor driving rules and safety standards. Although not explicitly a lens on the energy system at all, a social practices perspective understands the drivers of energy demand in very different terms from the individual economically-motivated actors that characterize the PTEM and, to some extent, the energy services perspective. It also underlines the importance of the socio-technical context of energy service demand and so indicates a way to develop new policies for changing demand-related practices as nexuses of things, routines, norms, and procedures, rather than through trying to affect individual decision-making piece by piece (Shove, 2010; Shove et al., 2012). For example, a policy programme for improving roof insulation in private housing could link an immediate price reduction for insulating materials to media programmes to enhance dwellers' practical knowledge, hands-on skills workshops for installers, as well as reconfiguring the meaning of roof insulation by describing it as a necessary characteristic of comfort and quality (instead of focusing on its economic or energy attributes).

Practice theories are also useful in helping explain why certain types of 'good ideas' do not diffuse as PTEM-type analysis might predict. For example, a study of improved cook stoves in India used three different frames to try to explain why rural people did not adopt stoves that were cheaper to operate and healthier for cooks (Mann, 2012). It began with an economic approach (we need to subsidize the stoves), then moved to an information deficit model (we need to tell people how good the stoves are for their health), and finally, used a practice theory approach to show that cooking is more about taste and tradition than economics and information. Rural cooks wanted their dhal to taste a certain way, and their adherence to traditional cooking methods indicated a complex and positive association with their food rather than the cost of cooking or ignorance of other cooking methods and their purported merits.

Although social practices evolve over time, the above example shows that social practices are fully capable of resisting the charms of standard policy instruments such as incentives and information. They are also resistant to regulations. Recent research has shown that the European Energy Performance of Buildings Directive has created Energy Performance Certificates but not yet a practice of energy-related home renovation (Bartiaux et al., 2014). The question going forward for both policymakers and social-practice researchers is whether practice theories can help create change as well as helping to explain why it does not happen as anticipated.

7.3.4 SOCIO-TECHNICAL TRANSITIONS

The socio-technical transitions perspective shifts the object of enquiry away from energy demand, energy services, and the social practices describing the rhythms of daily life, up to large-scale patterns and dynamics of social and technological change. A socio-technical transition is 'a major technological transformation in the way societal functions such as transportation, communication, housing, feeding are fulfilled . . . not only technological changes, but also changes in user practices, regulation, industrial networks, infrastructure and social meaning' (Geels, 2002: 1257). Demand for energy services is shaped, constrained, and stimulated by these 'systems of provision', and all the interests vested in them. A longstanding concern of evolutionary and institutional economists (Nelson and Winter, 1977; Nelson and Nelson, 2002), the socio-technical transitions lens on the energy system has been increasingly taken up by technology and innovation researchers and social theorists (Carlsson and Stankiewicz, 1991; Grubler, 1998; Geels, 2004).

Socio-technical transition theories are explicitly concerned with how complex and intermeshed systems change, whether endogenously or by design or exogenous influence. As the term suggests, socio-technical transitions are concerned less with the economic drivers of technological change and more with the roles, influences, and relationships of actors and social institutions at three different levels: micro, meso, and macro. This multi-level perspective is fundamental to this approach, as it depicts long-term systemic change as the outcome of dynamic interplay between these levels (Geels, 2002). The interplay involves, for example, exogenous, 'landscape' level forces (deregulation, crises, value shifts) opening spaces for innovations tested and proven in protected 'niches' to challenge dominant 'regimes' (Verbong and Geels, 2007). Socio-technical regimes are defined as: 'the rule-set of grammar embedded in a complex of engineering practices, production process technologies, product characteristics, skills and procedures, ways of handling relevant artefacts and persons, ways of defining problems; all of them embedded in institutions and infrastructures' (Rip and Kemp, 1998: 340). The regime is the meso-level of the multi-level perspective that also distinguishes the macro-level of landscapes, and the micro-level of niches (Rip and Kemp, 1998; Geels, 2002). Landscapes are exogenous, enduring political, environmental, and social aspects that guide the trajectory of socio-technical change. Examples of landscape-level forces include globalization, climate change, or cultural change. Niches are emergent novelties and incubators that bubble along until a destabilization of the regime opens space for their diffusion and—if successful—their dislodging of incumbent technologies, practices, and actors that constitute the regime (Schot and Geels, 2007).

Successful innovations can thus challenge dominant technologies, ways of thinking and doing, or social patterns. However, the principal characteristic of socio-technical systems is not change but stability (Wilson and Grubler, 2011). Lock-in (or inertia to change) in the energy system, with its associated patterns of energy demand, is conventionally explained as a function of increasing returns to scale, long-lived capital assets, dependence on infrastructure, and sunk costs (Unruh, 2000). These combine to give incumbent actors and technologies an enduring advantage. The rules embedded in a socio-technical regime offer a complementary explanation. These rules act as the deep structure or 'grammar' of the energy system (Geels, 2004). Cognitive rules—shared belief systems and expectations—make engineers and designers search, experiment, test, and learn in particular directions and not in others. Normative rules—perceptions and expectations of proper behaviour—establish mutuality and shared expectations, stabilizing social and organizational relationships. Formal rules—regulations, standards, contracts—further enshrine expectations about the direction and rate of change (Geels, 2004). Lock-in is thus also explained by vested interests, bounded or constrained thinking, and a deep-rooted habituation to the familiar.

From a socio-technical transitions theory perspective, the problem of understanding energy demand magnifies in complexity and scale, and the cast of explanatory variables becomes ever-wider. Interactions and relationships between different players in the energy system become particularly salient, as do the informal rules or expectations that govern these interactions. The demand for energy is not a short-term consequence of price-responsive or attitude-driven individual behaviours, but something that is 'systematically configured' over the long term under both social and technological influences (van Vliet et al., 2005). Because long timescales are an important part of the socio-technical transitions story, much of the research in this field has focused on historical case studies. These cases have been mainly located in Western Europe, particularly in the Netherlands, UK, and Germany. A survey of 540 articles on sustainability transitions also found that energy demand has not been a major focus of transitions theory to date (Markard et al., 2012). The survey found that the energy supply and distribution system has been the dominant empirical focus (36 per cent of all papers), followed by transportation (8 per cent). Empirical examples include low-carbon supply technologies like solar photovoltaics (Smith et al., 2014), public waste and hygiene infrastructure (Geels and Kemp, 2007), and electricity networks (Foxon et al., 2010). However, one of the new multi-year UK Energy Use Energy Demand Centres bases its work on an innovation and transitions theory perspective (CIED, 2014), so transitions research on energy demand is likely to increase.

7.4 **Implications for policy: how much energy is enough?**

This chapter has described the various trends in global energy demand between end-uses, over time, between countries and within countries. This chapter has also presented four different perspectives for understanding demand: (1) a physical, technical, and economic model (PTEM); (2) an energy services approach; (3) social practice theories; and (4) socio-technical transitions theory. Together, both the evidence and the multiple problem frames show that understanding energy demand is complex and contested.

How can policymakers seeking to reduce the adverse impacts of the energy system respond to competing and often conflicting explanations as to the nature, composition, and drivers of demand? The evident complexity of the problem is consistent with a broad socio-technical perspective rather than one that can be solved exclusively through technologies and prices. Yet the dominant policy frame in this field remains firmly rooted in a physical, technical, and economic model. This issue is taken up in detail in Chapter 9, which looks at various policy efforts to curb energy demand in the buildings sector. Policy considerations in other sectors are explored elsewhere in this book: personal mobility in Chapter 10; aviation and shipping in Chapter 11.

As a conclusion for this chapter, we consider how multiple explanations for the policy-relevant question of basic energy 'needs' could affect policy approaches. How much energy is needed, and for what, is a fundamental concern, both nationally and internationally. Is energy use a human right, or is it a privilege? How much energy is 'enough', and who decides?

In a broad historical context, increases in energy services have been proportional to increases in economic activity measured as gross domestic product (GDP). This historical context tends to identify increased energy use with economic growth. Decoupling energy consumption from GDP growth implies continuing improvement in service efficiencies (Haas et al., 2008). The potential for improving conversion efficiencies is inherently limited by physical and thermodynamic constraints (Cullen and Allwood, 2010b), although fundamental technological transitions in the means of providing energy services can result in substantial efficiency gains (e.g. tallow candles to electric lighting) (Fouquet, 2008).

Using alternatives to GDP—such as human well-being—as the denominator metric for system growth offers a more optimistic picture on the potential for decoupling. Returns to well-being or human development are closer to being saturated for additional services consumed (Haas et al., 2008). In other words, reductions in the quantity of services demanded (in developed economies) become feasible if seen from the point of view of human well-being rather than economic growth. The United Nations' Human Development Index (HDI) combines life expectancy, literacy, and income in an overall measure of human development. Steinberger and Roberts (2010) find that high HDI values can be achieved at moderate levels of energy consumption, and that trends in the HDI from 1975 to 2005 indicate a steady decoupling of development from energy demand growth. The authors find that 'for all the human development indicators we considered, the energy thresholds for high levels of human development are decreasing functions of time.... [Thus] achieving human well-being is becoming steadily more efficient' (Steinberger and Roberts, 2010: 429). Although Steinberger and Roberts' research shows that a given HDI value can be reached at ever lower levels of energy demand, they measure primary energy not final energy. This measurement choice masks differences in the structure of demand between countries. Moreover, their analysis does not explain why levels of energy services demanded and consumed continue to rise in developed economies (with income elasticities remaining positive).

An alternative approach is offered by Rao and Baer (2012) who estimate the energy requirements for a standard of 'decent living' corresponding to the Millennium Development Goals and greenhouse-gas emission reduction targets subject to 'equitable access to sustainable development'. Decent living is interpreted as household consumption of a set of basic goods and services comprising: food, safe water and sanitation, shelter, health care (basic needs); and education, clothing, television, refrigeration, and mobile phones (Rao and Baer, 2012). These services result directly in energy consumption, but associated energy used indirectly in the construction and maintenance of supporting infrastructure (physical and human resources) is also factored in. Druckman and Jackson (2010) have similarly estimated the energy and emissions requirements of a minimum income standard deemed to provide a 'decent life' in the UK, including not just basic services like nutrition, warmth, and shelter but also the means to participate effectively in society through, for example, mobility and communication.

Other social scientists argue that it is impossible to objectively establish what a decent life should be, and thus what the corresponding energy demand would be, as the answers and conceptions of a good life vary according to culture, time, and even individuals within a period or society. Ivan Illich (1977) argued that growth of new 'needs' is a key aspect of modern society, as what were once luxuries become necessities. There are many explanations for the increasing spiral of consumption, including social, cultural, psychological, and economic interpretations (Wilk, 2002). Karl Marx, for example, saw the consumption of goods as a socially constructed mechanism to justify a logic of overproduction in capitalist societies (Bartiaux et al., 2011: 67–8). Over time, many parts of a societal system may interact to produce 'needs' that become less and less negotiable (Bartiaux et al., 2011). Two such want–need spirals—air-conditioning and bigger houses—are considered further in Chapter 9.

As noted earlier, from an economic perspective the spiral of increasing consumption is readily explained. Technical improvements in energy conversion efficiency reduce the effective price of energy services and therefore increase the demand for these services—the direct rebound effect (Saunders, 1992). For example, better insulation leads to higher thermostat settings or larger homes (as they are cheaper to heat), more efficient engines lead to larger vehicles or more kilometres travelled (as they are cheaper to run), and so on (Haas et al., 2008). If the savings from lower energy service costs are not spent directly on more of those services, they can be spent on other energy-using services—the indirect rebound effect (Sorrell and Dimitropoulos, 2008). Numerous estimates and review studies of both direct and indirect rebound effects have found: (i) they vary widely between energy services; (ii) direct rebound effects for domestic energy services including mobility are typically less than 30 per cent; but (iii) they are difficult to estimate given the abundance of confounding factors (Sorrell et al., 2009; Ürge-Vorsatz et al., 2012: Table 10.3; Chitnis et al., 2013; Gillingham et al., 2013). Rebound effects can be mitigated by increasing energy prices to hold the effective price of energy services constant as service efficiencies improve, or according to Haas et al. (2008: 4019), through 'complete changes in customers' awareness, leading to a paradigm change influencing the elasticities of income, efficiency and individual power and quality of dwellings, light bulbs, cars, refrigerators, etc.' This is an echo of the 'behavioural and lifestyle changes' that are noted in but exogenous to PTEM-type analysis.

A more radical response to the rebound effect would be to concentrate on sufficiency rather than efficiency (Herring, 2006). Sufficiency sits within a broader philosophical rethinking of the purpose of development and the nature of progress (Princen, 2005). A sufficiency approach would address the broader social benefits of energy services delivered, rather than their quantities. It also brings into the discussion concepts of 'enoughness' or 'too muchness', which may be evident in other areas (like eating too much chocolate), but are not often discussed in the energy arena. The notion of setting limits for consumption, particularly in developed countries, is not a politically popular concept. Policy possibilities along this line of thinking include personal carbon trading (Fawcett and Parag, 2010; Parag and Eyre, 2010), which have been studied but not implemented.

Since the 1970s, energy efficiency gains have been outpaced by increases in the size, number, features, and use of energy consuming equipment (Moezzi, 1998). Other researchers have observed: 'when we were less efficient we used less energy' (Rudin, 2000: 8.331). This supersizing of expectations has led some energy efficiency advocates to recommend policy targets based on consumption levels rather than efficiency levels (Harris et al., 2006). Forward-thinking policymakers should thus consider limits to consumption. Much as the idea of 'limits to growth' framed economic and environmental debates in the 1970s, limits to consumption may be necessary to reach ambitious carbon reduction targets set for the twenty-first century.

This concluding section provides an initial exploration of the policy implications of contested theoretical and methodological approaches to the 'problem' of energy demand. Chapter 9 pushes this exploration further. Starting with the premise that energy demand is shaped both globally and locally, Chapter 9 considers the extent to which either technological or policy solutions in the buildings sector can be successfully shared—across borders, between groups, and over time.

▓ REFERENCES

BARTIAUX, F., FROGNEUX, N., & SERVAIS, O. 2011. Energy 'Needs', Desires, and Wishes: Anthropological Insights and Prospective Views. *In*: SHIOSHANSI, F. (ed.) *Energy, Sustainability, and the Environment: Technology, Incentives, and Behavior*, Burlington, MA and Oxford, UK, Elsevier: pp. 63–8.

BARTIAUX, F., GRAM-HANSSEN, K., FONSECA, P., OZOLIŅA, L., & CHRISTENSEN, T. H. 2014. A practice–theory approach to homeowners' energy retrofits in four European areas. *Building Research & Information*, 42, 525–38.

BLUMSTEIN, C. & STOFT, S. E. 1995. Technical efficiency, production functions and conservation supply curves. *Energy Policy*, 23, 765–8.

BOURDIEU, P. 1977. *Outline of a Theory of Practice*, London, Cambridge University Press.

BOURDIEU, P. 1990. *The Logic of Practice*, Cambridge, Polity Press.

CARLSSON, B. & STANKIEWICZ, R. 1991. On the nature, function and composition of technological systems. *Journal of Evolutionary Economics*, 1, 93–118.

CHITNIS, M., SORRELL, S., DRUCKMAN, A., FIRTH, S. K., & JACKSON, T. 2013. Turning lights into flights: Estimating direct and indirect rebound effects for UK households. *Energy Policy*, 55, 234–50.

CIED. 2014. *Centre for Innovation and Energy Demand* [Online]. Brighton: Sussex University. Available: <http://www.cied.ac.uk> [Accessed 28 May 2014].

CULLEN, J. M. & ALLWOOD, J. M. 2010a. The efficient use of energy: Tracing the global flow of energy from fuel to service. *Energy Policy*, 38, 75–81.

CULLEN, J. M. & ALLWOOD, J. M. 2010b. Theoretical efficiency limits for energy conversion devices. *Energy*, 35, 2059–69.

DRUCKMAN, A. & JACKSON, T. 2010. The bare necessities: How much household carbon do we really need? *Ecological Economics*, 69, 1794–804.

ENKVIST, P.-A., NAUCLER, T., & ROSANDER, J. 2007. A cost curve for greenhouse gas reduction. *The McKinsey Quarterly* 2007, Number 1: 35–45.

FAWCETT, T. & PARAG, Y. 2010. An introduction to personal carbon trading. *Climate Policy*, 10, 329–38.

FOUQUET, R. 2008. *Heat, Power, and Light: Revolutions in Energy Services*, Cheltenham, UK, Edward Elgar Publishing.

FOUQUET, R. 2010. The slow search for solutions: Lessons from historical energy transitions by sector and service. *Energy Policy*, 38, 6586–96.

FOUQUET, R. 2014. Long-run demand for energy services: income and price elasticities over two hundred years. *Review of Environmental Economics and Policy*, 8(2), 186–207.

FOXON, T. J., HAMMOND, G. P., & PEARSON, P. J. G. 2010. Developing transition pathways for a low carbon electricity system in the UK. *Technological Forecasting and Social Change*, 77, 1203–13.

GEELS, F. W. 2002. Technological transitions as evolutionary reconfiguration processes: A multi-level perspective and a case-study. *Research Policy*, 31, 1257–74.

GEELS, F. W. 2004. From sectoral systems of innovation to socio-technical systems: Insights about dynamics and change from sociology and institutional theory. *Research Policy*, 33, 897–920.

GEELS, F. W. & KEMP, R. 2007. Dynamics in socio-technical systems: Typology of change processes and contrasting case studies. *Technology in Society*, 29, 441–55.

GIDDENS, A. 1984. *The Constitution of Society: Outline of the Theory of Structuration*, Cambridge, Polity Press.

GILLINGHAM, K., KOTCHEN, M. J., RAPSON, D. S., & WAGNER, G. 2013. Energy policy: The rebound effect is overplayed. *Nature*, 493, 475–6.

GRAM-HANSSEN, K. 2010. Residential heat comfort practices: Understanding users. *Building Research & Information*, 38, 175–86.

GRAM-HANSSEN, K. 2011. Understanding change and continuity in residential energy consumption. *Journal of Consumer Culture*, 11, 61–78.

GRUBLER, A. 1998. *Technology and Global Change*, Cambridge, UK, Cambridge University Press.

GRUBLER, A., JOHANSSON, T. B., MUNDACA, L., NAKICENOVIC, N., PACHAURI, S., RIAHI, K., ROGNER, H.-H., & STRUPEIT, L. 2012. *Energy Primer. Global Energy Assessment*, Cambridge, UK, Cambridge University Press.

HAAS, R., NAKICENOVIC, N., AJANOVIC, A., FABER, T., KRANZL, L., MÜLLER, A. & RESCH, G. 2008. Towards sustainability of energy systems: A primer on how to apply the concept of energy services to identify necessary trends and policies. *Energy Policy*, 36, 4012–21.

HARGREAVES, T., NYE, M., & BURGESS, J. 2013. Keeping energy visible? Exploring how householders interact with feedback from smart energy monitors in the longer term. *Energy Policy*, 52, 126–34.

HARRIS, J., DIAMOND, R., IYER, M., & PAYNE, C. Don't Supersize Me! Toward a Policy of Consumption-Based Energy Efficiency. ACEEE Summer Study on Energy Efficiency in Buildings, 2006 Asilomar, CA., American Council for an Energy-Efficient Economy.

HERRING, H. 2006. Energy efficiency—a critical view. *Energy*, 31, 10–20.

IEA. 2007a. *Energy Balances of Non-OECD Countries*, Paris, France.

IEA. 2007b. *Energy Balances of OECD Countries*, Paris, France.

ILLICH, I. 1977. *Towards a History of Needs*, New York, Pantheon.

INTERTEK. 2012. Household Electricity Survey: A study of domestic electrical product usage, Intertek Report R66141. Milton Keynes, UK.

JOHANSSON, T. B., NAKICENOVIC, N., PATWARDHAN, A., & GOMEZ-ECHEVERRI, L. 2012. Summary for Policymakers. *In*: JOHANSSON, T. B., NAKICENOVIC, N., PATWARDHAN, A., & GOMEZ-ECHEVERRI, L. (eds) *Global Energy Assessment: Towards a Sustainable Future*, Cambridge, UK, Cambridge University Press.

LENOIR, A., CORY, S., DONN, M., & GARDE, F. Users' Behavior and Energy Performances of Net Zero Energy Buildings. 12th Conference of International Building Performance Simulation Association, 14–16 November 2011 Sydney, Australia.

LENZEN, M., WIER, M., COHEN, C., HAYAMI, H., PACHAURI, S., & SCHAEFFER, R. 2006. A comparative multivariate analysis of household energy requirements in Australia, Brazil, Denmark, India and Japan. *Energy*, 31, 181–207.

LEVINE, M., ÜRGE-VORSATZ, D., BLOK, K., GENG, L., HARVEY, D., LANG, S., LEVERMORE, G., MONGAMELI MEHLWANA, A., MIRASGEDIS, S., NOVIKOVA, A., RILLING, J., & YOSHINO, H. 2007. Residential and Commercial Buildings. *In*: METZ, B., DAVIDSON, O. R., BOSCH, P. R., DAVE, R., & MEYER, L. A. (eds) *Climate Change 2007: Mitigation*.

Contribution of Working Group III to the Fourth Assessment Report of the Intergovernmental Panel on Climate Change, Cambridge, UK and New York, USA, Cambridge University Press.

LUTZENHISER, L. 1992. A cultural model of household energy consumption. *Energy*, 17, 47–60.

LUTZENHISER, L. 1993. Social and behavioral aspects of energy use. *Annual Review of Energy and the Environment*, 18, 247–89.

LUTZENHISER, L. 2014. Through the energy efficiency looking glass: Rethinking assumptions of human behavior. *Energy Research and Social Science*, 1, 141–51.

LUTZENHISER, L. & LUTZENHISER, S. Looking at Lifestyle: The Impacts of American Ways of Life on Energy/Resource Demands and Pollution Patterns. ACEEE Summer Study on Energy Efficiency in Buildings, 2006 Asilomar, CA, American Council for an Energy Efficient Economy: 163–76.

MA, L., ALLWOOD, J. M., CULLEN, J. M., & LI, Z. 2012. The use of energy in China: Tracing the flow of energy from primary source to demand drivers. *Energy*, 40, 174–88.

MANN, P. A. G. 2012. *Achieving a mass-scale transition to clean cooking in India to improve public health*. PhD Dissertation, University of Oxford.

MARKARD, J., RAVEN, R., & TRUFFER, B. 2012. Sustainability transitions: An emerging field of research and its prospects. *Research Policy*, 41, 955–67.

MCKINSEY 2009. Pathways to a low-carbon economy. Version 2 of the Global Greenhouse Gas Abatement Cost Curve, London, UK, McKinsey and Company.

MEIER, A., WRIGHT, J., & ROSENFELD, A. 1983. Supplying energy through greater efficiency, Berkeley, CA, University of California: Berkeley.

MOEZZI, M. The Predicament of Efficiency. 1998 ACEEE Summer Study on Energy Efficiency in Buildings, 1998 Washington DC. American Council for an Energy-Efficient Economy: 4.273–4.283.

MOEZZI, M., IYER, M., LUTZENHISER, L., & WOODS, J. 2009. Behavioral assumptions in energy efficiency potential studies, Oakland, CA, California Institute for Energy and Environment.

MOEZZI, M. & JANDA, K. B. 2014. From 'if only' to 'social potential' in schemes to reduce building energy use. *Energy Research and Social Science*, 1, 30–40.

NELSON, R. R. & NELSON, K. 2002. Technology, institutions, and innovation systems. *Research Policy*, 31, 265–72.

NELSON, R. R. & WINTER, S. G. 1977. In search of useful theory of innovation. *Research Policy*, 6, 36–76.

PARAG, Y. & EYRE, N. 2010. Barriers to personal carbon trading in the policy arena. *Climate Policy*, 10, 353–68.

PRINCEN, T. 2005. *The Logic of Sufficiency*, Cambridge, MA, MIT Press.

RAO, N. D. & BAER, P. 2012. 'Decent living' emissions: a conceptual framework. *Sustainability Science*, 4, 656–81.

RECKWITZ, A. 2002. Toward a theory of social practices: a development in culturalist theorizing. *European Journal of Social Theory*, 5, 243–63.

RIAHI, K., DENTENER, F., GIELEN, D., GRUBLER, A., JEWELL, J., KLIMONT, Z., KREY, V., MCCOLLUM, D., PACHAURI, S., RAO, S., VAN RUIJVEN, B., VAN VUUREN, D. P., & WILSON, C. 2012. *Energy Pathways for Sustainable Development. The Global Energy Assessment*, Cambridge, UK, Cambridge University Press.

RIP, A. & KEMP, R. 1998. Technological Change. *In*: RAYNOR, S. & MALONE, E. (eds) *Human Choice and Climate Change 2: Resources and Technology*, Columbus, Ohio, Batelle Press.

ROSA, E. A. & DIETZ, T. 2013. Human drivers of national greenhouse-gas emissions. *Nature Climate Change*, 2, 581–6.

ROSENFELD, A., ATKINSON, C., KOOMEY, J., MEIER, A., MOWRIS, R. J., & PRICE, L. 1993. Conserved Energy Supply Curves for U.S. Buildings. *Contemporary Policy Issues*, 11, 45–68.

RUDIN, A. Why We Should Change Our Message and Goal from 'Use Energy Efficiently' to 'Use Less Energy'. ACEEE Summer Study on Energy Efficiency in Buildings, 2000 Asilomar, CA, American Council for an Energy-Efficient Economy, 8.330–8.340.

SAUNDERS, H. 1992. The Khazzoom-Brookes postulate and neoclassical growth. *The Energy Journal*, 13, 131–48.

SAUNDERS, H. 2013. Is what we think of as 'rebound' really just income effects in disguise? *Energy Policy*, 57, 308–17.

SCHATZKI, T. 1996. *Social Practices: A Wittgensteinian Approach to Human Activity and the Social*, Cambridge, Cambridge University Press.

SCHOT, J. & GEELS, F. W. 2007. Niches in evolutionary theories of technical change: a critical survey of the literature. *Journal of Evolutionary Economics*, 17, 605–22.

SCHWEBER, L. & LEIRINGER, R. 2012. Beyond the technical: a snapshot of energy and buildings research. *Building Research & Information*, 40, 481–92.

SHOVE, E. 1998. Gaps, barriers and conceptual chasms: theories of technology transfer and energy in buildings. *Energy Policy*, 26, 1105–12.

SHOVE, E. 2010. Beyond the ABC: climate change policy and theories of social change. *Environment and Planning A*, 42, 1273–85.

SHOVE, E., PANTZAR, M., & WATSON, M. 2012. *The Dynamics of Social Practice: Everyday Life and How it Changes*, London, Sage.

SMITH, A., KERN, F., RAVEN, R., & VERHEES, B. 2014. Spaces for sustainable innovation: Solar photovoltaic electricity in the UK. *Technological Forecasting and Social Change*, 81, 115–30.

SOCOLOW, R. 1978. *Saving Energy in the Home: Princeton's Experiments at Twin Rivers*, Cambridge, Ballinger.

SORRELL, S. 2009. Jevons' Paradox revisited: The evidence for backfire from improved energy efficiency. *Energy Policy*, 37, 1456–69.

SORRELL, S. & DIMITROPOULOS, J. 2008. The rebound effect: Microeconomic definitions, limitations and extensions. *Ecological Economics*, 65, 636–49.

SORRELL, S., DIMITROPOULOS, J., & SOMMERVILLE, M. 2009. Empirical estimates of the direct rebound effect: A review. *Energy Policy*, 37, 1356–71.

SOVACOOL, B. K. 2011. Conceptualizing urban household energy use: Climbing the 'Energy Services Ladder'. *Energy Policy*, 39, 1659–68.

STEINBERGER, J. K. & ROBERTS, J. T. 2010. From constraint to sufficiency: The decoupling of energy and carbon from human needs, 1975–2005. *Ecological Economics*, 70, 425–33.

SUNIKKA-BLANK, M. & GALVIN, R. 2012. Introducing the prebound effect: the gap between performance and actual energy consumption. *Building Research & Information*, 40, 260–73.

TRUNINGER, M. 2011. Cooking with Bimby in a moment of recruitment: Exploring conventions and practice perspectives. *Journal of Consumer Culture*, 11, 37–59.

UNRUH, G. 2000. Understanding carbon lock-in. *Energy Policy*, 28, 817–30.

ÜRGE-VORSATZ, D., EYRE, N., GRAHAM, P., HARVEY, D., HERTWICH, E., JIANG, Y., KORNEVALL, C., MAJUMDAR, M., MCMAHON, J. E., MIRASGEDIS, S., MURAKAMI, S., & NOVIKOVA, A. 2012. *Energy End-Use: Buildings. Global Energy Assessment*, Cambridge, UK, Cambridge University Press.

VAN VLIET, B., CHAPPELLS, H., & SHOVE, E. 2005. *Infrastructures of Consumption: Environmental Innovation in the Utilities Industries*, London, UK, Earthscan.

VERBONG, G. & GEELS, F. 2007. The ongoing energy transition: Lessons from a socio-technical, multi-level analysis of the Dutch electricity system (1960–2004). *Energy Policy*, 35, 1025–37.

WARDE, A. 2005. Consumption and theories of practice. *Journal of Consumer Culture*, 5, 131–51.

WILK, R. 2002. Consumption, human needs, and global environmental change. *Global Environmental Change*, 12, 5–13.

WILSON, C. & GRUBLER, A. 2011. Lessons from the history of technological change for clean energy scenarios and policies. *Natural Resources Forum*, 35, 165–84.

WOLFRAM, C., SHELEF, O., & GERTLER, P. 2012. How will energy demand develop in the developing world? *Journal of Economic Perspectives*, 26, 119–38.

ZHANG, T., SIEBERS, P.-O., & AICKELIN, U. 2012. A three-dimensional model of residential energy consumer archetypes for local energy policy design in the UK. *Energy Policy*, 47, 102–10.

8 Energy access and development in the twenty-first century

Xavier Lemaire

8.1 Introduction

This chapter first introduces some misconceptions that may have slowed down energy access in developing countries. It then describes the current situation in the developing world and strategies that have been successful. Finally, it presents some future trends in delivering energy access for all in developing countries.

8.1.1 MISCONCEPTIONS AROUND ENERGY ACCESS

Misconceptions from policymakers around energy access, combined with the lack of interest from utilities and politicians on this issue, have slowed down access to modern sources of energy in some developing countries.

8.1.1.1 Myth 1: Electrification is Just a Matter of Extending the Grid

The belief that the grid could reach all rural areas has slowed the expansion of distributed generation, leaving rural inhabitants to deal with costly and time-consuming products for inadequate services, like lighting from kerosene lamps, use of batteries, and traditional biomass; it seems nowadays rural inhabitants expect to receive a minimum level of comfort without waiting for the grid (Ellegård et al., 2004; Gustavsson & Ellegård, 2004). Even in peri-urban locations connected to the grid, end-users—when submitted to endless power cuts and the poor quality of services provided by the utilities—sometimes prefer to invest in small decentralized systems as a back-up. The tension between grid extension and decentralized generation still exists: there is among utilities the idea that in the long-term everyone shall be connected to the grid, even if it has been demonstrated that in remote locations, off-grid systems could be more economical, notably in less densely populated African countries (Szabo et al., 2011).

8.1.1.2 Myth 2: Access to Energy for all Implies Massive Public Subsidies and Donor Aid Funding

The dissemination of small renewable energy systems has often been done via donor-driven projects that—when no provision was made for after-sale maintenance—lasted only the duration of international funding. For instance, solar systems, instead of delivering electricity for the expected lifetime of the systems, were non-operational once batteries or electronic components were not maintained or replaced. Donor-given projects have tried to disseminate efficient cookstoves that have never been used (Clough, 2012; Kumar et al., 2013). Bio-digestion is another technology heavily promoted in Asian countries and to a lesser extent in some Sub-Saharan African countries where often a minority of bio-digesters function without defect (Bond and Templeton, 2011). Large-scale dissemination programmes of small wind and small hydro-power have also faced poor performance due to lack of proper management, notably in China (CREIA, 2009). In fact, the success of energy access implies the design of products adapted to local conditions of use, intensive marketing, and constant training.

8.1.1.3 Myth 3: Electricity is Enough to Guarantee the Development of an Area

Electricity has long been thought to be equal to development. But the evidence base for the link between rural electrification and development remains weak (World Bank, 2008). Far from just being a matter of increasing grid connections or the number of stand-alone systems, energy access is about providing energy services (Zerriffi, 2011). The focus on grid connection or on certain renewable electrification technologies, like photovoltaics, should not mask the great diversity of existing technologies to provide energy access; a successful strategy often relies on a mix of technologies (Karekezi and Kithyoma, 2002; Wamukonya, 2007). Other energy services like water pumping, heat, hot water, cooking, dryers, or maintaining the cold chain can be provided by using a number of RETs like water mills or passive solar technologies (e.g. solar dryers, solar water heaters). Furthermore, provision of access to some technologies may not lead to development without support measures in other sectors. 'The provision of energy services involves a more complex set of interdependent processes than "straightforward" transfer of technology' (Byrne et al., 2011: 1).

8.1.1.4 Myth 4: Abundance of Renewable Energy Sources can be Assimilated to Free Energy

The great availability of renewable energy sources has led to flaws in the design of programmes. While the world growth rate for grid-tied photovoltaic systems has been 40–80 per cent every year since 2000, off-grid growth rate has been only 5–15 per cent.

Paradoxically, solar systems, which were thought to be a 'technological fix' capable of quickly providing electricity to remote areas of developing countries, are still not disseminated widely and remain marginal compared to the number of un-electrified households. The idea that solar energy would be free was common in the 1970s and 1980s, but actually maintenance costs are far from negligible. This myth of free solar energy has repeatedly led to poorly designed solar programmes and a multiplication of bad experiences damaging the development of this market (Hoang-Gia, 1985; Lemaire, 2009). The same occurred for programmes of dissemination of biomass-based electricity generators where, apart from maintenance, the regular supply of biomass—even if abundant in some tropical countries—needs to be organized in vast areas. To some extent, for all renewable energy technologies (RETs) the abundance of resources has led policymakers to overlook technical and logistical constraints linked to the installation of systems in remote locations.

8.1.1.5 Myth 5: Small is Beautiful

Decentralized generation with stand-alone systems or mini-grids removes the need for a transmission network and can be cost-effective for rural electrification. But the financial system is still reluctant to finance projects using RETs that are more difficult to implement than centralized energy projects, where development banks can dispense large sums of credit in one transaction using well-identified procedures that lower transaction costs. Small systems have the disadvantage of being scattered in remote places, being difficult to monitor, and mobilizing limited funding. Nevertheless by implementing rural energy service companies managing a very large number of systems, small-scale decentralized RET projects can get access to international climate funding and development banks loans.

But national stakeholders can still prefer to favour short-term investments in conventional energies due to their lower capital cost and because of vested interests (Rehman et al., 2012). Furthermore, small non-interconnected systems require installation storage capacity or hybrid systems, which tend to drive costs up. Also small operators have difficulty accessing collateral for financing (N'Guessan, 2012).

8.1.2 ENERGY ACCESS IN DEVELOPING COUNTRIES AND RETS

8.1.2.1 Contrasting Situation across Continents and the Importance of Institutions

Access to electricity can vary among developing countries. Latin American countries, Northern African countries, and China have achieved a high level of electrification. Sub-Saharan African countries (apart from South Africa) have low electrification rates, sometimes reaching less than 10 per cent of the rural population (UNDP/WHO,

2009). The situation is more diverse in South East Asian countries. 'During the last 20 years, an additional 1.7 billion people (equivalent to the combined population of India and Sub-Saharan Africa) gained the benefits of electrification, while 1.6 billion people (equivalent to the combined population of China and the United States) secured access to generally less-polluting non-solid fuels' (Sustainable Energy for All, 2013: 11). But over 1.3 billion people around the world are without electricity and more than three billion people are without access to modern sources of fuel for cooking. These numbers are increasing because of the rising global population (UNDP/WHO, op. cit.).

The UN Energy Access for All initiative results from the awareness of the predicament of rural inhabitants who rely on traditional sources of energy. New international networks like the Global Alliance for Clean Cookstoves,[1] the Global Alliance for Productive Biogas,[2] or Lighting Africa[3] are capitalizing on experience of rural electrification with RETs and accelerating transfer of knowledge by promoting public–private partnerships. A number of rural electrification agencies have been established or re-activated in the 2000s, notably in Sub-Saharan African countries (REEGLE/SERN, 2013). Some countries have made important progress in a limited time period. 'Vietnam, for example, increased its electrification rate from less than 10% to 98% over three decades. Ghana increased the proportion of the total population with access to electricity from 45% in 2005 to 72% in 2010' (Scott and Seth, 2013). Success is linked to a long-term political commitment and strong institutional support combined with adequate rural electrification strategies (Barnes and Foley, 2004).

8.1.2.2 Centralized versus Decentralized RET Generation

Inhabitants of developing countries turn to distributed generation due to the weakness of electrical infrastructure and the quickly decreasing cost of now mature RETs (Moner-Girona et al., 2006). Even to be able to charge a mobile phone can make a sensible difference for isolated inhabitants (Mfuh, 2009).

Costs of most RETs have been constantly decreasing, often making them competitive with conventional stand-alone generation in rural areas. In an era not only of high prices of conventional energies, but of large fluctuations in fossil fuels prices, RETs can be used to reduce the risk linked to conventional energies. As demonstrated by Awerbuch (2004, 2005), adding RET projects to a portfolio of conventional generation assets serves to reduce overall portfolio cost and risk.

[1] <http://www.cleancookstoves.org/>. [2] <http://www.productivebiogas.org/>.
[3] <http://www.lightingafrica.org/>.

8.2 **Widening energy access**

Up-front cost and the lack of local expertise have been identified as the main barriers to the dissemination of RETs. But when financing and maintenance are guaranteed by local rural energy service companies (RESCOs), RETs can deliver for a lower cost than utilities an energy service in demand from end-users.

8.2.1 SUCCESSFUL RURAL ENERGY ACCESS STRATEGIES IN THE TWENTY-FIRST CENTURY

The potential of renewable energy sources is now relatively well identified due to extensive mapping but also due to powerful software modelling that enables the resource to be evaluated without a complete set of measurements over many years. The potential of hydro-power, wind, and biomass or geothermal is considerable in some countries; in most tropical countries, the potential of solar is high. But the location of these resources (except for solar, which is readily available almost everywhere) can be far from inhabited areas.

Rural electrification strategies rely on the right combination of institutions and appropriate technologies to meet the needs of the population. For mini-grids, this can be done while keeping in mind the possibility that the grid could be extended to this location. Stand-alone systems can provide electricity in remote places or on the outskirts of communities connected to the grid. Existing systems (solar systems for telecom antennas, local cogeneration systems) can be used as anchors to provide electricity for nearby inhabitants.

Rural electrification strategies can give priority to productive uses of electricity, like pumping water for agriculture. To provide electricity is not enough; productive use is also linked to the maximization of the impact of electrification by targeting measures to accompany inhabitants in their transition to the use of new forms of energy and helping them to increase their income (De Gouvello and Durix, 2008).

Another option can be to prioritize access for institutions like health centres or schools. A further option is to focus on lighting. Even if often presented as only a source of comfort, domestic lighting could in itself increase productivity by extending opening hours for small shops and enabling people to work at home later and children to study (van Campen et al., 2000; Gustavsson and Ellegård, op. cit.). Electricity is only one aspect of rural energy access, which needs to cover heat, cooking, and water supply as well—all needs that can be provided by a large variety of renewable and conventional energy products.

Power reforms and privatization are not conducive to rural electrification or to the uptake of RETs, unless they are accompanied by specific regulation to reduce the risk of investing in rural areas (Kozloff, 1998; Bacon and Besant-Jones, 2002; Moonga Haanyika, 2006). Consistent policies and regulations can give incentives to provide the remotest locations with access to energy services and can alter the natural bias in favour of electrifying dense urban areas.

8.2.2 THE RISE OF PUBLIC–PRIVATE PARTNERSHIPS

Main utilities often do not want to invest and maintain small-scale RET systems, which can instead be taken charge of by a network of small, specialized rural energy services enterprises. Public–private partnerships can accelerate the rate of electrification and energy access (Zomer, 2001; Barnes, 2007; Kammen and Kirubi, 2008). The implementation of local RET operators can not only reduce transaction costs, but can also offer a better quality of service. Rural energy services companies are indeed pivotal in the dissemination of new technologies, building networks of local retailers to maintain energy systems and guaranteeing the sustainable delivery of energy services; they can be cooperatives, associations or private entrepreneurs (Sanchez, 2006).

Rural energy service companies (RESCOs) need a customer base that is as large as possible in an area, and should include productive and less productive energy use in their services. They can extend their business to cover not just electrification but also other products like Liquefied Petroleum Gas (LPG), cookstoves, or solar water heaters. Diversification increases the stability of income. Business models for operators can rely on cash purchase, leasing, micro-credit, or fee-for-service (Scheutzlich et al., 2002; EDRC, 2003; Krause and Nordström, 2004; Schultem et al., 2003; Morris et al., 2007). Kenya is a well-known example of a country with widespread diffusion of several hundred thousand small solar systems relying on cash purchase (Jacobson, 2006). Bangladesh is an example of a country where micro-finance institutions have been able to fund a high number of solar home systems (Sovacool and Drupady, 2011; Komatsu et al., 2011). Currently, after a long period of maturation, one of the leading companies in Bangladesh has installed more than 1.5 million solar home systems and more than 0.5 million improved cookstoves.[4] South Africa is an example of a fee-for-service business model where rural companies collect a small fee every month and maintain several tens of thousands of solar systems, while also selling LPG (Lemaire, 2011).

For commercial banks this scheme is simpler than providing loans to each individual, as they need only to provide larger commercial loans to RESCOs. RESCOs help also to centralize subsidies from public authorities. Levels of subsidy are adjusted to the customer base that the public authorities want the RESCOs to reach. Targeted subsidies are matched to output and the effective delivery of an energy service. Financial support is also effective in giving indirect subsidies to rural energy service companies, with

---

[4] <http://www.gshakti.org/>.

financing assistance for surveys, business planning, capacity building, or market development. The sustainability of RESCOs relies on the capacity of rural inhabitants to cover operating costs and to keep RET systems running with an appropriate maintenance network. This implies a good design of the concession/area covered and good management by operators, but also a stable and transparent regulatory framework (Lemaire, 2014).

8.3 Trends for solar electrification at the beginning of the twenty-first century

In a number of developing countries, the dissemination of RET is gaining pace. This is particularly true for solar electrification.

8.3.1 FLEXIBILITY OF DECENTRALIZED SYSTEMS AND THE MOVE TO BIGGER SYSTEMS

Solar home systems have been presented as "pre-electrification systems", encouraging people to become familiar with electricity before moving to the grid (De Gouvello and Maigne, 2002). But RET-based systems can be conceived as the final source of electricity for remote locations.

In the past, solar programmes have tended to propose a standard 50-Watt peak solar home systems for everyone. Rural energy providers can today target more precisely the demand of end-users for affordable lighting with a wider range of systems and the combination of photovoltaic systems with a variety of technologies to satisfy energy needs other than basic lighting in rural areas. Leasing combined with micro-banking allows RESCOs to reach even the poorest.

The trend to adopt bigger systems in order to have a longer term off-grid solution will expand as the price of RET continues to decrease steadily. This is particularly true for photovoltaics. For instance, in Morocco, a 200-Watt peak system can today be provided for the cost of a 50-Watt peak system a few years ago (pers. interview, solar company, 2013).

8.3.2 THE PICO-PV MARKET

With the emergence of Light Emitting Diodes (LEDs), the provision of lighting can today be cost-effective with very small systems, which can also be used to charge a mobile phone. As the up-front cost of solar lanterns can be low (typically 20–50 USD), rural electrification with pico-PV could happen as quickly as the diffusion of mobile phones in

Africa. While it took decades to disseminate several million solar home systems world-wide, it took only few years to disseminate 4 million solar lanterns just in Africa, with projections of 20 to 28 million solar lanterns sold in Africa by 2015 (Lighting Africa, 2013). Pico-PV systems also sensitize customers to the possibilities offered by solar PV and incentivize them to buy bigger systems.

8.3.3 MOBILE PHONES: A DISRUPTIVE TECHNOLOGY?

The innovation introduced by the massive diffusion of mobiles phones into developing countries tends to create simultaneously new markets for small photovoltaic systems and the conditions for the diffusion of these systems. There were in 2012 more than 3 billion mobile phone subscribers.[5] In 2013 there were more than 38 times as many mobile phones as landlines in Sub-Saharan Africa.[6] Mobile phones help their users not only to communicate with their family but also with their business partners; they facilitate money generation by enabling even the most remote rural inhabitants to develop their own business by remaining connected with urban centres (Mfuh, op. cit.).

There are more than 500 million users of mobile phones without access to the grid (GSMA, 2011). Often car batteries are used to charge mobiles; users sometimes have to travel several miles to charge their mobiles and small businesses flourish by developing their activity around charging mobiles. The widespread use of mobiles generates a demand for small charging systems, where solar seems perfectly adequate. Mini solar panels can be used. Solar lamps also sometimes include a charging point. Solar handsets—where the panel is directly incorporated in the mobile phone—are also starting to be commercialized.

Another impact of mobile phones is the possibility opened up to solar companies to collect fees via mobile phones—small amounts of money can be charged to the user account at any time. The possibilities offered by mobiles to carry out money transfers via text messages are adapted to the situation of rural inhabitants of developing countries with low incomes who prefer to make small purchases regularly (as low as 0.03 USD to top-up their mobile) rather than bigger transactions.

Micro-banking with mobile phones can reduce transaction costs significantly for solar companies and enable end-users to spread their expenses and charge new units of electricity in their system at will. Micro-banking with mobile phones gives considerable flexibility both to utilities/rural energy services companies and end-users. 'Pay as you go' schemes are now implemented, notably in East Africa where end-users pay a small deposit for a small stand-alone solar system, get a system and then pay for the electricity and the micro-credit with their mobile; if they have no money, the solar system stops delivering electricity; when fully paid for the solar system belongs to end-user. Schemes

---

[5] <http://www.gsma.com>.
[6] *Africa Review*, Monday 13 November 2013. Website: <http://www.Africareview.com>.

using mobile phones are also used for metering micro-grids. Mobile telecoms can also facilitate the maintenance of RET systems by permanently monitoring their status and instantly providing headquarters of companies with detailed log-ins, reports of failures, and status of batteries.

8.4 **Conclusion**

Most technologies for decentralized rural energy access (like solar photovoltaics, small hydro-electricity, biogas, small wind) can be considered mature: some have been tested and implemented in the rural areas of a number of developing countries for decades. Business models have been developed which, combined with consistent policies in favour of rural areas, can accelerate electrification and complement grid extension. Decreasing costs of RETs—especially solar photovoltaics—and the increasing cost of conventional sources are a major drive in rural electrification, as it tends to make RETs such as solar cost-competitive with grid or off-grid diesel generation conventional generation, notably in major parts of rural Africa (Szabo et al., op. cit.). Micro-banking with mobile phones further reduces transaction costs for small energy companies. 'Pay as you go' is perfectly adapted to the use of electricity in the poorest segment of the population, where incomes are irregular. The apparition of pico-photovoltaic products like solar lanterns is also another game changer.

There are still gaps in the value chain and current programmes tend to put a strong emphasis on design, marketing, and after sales for all small-scale technologies like efficient cookstoves, pico-photovoltaic and small photovoltaic systems, and individual biogas systems, where end-user preferences are crucial.

Quality standards are essential due to the often poor quality of some imported products. The quality of the systems disseminated needs to be closely monitored (Duke et al., 2002; Otieno, 2003). Training of installers is also a prerequisite (Bates et al., 2003; Jacobson and Kammen, 2007). Well-managed rural energy service companies and a dense network of retailers are crucial for the installation and long-term maintenance of RETs in rural areas and for the relations they keep with end-users (Byakola et al., 2009). When standards and codes of practices are not enforced and the market develops in a totally unregulated way, low-quality products and poor installations hamper the creation of sustainable markets for RETs.

Energy access in developing countries is therefore still dependent on the policies in place in each country which can—when properly designed—greatly accelerate the dissemination of clean and efficient energy products. Furthermore—even if the massive-scale dissemination of small systems driven by sustained consumer demand seems to be able to make a difference in terms of rural development, subsidies and funding of infrastructures to give access to bigger loads are still a necessity for the development of productive uses of energy.

■ REFERENCES

Awerbuch, S., Jansen, J.C., & Beurskens, L.W.: *Building capacity for portfolio-based energy planning in developing countries—shifting the ground for debates*. Report. Paris, REEEP—UNEP, 2004.

Awerbuch, S. & Sauter, R.: *Exploiting the oil-GDP effect to support renewables deployment*. SPRU Paper no. 129. Science Policy Research Unit (SPRU), Brighton, University of Sussex, 2005.

Bacon, R.W. & Besant-Jones, J.: *Global electric power reform, privatization and liberalization of the electric power industry in developing countries*. Washington, World Bank —ESMAP, 2002.

Barnes, D. (ed.): *The challenge of rural electrification—strategies for developing countries*. Washington, Resources for the Future Press —ESMAP, 2007.

Barnes, D. & Foley, G.: *Rural electrification in the developing world—a summary of lessons from successful programs*. Washington, UNDP/World Bank— ESMAP, 2004.

Bates, J., Gunning, R., & Stapleton, G.: *PV for rural electrification in developing countries: a guide to capacity building requirements*. Report IEA-PVPS T9-03. Paris, International Energy Agency, 2003.

Bond, T. & Templeton, M.: History and future of domestic biogas plants in the developing world. *Energy for Sustainable Development*, 15:14, December 2011, pp. 347–54.

Byakola, T., Lema, O., Kristjansdottir, T., & Lineikro, J.: *Sustainable energy solutions in East Africa. Status, experiences and policy recommendations from NGOs in Tanzania, Kenya and Uganda*. Report. Oslo, Norges Naturvernforbund, 2009.

Byrne, R., Smith, A., Watson, J., & Ockwell, D.: *Energy pathways in low-carbon development: from technology transfer to socio-technical transformation*. STEPS Working Paper 46. Brighton, University of Sussex —STEPS Centre, 2011.

van Campen, B., Guidi, D., & Best, G.: *The potential and impact of solar photovoltaic systems for sustainable agriculture and rural development*. Environment and Natural Resources, Working Paper 2. Rome, Food and Agriculture Organization of the United Nations (FAO), 2000.

Clough, L.: *The improved Cookstove Sector in East Africa: Experience from the Developing Energy Enterprise Programme (DEEP)*. Report. London, GVEP, 2012. Site: <http://www.gvepinternational.org>. (Accessed April 2015).

CREIA/REN21: *Background paper: Chinese Renewable Status report*. Paris, REN21, October 2009. Site: <http://www.ren21.net>. (Accessed April 2015).

De Gouvello, C. & Maigne, Y.: *Decentralised Rural Electrification: An Opportunity for Mankind, Techniques for the Planet*. Paris, Systèmes Solaires, 2002.

De Gouvello, C. & Durix, L.: *Maximising the productive uses of electricity to increase the impact of rural electrification programs*. Report 332/08. Washington, World Bank —ESMAP, 2008.

Duke, R.D., Jacobson, A., & Kammen, D.M.: Photovoltaic module quality in the Kenyan solar home systems market. *Energy Policy*, 30:6 (2002), pp. 477–99.

Ellegård, A., Arvidson, A., Nordström, M., Kalumiana, O.S., & Mwanza, C.: Rural people pay for solar: experiences from the Zambia PV-ESCO project. *Renewable Energy*, 29:8 (2004), pp. 1251–63.

EDRC (Energy and Development Research Centre): *A review of international literature of ESCOs and fee-for-service approaches to rural electrification (Solar Home Systems)*. Cape Town, EDRC, 2003.

GSMA: *Green Power for Mobile Charging Choices*, 2011. London, GSMA. <http://www.gsma. com>. (Accessed April 2015).

Gustavsson, M. & Ellegård, A.: The impact of solar home systems on rural livehoods: experiences from the Nyimba Energy Service Company in Zambia. *Renewable Energy*, 29:7 (2004), pp. 1059–72.

Hoang-Gia, L.: L'utilisation de l'énergie solaire pour le développement du Tiers Monde. In Énergie populaire dans le Tiers-monde. *Environnement Africain. Cahiers d'étude du milieu et d'aménagement du territoire*, vol. V, 4 and vol. VI, 1–2 (1985), pp. 193–202.

Jacobson, A.: Connective power: solar electrification and social change in Kenya. *World Development*, 35:1 (2006), pp. 144–62.

Jacobson, A.D. & Kammen, M.: Engineering, institutions, and the public interest: evaluating product quality in the Kenyan solar photovoltaic industry. *Energy Policy*, 35:5 (2007), pp. 2960–8.

Karekezi, S. & Kithyoma, W.: Renewable energy strategies for rural Africa: is a PV-led renewable energy strategy the right approach for providing modern energy to the rural poor of Sub-Saharan Africa? *Energy Policy*, 30:11–12 (2002), pp. 1071–86.

Kammen, D. & Kirubi, C.: Poverty, energy, and resource use in developing countries—focus on Africa. *Annales New York Academic Science*, 1136 (2008), pp. 348–57.

Krause, M. & Nordström, S. (eds): *Solar photovoltaics in Africa—experiences with financing and delivery models*. Report. New York, UNDP and GEF, May, 2004.

Komatsu, S., Kaneko, S., & Ghosh, P.: Are micro-benefits negligible? The implications of the rapid expansion of solar home systems (SHS) in rural Bangladesh for sustainable development. *Energy Policy*, 39:7 (2011), pp. 4022–31.

Kozloff, K.: *Electricity Sector Reform in Developing Countries: Implications for Renewable Energy*. Policy Project Report, Washington DC: Renewable Energy Power Project, 1998.

Kumar, M., Kumar S., & Tyagi, S.K.: Design, development and technological advancement in the biomass cookstoves: A review. *Renewable and Sustainable Energy Review*, 26 (2013), pp. 265–85.

Lemaire, X.: De quelques mythes et réalités de l'énergie solaire photovoltaïque en milieu rural africain. In: M-J. Menozzi, F. Flipo and D. Pecaud (eds): *Energie et société: sciences, gouvernance et usages*. Saint-Remy-de-Provence, EDISUD (coll. Ecologie Humaine) (2009), pp. 122–30.

Lemaire, X.: Off-grid electrification with solar home systems: the experience of a fee-for-service concession in South Africa. *Energy for Sustainable Development (special issue on Rural Electrification)*, 15:3 (2011), pp. 277–83.

Lemaire, X.: Increasing energy access in rural areas of developing countries. In: J. Bundschuh & G. Chen (eds): *Sustainable Energy Solutions in Agriculture*. New York: CRC Press – Taylor & Francis Group (2014), Chap. 15, pp. 419–36.

Lighting Africa: *Lighting Africa Market Trends Report 2012—Overview of the Off-grid Lighting Market in Africa*, 2013. Nairobi, Lighting Africa. <http://www.lightingafrica.org> (Accessed April 2015).

Mfuh, W.: *The impact of mobile telephony services on performance outcomes of micro-businesses in developing economies: with evidence from micro-business communities in Afghanistan and Cameroon*. PhD thesis (2009), University of Warwick.

Moner-Girona, M., Ghanadan, R., Jacobson, A., & Kammen, D.M.: Decreasing PV costs in Africa—opportunities for rural electrification using solar PV in Sub-Saharan Africa. *Refocus*, 7:1 (January/February 2006), pp. 40–5.

Moonga Haanyika, C.: Rural electrification policies and institutional linkages. *Energy Policy*, 34:17 (2006), pp. 2977–93.

Morris, E., Winiecki, J., Chowdhary, S., & Cortiglia, K.: *Using micro-finance to expand energy services, summary of findings.* Washington DC: SEEP Network, November 2007. <http://www.seepnetwork.org/> (Accessed April 2015).

N'Guessan, M.: *Towards universal energy access particularly in rural and peri-urban areas of the ECOWAS region: approaches, opportunities and constraints*—final report, Nairobi, ECOWAS-UNDP, January 2012.

Nieuwenhout, F.D.J., van Dijk, Lassschuit, P.E., van Roekel, G., van Dijk, V.A.P., Hirsch, D., Arriaza H., Hankins, M., Sharma, B.D., & Wade, H.: Experience with solar home systems in developing countries: a review. *Progress in Photovoltaics: Research and Applications*, 9 (2001), pp. 455–74.

Otieno, D.: Solar PV in Kenya. *Refocus*, 4:5 (Sept–Oct. 2003), pp. 40–1.

REEGLE/SERN: *Country policy reviews.* Vienna, REEEP. Site: <http://www.reegle.org> (Accessed 18 October 2013).

Rehman, I.H., Kar, A., Banerjee, M., Kumar, P., Shardul, M., Mohanty, J., & Hossain, I.: Understanding the political economy and key drivers of energy access in addressing national energy access priorities and policies, *Energy Policy*, 47:1 (2012), pp. 27–37.

Sanchez, T.: *Electricity Services in Remote Rural Communities: The Small Enterprise Model.* Rugby, ITDG, 2006.

Scheutzlich, T., Pertz, K., Klinghammer, W., Scholand, M., & Wisniwski, S.: *Financing mechanisms for solar home systems in developing countries: the role of financing in the dissemination process.* Report IEA PVPS T9-01. Paris, International Energy Agency, 2002.

Schultem, B., van Hermert, B.H., & Sluijsc, Q.: *Summary of models for the implementation of solar home systems in developing countries.* Report IEA PVPS T9-02. Paris, International Energy Agency, 2003.

Scott, A. & Seth, P.: *The political economy of electricity distribution in developing countries—a review of the literature.* London, Overseas Development Institute (ODI)—Politics and Governance, 2013.

Sovacool, B.K. & Drupady, I.M.: Summoning earth and fire: The energy development implications of Grameen Shakti (GS) in Bangladesh. *Energy*, 36:7 (2011), pp. 4445–59.

Szabó, S., Bódis, K., Huld, T., and Moner-Girona, M.: Energy solutions in rural Africa: mapping electrification costs of distributed solar and diesel generation versus grid extension, *Environmental Resource Letters*, 6:3 (July–September 2011), pp. 1–8.

Sustainable Energy for All: *Global Tracking report* 2013. Vienna, United Nations. <http://www.se4all.org> (Accessed April 2015).

UNDP—WHO: *The energy access situation in developing countries—A review focusing on the Least Developed Countries and Sub-Saharan Africa*, 2009. New York, UNDP. Site: <http://www.who.int> (Accessed April 2015).

Wamukonya, N.: Solar home system electrification as a viable technology option for Africa's development, *Energy Policy*, 35:1 (2007), pp. 6–14.

World Bank: *The welfare impact of rural electrification: A reassessment of the costs and benefits—an impact evaluation.* Report. Washington D.C., The World Bank Independent Evaluation Group, 2008.

Zerriffi, H.: *Rural Electrification: Strategies for Distributed Generation.* Berlin, Springer, 2011.

Zomer, A.N.: *Rural electrification.* Ph.D. thesis, University of Twente, Twente, the Netherlands, 2001.

Part II
Global Energy: Options and Choices

9 Improving efficiency in buildings: conventional and alternative approaches

Kathryn B. Janda, Charlie Wilson, Mithra Moezzi, and Françoise Bartiaux

9.1 Introduction

This chapter describes selected technical, social, and policy approaches to reducing energy demand. It focuses in particular on efforts to improve energy efficiency in buildings and concentrates mostly on energy reduction in homes (including domestic technologies and appliances) rather than other building types, industry, or transport. Approaches to reducing energy demand in transport will be addressed in Chapters 10 and 11. Following from the discussion in Chapter 7, this chapter addresses the ways in which energy use is (and is not) problematized. As we show in this chapter, the framing of the problem affects the types of solutions that researchers and policymakers propose. A broader framing of the 'problem' of energy demand could contribute to a wider set of solutions. As explained further below, we argue that the dominant frame for energy research is based on positivist reasoning and that other epistemological approaches, such as interpretivism,[1] could contribute to further understanding and additional novel policy approaches.

In Chapter 7 we described energy demand as a problem that is global, cross-national, and contested. Environmentally, energy demand is a global problem because energy use damages the environment, including increasing the concentration of greenhouse gases in the atmosphere, which knows no geographic boundaries. Economically, globalization leads to some convergence in policy standards and/or cultural norms, and therefore energy use practices. Some energy prices (e.g. oil) are set on the world market and many end-use technologies are traded across borders (e.g. vehicles as discussed in Chapter 10; also appliances, consumer electronics, and heating and air-conditioning equipment). Growth of international trade and travel is also a significant driver of energy demand, as discussed in Chapter 11. As well as these global trends, energy

---

[1] We recognize that interpretivism is but one form of many non-positivist epistemologies. Our use of this term follows the framing used by Schweber and Leiringer (2012). We use it here as a single counterpoint to positivism to simplify the argument and reduce the number of possible 'isms' discussed in the chapter (e.g. antipositivism, constructivism, relativism, critical realism, post-positivism, historicism, etc.).

demand is also a cross-national problem because energy intensive processes are unequally distributed, creating different types of problems in different places. For example, cheap electricity in locations conducive to hydropower has attracted alumin- ium smelters to areas otherwise remote from industrialization; heavy industry has moved from locations with expensive labour to those with cheaper workers and processes. In addition, energy regimes, policies, and practices cluster differently between and within countries.

As discussed in Chapter 7, the precise reasons for how people use energy and its services are contested. The physical, technical, and economic model (PTEM) of rational energy use is the dominant policy and research model for understanding energy use. Other theoretical positions, however, question this model's ability to predict or encourage change for the future when it has failed to close the 'energy efficiency gap' over the past forty or so years. In particular, approaches like transitions theory, practice theories, and energy services focus on the importance of historical and social context, which reframe the 'problem' of energy use in relation to particular times, places, and purposes. There is therefore a tension in the field between a generalizable approach (such as that supported by the PTEM) versus approaches that take greater account of local characteristics, social variability, and cultural contexts.

We begin with a brief discussion of positivist versus interpretivist research meth- odologies, a longstanding discussion which has been taken up in many different fields. This discussion lays the groundwork for two contrasting visions at play in energy demand research according to a recent meta review of the energy and buildings literature by Schweber and Leiringer (2012). These authors found that almost 80 per cent of this literature takes a 'positivist' approach. The positivist approach uses methods drawn from the natural sciences and focuses on identifying generalizable patterns between variables. In contrast, an 'interpretivist' approach assumes 'that human behaviour is mediated by meaning and seeks to identify types of processes and their expression in particular contexts' (Schweber and Leiringer, 2012: 482). These authors found that an 'interpretivist' approach accounts for less than 20 per cent of the energy and building literature.

For the purposes of this chapter, we define 'positivism' as inclusive of a belief that there is a single, external objective reality that can be studied, predicted, and explained. This understanding of reality underpins ideas about the generalizability and transfer- ability of results in many fields. This chapter uses the term 'interpretivism' to mean that 'the study of social phenomena requires an understanding of the social worlds that people inhabit, which they have already interpreted by the meanings they produce and reproduce as a necessary part of their everyday activities together' (Blaikie, 2004: 509). Thus, interpretivism posits that social phenomena are based on perceptions of individ- uals and groups and are therefore multiple and varying in different contexts (Hudson and Ozanne, 1988). 'Reality' is therefore something that can be understood, but the generalizability of this understanding is tricky.

In this chapter, we take up the question of what these contested theoretical and epistemological approaches to the 'problem' of energy demand imply for policy. If energy demand is shaped globally, nationally, and locally, to what extent can either technological or policy responses be shared across borders? As a complete answer to this question is beyond the scope of this chapter, we instead explore it with reference to specific examples. We begin with a discussion of the conventional positivist approach to energy policy, in which general solutions are identified through engineering analyses known as conservation supply curves or marginal abatement cost curves. Challenges to this approach are noted, including the ways in which so-called 'barriers' to these solutions seem unassailable to policy instruments such as capital financing, energy labels, and regulations designed to overcome them. We then discuss a few housing trends which the conventional approach fails to address or explain. For example, different communities select different sets of energy solutions in response to similar goals. These experiences suggest that a more interpretative approach to questions of energy demand may provide additional policy opportunities that the dominant paradigm overlooks. The chapter concludes with possible implications for policies and policymakers resulting from restructuring the underlying assumptions on which most energy efficiency policy is based.

9.2 **Positivism in energy efficiency: the conventional approach**

A physical, technical, and economic model (PTEM) dominates the field of energy efficiency research (see Chapter 7). From this perspective, solutions to the 'problem' of energy demand are clearly identifiable through engineering–economic analyses. These are considered to be generalizable for similar physical, technical, and economic circumstances. Social and behavioural aspects are not included in these analyses, except as averages of expected levels of energy use.

9.2.1 SOLUTIONS THROUGH CONSERVATION SUPPLY CURVES (CSCS)

As discussed in Chapter 7, supply curves for energy conservation quantify the potential and annualized lifecycle cost of different technological and behavioural options for saving energy. The best solutions are the ones that have the lowest total costs, some of which may even be negative. Conservation supply curves (CSCs) were first introduced as an analytical tool by US researchers in the early 1980s (Meier, 1982; Meier et al., 1983; Blumstein and Stoft, 1995), and are now widely used in modelling and other energy analyses, and for visually summarizing and communicating the potential for energy productivity gains. Figure 9.1 depicts one of the earliest versions of this curve, which shows a micro energy conservation supply curve for heating improvements in a single hypothetical house. More recently, CSCs have been reworked as marginal abatement cost curves (MACCs) for CO_2 or greenhouse gases. MACCs similarly describe a supply curve ordered from low to high marginal cost, but for emission reductions rather than energy savings. McKinsey's MACCs have become particularly widely known and referenced (Enkvist et al., 2007; McKinsey, 2009). As most of the abatement options included on MACCs relate to energy, and many of the low-cost options relate to final energy use, MACCs are broadly similar to CSCs in terms of the information they convey.

CSCs and MACCs are useful both descriptively and analytically. They coherently summarize a large amount of information on discrete technological and behavioural options for reducing energy demand. They indicate which options are preferred, assuming that cost is always the dominant selection criterion as would be expected in competitive markets. They also quantify how much energy or emissions could be potentially reduced at a given cost, and so are widely used in 'bottom-up' modelling studies that, in the case of emission reductions, work upwards from an extensive catalogue of mitigation options to estimate sectoral or economy-wide outcomes of carbon pricing (Hanaoka and Kainuma, 2012). One particularly salient feature of both CSCs and MACCs is of interest here: the large potentials for negative cost energy saving or emission reduction opportunities. Negative marginal costs at market interest rates in terms of \$/MWh saved (CSCs) or \$/tCO$_2$ saved (MACCs) imply cost-effective opportunities that are not being pursued, which is often called the 'energy efficiency gap' (Jaffe and Stavins, 1994).

Both CSCs and MACCs have been widely criticized for a variety of methodological reasons, including unexpressed and substantial uncertainties around costs and saving

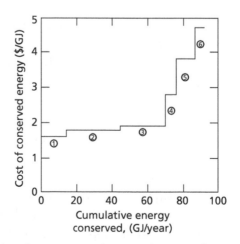

Figure 9.1 'Micro' space heating Conservation Supply Curve for a hypothetical house. Steps: (1) insulate ducts; (2) add wall insulation; (3) add attic insulation; (4) add intermittent ignition device; (5) weatherstrip; (6) tune up furnace

Source: Meier (1982: 24).

potentials, subjective assumptions on future energy prices and interest rates, choices of whose perspective is being adopted in the estimations, and summation errors resulting from the lack of accounting for interaction effects between options (e.g. Fleiter et al., 2009). Negative cost investment opportunities shown by CSCs and MACCs can be interpreted differently depending on one's perspective (Grubb, 2014). Economists argue that transaction costs or other hidden costs are omitted; if accounted for, these hidden costs could shift presumed negative costs above and beyond zero (Gillingham and Palmer, 2014). Building engineers argue that the technical potentials for efficiency gains are actually more plentiful and less costly; the MACCs are conservative, particularly if baseline assumptions about efficiency improvements are overly optimistic (Grubb, 2014). Other interpretations centre on the manifold 'co-benefits' of efficiency improvements in terms of thermal comfort, health, productivity, and so on that are not included in these calculations. These additional benefits imply returns on efficiency investments are even higher than suggested by the MACCs (e.g. Wilkinson et al., 2009). Despite critiques over scale and method, there is general agreement that energy efficiency can deliver more technical and economic potential than the market seems to provide.

9.2.2 OBSTACLES TO TECHNICAL AND ECONOMIC POTENTIAL

In technical and economic studies, the notion of 'market barriers' is widely used to explain why the 'energy efficiency gap' exists. Market barriers describe a wide variety of informational, cognitive, transactional, and other factors that prevent market actors from investing or behaving in a way that conventional policy models see as rational. 'If there are profits to be made, why do markets not capture these potentials? Certain characteristics of markets, technologies and end-users can inhibit rational, energy-saving choices...' (Levine et al., 2007: 418). These barriers are distinct from and in addition to a small number of market 'failures' with respect to an efficiency-centric view, particularly price externalities and split incentives or principal–agent problems (e.g. efficiency measures in rented properties for which landlords bear the costs and tenants enjoy the benefits).

From a technical and economic perspective, the pervasive interpretation of why low or negative cost energy savings are not exploited is inextricably linked to the cost-effective energy conservation opportunities portrayed in CSCs. (The same applies to emission reductions and MACCs.) The problem is therefore defined as one of market barriers, and how these barriers can be targeted, removed, or overcome. The focus on barrier reduction thus generally ignores interpretations and opportunities that lie outside of the PTEM framework. Alternative approaches, which we discuss in a later section, incorporate a broader range of options, particularly for human, organizational, and social behaviour. Nevertheless, this conventional framing dominates all the major energy and climate assessments including the IPCC reports (Levine et al., 2007) and the Global

Energy Assessment (Ürge-Vorsatz et al., 2012) as well as major policy programmes and institutional initiatives (e.g., Brown et al., 2009; DECC, 2012). As noted earlier, CSCs and MACCs are also widely used in bottom-up modelling assessments, typically at a sectoral or regional scale. For example, a comparison of supply curves for GHG emission reductions to 2030 with marginal abatement costs up to $200/tCO$_2$-eq found great differences both regionally and between the models used to generate and analyse these abatement opportunities (Hanaoka and Kainuma, 2012).

CSCs and MACCs underwrite generalizable assessments and results in the framing of both the energy efficiency problem and its solution. CSCs (and MACCS) are designed to show that cost-effective energy efficiency measures exist, and the solution is to adopt all of the measures that are at or below the current cost of energy (or emissions). The market barriers that explain these unexploited energy efficiency options are heterogeneous. Prescribed policy solutions for removing these obstacles are similarly wide ranging in terms of specific instruments, targets, and actors. As an example, Ürge-Vorgatz and colleagues find:

The market-based realization of significant, mostly cost-effective efficiency opportunities in buildings is hampered by a wide range of strong barriers... due to the large number and diversity of barriers... policy portfolios, tailored to different target groups and tailored to a specific set of barriers, are necessary to optimize results (Ürge-Vorsatz et al., 2012: 655).

Their analysis provides a detailed characterization of barriers with associated quantitative impacts on energy use (Ürge-Vorsatz et al., 2012: Table 10.12). They also note issues specific to developing countries that include lack of qualified personnel, insufficient levels of energy services, subsidized energy prices, and the influence of Western-originated architectural designs that may be inappropriate for local climatic conditions (e.g. steel and glass high rises). Ürge-Vorsatz et al. (2012: 698) also state that solutions to these barriers 'must address many principal actors and their intermediaries and include increased education and training of professionals and consumers, improved information, pricing policies, and regulations'. These policy solutions are needed in combination not in isolation, both because policies have interacting effects and because efficiency gains require the coordination of many different small activities throughout the household sector and the wider economy (Brophy Haney et al., 2011).

Beyond the technical and economic energy efficiency literature, it is important to note that scholars from other traditions have criticized both the PTEM and the existence of market barriers. Shove (1998) argued that the conventional framing of 'barriers' to cost-effective energy efficiency defines the problem technically, to be resolved universally by technologies and practices supported or incentivized by policies and standards (Shove, 1998). The role of social scientific research is accordingly closed down to explaining and filling the 'energy efficiency gap' identified by technical analysis, creating a false distinction between the technical and social, and precluding a deeper understanding of the deeply embedded nature of energy use in many contexts, particularly the home (Shove, 1998; Lutzenhiser and Shove, 1999; Lutzenhiser, 2002). The barriers approach is also critiqued for having 'an individualistic view of action' instead of recognizing that action is embedded in a social context (Shove et al., 1998: 301). Guy and Shove (2000: 131) note that '[r]epeated calls for "additional research to understand barriers" (...) reinforce the belief (...) that such barriers are real'. Further, as noted by Bartiaux (2009: 2), the term 'barriers' suggests that they can be 'jumped' over one at the time, which does not reflect the interactions between social–technical factors or their co-evolution. Therefore, several authors propose alternatives: the 'inertia model' (Jensen, 2005), the term of 'brakes' (Bartiaux et al., 2006), and more generally, a broader view of the socio-technical context in which energy-related practices and changes take place (as discussed in Chapter 7).

Whether 'market barriers' to energy efficiency uptake are real things or not is therefore contested. A conventional PTEM approach declares that market barriers are real things, since they explain irrational behaviour for which there is no other economically logical explanation. From other perspectives, PTEM itself is flawed because it rests upon de-contextualized economic logic. From this perspective, the 'barriers' themselves are imagined things, real only in the context of the PTEM. No one disputes that there is less energy efficiency uptake than would be ideal for the planet, but how to solve this problem (e.g. overcome barriers or change social practices?) depends on how the 'problem' of improving efficiency itself is framed and what interventions are considered to be appropriate. This point underscores the importance and effect of different epistemological approaches, which we discussed briefly earlier in terms of positivism and interpretivism, and to which we will return in the final section.

This discussion raises these questions for the reader rather than advocating for one position or the other. In the next section, we discuss how various policy tools have been used to address market barriers, which are generally assumed to be real things and therefore assailable through financing, information, and regulation.

9.3 Positivism's limited results in assailing market barriers

Capital cost financing, greater levels of information, and regulatory standards are all viable policy tools. However, when applied in practice to overcome market barriers to energy efficiency uptake, some unexpected results can occur. In this section, we review several unanticipated effects of these policy tools in the energy efficiency field. Most importantly, they have not measurably contributed to mending the 'energy efficiency gap' or proving the positivist approach to generalizable solutions. The limited results of the conventional approach suggest that new policy approaches, based on a broader range of epistemologies, may be useful. This section focuses on (1) financing, (2) labelling, and (3) standards.

9.3.1 CAPITAL COST FINANCING: THE CASE OF THE UK'S 'GREEN DEAL'

In the domestic sector, the largest technical potential for residential energy savings is achievable through structural efficiency improvements to the building envelope and upgrades to heating or cooling systems (Dietz et al., 2009). In developed economies with stable populations and a mature housing stock, such improvements are available primarily by retrofit, renovation, or other home improvement activities. In rapidly urbanizing economies, building standards and codes for new residential buildings can play a more influential role and are discussed later in this section.

Research following the dominant PTEM framing of the energy efficiency problem repeatedly identifies upfront cost and access to capital as the main barrier that prevents cost-effective efficiency investments in better insulation, windows, doors, and boilers (Emmert et al., 2010; Weiss et al., 2012). Many policy instruments have been proposed and tested to overcome this financial barrier by establishing a mechanism for households to amortize the cost of efficiency measures over time.

Property Assessed Clean Energy (PACE) financing allows upfront costs to be repaid over time through property taxes. Mortgage financing similarly uses loans secured against the property as the repayment mechanism. Energy service companies can invest in efficiency measures as part of bundled service contracts, although transaction costs to energy service contracting in the residential market have limited this approach compared with the institutional sector (Vine, 2005). Upfront costs can also be financed through utility-run demand-side management programmes, and repaid through monthly bills (Brown et al., 2010).

In the UK, the 'Green Deal' was introduced in January 2013 as a major regulatory innovation to instigate 'a revolution in British property' which will put 'consumers back in control' (DECC, 2011: 10). Described as 'the biggest shift in the history of energy efficiency policy in the UK since the oil crises' (Rosenow et al., 2013: 439), the Green Deal allows the upfront costs of energy efficient renovations to be financed by a third party and repaid over time through the electricity bill for the property. To address uncertainty over future energy savings and contractor reliability, Green Deal financing is only available from accredited providers and subject to a prior technical assessment by certified home energy experts. These assessments include a requirement, known as the 'Golden Rule', to recommend only those efficiency measures whose energy savings should exceed monthly repayments (DECC, 2010).

In targeting the upfront cost barrier, the Green Deal is the latest in a long line of policy initiatives to provide or enable the upfront costs of efficiency measures in homes to be amortized. A 2008 review of residential efficiency financing (with an emphasis on on-bill financing) in the USA and Canada found eighteen programmes (Fuller, 2009). This review revealed a number of programme limitations, including: limited applicability of the programmes to households most in need, low participation rates, difficulty assuring that savings will exceed payments, limited support for comprehensive energy retrofits, the inability of many programmes to cover their costs, and issues particular to on-bill financing. A 2011 review in the USA alone included thirty-one programmes from twenty different states of which nineteen were on-bill financing (Bell et al., 2011). The expansion of programmes in the USA was enabled in 2009 by the federal government providing fiscal stimulus money to state-level energy agencies to cover the capital costs of the financing programmes (Brown et al., 2010).

Such financing mechanisms are potentially attractive in that: (1) they allow investment costs to be repaid through energy cost savings; (2) they overcome split incentive problems if those occupying a property repay the costs of improvements to that property; (3) they offer lower credit risk than unsecured loans (Bell et al., 2011). However, these mechanisms seem more attractive to policymakers than to intended participants. The Green Deal, like the programmes reviewed in the USA, has had a very low number of participants. Between January and October 2013 (the most recent period for which data were available at the time of writing), there had been 101,851 Green Deal home assessments but only 1,173 (~1 per cent) had 'converted' into Green Deal financing plans (DECC, 2013). An earlier UK trial of 'Pay-As-You-Save' financing through property taxes reported that many households dropped out of the trial after having initially expressed interest. Their reasons for leaving included uncertainty about expected energy savings (Bioregional, 2011). Low participation rates in financing programmes are attributed to a lack of awareness of potential efficiency measures and of the available financing mechanisms, and to the hassle of searching for information and applying for financing (Bell et al., 2011). Informational and transaction cost barriers thus reinforce the financial barriers to energy efficient investments.

This interpretation of the issues with on-bill and loan financing programmes is entirely consistent with the barriers framing of the problem that the financing programmes are designed to address. Yet as Borgeson et al. conclude: '[Although the] idea that financing can deliver the long-heralded low hanging fruit of energy efficiency in buildings is intellectually appealing, financing as the most important element of program design strategy has not been widely substantiated in over 25 years of experience with financing programs' (Borgeson et al., 2012). Financing policies have under-delivered because lack of financing is rarely the *primary* obstacle to efficiency investments. Other barriers including 'information and hassle costs, split incentives, performance uncertainty' are more important at scale (Borgeson et al., 2012).

There are alternative interpretations that require wholly different problem framings from the conventional approach, which assumes that homeowners are motivated to invest in energy efficient improvements in their properties but are prevented from doing so by barriers. An alternative framing suggests that homeowners adapt and improve their home to better meet the demands and challenges of everyday life (Wilson et al., 2013a). These challenges might range from increasing thermal comfort to competing uses for shared space (e.g. work vs play), accommodating different household member's interests and needs, or modifications for people with limited mobility. As Karvonen (2013: 569) describes: 'domestic retrofit is not an activity of changing a house . . . from poor energy performance to exceptional energy performance, but an intervention into the rhythms of domestic habitation'. The processes of including efficiency measures as part of major home improvements are therefore situated in the domestic environment and its everyday practices (McCormack and Schwanen, 2011). The barriers identified to the uptake of financing mechanisms are misaligned with the ultimate reasons why people might decide to redesign or structurally change a particular part of their domestic environment (Wilson et al., 2013b). If people don't know or care about energy efficiency, overcoming financial barriers will not solve the underlying problem with low energy efficiency adoption rates (Moezzi and Janda, 2014). To overcome the lack of knowledge and caring, the market barriers approach turns to information programmes for help.

9.3.2 INFORMATION: LABELLING AND THE ENERGY PERFORMANCE OF BUILDINGS DIRECTIVE

Lack of information among consumers and contractors is another type of barrier that is often underlined in the PTEM. To overcome this barrier, the European Union adopted the Energy Performance of Buildings Directive (EPBD) in 2002 and recast it more strongly in 2010. It is an information-based policy instrument, aimed at educating potential new owners (or new tenants) about the energy performance of the dwelling they could buy (or rent) by a mandatory energy performance certificate (EPC). As for each European Directive, its translation into national or regional law and its implementation are the responsibility of each member state. For the EPBD, these have been done in (very) different ways and at various paces in different countries (Andaloro et al., 2010).

Denmark has been one of the frontrunners of the EPBD because it has had a mandatory energy labelling scheme for buildings in place since 1997. However, in 2004–5, this scheme was not a strong incentive for energy-efficient refurbishment (Gram-Hanssen et al., 2007). Adjei et al. (2011) later confirm the rather weak influence of the energy labels on refurbishment for Denmark and other European countries through a web-based survey of almost 3200 homeowners from Denmark, Germany, the Netherlands, England, and Finland. To compare and quantify the relative importance of different factors in influencing energy-efficient refurbishment, the authors specified a binary logistic regression model on 1804 respondents (see Adjei et al. (2011: 95–98)). Results show that the relative effect of the Energy Performance Certificate (EPC) is one of the least influential in refurbishment actions. Higher factors on the list, *ceteris paribus*, include the overall condition and age of the house, which suggest that refurbishment may be undertaken more commonly for structural or aesthetic reasons rather than for energy or environmental purposes (see Table 9.1).

The same conclusion was reached from qualitative interviews carried out as part of a comparative study of five different European countries: Portugal, Latvia, the Czech

Table 9.1 Factors influencing energy-efficient refurbishment (Denmark, Germany, Netherlands, England)

Factors influencing refurbishment	Odds ratio (95% confidence interval)
Very poor condition	18.0
Poor condition	5.1
1946–70	4.1
1971–80	4.1
Neither poor nor good condition	3.4
1981–90	3
1991–2000	2.6
1919–45	2.2
Good condition	2
Existing energy problem	1.5
EPC with recommendations	1.5
Environmental activity	1.1
Time since dwelling purchased	1.0

Source: adapted from (Adjei et al., 2011: 98).

Republic, Bulgaria, and the Belgian region of Wallonia (Bartiaux, 2011). The results obtained in Denmark, Latvia, Portugal, and Wallonia were later revisited with a social-practice perspective (Bartiaux et al., 2014, also see Chapter 7 for a discussion of social practices). This investigation helps explain why the EPBD has had little effect: 'the aim of the EPBD is to label buildings, (…) [whereas] [a]t the same time there is an aspiration to compel persons (e.g. homeowners) into a (still non-existing, as argued here) practice of energy retrofitting'(Bartiaux et al., 2014: 10). Another reason given is that in 2009–10, the EPC was 'an isolated market instrument' that was 'not related to other institutionalized rules in the four studied countries/regions, which lowers their potential impact' (Bartiaux et al., 2014: 11). The weak influence that EPCs have on energy retrofits in many European countries is not surprising. Indeed, sociologists have shown earlier that 'from a social perspective, labels have no meaning in their own right – they must be noticed and interpreted by actors in socially relevant terms. The relevant process is not so much the overcoming of ignorance, but the active creation of new shared understandings' (Shove et al., 1998: 306). The labels were intended to activate a presumed system of latent relationships between people and the energy performance of buildings. However, the existence of these relationships is highly questionable, as well as whether labels activate them or not.

9.3.3 REGULATORY APPROACHES: WHAT DO STANDARDS STANDARDIZE?

Energy standards for buildings and appliances, whether mandatory or voluntary, have been one of the most important policy mechanisms for defining, fostering, and diffusing energy efficiency. Unlike new forms of financing or information, which aim to fix market barriers, economic theory allows regulatory intervention in the market itself when it fails, for example, to capture environmental externalities.

Once developed, promulgated, and enforced, energy standards can have a powerful influence: they serve to make the most inefficient buildings and appliances illegal. Although standards aim to be just and rational in this process, they also reflect (and can affect) social norms, power dynamics, and historical patterns. New standards are rarely designed from scratch; they are often based on protocols and methods from previous practice or from other countries (Janda and Busch, 1994; Janda, 2009). Standards set benchmarks and assure certain kinds of comparability between the energy performance of appliances and buildings. For example, energy standards may group televisions of the same size and buildings of the same type together to afford some level of generalizability across a varied landscape. But they can also have unintended consequences that can reshape how societies use energy (Shove and Moezzi, 2004). Besides cultural and market changes, they can even become a mechanism that drives consumption higher by forcing out or marginalizing lower energy-use alternatives. This may happen especially when standards are transferred from more developed, higher-consuming countries to lower-consuming and lesser-developed countries, but it can happen even within a country. For example, policies supporting the uptake of efficient air conditioners have been shown to slightly increase the adoption of air conditioning itself (Samiullah et al., 2002). Subsidies for energy-efficient air conditioner replacement in Mexico were found to increase energy consumption rather than reduce it (Davis et al., 2013).

Energy standards contain definitions of energy efficiency that are rooted in physical phenomena, but they also incorporate assumptions about the purpose of the building, equipment, or appliance at issue, how it will be used, and what features it must possess. These assumptions may be relatively hidden in test procedures or simulation software, or they may be an obvious element in product specifications. Products that satisfy these criteria will be defined as efficient within the context of policy instruments, but if they are not used as designed, the intended savings from efficiency may not be realized. Though this shortfall is often difficult to see, the case of programmable thermostats in the USA is one documented example (Peffer et al., 2013). In particular, the programmable features of these thermostats are often not used as envisioned, and even when they are, they may be less energy conserving than the manual thermostat management practices they replace.

Products developed to meet particular standards can also lead to conceptual, institutional, or material shifts (Healy, 2008) that escalate consumption. For example, if a standard that is developed in one country is used as the basis for a standard in another, some patterns of use may be implicitly transferred as well. In China, clothes washing was traditionally with cold water, but newer front-loading washing machines introduced from Europe use hot water as the only option, amounting to much higher total energy use than the cold water wash it starts to replace (Lin and Iyer, 2007; Harvey, 2010). Standards can be and often are adapted to local conditions, but concerns about trade barriers and the high cost of standards development encourage some degree of harmonization (Lin and Fridley, 2006).

Some of the most interesting examples of the unintended consequences of standards speak to questions of what a building should be or do. For example, home energy rating schemes (e.g. the US 'Home Energy Rating System' (RESNET, 2014)) typically assume that a house will use electricity or fuels for heating or cooling. In so doing, their specifications for efficiency and function can penalize and otherwise work against the construction of 'free-running' homes that do not rely on mechanical systems for heating or cooling, even where they are quite easy to build (Kordjamshidi and King, 2009). In the long run such standards can help institutionalize air conditioning (Cooper, 1998), which we discuss in greater detail in the next section on problems that are particularly intractable for a positivist approach.

9.4 **Challenges to positivism in energy demand research: towards other approaches**

Defining the problem of improving efficiency in terms of barriers and so framing the objectives for energy efficiency policy as barrier removal has a number of benefits (Wilson et al., 2013b). It offers a clear and tractable analytical framework for understanding energy efficient behaviour and investments. It speaks directly to policy concerns for energy savings or emission reductions. It identifies both options and potentials for low-cost information-based, financial, and regulatory interventions. It provides a clear and relatively finite set of explanatory variables for energy efficient technology adoption that harmonize well with stable of policy instruments available. Its repetitive use also ensures that it is easy to recall or cognitively 'available' (Wilson et al., 2013b).

However, the generalizable problem definition is itself problematic as it offers only a partial view of how energy use can be minimized and precludes interpretations and insights from a wealth of other perspectives. This argument has been made forcefully in relation to the dominant individualist and behavioural framing of UK climate change policy (Shove, 2010), but applies equally to the closely related CSC and MACC interpretations of energy efficiency (e.g. DECC (2012)). Chapter 7 illustrated how different perspectives on, and problem definitions associated with, energy use lead to very different analyses. Energy services, social practices, and socio-technical transitions were offered as examples to contrast with the PTEM perspective associated with CSCs and market barriers. Identifying and removing barriers to cost-effective energy-saving technologies is not an effective solution if the problem is reframed, for instance, in terms of energy providing useful services that enable normal and socially acceptable activities to be carried out as part of routine domestic life (Wilhite et al., 2000). In the next section, we consider some issues that a more complete framing of energy demand reduction could grapple with that the conventional approach cannot.

9.4.1 SOME LESS TRACTABLE PROBLEMS: COOLING AND SUPERSIZING

The use of space conditioning technologies to make homes comfortable is a good example of how energy consumption has become configured and embedded in expectations and routines. These expectations and routines resist the instrumental reach of policies intended to remove market barriers. Households' desires and expectations for thermal comfort have evolved over time (Shove, 2003). So too have the design of houses (e.g. room sizes, window area, opportunities for passive cooling), energy technologies (e.g. furnaces, thermostats), supporting infrastructure and institutions (e.g. electricity grids, utility tariffs, and services), as well as social norms (e.g. indoor temperatures, room occupancy profiles) (Wilhite et al., 2000; Wilhite, 2008). These changes in norms and technologies affect one another and drive further change.

In the three decades from 1962–92, use of air-conditioning spread from 12 per cent to 64 per cent of US homes (Kempton and Lutzenhiser, 1992), and by 2001 had risen to 75 per cent of households (EIA, 2001). In 2009, central air conditioning was a standard feature in 88 per cent of new single family homes (U.S. Census Bureau, 2012: Table 971, p. 610). The penetration of AC units into Australian homes has similarly doubled to about 65 per cent in the past decade (Strengers, 2012). Nor are these trends confined to developed economies. Citing unpublished work by Aufhammer, Wolfram, and colleagues note that in urban China there were 8 air-conditioning (AC) units per 100 households in 1995, but 106 units per 100 households in 2009 (Wolfram et al., 2012).

The availability and adoption of air-conditioning technologies has led to changes in the way homes are designed and constructed: verandas, eaves, thermal mass, and other means of passive cooling, have ceased to be integral features (Wilhite, 2008). These changes in design have had a 'ratchet' effect on the need for air-conditioning as part of a broader normalization of the ability to control and customize the indoor thermal environment (Shove, 2003).

Prior to the availability and affordability of mechanical or electrical heating and cooling, passive building design was commonplace. Passive design uses building architecture and construction to collect or deflect solar energy as heat and so does not rely on mechanical or electrical heating and cooling systems (Anderson and Michal, 1978). Examples of passive designs that have high impact in terms of energy savings include: long east/west building axes to maximize solar gain on the south-facing side; sizing and shading windows to face the midday sun in the winter and be shaded in the summer; using shading elements (e.g. shrubbery, trees, trellises) to protect against solar heat gain in the summer while promoting it in the winter; and insulating building envelopes to reduce heat gain or loss (Lechner, 2008). Longitudinal studies of the decline in passive design strategies are not available, neither within country nor as cross-country comparisons. However, Kruzner et al. (2013) use a nationally-representative snapshot of 1,000 existing homes in the USA to evaluate whether passive design strategies in both warm and cool climate regions were widely used. No significant national trends were found towards passive design in terms of orientation, roof colour, or level of shading.

The ready availability and relative affordability of mechanical or electrical heating and cooling systems helps explain this current snapshot of the existing building stock: 'competing design considerations (e.g. current styles, the status-quo) may receive more priority...rather than optimize the passive design of a home in a hot climate, for example, a designer can just specify a more powerful air conditioning unit' (Kruzner et al., 2013: 83). The loss of passive design techniques in home design and construction may be particularly marked in the USA as retail electricity and natural gas costs are very roughly half those of Europe or Japan (EIA, 2010, 2012).

The relatively recent rise of air-conditioning in homes usefully illustrates the inter-twining or 'co-evolution' of technological and social influences on energy demand. Changing practices of household cooling may be conventionally seen as the immediate drivers of energy demand. But these practices can only be understood as part of the co-evolution of technologies and material infrastructures, expectations for comfort, health, and well-being associated with indoor climate and temperature, and practical knowledge about how to cool the body and home with or without air-conditioning (Shove, 2003; Strengers, 2012). As described above, efficiency standards and policies can be complicit in these evolutions by inadvertently promoting air-conditioning itself while directly promoting efficient air-conditioners.

Air-conditioning is just one example of the complex socio-technical relationships in the built environment. Another US study notes the diminishing craftsmanship in house construction and design leading not just to the loss of passive design features but also to ever-larger homes with ever-higher energy service demand under a 'bigger is better' rubric (Wilson and Boehland, 2005). Between 1970 and 2009, the size of the average new American home has climbed more than 50 per cent (U.S. Census Bureau, 1995, 2012). Over the last 40 years, efficiency gains have been outpaced by increases in the size, number, features, and use of energy-consuming equipment (Moezzi, 1998; Rudin, 2000). This supersizing of expectations has led some energy efficiency advocates to recommend policy targets based on absolute levels of energy consumption rather than on levels of efficiency (Harris et al., 2006). However, the difficulty remains that the conventional approach to energy efficiency does not have a wide-angle lens that can see beyond 'modest, incremental, widget-improving solutions' to look at 'important problems [...] and significant struggles between interests and ideas, with money, power, and the environment at stake' (Lutzenhiser, 2014: 149; Sovacool, 2008).

9.4.2 ALTERNATIVE APPROACHES

If the conventional approach to energy efficiency cannot address broader social trends, are there alternative approaches that can? What if policy was to take a non-positivist approach that recognizes and supports a broader range of practices rather than relying on 'average behaviour' as a 'rule of thumb' or assuming the best choice is always the most cost-effective as calculated by an energy expert? Different community energy groups, for instance, can hold quite different conceptions of 'sustainable energy'. Some community groups install photovoltaic (PV) panels; some retrofit houses; others change their behaviours; or some combination of these actions (Seyfang et al., 2013). A PTEM perspective informed by a CSC cannot easily make sense of these multiple approaches, since behaviour change is the most economically viable option, and housing retrofits should pay back faster than installing PV (in the absence of feed-in-tariffs). In practice, however, PV is seen as a worthy upgrade; insulation is boring; and behaviour change is a nuisance. The social meanings do not often correlate with the costs. Previous work has shown that even within energy efficiency projects where incremental costs of efficient technologies are covered, some people do not always choose the 'most efficient' solution (Janda, 1998). Instead, some people may choose what they believe to be marketable, reliable, or comfortable. In some cases, people might also choose what is sustainable and sufficient instead of 'efficient' (Princen, 2005). These examples illustrate some of the many ways in which what people do (and why) can 'overflow' the conventional economic framework (Callon, 1998) provided by PTEM.

Can social practices be changed to be better aligned with economic analyses? Possibly, but it would take an approach that is broader and deeper than either the market 'barrier' approaches discussed above or even the 'behaviour change' efforts that are currently underway. Many of the behaviour change efforts take their cue from the PTEM and try to get people to use technologies 'properly'. These approaches are generally focused on getting people at home to know and care about energy in ways that are predefined by energy experts (Moezzi and Janda, 2014; Janda and Moezzi, 2014). This is not the highest level of citizen participation, as defined by Arnstein (1969) in an early influential essay on the topic. On Arnstein's ladder of citizen participation (see Figure 9.2) this 'informing' approach is only the third rung out of eight. It falls into the 'tokenism' area, ranking just above 'manipulation' and 'therapy'. In the UK, interventions designed to promote 'behavioural change' have been subject to 'an increasingly centralised, top-down push towards a universalist approach' (Chatterton and Wilson, 2013: 2). The focus on 'subconscious, automatic decision-making processes and heuristics' (Chatterton and Wilson, 2013: 4), suggests some interventions aim at the lowest rung of the citizen participation ladder (e.g. Thaler and Sunstein's 'Nudge' (2008)). There are, however, alternatives. Options for encouraging higher levels of citizen participation include increasing the level of building literacy within the general population (Janda, 2011); treating energy use as an opportunity for creative participation from citizens (Moezzi and Janda, 2014); encouraging demand-side participation (Devine-Wright and Devine-Wright, 2004); and building a cohesive energy training and support infrastructure at the community level, modelled after public investment in community sports teams and infrastructure (Hamilton and Berry, 2013). Some additional promising research strands have been noted for citizen science (Moezzi and Janda, 2014), building communities (Axon et al.,

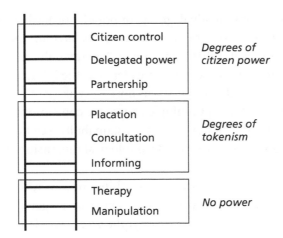

Figure 9.2 Ladder of citizen participation

Source: based on Arnstein (1969).

2012), and a 'middle-out' approach to energy transitions (Janda and Parag, 2013; Parag and Janda, 2014). A complementary concept of 'social potential' for energy savings could serve as a focal point for developing new tools and frameworks that invite a more active engagement of people—particularly in communities and groups—in helping define and address energy problems (Janda, 2014).

Policymakers like to have an evidence base from which to work, and to be able to see the policies they develop as potentially spreading positive impacts across the general population. However, in practice, voluntary participation is usually quite segmented, and the outcomes are difficult to predict. From a policy perspective, is there an efficient way to foster niches without over-favouring some particular group or groups? How can policymakers justify support for social processes that have less definable, more obviously uncertain outcomes as compared to ostensibly predictable technical ones? Some of these tensions are illustrated in a fictional conversation between a social scientist and a policymaker (Shove, 2012). One of the conclusions from this dialogue is that policymakers will also have to learn to see things in a different way, one that uses a broader range of the perspectives that social science has to offer.

To further open up thinking beyond conventional approaches, policy practices and efficiency goals could be reshaped to include different evidence and ideas from beyond the economic social sciences. A much richer involvement of social scientists in understanding how people use energy and in shaping relevant policies (not to mention buildings, things, discourses, and information) is required to recast energy problems and to find more effective ways of addressing these problems (Lutzenhiser, 2014; Sovacool, 2014; Stern, 2014). For example, Chatterton and Wilson (2013) suggest a 'four dimensions of behaviour' framework designed to facilitate understanding and discussion of observable behaviours across different disciplines. This framework focuses on who is behaving (e.g. an individual, group, or population); what shapes the behaviour; how the behaviour relates to time; and how the behaviour inter-relates with other behaviours. Stephenson et al. (2010) propose an 'energy cultures' framework that suggests energy behaviour could usefully be considered as the interactions between material culture, energy practices, and cultural/social norms. Wilk (2002) proposes a 'heterodox multigenic theory' that can accept multiple causes for consumption at different levels of analysis. More broadly, the newer field of sustainability science is exploring similar questions about the social, political, and technological dimensions of linking knowledge and action (Miller et al., 2013).

Alternative approaches need not be over-arching, however, to be useful. Integrating conventional and alternative approaches may allow us to gather richer insights from practical case studies. For example, a 'realist synthesis' approach, borrowed from the field of evidence-based health care (Pawson et al., 2004), is being applied to a study of the smart meter rollout in the UK (Darby, 2014). This approach aims to find out what works in different contexts, for whom, and how by integrating evidence that relates to context as well as to the actual intervention—a complex change introduced into a complex situation. Another example involves the explicit use of different narrative forms at different stages of energy efficiency projects (Janda and Topouzi, 2013). The positivist 'hero story' wins the funding, but an interpretivist 'learning story' helps mediate the lessons learned in practice, as well as valuing the multiple perspectives of building users, rather than just the goals of the design or research teams. A similar story-telling approach is being used in an International Energy Agency demand side management research programme on behaviour change to help relate theory to practice (Mourik and Rotmann, 2014).

Policies and research that focus either on technologies or behaviours are often implemented, while the full creative power of people and resulting diversity of possible solutions may be overlooked. Social processes are gradually receiving more attention in technical realms, but there is far less attention paid to social process than behaviour, and far less attention paid to behaviour than to technology. For example, in the recent Global Energy Assessment, there is a section on 'Social, Professional, and Behavioural Opportunities and Challenges'. However, this section represents only 4 pages out of the 125 devoted to technical and economic energy and building issues (Urge-Vorsatz et al., 2011). The energy demand research agenda in the UK also shows some indicative change in expanding its approach beyond the PTEM. Two out of the six 'end use energy demand centres' supported by major grants from the UK Research Councils in 2012 draw upon theoretical traditions—practice theories and transitions theory (see Chapter 7)—that are largely outside the PTEM. Although the links between research agendas and policy practices are neither seamless nor entirely direct, these centres aim to contribute to both research and policy. Hence, even though it is difficult to know precisely what insights these alternative theoretical approaches will deliver, it is reasonable to assume the existence and work of these differently-oriented centres will affect both research and policy practices.

The examples outlined above provide brief illustrations of how policy and research can expand beyond the PTEM to include a broader view of people and their potential contributions to deeper, broader, changes in societal energy use. These movements are still unwieldy with respect to typical evaluation frameworks, the funding mechanisms that are available to understand and promote them, and the formal scalability of perceived solution sets. But they also help capture a revised view of what people might do, and why they might do it, that can enrich and expand beyond more consumer-oriented viewpoints on energy use. In better being able to see, acknowledge, study, and debate these possibilities, energy policy, programmes, and research may be better able to support a more realistic and more powerful view of where people fit.

9.5 Conclusions and Next Steps

In this chapter, we addressed the question of providing solutions to policy problems associated with energy demand in the built environment from different perspectives. The dominant framing of the problem uses a Physical, Technical, and Economic Model (PTEM), and is associated with a positivist research epistemology. Alternative framings of the problem are growing in significance, employ various theoretical perspectives, and draw upon alternative, non-positivist epistemologies (which this chapter groups together under the term 'interpretivism' for convenience).

In other fields where positivism and other epistemologies are under discussion, there are a number of scholars who argue that positivism and its alternatives, such as inter-pretivism, are mutually exclusive. Some of these scholars argue for the supremacy of one approach over the other, others simply say that they are incompatible (Schweber and Leiringer, 2012). While they may be from incompatible worldviews, in our view research from both approaches usefully contributes to the energy demand field. As the positivist approach dominates the field, we would like to see it more fully complemented by alternative approaches. And we would welcome healthy programmes of such alternative research to help build its utility in the policy sphere.

Moving from a singular reality populated by average individuals to a reality populated with multiple perspectives need not mean moving from one solution to an infinite number of possibilities. Non-positivist approaches could help develop new categories for policy segmentation models, namely for actors that are understudied in the energy system, such as middle agents (Janda and Parag, 2013) and organizations (Cooremans, 2011, 2012; Janda, 2014). Non-positivist approaches also point at alternative framings of the topics studied as well as at novel policy instruments. More importantly, it could help re-engage populations (both researchers and respondents) in broader and more participatory discussions on how to improve both energy efficiency and broader forms of social, economic, and ecological sustainability in the built environment.

■ REFERENCES

ADJEI, A., HAMILTON, L., & ROYS, M. 2011. A study of homeowners' energy efficiency improvements and the impact of the Energy Performance Certificate, Hertfordshire, UK, BRE.

ANDALORO, A. P. F., SALOMONE, R., IOPPOLO, G., & ANDALORO, L. 2010. Energy certification of buildings: A comparative analysis of progress towards implementation in European countries. *Energy Policy*, 38, 5840–66.

ANDERSON, B. & MICHAL, C. 1978. Passive solar design. *Annual Review of Energy and the Environment*, 3, 57–100.

ARNSTEIN, S. R. 1969. A ladder of citizen participation. *Journal of the American Institute of Planners*, 35, 216–24.

AXON, C. J., BRIGHT, S. J., DIXON, T. J., JANDA, K. B., & KOLOKOTRONI, M. 2012. Building communities: reducing energy use in tenanted commercial property. *Building Research & Information*, 40, 461–72.

BARTIAUX, F. 2009. Changing energy-related practices and behaviours in the residential sector: Sociological approaches. EFONET Paper WS7.2. Energy Foresight Network (EFONET).

BARTIAUX, F. 2011. A qualitative study on home energy-related renovation in five European countries: homeowners' practices and opinions, Louvain-La-Neuve (Belgium). IDEAL EPBD: Louvain-La-Neuve (Belgium).

BARTIAUX, F., GRAM-HANSSEN, K., FONSECA, P., OZOLIŅA, L., & CHRISTENSEN, T. H. 2014. A practice–theory approach to homeowners' energy retrofits in four European areas. *Building Research & Information*, 42, 525–38.

BARTIAUX, F., VEKEMANS, G., GRAM-HANSSEN, K., MAES, D., CANTAERT, M., SPIES, B. T., & DESMEDT, J. 2006. Socio-technical factors influencing Residential Energy Consumption (SEREC), D/2005/1191/9. Final Report - SEREC Project CP-52. Belgian Science Policy: Brussels.

BELL, C. J., NADEL, S., & HAYES, S. 2011. On-Bill Financing for Energy Efficiency Improvements: A Review of Current Program Challenges, Opportunities, and Best Practices. American Council for an Energy Efficient Economy (ACEEE): Washington, DC.

BIOREGIONAL 2011. Helping to inform the Green Deal: green shoots from Pay As You Save. Bioregional, with B&Q and the London Borough of Sutton: Wallington, Surrey.

BLAIKIE, N. 2004. Interpretivism. *In*: LEWIS-BECK, M., BRYMAN, A. & LIAO, T. (ed.) *Encyclopedia of Social Science Research Methods*. Thousand Oaks, CA, SAGE Publications, Inc.

BLUMSTEIN, C. & STOFT, S. E. 1995. Technical efficiency, production functions and conservation supply curves. *Energy Policy*, 23, 765–8.

BORGESON, M., ZIMRING, M., & GOLDMAN, C. The Limits of Financing for Energy Efficiency. ACEEE Summer Study on Energy Efficiency in Buildings, 2012 Asilomar, CA, American Council for an Energy Efficient Economy (ACEEE).

BROPHY HANEY, A., JAMASB, T., PLATCHKOV, L. M., & POLLITT, M. G. 2011. Demand-side Management Strategies and the Residential Sector: Lessons from the International Experience. *In*: JAMASB, T. & POLLITT, M. G. (eds.) *The Future of Electricity Demand: Customers, Citizens and Loads*. Cambridge, UK, Cambridge University Press.

BROWN, M., CHANDLER, J., & LAPSA, M. V. 2010. Adding a Behavioral Dimension to Utility Policies that Promote Residential Efficiency. *In*: EHRHARDT-MARTINEZ, K. & LAITNER, J. A.

(eds) *People-Centred Initiatives for Increasing Energy Savings.* Washington, DC, American Council for an Energy Efficient Economy (ACEEE).

BROWN, M., CHANDLER, J., LAPSA, M. V. & ALLY, M. 2009. Making Buildings Part of the Climate Solution: Policy Options to Promote Energy Efficiency. Oak Ridge, TN, Oak Ridge National Laboratory.

CALLON, M. 1998. An Essay on Framing and Overflowing: Economic Externalities Revisited by Sociology. *In:* CALLON, M. (ed.) *The Laws of the Markets.* Oxford & Keele, Blackwell Publishers & Sociological Review.

CHATTERTON, T. & WILSON, C. 2013. The 'Four Dimensions of Behaviour' framework: a tool for characterising behaviours to help design better interventions. *Transportation Planning and Technology*, 37, 38–61.

COOPER, G. 1998. *Air-conditioning America: Engineers and the Controlled Environment, 1900–1960.* Baltimore, John Hopkins University Press.

COOREMANS, C. 2012. Investment in energy efficiency: do the characteristics of investments matter? *Energy Efficiency*, 5, 497–518.

COOREMANS, C. 2011. Make it strategic! Financial investment logic is not enough. *Energy Efficiency*, 4, 473–92.

DARBY, S. 2014. What does a successful technology introduction look like? Developing a consistent means of evaluating smart meter rollout *Behave.* Oxford, The UK Energy Research Centre.

DAVIS, L., WRIGHT, A. & GERTLER, P. 2013. Cash for Coolers: Evaluating a Large-Scale Appliance Replacement Program in Mexico, Berkeley, Energy Institute at Haas: <http://cega. berkeley.edu/assets/cega_research_projects/71/Cash_for_Coolers_Evaluating_the_Impact_ of_an_Applicance_Replacement_Program_in_Mexico. > (Accessed April 2015).

DECC 2010. The Green Deal: A Summary of the Government's Proposal. Department of Energy and Climate Change (DECC), London, UK.

DECC 2011. The Green Deal and Energy Company Obligation: Consultation Document. Department of Energy and Climate Change (DECC), London, UK.

DECC 2012. The Energy Efficiency Strategy: The Energy Efficiency Opportunity in the UK. Department of Energy and Climate Change (DECC), London, UK.

DECC 2013. Domestic Green Deal and Energy Company Obligation in Great Britain. Department of Energy and Climate Change (DECC), London, UK.

DEVINE-WRIGHT, H. & DEVINE-WRIGHT, P. 2004. From demand side management to demand side participation: tracing an environmental psychology of sustainable electricity system evolution. *Journal of Applied Psychology*, 6, 167–77.

DIETZ, T., GARDNER, G. T., GILLIGAN, J., STERN, P. C., & VANDENBERGH, M. P. 2009. Household actions can provide a behavioral wedge to rapidly reduce US carbon emissions. *Proceedings of the National Academy of Sciences*, 106, 18452–6.

EIA. 2001. Residential Energy Consumption Survey. US Energy Information Administration (EIA): Washington, DC.

EIA. 2010. Electricity Prices for Households for Selected Countries. US Energy Information Administration (EIA), Washington, DC.

EIA. 2012. Natural Gas Prices for Households. US Energy Information Administration (EIA), Washington, DC.

EMMERT, S., VAN DE LINDT, M., & LUITEN, H. 2010. BarEnergy. Barriers to change in energy behaviour among end consumers and households. Integration of Three Empirical Studies. Organisation for Applied Scientific Research (TNO), Delft, the Netherlands.

ENKVIST, P.-A., NAUCLER, T., & ROSANDER, J. 2007. A cost curve for greenhouse gas reduction. *The McKinsey Quarterly*, 1, 35–45.

FLEITER, T., EICHHAMMER, W., & WIETSCHEL, M. Costs and potentials of energy savings in European industry—a critical assessment of the concept of conservation supply curves. ECEEE Summer Study, 2009 Toulon, France. European Council for an Energy Efficient Economy (ECEEE), 1261–72.

FULLER, M. 2009. Enabling Investments in Energy Efficiency: A study of energy efficiency programs that reduce first-cost barriers in the residential sector. University of California, Berkeley Berkeley, CA.

GILLINGHAM, K. & PALMER, K. 2014. Bridging the energy efficiency gap: policy insights from economic theory and empirical evidence. *Review of Environmental Economics and Policy*, 8, 18–38.

GRAM-HANSSEN, K., BARTIAUX, F., MICHAEL JENSEN, O., & CANTAERT, M. 2007. Do homeowners use energy labels? A comparison between Denmark and Belgium. *Energy Policy*, 35, 2879–88.

GRUBB, M. 2014. Why so Wasteful? *In*: GRUBB, M., HOURCADE, J.-C., & NEUHOFF, K. (eds.) *Planetary Economics: Energy, Climate Change and the Three Domains of Sustainable Development*. London, UK, Earthscan.

GUY, S. & SHOVE, E. 2000. *A Sociology of Energy, Buildings, and the Environment*, London, Routledge.

HAMILTON, J. & BERRY, S. 2013. Rethinking energy efficiency delivery—what can we learn from sport? ECEEE Summer Study, 3–8 June 2013 Presqu'île de Giens, France. European Council for an Energy-Efficient Economy, 323–31.

HANAOKA, T. & KAINUMA, M. 2012. Low-carbon transitions in world regions: comparison of technological mitigation potential and costs in 2020 and 2030 through bottom-up analyses. *Sustainability Science*, 7, 117–37.

HARRIS, J., DIAMOND, R., IYER, M., & PAYNE, C. 2006. Don't Supersize Me! Toward a Policy of Consumption-Based Energy Efficiency. ACEEE Summer Study on Energy Efficiency in Buildings, 2006 Asilomar, CA, American Council for an Energy-Efficient Economy.

HARVEY, L. D. D. 2010. *Energy Efficiency and the Demand for Energy Services*, London, Earthscan Ltd.

HEALY, S. 2008. Air-conditioning and the 'homogenization' of people and built environments. *Building Research & Information*, 36, 312–22.

HUDSON, L. A. & OZANNE, J. L. 1988. Alternative ways of seeking knowledge in consumer research. *Journal of Consumer Research*, 14, 508–21.

JAFFE, A. B. & STAVINS, R. N. 1994. The energy efficiency gap: what does it mean? *Energy Policy*, 22, 804–10.

JANDA, K. B. & BUSCH, J. F. 1994. Worldwide status of energy standards for buildings. *Energy*, 19, 27–44.

JANDA, K. B. 1998. *Building change: effects of professional culture and organizational context on energy efficiency adoption in buildings*. Dissertation, University of California at Berkeley.

JANDA, K. B. 2009. Worldwide Status of Energy Standards for Buildings: A 2009 Update. ECEEE Summer Study, June 1–6, 2009 Colle Sur Loop, France. European Council for an Energy-Efficient Economy, 485–91.

JANDA, K. B. 2011. Buildings don't use energy: people do. *Architectural Science Review*, 54, 15–22.

JANDA, K. B. 2014. Building communities and social potential: between and beyond organisations and individuals in commercial properties. *Energy Policy*, 67, 48–55.

JANDA, K. B. & MOEZZI, M. 2014 Broadening the Energy Savings Potential of People: From Technology and Behavior to Citizen Science and Social Potential. Proceedings of ACEEE Summer Study on Energy Efficiency in Buildings, August 17–22, 2014 (Asilomar, CA). Vol. 7, pp. 133–46. American Council for an Energy Efficient Economy.

JANDA, K. B. & PARAG, Y. 2013. A middle-out approach for improving energy performance in buildings. *Building Research & Information*, 41, 39–50.

JANDA, K. B. & TOPOUZI, M. 2013. Closing the loop: using hero stories and learning stories to remake energy policy. ECEEE Summer Study, 3–8 June 2013 Presqu'île de Giens, France. European Council for an Energy-Efficient Economy.

JENSEN, O. M. 2005. Consumer inertia to energy saving. ECEEE Summer Study, 30 May–4 June, 2005 Mandelieu, France. European Council for an Energy-Efficient Economy, 1327–34.

KARVONEN, A. 2013. Towards systemic domestic retrofit: a social practices approach, Building Research & Information, 41:5, 563–74.

KEMPTON, W. & LUTZENHISER, L. 1992. Introduction. *Energy and Buildings*, 18, 171–6.

KORDJAMSHIDI, M. & KING, S. 2009. Overcoming problems in house energy ratings in temperate climates: A proposed new rating framework. *Energy and Buildings*, 41, 125–32.

KRUZNER, K., COX, K., MACHMER, B., & KLOTZ, L. 2013. Trends in observable passive solar design strategies for existing homes in the U.S. *Energy Policy*, 55, 82–94.

LECHNER, N. 2008. *Heating, Cooling and Lighting: Sustainable Design Methods for Architects*, Hoboken, NJ, John Wiley & Sons.

LEVINE, M., ÜRGE-VORSATZ, D., BLOK, K., GENG, L., HARVEY, D., LANG, S., LEVERMORE, G., MONGAMELI MEHLWANA, A., MIRASGEDIS, S., NOVIKOVA, A., RILLING, J., & YOSHINO, H. 2007. Residential and commercial buildings. *In*: METZ, B., DAVIDSON, O. R., BOSCH, P. R., DAVE, R., & MEYER, L. A. (eds) *Climate Change 2007: Mitigation. Contribution of Working Group III to the Fourth Assessment Report of the Intergovernmental Panel on Climate Change*. Cambridge, UK and New York, USA, Cambridge University Press.

LIN, J. & FRIDLEY, D. 2006. Harmonization of energy efficiency standards: searching for common ground. ACEEE Summer Study on Energy Efficiency in Buildings, 2006 Asilomar, CA. American Council for an Energy-Efficient Economy, 9.178–9.186.

LIN, J. & IYER, M. 2007. Cold or hot wash: Technological choices, cultural change, and their impact on clothes-washing energy use in China. *Energy Policy*, 35, 3046–52.

LUTZENHISER, L. 2002. Marketing Household Energy Conservation: The Message and the Reality. *In*: DIETZ, T. & STERN, P. C. (eds) *New Tools for Environmental Protection: Education, Information, and Voluntary Measures*. Washington, DC, National Academy Press.

LUTZENHISER, L. 2014. Through the energy efficiency looking glass: Rethinking Assumptions of Human Behavior. *Energy Research and Social Science*, 1, 141–51.

LUTZENHISER, L. & SHOVE, E. 1999. Contracting knowledge: the organizational limits to interdisciplinary energy efficiency research and development in the US and the UK. *Energy Policy*, 27, 217–27.

MCCORMACK, D. P. & SCHWANEN, T. 2011. The space–times of decision making. *Environment and Planning A*, 43, 2801–18.

MCKINSEY 2009. Pathways to a low-carbon economy. Version 2 of the Global Greenhouse Gas Abatement Cost Curve. London, McKinsey and Company.

MEIER, A. K. 1982. *Supply curves of conserved energy*. PhD, University of California at Berkeley.

MILLER, T., WIEK, A., SAREWITZ, D., ROBINSON, J., OLSSON, L., KRIEBEL, D., & LOORBACH, D. 2013. The future of sustainability science: a solutions-oriented research agenda. *Sustainability Science*, 9, 239–46.

MOEZZI, M. 1998. The Predicament of Efficiency. 1998 ACEEE Summer Study on Energy Efficiency in Buildings, 1998 Washington DC. American Council for an Energy-Efficient Economy, 4.273–4.283.

MOEZZI, M. & JANDA, K. B. 2014. From 'if only' to 'social potential' in schemes to reduce building energy use. *Energy Research and Social Science*, 1, 30–40.

MOURIK, R. & ROTMANN, S. 2014. *Task 24: The Monster Storybook Subtask 1 analysis of IEA DSM Task 24: Closing the Loop: Behaviour Change in DSM—From Theory to Practice*, Paris, IEA.

PARAG, Y. & JANDA, K. B. 2014. More than filler: middle actors and socio-technical change in the energy system from the 'middle-out'. *Energy Research and Social Science*, 3, 102–12.

PAWSON, R., GREENHALGH, T., HARVEY, G., & WALSHE, K. 2004. Realist Synthesis: An Introduction, RMP Methods Paper 2/2004. Manchester, University of Manchester.

PEFFER, T., PERRY, D., PRITONI, M., ARAGON, C., & MEIER, A. 2013. Facilitating energy savings with programmable thermostats: evaluation and guidelines for the thermostat user interface. *Ergonomics*, 56, 463–79.

PRINCEN, T. 2005. *The Logic of Sufficiency*, Cambridge, MA, MIT Press.

RESNET. 2014. *What is the HERS Index?* [Online]. Oceanside, CA: Residential Energy Services Network (RESNET). Available: <http://www.resnet.us/hers-index> [Accessed 24 July 2014].

ROSENOW, J., CROFT, D., & EYRE, N. 2013. Energy policy in transition: evidence from energy supply and demand in the UK. ECEEE Summer Study (European Council for an Energy Efficient Economy), 2013 Toulon, France, pp. 439–47.

RUDIN, A. 2000. Why We Should Change Our Message and Goal from 'Use Energy Efficiently' to 'Use Less Energy'. ACEEE Summer Study on Energy Efficiency in Buildings, 2000 Asilomar, CA. American Council for an Energy-Efficient Economy, 8.330–8.340.

SAMIULLAH, S., HUNGERFORD, D., & KANDEL, A. 2002. Do Central Air Conditioner Rebates Encourage Adoption of Air Conditioning? American Council for an Energy-Efficient Economy 2002 Summer Study, 2002 Asilomar, CA. ACEEE, 8.253–8.263.

SCHWEBER, L. & LEIRINGER, R. 2012. Beyond the technical: a snapshot of energy and buildings research. *Building Research & Information*, 40, 481–92.

SEYFANG, G., PARK, J. J., & SMITH, A. 2013. A thousand flowers blooming? An examination of community energy in the UK. *Energy Policy*, 61, 977–89.

SHOVE, E. 1998. Gaps, barriers and conceptual chasms: theories of technology transfer and energy in buildings. *Energy Policy*, 26, 1105–12.

SHOVE, E. 2003. *Comfort, Cleanliness, and Convenience: The Social Organisation of Normality.* Oxford, UK, Berg.

SHOVE, E. 2010. Beyond the ABC: climate change policy and theories of social change. *Environment and Planning A*, 42, 1273–85.

SHOVE, E. 2012. Putting practice into policy: reconfiguring questions of consumption and climate change. *Contemporary Social Science*, 9 (4), 415–29.

SHOVE, E. & MOEZZI, M. 2004. What do standards standardize? American Council for an Energy-Efficient Economy 2004 Summer Study, August 22–27, 2004 Asilomar, CA. Washington, DC, ACEEE.

SHOVE, E., LUTZENHISER, L., GUY, S., HACKETT, B., & WILHITE, H. 1998. Energy and Social Systems. *In*: RAYNER, S. & MALONE, E. L. (eds) *Human Choice and Climate Change.* Columbus, Ohio, Battelle Press.

SOVACOOL, B. K. 2008. *The Dirty Energy Dilemma: What's Blocking Clean Power in the United States.* Westport, CT, Praeger.

SOVACOOL, B. K. 2014. What are we talking about? Analyzing fifteen years of energy scholarship and proposing a social science research agenda. *Energy Research and Social Science*, 1, 1–29.

STEPHENSON, J., BARTON, B., CARRINGTON, G., GNOTH, D., LAWSON, R., & THORSNES, P. 2010. Energy cultures: A framework for understanding energy behaviours. *Energy Policy*, 38, 6120–9.

STERN, P. C. 2014. Individual and household interactions with energy systems: toward integrated understanding. *Energy Research and Social Science*, 1, 41–8.

STRENGERS, Y. 2012. Peak electricity demand and social practice theories: Reframing the role of change agents in the energy sector. *Energy Policy*, 44, 226–34.

THALER, R. & SUNSTEIN, C. R. 2008. *Nudge: Improving Decisions about Health, Wealth, and Happiness*, New Haven, CT, Yale University Press.

U.S. CENSUS BUREAU. 1995. Statistical Abstract of the United States: 1995. Washington DC. <http://www.census.gov/compendia/statab/2012/tables/12s0971.pdf>. (Accessed April 2015).

U.S. CENSUS BUREAU. 2012. Statistical Abstract of the United States: 2012. Washington DC. <http://www.census.gov/compendia/statab/2012/tables/12s0971.pdf>. (Accessed April 2015).

ÜRGE-VORSATZ, D., EYRE, N., GRAHAM, P., HARVEY, D., HERTWICH, E., JIANG, Y., KORNEVALL, C., MAJUMDAR, M., MCMAHON, J. E., MIRASGEDIS, S., MURAKAMI, S., & NOVIKOVA, A. 2012. *Energy End-Use: Buildings. Global Energy Assessment.* Cambridge, UK, Cambridge University Press.

URGE-VORSATZ, D., EYRE, N., GRAHAM, P., HARVEY, L. D. D., HERTWICH, E., KORNE-VALL, C., MAJUMDAR, M., MCMAHON, J., MIRASGEDIS, S., MURAKAMI, S., NOVI-KOVA, A., & JANDA, K. B. 2011. *Knowledge Module 10: Energy End Use: Buildings. The Global Energy Assessment (GEA).* Laxenburg, Austria, IIASA.

VINE, E. 2005. An international survey of the Energy Service Company (ESCO) industry. *Energy Policy*, I33, 691–704.

WEISS, J., DUNKELBERG, E., & VOGELPOHL, T. 2012. Improving policy instruments to better tap into homeowner refurbishment potential: Lessons learned from a case study in Germany. *Energy Policy*, 44, 406–15.

WILHITE, H. 2008. New thinking on the agentive relationship between end-use technologies and energy-using practices. *Energy Efficiency*, 1, 121–30.

WILHITE, H., SHOVE, E., LUTZENHISER, L., & KEMPTON, W. 2000. The Legacy of Twenty Years of Energy Demand Management: We Know More About Individual Behaviour But Next to Nothing About Demand. *In*: JOCHEM, E., SATHAYE, J., & BOUILLE, D. (eds) *Society, Behaviour, and Climate Change Mitigation*. Dordrecht, The Netherlands, Kluwer Academic Publishers.

WILK, R. 2002. Consumption, human needs, and global environmental change. *Global Environmental Change*, 12, 5–13.

WILKINSON, P., SMITH, K. R., DAVIES, M., ADAIR, H., ARMSTRONG, B. G., BARRETT, M., BRUCE, N., HAINES, A., HAMILTON, I., ORESZCZYN, T., RIDLEY, I., TONNE, C., & CHALABI, Z. 2009. Public health benefits of strategies to reduce greenhouse-gas emissions: household energy. *The Lancet*, 374, 1917–29.

WILSON, A. & BOEHLAND, J. 2005. Small is beautiful: US house size, resource use, and the environment. *Journal of Industrial Ecology*, 9, 277–87.

WILSON, C., CRANE, L., & CHRYSSOCHOIDIS, G. 2013a. The conditions of normal domestic life help explain homeowners' decisions to renovate. ECEEE Summer Study (European Council for an Energy Efficient Economy), 2013 Toulon, France, pp. 2333–47.

WILSON, C., CRANE, L., & CHRYSSOCHOIDIS, G. 2013b. Why do people decide to renovate their homes to improve energy efficiency? Working Paper. Norwich, UK.

WOLFRAM, C., SHELEF, O., & GERTLER, P. 2012. How will energy demand develop in the developing world? *Journal of Economic Perspectives*, 26, 119–38.

WRIGHT, J., & ROSENFELD, A. 1983. Supplying energy through greater efficiency, University of California at Berkeley Los Angeles, CA.

10 Challenges and options for sustainable travel: mobility, motorization, and vehicle technologies

Hannah Daly, Paul E. Dodds, and Will McDowall

10.1 Introduction

The scale of the global transport system has expanded immensely with the growth of the world economy: The average daily motorized travel distance has increased from 3.9 to 16.5 PKT (passenger kilometres travelled) per person between 1950 and 2005, and is projected to almost double to 31.2 PKT by 2050 (Schäfer et al., 2009). While increased mobility has brought many benefits—affording people greater access to goods, services, and opportunities, improved communication and being a necessary condition for free trade and globalization—the hugely expanding transport system has brought pervasive, profound, and many negative impacts on society on local, regional, and global scales.

The energy and transport systems are deeply enmeshed. Transport accounts for 29 per cent of global final energy demand, making it the most significant end-use sector in terms of primary energy, and this demand is projected to increase up to 1.3–2 per cent annually between 2010 and 2035 (Kahn Ribeiro et al., 2007; IEA, 2012: 63). The growth in mobility has been made possible by cheap oil: this single commodity fuels 97 per cent of world transport by energy content. This reliance on oil is the cause of two major concerns, the security of energy supply and the rising contribution of transport to global climate change.

This chapter examines the challenges that are faced in mitigating the unsustainable pathway of transport demand, focusing on land passenger transport. We have not addressed freight transport, though trucks are projected to be an important component of oil demand growth (IEA, 2012). Aviation and shipping are dealt with separately, in Chapter 11.

Large-scale reductions in transport-related CO_2 emissions are required to meet ambitious climate targets. In the most ambitious scenario from the IEA's Mobility Model (MoMo), for example, transport emissions in 2050 are reduced by 25 per cent on 2005 levels (70 per cent lower than the baseline), but achieving this requires great changes to vehicle technologies, including 50 per cent or greater energy efficiency improvement in light duty vehicle (LDV) efficiency, 30–50 per cent improvement of efficiency of other motorized transport modes, 25 per cent substitution of fossil fuels by biofuels, and large-scale penetration of electric and/or fuel-cell drivetrains (IEA, 2009; Fulton et al., 2013). Other studies have produced comparable portfolios of transport technologies and fuels to meet decarbonization targets in transport (Schäfer and Jacoby, 2006; Girod et al., 2012, 2013). Section 10.3 looks at the choices for energy efficient and low-carbon drivetrain technologies and transport fuels and outlines the policy options and challenges for their wide-scale diffusion.

The dominant emphasis in transport energy studies has been on technology solutions (the PTEM approach outlined in Chapter 7). The role that behaviour, demand reduction, and modal shift can play in mitigating GHG emissions are frequently not addressed in transport energy models and studies, nor are the external and intangible costs of mobility and transport technologies. Section 10.2 addresses the role of travel demand and modal choice, and particularly the impact of growing car ownership, factors which are frequently not examined in detail by global studies. Section 10.4 focuses on the societal and policy drivers for sustainable transport, given the negative externalities, market failures, interdependencies, and co-benefits of different policies and choices, and considers how the transport system is evolving as a result of important trends in economic development and urbanization.

10.2 Travel demand and travel intensity

CO_2 emissions from passenger transport can be decomposed into four main drivers: *Total mobility*, (PKT); *travel intensity*, which is the number of vehicle kilometres travelled (VKT) per passenger kilometre and depends on the structure of modal demand and vehicle occupancy; the technology-specific *energy intensity*; and finally the fuel specific CO_2 *intensity*. This section looks at the first two drivers, specifically mobility and modal choice, which are interdependent and frequently overlooked in transport energy studies.

10.2.1 TRENDS IN DEMAND GROWTH AND MOTORIZATION

Motorized travel demand is projected to grow by an order of magnitude in the century between 1950 and 2050. The level of mobility is coupled strongly with income: the motorized distance travelled by citizens of industrialized countries in 2005 was about five times greater than citizens of developing countries, and this inequality is forecast to grow to 6.4 times by 2050 (Schäfer et al., 2009). People have a limited amount of time to travel, and it has been observed that across time and regions the average time a population spends travelling tends to be held constant at 1.1–1.2 hours per day, known as a travel time budget (Metz, 2010). With increased income, people take advantage of faster transport, and this constant travel time budget has been used to explain the historical shift away from slower non-motorized modes and public transport towards private car and air travel. Therefore, growth in passenger travel demand and in motorization have gone hand-in-hand to a large extent and should not be treated in isolation.

Several studies have projected a saturation of car demand in developed countries because of the diminishing returns for faster travel (Metz, 2010) and fuel price rises, constrained land use, and demographics (Millard-Ball and Schipper, 2010). However, these studies do not consider the contribution of international aviation: a fixed travel time and the desire for greater mobility will be met with ever-faster modes of transport. Three stages of mobility have been defined by Schäfer et al. (2009): below an annual mobility level of 1,000 PKT per person, low-speed public transport modes dominate. For annual mobility levels of between 1,000 and 10,000 PKT per person there is a strong rise in car use and public transport declines to between 10 and 30 per cent of total motorized travel. A third stage of mobility evolution is underway in North America, where air travel constitutes almost a fifth of transport demand, and is projected to continue to grow at a rapid rate.

The level of motorization and structure of travel demand matters greatly to energy and CO_2: the CO_2 intensity profile of different travel modes differ greatly. Walking and cycling do not contribute to fossil fuel use, while cars emit up to 5 times the CO_2 per PKT of public transport (Scholl et al., 1996). The shift away from public and non-motorized modes has caused a rise in total travel demand and the energy intensity of mobility, both of which have contributed to increases in CO_2 emissions.

Growing private car ownership leads to an increase in energy demand in several ways. Access to faster speeds allows people to travel greater distances within their time budget. Private car ownership is a capital intensive investment and is only available to households above a certain income. Once a household buys a car it tends to get locked into this mode, as the marginal cost of car travel is low once the initial investment is made. Growing car ownership decreases ride sharing and so causes a decrease in the load factor of cars, and so an increase in the energy use per PKT. Such an effect has dampened improvements in energy per PKT from the increased engine efficiency of cars (Scholl et al., 1996).

Increased motorization has largely been attributed to income growth. Chamon et al. (2008) describe a non-linear relationship between per-capita income and car ownership. Ownership rates are typically minimal in low-income countries, but increase rapidly as per-capita income rises above a threshold of US$5,000 (2,000 prices). Ownership rises with income also among developed countries, though saturation in car ownership is reached and air travel increases. This threshold is being reached by key emerging markets, in particular India and China (Wang et al., 2011), which together account for over a third of the world's population and at present 14 per cent of cars. All other variables being equal, if these two countries attained the per-capita car ownership levels of the USA, the global population of cars would triple to 3 billion. Indeed, Chamon et al.

(2008) calculate that the number of cars will increase by 250 per cent by 2050 with the majority of growth occurring in China and India. To meet a target of a net reduction in transport GHG emissions in that timeframe would require a massive shift to alternative transport fuels and drivetrains, even with great improvements to energy efficiency. The potential of technologies is explored in Section 10.3, and the social and political implications of this unchecked growth in car use is discussed in Section 10.4.

Beyond income, land-use and transport infrastructure also strongly influence travel patterns and motorization (Cervero and Kockelman, 1997). Urbanization is a trend which will affect the patterns of mobility demand: most of the world's population live in urban areas, and most future population growth will occur in urban areas. The dominant trends in cities across the world have been reductions in population densities and public transport mode share, and increases in home to work distance and automobile use (Lefèvre, 2009). Cities that are dense, mixed use, and highly structured through land use and transport planning have far lower levels of CO_2 emissions due to transport than sprawling unplanned cities. Transport plays a fundamental part in the development of urban areas: access to housing, employment, and education are all dependent on the transport system. Without containment of urban sprawl, increases in city size and a reduction in population density result in motorization and an increase in energy consumption (Newman and Kenworthy, 1989). Making cities 'liveable'—economically efficient, socially sound, and environmentally friendly—can be a driver for reducing the negative impacts of transport (Vuchic, 1999).

Rapid urbanization in developing countries leaves local governments unable to keep up with transport infrastructure, leading to insufficient public transport investment, and increases in motor use and congestion. The rate of car ownership grows faster than road space, decreasing the quality of the city space and increasing pollution, accident rates, noise, and journey times.

10.2.2 POLICIES AND DRIVERS FOR SUSTAINABLE TRAVEL PATTERNS

10.2.2.1 Facing up to Mobility Management

Growth in mobility and a high quality transport system are seen as essential prerequisites for economic growth, and consequently there is little political appetite for curbing mobility. The White Paper of the European Commission (2011) on transport, for example, recognizes the need to reduce the dependence of the transport system on oil, but considers growth in transport as vital for economic growth and the wellbeing of citizens.

However, the convention that the unbounded increase in mobility is a universal necessity and benefit has been questioned (Coombe, 1995). Not only is unbounded mobility not necessarily needed, but studies are increasingly showing that meaningful improvements in the sustainability of transport will not be met by technological solutions alone, but will require changes to mobility and patterns in transport demand (Chapman, 2007; Wee, 2014; Hickman and Banister, 2007).

Models and analysis used to look at long-term energy decarbonization generally fail to look at the true costs of mobility, and the potential for behaviour and demand measures are missing or very limited (for example, Fulton et al. 2009). The sole focus on technology results in significant oversights in transport energy models and hence advice to policymakers for reducing transport's harmful impacts. By making exogenous assumptions about mobility demand and mode choice, studies neglect to address the huge contribution that these variables play in determining energy demand in transport and the potential for mitigating emissions by investing in changing mobility practices (Daly et al., 2014). Further, mobility and transport technologies have vast effects on society, particularly on land use, safety, health, pollution, economic growth, and social cohesion and inclusion: by only focusing on CO_2 and direct costs, analyses of the values and barriers to different approaches are limited.

10.2.2.2 Planning, Land Use, and Infrastructure

A major driver of travel demand and motorization is the building of road infrastructure to enable car use. Traditional transport planning in developed countries has largely been based on accommodating the growth of the private car. The expansion of roads in the United Kingdom and the United States in the twentieth century coincided with the so-called 'predict and provide' model of transport planning, where the orthodoxy was to forecast future traffic and then build the road infrastructure to accommodate it and avoid traffic congestion. This approach in the UK has largely been discredited (Coombe, 1995): it is now accepted that it is not possible to match road capacity to forecasted demand and other solutions have to be found. Further, it has been found that building roads to relieve congestion is not effective, as over the long term people use the savings in travel time to travel greater distances, so in fact road building schemes induce travel (Metz, 2008). Transport planners in many countries continue to justify building road infrastructure largely on the basis of travel time savings, reasoning which has been criticized (Santos and Bhakar, 2006).

Managing land use is vital for reducing the long-term impact of transport. Allowing urban sprawl and one-off rural developments without access to services within walking or cycling distances will lock people in to years of motorized travel with no alternative options. The design of neighbourhoods plays a very strong part in mode choice and travel distance: measures for improving the safety and attractiveness of walking and cycling and for mixed land use can successfully reduce car travel. Land use regulations for promoting urban sustainable transport are outlined by Lefèvre (2009). Energy, transport, and land use goals and policies should be coordinated across government, for example by incorporating transport energy into urban planning tools (Saunders et al., 2008). Transport, urban, and land use planners must play a strong role in designing the built environment to enable lower levels of mobility and car use without sacrificing accessibility.

10.2.2.3 Influencing Travel Demand and Mode Choice

Like other final energy use, passenger transport is mainly a derived demand, a means of fulfilling the need for socializing, travelling to work, purchasing goods and services, and other ends. From this perspective, actions that currently are fulfilled through mobility can be met by other means. Substitutions for travel, particularly modern information and communications technologies, can frequently be a substitute for physical transport, with social networking and online communication, home working and online shopping. The promise of telecommunications as a substitute for mobility, however, has not decreased the demand for mobility as has been predicted in the past (Mokhtarian, 2009).

Low-occupancy car use is very inefficient in terms of energy use compared with high-occupancy public transport, high-occupancy car use, or non-motorized travel, but is most popular because of its convenience. Measures causing a decrease in 'travel intensity' (VKT/PKT) range from increasing the load factor of all modes to encouraging a shift from private car use to high-occupancy or non-motorized modes. Increasing the occupancy of private cars can be incentivized with high-occupancy vehicle lanes, workplace measures to encourage ride sharing, and the use of digital ride sharing, enabled by the internet (ITDP, 2010).

Changing price signals can be an effective way to reduce car use. Making the full external costs of transport internal to the cost to the user is an economically-efficient way of mitigating the damages of transport. The economic orthodoxy is to set a tax on the pollution or damage equal to the marginal damage cost, so that the level at which the costs are imposed by the marginal unit of damage is equal to the costs of its abatement (Owens, 1995). This exists currently in transport with fuel and (to a limited extent) CO_2, where these are significantly taxed, but the external costs associated with mobility in most countries are not internalized, except in limited cases with congestion charging. The outright increase in the charges of private cars and fuel prices can be regressive, however, particularly in developed countries where poor people are often locked into car ownership and dependent on private cars for access to employment. This can be addressed by making car charges and taxes based on use as opposed to ownership: a market failure of car pricing is the high ownership cost relative to usage cost. Once cars are bought they tend to be used in preference to other available modes, as the sunk capital investment cost and annual tax and insurance are not dependent on travel distance, therefore the marginal cost of car use is lower than the real per-kilometre cost. Measures to price and tax car ownership and use on a per-kilometre basis can be an economically efficient and progressive means of mitigating this effect.

Use-based pricing can take a number of forms: road user charging, which can take the form of urban congestion charges and road tolls, can be an economically efficient way of reducing congestion (World Bank, 1968). Congestion charges in cities have successfully reduced urban car use. Cost–benefit analyses have shown that congestion charges have yielded significant societal surpluses through reduced traffic delays, improved journey time reliability, reduced waiting time at bus stops, lower fuel consumption, less pollution and fewer accidents, and a more pleasant environment (Transport for London, 2007; Eliasson, 2009). It is possible that in the future, in-vehicle or on-road sensors could be used to calculate tax and insurance on the basis of car use, and mileage based usage fees (MBUF) or distance-based charging is relatively widespread for heavy vehicles based on truck weight and distance travelled.

Car sharing schemes can also be beneficial for pricing car use, removing the need for car ownership and so improving the marginal attractiveness of public and non-motorized transport, while giving access to low-income households where ownership is not affordable.

With growing incomes, the demand for mobility and comfort will continue to grow irrespective of the environmental impact. This will likely be met through private car use unless the competitiveness of alternative modes is improved by creating more attractive environments for walking and cycling and creating high quality public transport. However, there has been limited success in reducing urban car use: in a study of twelve major cities' attempts to curb private vehicle use, each city has experienced a growth in transport energy consumption and GHG emissions and/or a growth in private car ownership despite policies in place (Poudenx, 2008).

Reducing car use in an equitable way can also be achieved by providing high-quality public transport, which carries a large proportion of passenger trips in dense urban areas. In many developing cities, the modal share of buses can be in the range of 56–85 per cent and are frequently provided for by hundreds of separate bus companies. While there is often little political will to divert road space away from cars given the relative political strength of motorists, dedicated bus lanes and bus rapid transit (BRT) systems can move passengers very efficiently and at lower cost than private cars (GEA, 2012: chapter 9). Light rail transit can work in a similar way, separated from traffic but using existing road space. Subway systems have the highest cost per PKT but have the potential for higher capacities. Without dedicated road space, conventional buses tend to be subjected to significant congestion.

The most energy efficient mode for long distance travel is rail. Inter-city and high-speed railways can be comparable in time with cars and plane trips respectively, making inter-urban railways attractive for reducing car trips. This requires integrated public transport at either end with clear ticketing and information. Suburban and rural areas are, however, more difficult to provide with public transport because of low population densities.

10.3 **Energy efficient vehicles and low carbon fuels**

Road transport has mostly relied on the same petroleum fuels and the same basic internal combustion engine (ICE) designs since the advent of the passenger car more than 100 years ago. Petroleum fuels have high energy densities and are cheap to produce, easy to handle, and fast to refill; moreover, ICE powertrains are also cheaper to manufacture than alternative technologies. However, road vehicles produce a large amount of CO_2, NO_X, and other emissions that affect air quality and there is strong pressure on the industry to reduce emissions from cars, particularly in Europe, parts of North America, and more recently Asia. Three strategies are currently being employed to reduce emissions: efficiency, alternative fuels, and new drivetrains. This section examines the status and outlook for these vehicle technologies and fuels, then considers how transitions to alternative technologies can occur and the types of policy that can achieve these transitions.

10.3.1 STATUS AND OUTLOOK FOR VEHICLE DRIVETRAINS AND FUELS

Vehicle manufacturers are combining low-carbon fuels and new powertrain types in a wide array of possible configurations in order to find the best performing low-carbon vehicles in response to policy drivers (Plotkin et al., 2009). Hydrocarbon fuels with lower carbon intensities (e.g. liquefied petroleum gas (LPG) and compressed natural gas (CNG)), biofuels, electricity, and hydrogen are all in use or under consideration. Different conversion devices (engines, fuel cells) and powertrains (parallel and series hybrid, plug-in hybrid) can be combined with these fuels in an array of configurations. The move towards hybridization represents a major transition from the last century.

10.3.1.1 Conventional ICE

ICE vehicles powered by petrol or diesel account for the vast majority of the global vehicle fleet. While there have been significant breakthroughs in the past to reduce the air quality impact of road vehicles, until recently there was little pressure to improve fuel efficiency and the gradual move to using lighter materials has been offset in most countries by an increase in the average engine size of the car fleet (Daly and Ó Gallachóir, 2011). More recently, a combination of higher fuel prices, higher taxes, and legislation that aims to reduce average new car CO_2 emissions below 95 gCO_2/km (EC, 2009) have encouraged manufacturers to substantially improve vehicle performance. These improvements have been achieved by reducing vehicle weight, reducing aerodynamic and mechanical drag, and using more efficient ancillary devices and components (SMMT, 2013).

One strategy to reduce CO_2 emissions from ICE vehicles is to switch to fuels with lower carbon intensities. LPG is available in a number of countries but has not achieved a significant market share anywhere. Several countries, notably Argentina and Italy, have introduced compressed natural gas (CNG) in government-led initiatives. These transitions have been driven by pollution and energy security as opposed to GHG objectives. In Section 10.3.2, we identify some useful lessons from attempted transitions to CNG that can inform future transitions.

10.3.1.2 Biofuels

Biofuels can be low-carbon alternatives to petroleum fuels that are produced from biomass (Melillo et al., 2009). The principal advantage of biofuels is that they can be blended in small quantities with petroleum-derived fuels and used in existing vehicles, therefore achieving instant GHG reductions. With minor engine changes to adjust for the lower energy density of biofuels, they can be used in much greater proportions. For example, in Brazil, ethanol from sugar cane has been mixed with petrol since 1976, in response to the first oil crisis, and the mandatory proportion of ethanol is currently 25 per cent[1] (Ministério da Agricultura, 2007). The wider issues for biofuel use are explored in Chapter 16.

10.3.1.3 Hybrid Electric Vehicles

Hybrid vehicles use two or more distinct power sources, typically an ICE and an electric motor, and have become widespread following the introduction of Toyota's *Prius*. Vehicle efficiency is increased because the engine can run closer to peak efficiency, with fluctuations in power demand augmented by the battery. Furthermore, regenerative braking captures some of the kinetic energy that is otherwise lost during braking, and start–stop systems prevent inefficient idling. The increase in efficiency is achieved through a higher capital cost.

Plug-in hybrid vehicles, first launched commercially in December 2010, are similar but have a greater electrical storage capacity and a more powerful electric motor. The batteries can be charged when the vehicle is parked and short journeys can be made using only the electric drive. The principal disadvantage is the increased weight and cost of the electric powertrain and particularly the batteries.

10.3.1.4 Battery Electric Vehicles

Battery electric vehicles (BEVs) use batteries as the sole power source, and they offer the potential of decoupling road transport from fossil fuels. As a result, BEVs have been seen as a key step to fully decarbonizing the global energy system (Weiss et al., 2012; Bahn et al., 2013), and many governments have developed targets and subsidies for their adoption. Yet BEVs face a number of barriers. Batteries are expensive; they require a long recharging time; and their relatively low energy density and resulting shorter range

---

[1] The mandatory proportion of ethanol is occasionally reduced in Brazil if there are supply issues. Most Brazilian vehicles are flex-fuel, meaning they are adaptable to different proportions of ethanol.

limits the utility of the car. The number of batteries cannot simply be increased to extend the range because mass compounding means that larger cars require proportionally larger batteries (Thomas, 2009a). This means that battery electric drivetrains could be most suitable for small urban cars and may not be a direct replacement for all ICE cars. Urban delivery vehicles are another potential market; in the UK, for example, BEVs have been used for overnight milk deliveries for decades (Høyer, 2008).

The cost of providing infrastructure for electric vehicles is unclear, with uncertainties about technology cost, likely user behaviour, and the potential scale of adoption. If consumers charge their cars overnight, this would not influence peak electricity demand but could require reinforcement of the local electricity distribution networks if BEVs were to be used on a large scale. Batteries could also be charged at public charge points both day and night, possibly using fast-charging facilities, which would require the deployment of substantial expensive infrastructure, and which may influence peak electricity demand with implications for the wider energy system.

Electric vehicles do not produce tail-pipe emissions, and the well-to-wheel reduction in CO_2 emissions depends on the carbon intensity of the electricity used to charge the batteries.

10.3.1.5 Hydrogen Fuel Cell Vehicles

Hydrogen is attractive as a vehicle fuel because it can be made from a wide variety of sources and is clean at the point of use, with well-to-wheel emissions dependent on the mode of hydrogen production. Hydrogen can be used in an internal combustion engine, but fuel cells are preferred as they have higher fuel conversion efficiencies. Fuel cells generate electricity from hydrogen, and most hydrogen vehicle prototypes use fuel cells in combination with a battery in a hybrid electric powertrain. Several automakers have announced their intention to begin mass production of fuel cell vehicles between 2015 and 2017.

Like battery electric vehicles, fuel cell vehicles face a number of barriers to their adoption. First, onboard storage of hydrogen remains problematic because hydrogen has low volumetric energy density, limiting the amount that can practically be stored on board with implications for vehicle range. Liquid, compressed gas, and various solid-state storage options have been explored. In recent years automakers have focused on compressed gaseous hydrogen, typically stored at 700 bar, which enables adequate vehicle range and rapid refuelling. Research efforts into onboard storage are ongoing as the hydrogen tank is one of the most significant costs in a fuel cell vehicle.

A second major barrier to the adoption of hydrogen vehicles is the lack of fuel infrastructure. A network of production facilities, refuelling stations, and delivery mechanisms would need to be created to support the first adopters and this would initially be underutilized during a transition to large-scale hydrogen vehicle deployment. The costs associated with establishing this infrastructure, and bringing it to maturity, are substantial. Early investors in infrastructure do not know whether there will be sufficient demand in the future to make the facilities viable so there is a substantial risk attached to such investments. Agnolucci (2007) identifies several strategies that have been proposed to deal with this conundrum.

Finally, fuel cell vehicles have high capital costs, despite considerable success in reducing costs in recent years (Fuel Cell Technologies Office, 2013).

10.3.2 ON-ROAD EFFICIENCY OF THE CAR STOCK

Estimating the fuel consumption of a vehicle in use is not as straightforward as looking at the engine profile: there is an increasing discrepancy between the fuel consumption of vehicles under real-world driving conditions and from type approval tests (Zachariadis, 2006), leading to a significant source of uncertainty in estimating actual fuel consumption. While the efficiency of individual car models has improved, this has not always lead to an improvement in the efficiency of the overall car stock (Van den Brink and Van Wee, 2001; Meyer and Wessely, 2009; Daly and Ó Gallachóir, 2011), where purchasing trends towards larger and heavier cars have offset efficiency gains from standards and taxes; pollution and safety features of cars have also led to increased vehicle weight.

Immediate reductions in energy demand can be achieved through educating drivers on efficient driving styles. This can firstly be achieved by lessening strong acceleration and breaking, which lead to losses in efficiency and also contribute more to locally polluting emissions, which are proportional to acceleration as opposed to average speed. Smokers et al. (2006) calculate that driver awareness campaigns can achieve a 3 per cent energy saving. Further, a mandate that new vehicles be fitted with gearshift indicators could give a 1.5 per cent on-road energy saving for new cars (Smokers et al., 2006).

10.3.3 TRANSITIONS TO ALTERNATIVE TECHNOLOGIES

Techno-economic analysis often embeds the assumption that a transition to a new type of vehicle and fuel will occur when the costs and performance of the new technology are preferable to those of incumbent technologies. However, analysis of previous attempts to foster transitions to new fuels suggests a more complex picture.

First and most obviously, cars and refuelling infrastructure are complementary goods, with the 'chicken-and-egg' dynamics that this entails (no incentive to buy a vehicle until the infrastructure is there, no incentive to build infrastructure until the vehicles are there). Beyond this, it is clear that ICE cars also benefit from networks of other complementary goods and services (mechanics, for example), and from a variety of socio-technical factors, such as social norms (for example, what a desirable car 'should sound like'), and an appropriate regulatory environment (Geels, 2002). Moreover, studies of the diffusion of innovations show that potential consumers of a new technology must gain some degree of familiarity with it before they have a willingness even to consider adopting it. The visibility and 'trialability' of the new technology is thus important, yet is clearly limited in the early stages of a transition.

Modellers examining these issues using system dynamics approaches typically find that their inclusion or exclusion has a decisive effect on results, with transitions that appear to be optimal from a techno-economic perspective failing to occur in model runs in which such dynamics are captured (Struben and Sterman, 2008). Furthermore, historical precedent suggests that transitions to alternative fuels are difficult and slow, except where rather extreme political circumstances drive rapid change (as has happened recently in the case of Iran, where sanctions have resulted in rapid conversion of vehicles to compressed natural gas). Transitions have tended to occur most rapidly where it is possible for a new fuel to be used alongside conventional fuels (as with 'bi-fuel' or 'flex fuel' vehicles) and where conversion of vehicles to use of the new fuel is a possibility (McDowall, forthcoming).

10.3.4 POLICIES AND DRIVERS FOR LOW-CARBON VEHICLES

Measures for reducing the energy and carbon intensity of cars broadly fall into two categories, fiscal incentives through taxation and subsidies, and mandates, and the following sections characterize different measures accordingly.

10.3.4.1 Financial Incentives

- *Fuel and CO$_2$ taxes*: Fuel price increases lead to reductions in overall travel demand, shifts to more energy efficient technologies and modes, and promote more energy efficient driving styles (Gross et al., 2009; Clerides and Zachariadis, 2008). Carbon taxes, moreover, incentivize biofuels and other less carbon intensive fuels.
- *Vehicle tax*: Changing the structure of car purchase incentives has worked historically to reduce new-car emissions intensity (Rogan et al., 2011). Vehicle ownership tax based on CO$_2$ emissions helps to offset the high discount rate of consumers, which makes short-term savings in capital cost favourable over long-term fuel savings from energy efficient cars (Schäfer and Jacoby, 2006).
- *Subsidies for alternative fuel vehicles (AFV)*: Subsidies can reduce the purchase cost of otherwise expensive technologies to encourage early investment. Subsidies can be seen as unfair, giving the rich easier access to very fuel efficient cars, as is the case with BEVs.

10.3.4.2 Targets and Mandates

Voluntary agreements between governments and vehicle manufacturers set vehicle efficiency targets. These have not been a major stimulus for achieving efficiency gains.

For example, the European Commission's voluntary agreement with the automobile industry did not meet its target (Fontaras and Samaras, 2007; Zachariadis 2006).

Mandatory CO_2 emission standards set by EU legislation now compel car manufacturers to reach average new car emissions of 130 gCO_2/km by 2015 (EC, 2009). A further target of 95 gCO_2/km is being resisted by German lobbying, whose car manufacturers produce heavier and more fuel intensive vehicles and have more to lose from the target (Lewis, 2013). US Corporate Average Fuel Economy (CAFE) standards have imposed fuel performance standards on manufacturers since 1975.

Renewable fuel obligations in the European Union have mandated that renewable energy, including biofuels, shall fuel 10 per cent of the transport sector in 2020 (Van Noorden, 2013).

10.4 Beyond energy and CO_2: conflicting and complementary drivers for transport policy

10.4.1 LOCAL IMPACTS: AIR QUALITY, HEALTH, AND SAFETY

The drivers of change and impacts of road transport are complicated and often interdependent. Transport happens in the public sphere and has strong negative externalities, especially for people who do not have access to motorized mobility. Frequently, the negative impacts of transport are drivers of change, leading to measures which have unintended and unexpected consequences for energy and GHG emissions, both positive and negative. Conversely, measures for mitigating transport CO_2 have the potential for strong local co-benefits, which can help in justifying taxation and investment. This section explores some of these interactions.

Road transport is a major cause of deaths. Road traffic injuries are the eighth leading cause of death in the world and are the leading cause of death among those aged 15–29 years. Approximately 1.24 million deaths occurred on the world's roads in 2010, with more than half the death toll being borne by vulnerable road users (motorcyclists, pedestrians, and cyclists) who had almost no responsibility for the incidents (WHO, 2013). In Africa, 38 per cent of road deaths are pedestrians (WHO, 2013). The main response to the danger of driving has been focused on protecting car drivers: cars are increasingly heavy and have safety features that add to weight, leading to lower energy efficiency. This exacerbates the burden of risk to more vulnerable road users, encouraging a further modal shift away from walking and cycling, further exacerbating the energy demand of transport. Road safety is also promoted through highway engineering which has also focused on protecting private car users, leading to the same modal shift impact (Adams, 1985) as well as limiting the access of pedestrians to roads. Not only is the burden of risk unfairly borne by vulnerable road users, it encourages the use of private cars and it increases the weight of cars, causing a self-reinforcing cycle.

In contrast, measures promoting the safety of walkers and cyclists can reduce energy demand, increase access, and decrease unfair risk. Furthermore, lowering speed limits on motorways has the potential to reduce GHG emissions by 10–20 per cent (Barrett, 2007), along with air pollution and road deaths.

Air pollution is also a major externality of road transport. The health effects of air pollution are more uncertain than traffic accidents, but some studies indicate that the number deaths from transport-related air pollution is similar to those from traffic accidents (Krzyzanowski et al., 2005). Climate change mitigation in transport will benefit public health considerably: Woodcock et al. (2009) simulate the impact of lower-emission vehicles along with reduced use of motor vehicles in cities, finding a reduction in the number of years of life lost from heart disease of 10–19 per cent in London and 11–25 per cent in Delhi as a result of active travel and lower pollution. Increasing distances walked and cycled, which is associated with greater use of public transport, leads to large health benefits (Lachapelle and Frank, 2009).

Tackling local air pollution has led to fuel-switching measures with co-benefits for GHG emissions: Section 10.3.1 described transitions to low carbon fuels (LPG and CNG) which were not prompted from climate change concerns. The promotion of alternative powertrains, energy efficiency, mobility demand, and mode-switching measures can all have positive spillovers for the mitigation of air pollution.

Similarly to air quality and traffic accidents, tackling the issues of congestion, noise, land use, and environmental and ecosystem degradation can have positive spillovers for energy and GHG emissions.

10.4.2 ECONOMIC GROWTH, WELFARE, AND EQUITY

The inherent utility of mobility is the historical reason for the massive growth in private travel demand. Mass car ownership has historically been an integral component of the transition to an advanced economy. Workers can cover longer distances in their daily commutes, increasing the size of the labour market and increasing economic efficiency. Greater mobility is a utility that increases access to goods and services and increases communication. Moreover, greater demand in transport has positive spillovers on the production side via demands for raw materials, car manufacturing, and road infrastructure. While road investment projects are frequently carried out in the name of development, it is not clear whether this promotes economic growth (Banister and Berechman, 2001). Section 10.2 discusses means for maintaining this utility via provision of public transport and substitutes for mobility, such as home working.

Increasing mobility for the poor can have great benefits for welfare, equity, and economic growth. In this respect, fuel taxes and other charges on private cars are problematic and can be regressive in advanced economies where the poor are locked into car transport, and spend a greater proportion of disposable income on transport than the rich (Aasness and Larsen, 2003). This can be overcome by the provision of high-quality public transport as an alternative to car travel. However, sprawling cities with high levels of vehicle ownership are a great cause of environmental degradation and social exclusion (Power, 2001): the untamed growth of private cars overwhelms the operation of urban transport systems and makes the journey to work prohibitively long, unpleasant, and expensive for the poor.

10.4.3 REBOUND AND THE VALUE OF TIME

Within the suite of measures described in Sections 10.2.2 and 10.3.4 there can be unexpected contradictions and incompatibilities. The 'rebound effect' happens on different scales and in unexpected ways: efficiency measures, particularly fuel economy standards, can decrease the cost of travel and so increase the level of demand (Sorrell et al., 2009). Brännlund et al. (2007) estimated that in Sweden, a 20 per cent increase in the efficiency of cars would result in an increase in vehicle activity of 7.5 per cent. This rebound can be overcome with fuel tax and road pricing.

Rebound in activity can also happen as a result of measures which reduce the travel time of journeys: there is evidence of a fixed travel time budget, where people on average reorganize their travel plans in order to reinvest the travel time savings into moving greater distances (Metz, 2008). This can explain why road building aimed at relieving congestion frequently fails to have long-term benefits. This also applies when creating a modal shift away from private cars without reducing road space: for example, it is estimated that induced traffic could offset around 33 per cent of the modelled reduction in traffic as a result of the London congestion charge (Transport for London, 2007).

10.5 **Conclusion**

The pervasive influence of the transport system has led to profoundly negative impacts on the environment through resource use and climate change, on human health through pollution and lifestyle change, and on societies through inequality and the nature of the built environment. There is a huge implementation gap that results from the attractiveness of private transport and the many benefits of mobility. The distance between the trajectory of transport in 'business as usual' scenarios and scenarios consistent with GHG stabilization is vast, and the window for action is closing. It is vital to address the issue of large-scale and growing fossil energy demands and GHG emissions as a result of passenger transport on many levels with consideration for the significant interdependencies and uncertainties associated with mobility and technology.

Substantial reductions in the climate impact of transport can be made but only if a substantial change in mobility patterns is achieved, along with technological innovation. There is growing evidence that achieving reductions in GHG emissions from transport consistent with global climate targets requires integrated packages of measures to bring about change in transport behaviour and technologies: relying on one aspect alone is not likely to be sufficient (Owens, 1995; Hickman and Banister, 2007; Banister, 2008; Leduc et al., 2010). New technologies and fuels that reduce the energy and GHG intensity of vehicles must be pursued. Transport technologies have improved greatly in the past; however, while individual engines have become cleaner, transport has become more polluting. Similarly, while engines are more energy efficient, average cars have not achieved significant on-road savings. Thus far, savings from technological advances have been consumed in greater distances travelled and greater weight and performance of cars.

A portfolio of technology options is required to meet the societal goals of mitigating climate change, air pollution, and oil dependence, beginning with hybrids, plug-in hybrids, and biofuels, and transitioning to full battery electric and fuel cell vehicles in the long term (Thomas, 2009b). There is a very wide range of technology and fuel options with different strengths and weaknesses, and it must be recognized that consumer preference will determine drivetrain adoption, as opposed to necessarily the most suitable or cost-effective technology from an energy system perspective. Therefore, an understanding of behaviour and technology use is needed to focus policy.

There is a growing consensus that stabilization of GHG emissions from transport is unlikely to come about without so-called behaviour change (Chapman, 2007; GEA 2012), resulting in reductions in the demand for transport services. Substantial targets for GHG reductions are not likely to be met if demand grows under a business-as-usual scenario (Hickman and Banister, 2007). Relying on technology options alone is risky: deployment of AFVs may be more challenging than anticipated, particularly if costs do not sufficiently decrease, and, furthermore, relying on technology for decarbonization fails to deal with other negative externalities caused by high levels of mobility (Johansson, 2009).

Several market failures are apparent in transport, which inhibit effective change: private car transport is not priced on a per-kilometre basis; consumers have high observed discount rates (Sutherland, 1991), making efficiency measures difficult to see through without incentives; and the burdens of risk, climate change, and pollution are greatly borne by those with the least access to transport.

■ REFERENCES

Aasness, J. & Larsen, E. R. 2003. Distributional effects of environmental taxes on transportation. *Journal of Consumer Policy*, 26(3), 279–300.

Adams, J. G. U. 1985. *Risk and Freedom: The Record of Road Safety Regulation*, Brefi Press.

Agnolucci, P. 2007. Hydrogen infrastructure for the transport sector. *International Journal of Hydrogen Energy*, 32, 3526–44.

Bahn, O., Marcy, M., Vaillancourt, K., & Waaub, J.-P. 2013. Electrification of the Canadian road transportation sector: A 2050 outlook with TIMES-Canada. *Energy Policy*, 62, 593–606.

Banister, D. 2008. The sustainable mobility paradigm. *Transport Policy*, 15, 73–80.

Banister, D. & Berechman, Y. 2001. Transport investment and the promotion of economic growth. *Journal of Transport Geography*, 9, 209–18.

Barrett, M. 2007. *Energy Scenarios for Europe: Strategies for the control of emissions of carbon dioxide and air pollutants and to enhance energy security*. For the Swedish Environmental Protection Agency, Available at <http://www.unece.org/fileadmin/DAM/env/lrtap/TaskForce/tfiam/33meeting/tfiam33_Barrett_EnergyScenariosEu.pdf> (Accessed April 2015).

Brännlund, R., Ghalwash, T., & Nordström, J. 2007. Increased energy efficiency and the rebound effect: Effects on consumption and emissions. *Energy Economics*, 29, 1–17.

Cervero, R. & Kockelman, K. 1997. Travel demand and the 3Ds: density, diversity, and design. *Transportation Research Part D: Transport and Environment*, 2, 199–219.

Chamon, M., Mauro, P., & Okawa, Y. 2008. Mass Car Ownership in the Emerging Market Giants. *Economic Policy*, 23, 243–96.

Chapman, L. 2007. Transport and climate change: a review. *Journal of Transport Geography*, 15, 354–67.

Clerides, S. & Zachariadis, T. 2008. The effect of standards and fuel prices on automobile fuel economy: An international analysis. *Energy Economics*, 30, 2657–72.

Coombe, D. 1995. Trunk roads and the generation of traffic. *Highways and Transportation*, 42(4), 2.

Daly, H. E. & Ó Gallachóir, B. P. 2011. Modelling private car energy demand using a technological car stock model. *Transportation Research Part D: Transport and Environment*, 16, 93–101.

Daly, H. E., Ramea, K., Chiodi, A., Yeh, S., Gargiulo, M., & Ó Gallachóir, B. P. 2014. Incorporating travel behaviour and travel time into TIMES energy systems models. *Applied Energy*, 135 (0), 429–39.

Eliasson, J. 2009. A cost–benefit analysis of the Stockholm congestion charging system. *Transportation Research Part A: Policy and Practice*, 43, 468–80.

EC 2009. Regulation (EC) No 443/2009 of the European Parliament and of the Council of 23 April 2009. *Official Journal of the European Union* L, 140.

European Commission 2011. WHITE PAPER: Roadmap to a Single European Transport Area—Towards a competitive and resource efficient transport system, 144 final, Brussels, 28.3.2011.

Fontaras, G. & Samaras, Z. 2007. A quantitative analysis of the European Automakers' voluntary commitment to reduce CO2 emissions from new passenger cars based on independent experimental data. *Energy Policy*, 35, 2239–48.

Fuel Cell Technologies Office 2013. Progress and Accomplishments in Hydrogen and Fuel Cells. EERE Fuel Cells Technologies Office, U.S. Department of Energy. March 2013.

Fulton, L., Cazzola, P., & Cuenot, F. S. 2009. IEA Mobility Model (MoMo) and its use in the ETP 2008. *Energy Policy*, 37, 3758–68.

Fulton, L., & Lah, O., Cuenot, F. 2013. Transport pathways for light duty vehicles: towards a 2° scenario. *Sustainability*, 5, 1863–74 (DOI: 10.3390/su5051863).

GEA 2012. *Global Energy Assessment—Toward a Sustainable Future*, Cambridge University Press, Cambridge, UK and New York, NY, USA and the International Institute for Applied Systems Analysis, Laxenburg, Austria.

Geels, F. W. 2002. Technological transitions as evolutionary reconfiguration processes: a multi-level perspective and a case-study. *Research Policy*, 31(8–9), 1257–74.

Girod, B., Van Vuuren, D. P., & Deetman, S. 2012. Global travel within the 2C climate target. *Energy Policy*, 45, 152–66.

Girod, B., Vuuren, D., Grahn, M., Kitous, A., Kim, S., & Kyle, P. 2013. Climate impact of transportation A model comparison. *Climatic Change*, 118, 595–608.

Gross, R., Heptonstall, P., Anable, J., Greenacre, P., & E4TECH 2009. *What policies are effective at reducing carbon emissions from surface passenger transport? A review of interventions to encourage behavioural and technological change*. A report produced by the Technology and Policy Assessment Function of the UK Energy Research Centre. UKERC, London.

Hickman, R. & Banister, D. 2007. Looking over the horizon: Transport and reduced CO2 emissions in the UK by 2030. *Transport Policy*, 14, 377–87.

Høyer, K. G. 2008. The history of alternative fuels in transportation: The case of electric and hybrid cars. *Utilities Policy*, 16, 63–71.

IEA. 2009. *Transport Energy and CO_2—Moving Towards Sustainability*. Paris, International Energy Agency.

IEA. 2012. *World Energy Outlook 2012*. Paris, France, International Energy Agency.

ITDP. 2010. *Promoting Sustainable and Equitable Transportation Worldwide*. Institute for Transportation and Development Policy (ITDP).

Johansson, B. 2009. Will restrictions on CO2 emissions require reductions in transport demand? *Energy Policy*, 37, 3212–20.

Kahn Ribeiro, S., Kobayashi, S., Beuthe, M., Gasca, J., Greene, D., Lee, D. S, Muromachi, Y., Newton, P. J., Plotkin, S., Sperling, D., Wit, R., Zhou, P. J. 2007. Transport and its Infrastructure. In: *Climate Change 2007: Mitigation. Contribution of Working Group III to the Fourth Assessment Report of the Intergovernmental Panel on Climate Change* (B. Metz, O. R. Davidson, P. R. Bosch, R. Dave, L. A. Meyer (eds)), Cambridge University Press, Cambridge, United Kingdom and New York, NY, USA.

Krzyzanowski, M., Kuna-Dibbert, B., Schneider, J. 2005. *Health effects of transport-related air pollution*, WHO Library Cataloguing in Publication Data. World Health Organisation, Copenhagen, Denmark.

Lachapelle, U. & Frank, L. D. 2009. Transit and health: mode of transport, employer-sponsored public transit pass programs, and physical activity. *Journal of Public Health Policy*, 30, S73–S94.

Leduc, G., Mongelli, I., Uihlein, A., & Nemry, F. 2010. How can our cars become less polluting? An assessment of the environmental improvement potential of cars. *Transport Policy*, 17, 409–19.

Lefèvre, B. 2009. Urban transport energy consumption: Determinants and strategies for its reduction an analysis of the literature. *Sapiens*, 2.

Lewis, B. 2013. Germany wins backing of EU ministers to block car emissions law. *Reuters* [online], 14 October. Available at <http://www.reuters.com/article/2013/10/14/us-eu-cars-emissions-idUSBRE99D0JV20131014> (Accessed 8 November 2013).

McDowall, W. Forthcoming. Are scenarios of hydrogen vehicle adoption optimistic? A comparison of hydrogen scenarios with historical analogies. *Environmental Innovation and Societal Transitions*.

Melillo, J. M., Reilly, J. M., Kicklighter, D. W., Gurgel, A. C., Cronin, T. W., Paltsev, S., Felzer, B. S., Wang, X., Sokolov, A. P., & Schlosser, C. A. 2009. Indirect emissions from biofuels: how important? *Science*, 326, 1397–9.

Meyer, I. & Wessely, S. 2009. Fuel efficiency of the Austrian passenger vehicle fleet—Analysis of trends in the technological profile and related impacts on CO2 emissions. *Energy Policy*, 37, 3779–89.

Metz, D. 2008. The myth of travel time saving. *Transport Reviews*, 28, 321–36.

Metz, D. 2010. Saturation of demand for daily travel. *Transport Reviews*, 30, 659–74.

Millard-Ball, A. & Schipper, L. 2010. Are we reaching peak travel? Trends in passenger transport in eight industrialized countries. *Transport Reviews*, 31, 357–78.

Ministério Da Agricultura, Pecuária E Abastecimento. 2007. Portaria Nº 143, de 27 de Junho de 2007. Brazil: <http://extranet.agricultura.gov.br/sislegis-consulta/consultarLegislacao.do?operacao=visualizar&id=17886>. (Accessed April 2015).

Mokhtarian, P. 2009. If telecommunication is such a good substitute for travel, why does congestion continue to get worse? *Transportation Letters*, 1(1), 1–17.

Newman, P. & Kenworthy, J. R.1989. *Cities and Automobile Dependence: An International Sourcebook*. Aldershot: Gower.

Owens, S. 1995. From "predict and provide" to "predict and prevent"?: Pricing and planning in transport policy. *Transport Policy*, 2(1), 43–9.

Plotkin, S., Singh, M., Patterson, P., Ward, J., Wood, F., Kydes, N., Holte, J., Moore, J., Miller, G., Das, S., & Greene, D. 2009. *Multi-path Transportation Futures Study: Vehicle Characterization and Scenario Analyses*. USA, Argonne National Laboratory.

Poudenx, P. 2008. The effect of transportation policies on energy consumption and greenhouse gas emission from urban passenger transportation. *Transportation Research Part A: Policy and Practice*, 42, 901–9.

Power, A. 2001. Social exclusion and urban sprawl: Is the rescue of cities possible?. *Regional Studies*, 35(8), 731–42.

Rogan, F., Dennehy, E., Daly, H., Howley, M., & Ó Gallachóir, B. P. 2011. Impacts of an emission based private car taxation policy—First year ex-post analysis. *Transportation Research Part A: Policy and Practice*, 45, 583–97.

Santos, G. & Bhakar, J. 2006. The impact of the London congestion charging scheme on the generalised cost of car commuters to the city of London from a value of travel time savings perspective. *Transport Policy*, 13, 22–33.

Saunders, M. J., Kuhnimhof, T., Chlond, B., & Da Silva, A.N.R. 2008. Incorporating transport energy into urban planning. *Transportation Research Part A: Policy and Practice*, 42, 874–82.

Schäfer, A., Heywood, J. B., Jacoby, H,. & Waitz, I. 2009. *Transportation in a Climate Constrained World*, The MIT Press. Cambridge, MA.

Schäfer, A. & Jacoby, H. D. 2006. Vehicle technology under CO2 constraint: a general equilibrium analysis. *Energy Policy*, 34, 975–85.

Scholl, L., Schipper, L., & Kiang, L. 1996. CO2 emissions from passenger transport: a comparison of international trends from 1973 to 1992. *Energy Policy*, 24, 17–30.

Sorrell, S., Dimitropoulos, J., & Sommerville, M. 2009. Empirical estimates of the direct rebound effect: A review. *Energy Policy*, 37, 1356–71.

SMMT 2013. *New Car CO2 Report 2013*. The Society of Motor Manufacturers and Traders. <http://www.smmt.co.uk/wp-content/uploads/sites/2/SMMT-New-Car-CO2-Report-2013-web.pdf>. (Accessed April 2015).

Smokers, R., Vermeulen, R., Van Mieghem, R., Gense, R., Skinner, I., Fergusson, M., Mackay, E., Ten Brink, P., Fontaras, G., & Samaras, Z. 2006 Review and analysis of the reduction potential and costs of technological and other measures to reduce CO2-emissions from passenger cars. Final Report. *TNO Report 06.OR.PT.040.2/RSM. Report to the European Commission.* <http://ec.europa.eu/enterprise/sectors/automotive/files/projects/report_co2_reduction_en.pdf>. (Accessed April 2015).

Struben, J. & Sterman, J. 2008. Transition challenges for alternative fuel vehicle and transportation systems. *Environment and Planning B: Planning and Design*, 6, 1070–97.

Sutherland, R. J. 1991. Market barriers to energy-efficiency investments. *The Energy Journal*, 12 (3), 15–34.

Thomas, C. E. 2009a. Fuel cell and battery electric vehicles compared. *International Journal of Hydrogen Energy*, 34, 6005–20.

Thomas, C. E. 2009b. Transportation options in a carbon-constrained world: Hybrids, plug-in hybrids, biofuels, fuel cell electric vehicles, and battery electric vehicles. *International Journal of Hydrogen Energy*, 34, 9279–96.

Transport For London. 2007. *Central London Congestion Charging Scheme: ex-post evaluation of the quantified impacts of the original scheme.* <https://www.tfl.gov.uk/cdn/static/cms/documents/ex-post-evaluation-of-quantified-impacts-of-original-scheme.pdf>. (Accessed April 2015).

Van Den Brink, R. M. M. & Van Wee, B. 2001. Why has car-fleet specific fuel consumption not shown any decrease since 1990? Quantitative analysis of Dutch passenger car-fleet specific fuel consumption. *Transportation Research Part D: Transport and Environment*, 6, 75–93.

Van Noorden, R. 2013. EU debates U-turn on biofuels policy. *Nature*, 499, 13–14.

Vuchic V. R. 1999. *Transportation for Livable Cities*. Rutgers Center for Urban Policy Research, New Jersey, USA.

Wang, Y., Teter, J., & Sperling, D. 2011. China's soaring vehicle population: Even greater than forecasted? *Energy Policy*, 39, 3296–306.

Wee, B. 2014. The Unsustainability of Car Use. In: Gärling, T., Ettema, D., & Friman, M. (eds) *Handbook of Sustainable Travel*. Springer, Netherlands.

Weiss, M., Patel, M. K., Junginger, M., Perujo, A., Bonnel, P., & van Grootveld, G. 2012. On the electrification of road transport—Learning rates and price forecasts for hybrid-electric and battery-electric vehicles. *Energy Policy*, 48, 374–93.

WHO, 2013. Global Health Observatory (GHO) data repository [online]. Available at <http://www.who.int/gho/database/en/>. (Accessed 8 November 2013).

Woodcock, J., Edwards, P., Tonne, C., Armstrong, B. G., Ashiru, O., Banister, D., Beevers, S., Chalabi, Z., Chowdhury, Z., Cohen, A., Franco, O. H., Haines, A., Hickman, R., Lindsay, G., Mittal, I., Mohan, D., Tiwari, G., Woodward, A., & Roberts, I. 2009. Public health benefits of strategies to reduce greenhouse-gas emissions: urban land transport. *The Lancet*, 374, 1930–43.

World Bank. 1968. *The Economics of Road User Charges*. Washington, DC: World Bank. <http://documents.worldbank.org/curated/en/1968/01/16811589/economics-road-user-charges>. (Accessed April 2015).

Zachariadis, T. 2006. On the baseline evolution of automobile fuel economy in Europe. *Energy Policy*, 34, 1773–85.

11 **Shipping and aviation**

Antony Evans and Tristan Smith

11.1 **Introduction**

International transport, particularly passenger transport, was revolutionized by the development of the powered aircraft in 1903. For the first time, the possibility arose for passengers and freight to be transported by air instead of over land or sea, making previously impossible trips not only possible, but practical and even attractive. During the same time period, the modern fuel oil and oil-derivative powered 'motor' ship took over from sailing and steam ships and revolutionized trade and facilitated globalization by providing reliable and low-cost international freight transport services. These two sectors present similar challenges for policymakers, particularly with regard to GHG emission control, including through the regulation of energy efficiency. Their inter-national nature left them outside of national responsibilities in the Kyoto Protocol with the development of an appropriate mitigation strategy instead delegated to their respective UN agencies (the International Civil Aviation Organisation and Inter-national Maritime Organisation). Both sectors have seen substantial growth in recent years and forecast continuations of these trends. And both sectors remain wedded to the (comparatively) low-cost, energy-dense, and portable energy sources (aviation and marine fuels) derived from oil. This chapter describes some of the fundamentals driving the recent and future evolution of shipping and aviation's transport demand, the possible pathways for each sector to lower carbon and energy intensity in their transport supply, and the foreseeable policy and wider contexts that could influence their futures. The description is derived from a number of different theoretical viewpoints, mirroring the approach taken in Chapter 7. These include techno-economic modelling theory, energy services theory, and socio-technical transitions theory, and demonstrate the important contributions different theoretical frameworks can make, but also the benefit of combining the approaches in interdisciplinary research.

11.2 **Energy demand and climate impacts**

Aviation was estimated to be responsible for the demand of 214 million tonnes of fuel (dominated by Jet-A1) in 2012, which equates to approximately 2 per cent of total anthropogenic CO_2 emissions (or about 12 per cent of CO_2 emissions from all transportation sources) (ATAG, 2012). This is consistent with estimates from the International Panel on Climate Change (IPCC) (Penner, 1999). By 2050, the contribution of aviation to anthropogenic CO_2 emissions is expected to increase to around 3 per cent. However, because of the significant non-CO_2 effects of aviation, described in this chapter, aviation is estimated to have accounted for around 3.5 per cent of global anthropogenic radiative forcing (RF) in 2005, increasing to around 4.9 per cent if recent estimates for the uncertain effect from aviation induced cirrus clouds are considered (Lee et al., 2010).

Shipping was estimated to be responsible for the demand of 333 million tonnes of fuel (dominated by heavy fuel oil and distillate (crude oil derived) fuels diesel and gasoil), which equates to approximately 3.3 per cent of global anthropogenic CO_2 emissions in 2007 (Buhaug et al., 2009). These figures are derived from activity models and are inconsistent with estimates produced by the International Energy Agency, 2007, which are derived from marine bunker fuel sales data. The inconsistency can be attributed to uncertainties that occur in the reporting and classification of marine bunker fuel sales data, and, as there are also large and unquantified uncertainties in the activity model calculation, remain unresolved.

Energy demand for both shipping and aviation can be decomposed as the following

$$E = T.1 \tag{11.1}$$

Where T is the transport demand in passenger km or tonne km, and I is the energy intensity of the transport supply (e.g. joules (J) per tonne km or J per passenger km). This decomposition is also a convenient framing to the different theoretical perspectives: energy services theory is well suited to the study of transport demand, whereas techno-economic and socio-technical transitions theory are particularly valuable for the study of energy intensity. This chapter starts by following this decomposition into demand and energy intensity, before bringing the two subsets together for the consideration of cross-cutting issues around policy.

11.3 **Transport demand**

Shipping and aviation, like most forms of transport (and with some minor exceptions, e.g. cruises) are both derived demands.[1] This description is consistent with an energy services theoretical perspective that characterizes a ship or an aircraft as end-use technology that provides or enables useful services such as business travel, leisure travel, or trade of commodities or manufactured goods. Therefore energy demand and energy

---

[1] This term is applied when the demand is not so much for the commodity or product actually purchased as for something else (in this case an energy service), which may be obtained through the purchase.

demand growth in shipping and aviation are importantly linked to these demands for services, the fundamentals of which are described further here in their historical, current, and future contexts.

International trade has existed for centuries, consisting largely of the transport of specialist goods and commodities for long distances at sea (e.g. cotton, wool, spices, and other agricultural products). In some instances, that trade was driven by the demand for a good that could not be sourced locally (e.g. spices due to global climate and agricultural productivity variations). While this remains an explanation for the existing patterns of trade (particularly for extracted commodities, which for geological reasons are distributed unevenly around the world and not necessarily collocated with their consumption location), a more recent trend in demand growth has been associated with one of the last fifty years' most notable economic phenomena—globalization.

Globalization refers to the increasingly global nature of the market in which goods and services are produced and consumed. The principle can be summarized, for a good consumed in country 'b', by (11.2):

$$CP_a + CT_{ab} \leq CP_b, \tag{11.2}$$

where CP is the cost of production and CT is the cost of transport. Provided that the combined cost of production and cost of transport from country 'a' to 'b' is less than the cost of local production (in 'b') then it is economically viable to locate production in country 'a'. Wage and skills differentials between countries can be one explanation for a geographical differential in cost of production; however, these have always existed. What has not always existed is a supply of safe, reliable, and cheap transport connecting the world's consumers with the world's raw materials and skilled, low-cost labour markets.

That connection for the majority of the world's trade is provided by the shipping industry, and as the industry and global transport infrastructure have developed and become more efficient, distance-to-market as a parameter that affects the competitiveness of many goods and commodities has reduced in significance.

An important trend, however, is the rising share of high-value commodities and the associated need to transport these commodities quickly. This has led to an increase in the importance of aviation in freight transport. And while airfreight represents less than 2 per cent of world trade by weight, it represents nearly 40 per cent by value (Kasarda and Green, 2005).

Dedicated freight aircraft represent less than 10 per cent of the global aircraft fleet (Boeing, 2012, 2013). The vast majority of aircraft serve primarily passenger travel. It was not until after the Second World War that long-distance intercity passenger travel ceased to be dominated by rail and shipping and shifted to aviation. It took technological innovations during and after the Second World War, most notably the development of the jet engine, for aircraft to become sufficiently cost-effective for aviation to compete with surface transport. By the 1950s, advancements in aircraft technology along with government investment in aviation infrastructure (particularly airports) led to aviation taking over from rail to become the most important mode of transport for commercial intercity passenger travel in the United States (Schäfer et al., 2009). The growth of aviation in other industrialized countries and regions followed similar trends. Between 1960 and 2011, worldwide scheduled passenger air travel grew from 109 billion passenger-kilometres travelled to 3.7 trillion—an average growth rate of over 7 per cent per year (ICAO, 2006a; UN, 2013).

Like freight, demand for passenger air travel also has strong links to globalization, although for different reasons. As economies around the world have grown, business is increasingly done between companies in different countries, and even on different continents. This requires personal interaction as well as the transfer of goods and information. The increase in personal interaction for business purposes between different countries is one contributor to the rapid growth in aviation. A further impact of, and contributor to, globalization is the migration of people around the world to live and work in a different country from where they grew up. This has led to increased travel for visiting friends and relatives (termed 'VFR'). This migration has been enabled by globalization, but is often driven by disparity in earning potential in different countries. While in many cases this represents a 'brain drain' for countries, the VFR travel back to the country can bring much needed economic benefits, as can cash remittances from migrants. Similarly, vacation travel can be a significant economic input into a country. As income levels have risen, people have also become more willing to spend money on vacation travel, and particularly travel over greater distances. This has also contributed to the increase in demand for air travel. This effect has been further stimulated by the relatively recent development of the low-cost airline business model. This has led to a decline in the cost of air travel, as shown for the domestic US market in Figure 11.1, from both the dedicated low-cost carriers and the existing network carriers, because of the need to compete. This has made air travel for vacation purposes more accessible, and has contributed to the rise in air traffic over the last decade. A further, longer-term, contributor to the increase in demand for air travel is the continuing trend to shift from slower to faster modes of transport across all distances (Schäfer et al., 2009).

For global maritime transport demand, Figure 11.2 shows the historical trend in the transport demand (in units of billion tonne miles: $1 \times 10^9 \times$ mass in tonnes $\times$ distance in nautical miles) for some of the main commodity groups of the shipping industry. The graph shows that, at least for global aggregations of flows of commodities, there is a high level of correlation between GDP and transport demand. Figure 11.3 shows similar results within the aviation sector, both for passenger and airfreight traffic.

11.4 **Trends/expectations in demand growth**

Despite the uncertainties involved in making extrapolations from historical trends to estimate what might happen in the future, some scenarios use correlation with GDP

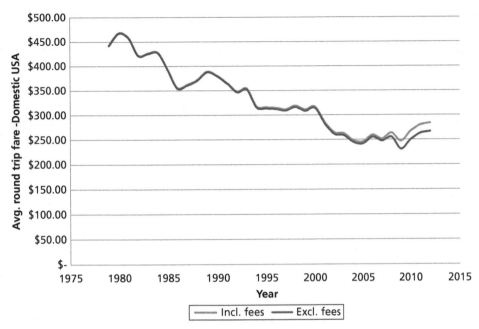

Figure 11.1 Average real round-trip airfares in the domestic United States from 1979 to 2012 (Airlines for America, 2014)

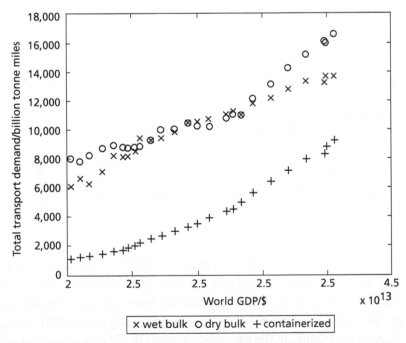

Figure 11.2 Relationship between world GDP (1985–2009) and transport demand for different marine transport commodities

(*Source*: UNCTAD).

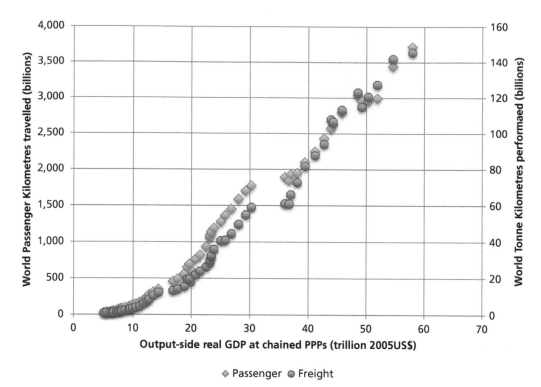

Figure 11.3 Relationship between global output-side GDP at chained PPPs,[2] against global air passenger (left axis) and airfreight (including mail) (right axis) traffic, 1950–2005. Note that passenger and freight air traffic from 1950 to 1970 does not include data from the Union of Soviet Socialist Republics (USSR)

(*Source*: GDP data from Feenstra et al., 2013; passenger and freight air traffic data from ICAO, 2006).

Table 11.1 Transport demand in 2050

SRES scenario	A1B	A2	B1	B2
Ocean-going shipping	320	240	220	180
Coastwise shipping	320	270	220	220
Container	1,230	960	850	690
Average (all ships)	540	421	372	302

forecasts (e.g. those available for IPCC scenarios, SRES (Special Report on Emissions Scenarios)) to estimate future global marine transport demand. This method can produce dramatic growth rates. Other approaches have applied resource constraints (e.g. finite oil) or wider CO_2 mitigation scenarios (e.g. Buhaug et al., 2009). Using such an approach, Buhaug et al. estimate the transport demand in 2050 as a function of different IPCC SRES scenarios, to be represented by the values in Table 11.1. The values are

[2] PPP stands for Purchasing Power Parity, which consists of GDP in market prices adjusted by an exchange rate between two currencies that reflects the relative purchasing power of the currencies.

indexed to the 2007 transport demand (100). The range of values shows that total maritime transport demand is expected to increase by between a factor of 3 and 5 over the next forty years.

The theoretical perspective used to derive such forecasts, whilst not explicitly listed as such, could be thought of as being predominantly techno-economic, for example defined by the location of resource constraints and the correlation of freight weight–distance with GDP. A social practice theory approach would attempt to consider influences of end-user attitude change and policy on transport demand. For example, popular movements such as the 'slow food movement' or trends for consumer goods from overseas 'sweat shops' no longer to be considered ethical may shift demand to more locally produced and transparent sources or drive substitution by alternative goods.

Mander et al. 2012 apply an energy services perspective to the derivation of future transport demand scenarios for the UK. They envision a number of different scenarios for the UK's future energy commodity mix, all compatible with the UK Climate Change Act legislation. Consistent with the Act's incentivization of a shift away from the current mix of energy commodity imports dominated by fossil fuels, the future scenarios include reductions in demand for the import of oil, gas, and coal and increased demand for imports of biofuel, biomass, and so on. The results are substantially different from the forecasts of transport demand that can be generated through a simple extrapolation of historical trends, and show the importance of aligning freight transport demand scenarios with assumptions about both national and international mitigation and adaptation policies.

Many forecasts for future growth in passenger air transport are, similar to Table 11.1, high. The Airbus Global Market Forecast from 2013 to 2032 (Airbus, 2013) and the Boeing Current Market Outlook from 2013 to 2032 (Boeing, 2013) both predict growth rates for global air transport of around 5 per cent per year. By 2050 conservative estimates predict a 30–110 per cent growth in passenger kilometres travelled over 2005 levels (Berghof et al., 2005), while other estimates project an increase of an order of magnitude (Schäfer, 2007). The forecast growth, however, differs significantly by region. Airbus (2013) predict that developing regions such as China, India, the Middle East, Latin America, and Eastern Europe, will grow at around 6 per cent per year, while developed regions such as Western Europe and North America are predicted to grow at closer to 4 per cent per year. Boeing (2013) makes similar forecasts.

Passenger transport has also seen some mode shift. While there is a general shift to faster modes of transport, this has not been exclusive to aviation. Where there has been significant investment in high-speed rail, such as in Spain, France, Japan, Korea, and China, there has in fact been a shift in passenger traffic from air to rail. For example, following the introduction of a high-speed rail service between Seoul and Daegu in Korea in 2004, there was a reduction in domestic aviation demand by between 34 per cent and 75 per cent (Park and Ha, 2006). Similarly, with the introduction of a high-speed rail service between Madrid and Seville in Spain in 1992, the market share of aviation on the route dropped by 33 per cent.

11.5 **Historic/existing carbon/energy intensity of transport supply**

Useful insights into aircraft energy intensity are provided by techno-economic theoretical perspectives: contrary to surface modes of transport, where increases in a vehicle's carrying capacity lead to a decline in energy intensity; for a given load factor, aircraft energy intensity is largely independent of size. This is because aircraft energy use is proportional to total vehicle weight because of the lift required to keep the aircraft in the air. Hence scale effects often observed in other modes are not observed in aviation. Instead, energy intensity in aircraft is largely a function of stage length (the distance between take-off and landing) and technology age. Take-off and climb are the most energy-intensive portions of a flight. For short-haul flights, which are typically operated by small aircraft, take-off and climb fuel may account for a comparatively large share of fuel burn, increasing the energy use per unit distance (and therefore energy intensity). For very long-haul flights, which are typically operated by larger aircraft, take-off and climb fuel may become proportionately negligible, but the fuel weight required for the flight becomes a significant portion of the total vehicle weight. This reduces the available payload weight, increasing energy use per revenue passenger or tonne of freight (and therefore energy intensity). The lowest energy intensity (E_U) exists for stage lengths between these two extremes, of around 2,000 km, as shown in Figure 11.4 from Babikian et al. (2002).

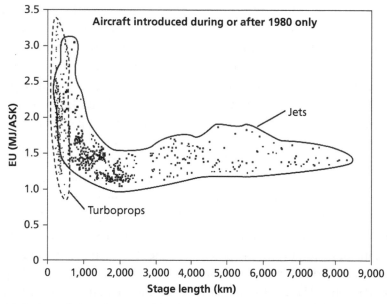

Figure 11.4 Variation of aircraft energy intensity (measured in Mega-Joules per Available Passenger Seat Kilometre) with stage length (Babikian et al., 2002)

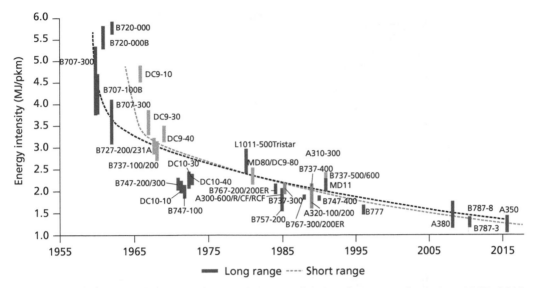

Figure 11.5 Historical and forecast improvements in aircraft energy intensity, 1955–2015 (International Energy Agency, 2009)

Note: The range of points for each aircraft reflects varying configurations, connected dots show estimated trends for short- and long-range aircraft.

Even more significant, however, is technology level. Figure 11.5 from the International Energy Agency (2009) shows the evolution of aircraft energy intensity from the 1950s to predicted levels in 2015. Clearly, aircraft—both small single aisle aircraft (short range) and large twin aisle aircraft (long range)—have experienced a strong decline in energy intensity over this time. Between 1959 and 1995, average new aircraft energy intensity declined by nearly two-thirds, with nearly 60 per cent of this being due to improvements in engine efficiency, just over 20 per cent due to increases in aerodynamic efficiency, and just over 15 per cent due to increased load factors (Schäfer et al., 2009). Technology has therefore been the dominant driver behind this improvement. Because fuel has always been a significant contributor to airline operating costs, airlines, and therefore manufacturers, have always had an incentive to improve fuel burn performance. This has increased with increasing fuel prices. From 2003 to 2013 alone, fuel has doubled as a percentage of airline costs (Boeing, 2013). So reducing energy intensity will continue to be an important objective in the future.

Similarly to the aviation sector, fuel cost is an important component of shipping industry total costs (Stopford, 2009) and creates an incentive to reduce costs through increased energy efficiency. With rising oil prices, as in aviation, fuel cost has taken an increased share of total cost particularly over the last decade. Shipping has a number of mechanisms at its disposal to modify the carbon intensity of transport supply. Unlike aviation, scale effects are important and larger ships can have energy intensities per unit of transport supply that are an order of magnitude lower than smaller ships. This is shown in Figure 11.6, with tankers in the size 1 category (0–10,000 tonnes deadweight) shown to have carbon intensity an order of magnitude greater than the largest size category (200,000 tonnes deadweight and above). Speed is also an important influence on energy intensity, with the consequence of the physics

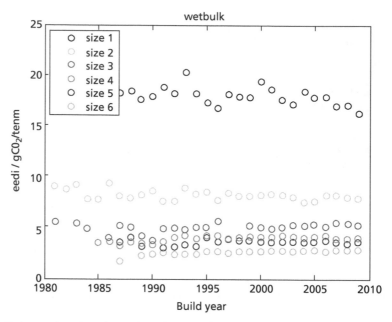

Figure 11.6 Estimated carbon intensity of newbuild tankers in size categories from build year 1980 to 2010

governing ship hydrodynamics meaning that speed reductions of 10 per cent result in an energy intensity reduction of 20 per cent (Smith, 2012a). There are also technology and engineering solutions that influence energy intensity—modifications to ship's engines, propulsors, hullform, and coatings can all reduce energy intensity. In light of these opportunities and the backdrop over the last decade of rising oil prices, the apparent stagnation of shipping's energy intensity over the last 30 years, at least within size range categories (shown in Figure 11.6), raises some questions. These questions cannot be answered just by techno-economic theory, which would expect that the trend of increasing fuel cost would result in a reduction in energy intensity over this period. Socio-technical literature has used shipping, and the transition from sail power to steam in the nineteenth century, as a case-study to understand technological transitions (Geels, 2002). However, there has been little literature to date that has created insight into the trend observed in Figure 11.6. This is partly due to the paucity and comparative unreliability of data characterizing the existing fleet. Due to emerging regulation the quality of information should now be improving and this in turn should enable researchers to address these questions.

11.6 **Planning, land use, and infrastructure influences on transport supply and its energy intensity**

Growth in air traffic is constrained by both airport capacity and airspace capacity. Airport capacity, and particularly runway capacity, is, however, generally the dominant constraint. Airport capacity expansion has become difficult in more industrialized nations because of opposition from local communities and limited space availability. While many airports were originally developed well away from city centres, businesses have grown around them because of the improved access to intercity transport. Hence communities serving these businesses, and the aviation business itself, have grown around the airports accordingly. This is less of a problem in many developing economies, however, where there is greater availability of land, and less political and social resistance to constructing new airports or adding capacity to existing airports. For example, only one new airport (with a runway longer than 10,000 ft) was constructed in the United States in the ten years from 2003 to 2014. In contrast, twenty-two and eight new airports (with a runway longer than 10,000 ft) were constructed in China and India over this time period respectively (CIA, 2013), although this difference is partly due to the fact that there are already a high number of airports in the United States, while there are relatively few in China and India.

Airspace congestion is currently only a significant issue in the world regions with the highest traffic levels, that is, North America and Western Europe. Hence there are significant efforts to improve the efficiency of both these airspace systems, to allow aircraft to fly closer to the optimal flight paths than is currently the case. The Single European Sky programme, launched in 1999, is an initiative to organize the European airspace into functional blocks, according to traffic flows rather than to national borders, while also modernizing and optimizing the future European Air Traffic Management network through advanced technologies and procedures (Eurocontrol, 2013). Similarly, in the United States, the Next Generation Air Transportation System (NextGen), which is intended to make air travel more convenient, predictable, and environmentally friendly, will be enabled by a shift to satellite-based and digital technologies and new procedures (FAA, 2013).

As demonstrated in Figure 11.6, large ships are more energy-efficient per unit of transport supply than smaller ships. Therefore an increase in energy efficiency could be achieved through an increase in the average size of ships—both by increasing the size of the largest ships and shifting transport supply from small ships to large ships. Constraints to the design and operation of larger ships can be divided into three categories: technology, market, and infrastructure. From the infrastructure perspective, a ship can only be deployed on a trade route if it can be safely navigated at either end of its voyage, and pass through any constrictions during its voyage (e.g. the canals in Panama and Suez), and the port or terminal can support the vessel when alongside and load or unload its cargo.

The cost of dredging (to enable access for larger ships) and development of port infrastructure mean that port access is still a significant obstacle in many countries, particularly in the developing world, to the operation of larger ships. This also presents an interesting trade-off between a ship's design for minimal resistance (which occurs at a certain combination of the main dimension parameters length, beam, and draught) and design for operational flexibility (the ability to operate on the largest range of trade routes and therefore selection of values of beam and draught to accommodate physical restrictions), which can be missed in more simplistic techno-economic models that ignore such constraints (for example, IMO (2010a)).

11.7 **Potential for future energy efficiency gains and carbon intensity reduction fuel**

Biofuel is seen as one of the most important technologies for reducing greenhouse gas emissions in the aviation sector. This is because of the very high gravimetric and volumetric energy densities (MJ/kg and MJ/L respectively) of synthetic paraffinic kerosene, which is comparable to those of conventional jet fuel (Jet A). Both densities are important because the aircraft energy use is not only to overcome drag (for which volumetric energy density is important), but also to generate the lift required to carry the vehicle, cargo, and fuel (high gravimetric density reduces the weight of fuel required for a particular journey). Because of this, and because of the lack of other alternative fuels for aviation, it has been proposed that aviation have priority as a biofuel user (Sgouridis, 2012). Indeed, based on supply curves from the US Department of Energy's 'Billion-Ton Update' study (Department of Energy, 2011) and using a 50 per cent conversion efficiency of biomass to jet fuel, waste feedstocks for cellulosic biomass in the USA in 2012 were theoretically enough to supply all domestic operations of the US commercial airline fleet in that year. However, this would require prioritization for aviation, and would be unlikely to apply in the future, as demand for air travel is likely to grow faster than the supply of waste feedstock.

Hydrogen has been suggested as a source of energy for aircraft in the past, and in fact Tupolev flew a modified Tu-154B with hydrogen fuel more than twenty-five years ago. However, while it is technologically feasible, the weight and size of the liquid hydrogen tanks make it highly uneconomical, and this challenge has not been overcome. Hydrogen is not therefore seen as a compelling alternative fuel for aviation.

Shipping fuels used in the last decades are predominantly residual fuels derived from crude oil, particularly heavy fuel oil (HFO). Tightening regulations on sulphurous and nitrogen oxide emissions, and increased pressure on CO_2 emission regulations have led to particular attention being paid to the potential of liquid natural gas as a new fuel in the industry. To date, the lack of refuelling infrastructure means that this remains the subject of pilot projects only. Further alternatives exist, including biofuels and hydrogen, and studies have been undertaken to consider the possible future landscape and market shares of these different fuels. One of these studies, Lloyd's Register and UCL Energy Institute (2014), suggests that without further changes in CO_2 emission regulations, and considering expectations of bioenergy availability, the industry will continue to use similar fuels, with some moderate penetration (depending on relative oil and gas prices) of liquid natural gas. However, under certain specifications of a hypothetical global shipping emission trading scheme, hydrogen becomes a viable shipping fuel.

11.8 **Energy efficiency technologies**

A number of technologies are under development that may have a significant impact on aircraft energy use. Already, the use of composite materials and improved electrical systems has allowed manufactures to reduce aircraft weight significantly, leading to reductions in fuel burn. The Boeing 787, for example, is 20 per cent more fuel efficient than the B767, and nearly half of this is achieved through weight reduction. The other half is achieved through improved engines and aerodynamics. Engines, particularly, are expected to improve significantly in the future. The most notable design development is likely to be the open rotor or propfan, which uses a gas turbine to drive an unshielded propeller like a turboprop, but the propeller itself is designed with a large number of short, highly twisted blades, similar to a turbofan's fan. The engine is hoped to gain the efficiency benefits of a turboprop, at the speed of a turbofan, giving between 35 per cent and 45 per cent fuel burn improvement over existing narrow-body aircraft depending on assumptions about operating speed (Vera Morales et al., 2011). Aircraft design changes are also envisioned to make significant improvements to energy use. The most notable of these is the blended-wing-body. While this flying-wing design, which has significantly improved lift-to-drag ratios over traditional 'tube-and-wing' aircraft, has existed since the Second World War, it has not as yet been applied to commercial airliners. A number of studies have, however, been done, predicting fuel burn improvements over existing types of between 19 per cent and 54 per cent, depending on assumptions about operations, engines and so on (Graham et al., 2014).

For shipping, the technology options can be categorized under a number of sub-headings:

- Drag reduction: total drag reduction (e.g. devices for wave-making drag reduction); friction drag reduction (e.g. hull coatings and air lubrication); appendage drag reduction (e.g. high lift rudders); parasitic drag reduction (e.g. fairing of hull openings); aerodynamic drag reduction (e.g. fairing of superstructure).
- Propulsion energy efficiency increase: propulsor efficiency increase (e.g. alternative propulsors); propeller inflow and outflow efficiency increase (e.g. pre-swirl ducts and vane wheels).
- Engine and machinery efficiency increase (e.g. engine tuning or control).
- Auxiliary energy efficiency increase: heat recovery systems; electrification and control of auxiliary systems.

11.9 **Operations**

Because of the high share of fuel costs in airline direct operating costs, airline operations are already carefully optimized. However, with the development of new technology, opportunities are arising for improvements. New, lighter seats and other cabin accessories are leading to cabin retrofits, which can reduce weight, although in many cases they just offset the increase in weight due to improved audio-visual equipment. Improved revenue management techniques mean that passengers that are spilled—not finding tickets for the flight they want to fly on, and therefore choosing a different airline or not flying at all—are typically low-value customers. Hence it has become economic to increase load factors. Better matching of aircraft type to mission, particularly using smaller aircraft for 'thin' routes with low load factors, could also reduce fuel burn, but because of airline requirements for flexibility in their fleets, this change is unlikely to be adopted soon. Improved engine and aerodynamic maintenance has also been adopted, reducing fuel burn further. Furthermore, fuel burn can be reduced by flying with less contingency fuel, although there are arguments from some pilots that this has a negative impact on safety (USA Today, 2008). Other operational changes exist that would only be implemented were airline costs to change, due to, for example, policy intervention. These include flying slower, that is, closer to the minimum fuel burn speed as opposed to the minimum cost speed; not tankering fuel (tankering refers to fueling an aircraft with more fuel than required because of lower fuel costs at the origin airport than at the destination); and early retirement of older, less fuel-efficient aircraft, replacing them with newer, more fuel-efficient aircraft.

In the shipping industry, ship speed reductions have been the explanation for reduced energy demand and emissions over the period since the financial crisis (2008 to 2012). In this instance it was not regulation, but the particular market conditions (an oversupply of ships, low revenues and high energy costs) that caused the phenomenon. However, this episode has illustrated the large potential this particular operational 'lever' has for influencing energy and emissions intensity in the future.

11.10 Transitions and take-up of alternative technologies

Because of the long life of most civil aircraft (in the order of thirty years), projected transition to new technology is generally slow (although less so in high-growth regions), and dependent on fuel price, policy measures, the range of technologies available, and when they will become available. Dray et al. (2010) perform a detailed study of technology uptake in Europe under different carbon prices, with aviation included in the European Emissions Trading System (ETS). They observe that some options are rapidly taken up under all scenarios (e.g. improved air traffic management (ATM)), others are taken up more slowly by specific aircraft classes depending on the scenario (e.g. biofuels) and others have negligible impact in the cases studied. High uptake of one mitigation option may also reduce the uptake of other options.

The literature on the potential for future technology take-up in shipping is dominated by techno-economic and physical technical economic models (PTEM), for example

Buhaug et al. (2009), IMO (2010a), and Eide et al. (2012). An exception which has taken a social-technical perspective is Kohler and Senger (2012), which considered the application of agent-based models to understand the diffusion of technology between shipping's stakeholders, particularly how the implementers of innovation (shipbuilders and equipment suppliers) interact with the professional societies, classification societies, and investment mechanisms. This reveals the need to understand these interactions and the potential barriers or opportunities for enabling they represent which may not be being captured in PTEM literature. One barrier that has been observed to be significant in other industries is the principal–agent relationship that exists in particular between the ship-owner and the charterer (the entity operating a ship and often paying fuel costs for the ship). A social-practice theory analysis of this relationship is presented in Rehmatulla et al. (2013), with quantifications of the barrier applied to the shipping system model GloTraM to simulate the consequence of principal–agent relationships on technology take-up and transitions in the shipping industry. Unsurprisingly, the occurrence of market barriers is shown to significantly reduce technical energy intensity improvements in the industry, although the study acknowledges that further work is required to assess and quantify the barriers.

11.12 **Policies and drivers for transitions**

The only existing legislation driving reduction in emissions, and hence energy use, in aviation is the European Emissions Trading System (ETS), and at the time of writing its implementation is still a matter of debate. While aviation was to be included in the ETS from 2012, there has been significant political debate internationally surrounding how it is to be included, and whether or not it violates international agreements (particularly the Chicago Convention), which prevent the taxation of commercial aviation fuel (ICAO, 2006b). One consequence of the proposal to include aviation in the EU ETS has been pressure to instead develop a global scheme through The International Civil Aviation Organization (ICAO)—a specialized agency of the United Nations. Recently, there has been an agreement that such a scheme should be implemented by 2020, with a detailed plan by ICAO's next general assembly in 2016 (Alcock, 2013; theguardian.com, 2013). Despite the limited policy intervention in this area, some organizations and countries have emissions targets especially for aviation. The International Air Transport Association (IATA), the trade association for the world's airlines, has pledged to improve fuel efficiency by an average of 1.5 per cent per year to 2020; to stabilize carbon emissions from 2020 with carbon-neutral growth; and to reduce carbon emissions by 50 per cent by 2050 compared to 2005 (IATA, 2009). The UK government has adopted a slightly less stringent target to reduce UK aviation emissions to 2005 levels by 2050 (Committee on Climate Change, 2009). While not binding, these targets do indicate that there is an awareness of the problem of aviation emissions linked to energy use.

On 1 January 2013, shipping entered a new regime with the entering into force of its first global regulation on energy efficiency, the Energy Efficiency Design Index (EEDI), mandating that new-build ships meet a maximum carbon intensity (used in this instance as a proxy for energy intensity). At the same time, the International Maritime Organisation (IMO) introduced the Ship Energy Efficiency Management Plan (SEEMP). The IMO forecast that these policies would result in modest emissions reductions in the global fleet (IMO, 2012), using a PTEM theoretical approach. Even when allowing for uncertainty in the analysis of the emissions forecasts and reductions, these reductions are not aligned with the UNFCCC and COP rhetoric and significantly more will need to be done (Anderson and Bows, 2012). For such purposes, the IMO has progressed discussion on Market Based Measures (MBM) over a number of years (IMO, 2010b). No consensus has formed on whether a MBM (of a number of candidate forms), or a command and control approach is acceptable, although much of this is to do with obstructions to progress created by the misalignment between the UNFCCC principle of Common But Differentiated Responsibilities, and the IMO principle of No More Favourable Treatment.

Regulation of shipping's energy efficiency has also been considered at the EU level, and CE Delft (2009) provided a number of potential mechanisms that the EU could use to apply unilateral regulation to shipping. To date, the EU has implemented none of these mechanisms, although it has produced a proposal for the implementation of Measuring, Reporting, and Verification (EC, 2013), which could be a first step to a regulation requiring progressive reductions in emissions or emissions intensity over time. Whether shipping will find itself in a complicated regional and national patchwork of GHG and energy efficiency regulation, or whether a homogeneous international policy landscape defined by the IMO will prevail, remains unknown and will be the subject of ongoing debate for at least the remainder of this decade.

11.13 Other climate impacts

As mentioned above, non-CO_2 climate effects are significant in aviation. These include particularly NO_x effects, and the effect of contrails and contrail induced cirrus, as described by Penner (1999), Sausen et al. (2005), and Lee et al. (2010). Krammer et al. (2013) shows that widespread use of aviation biofuel may lead to a scenario in which aviation growth is accompanied by flat or decreasing carbon emissions, but an increasing total climate impact. No policy measures have so far been considered to account for these non-CO_2 effects, although an 'uplift factor' has been proposed, increasing the accounted-for CO_2 emissions under an ETS, so as to capture non-CO_2 effects. However, the scientific community is in agreement that problems exist with the use of uplift factors. The different impacts due to different aircraft emissions act over different time and length scales, so basing any calculations on the time and length scales associated with

CO_2 emissions can produce misleading results, and may not drive responses that reduce non-CO_2 effects significantly (Bows et al., 2005; Dray et al. 2009).

11.14 **Unintended consequences**

The rebound effect in the aviation sector was estimated by Evans and Schäfer (2013), for a relatively small network (twenty-two airports), covering the busiest airports in the United States. Their results indicate that the average rebound effect in this network is about 19 per cent, for the range of aircraft fuel burn reductions modelled. This is the net impact of an increase in air transport supply to satisfy the rising passenger demand, airline operational effects that further increase supply, including particularly frequency competition effects, and the mitigating effects of an increase in flight delays. Although the magnitude of the rebound effect seems relatively small, it can be significant for a sector that has comparatively few options for reducing greenhouse gas emissions.

Shipping has an unintended consequence risk from the interaction between technical energy efficiency, operational energy efficiency, and profit maximization. Smith (2012b) illustrates how improvements in energy efficiency achieved through fitting technology to ships can result, through the rational application of profit maximization by firms considering both revenues and costs, in increased operational speeds. These increased operational speeds in turn result in higher operational emissions that negate emissions reductions that might have otherwise been achieved.

11.15 **Conclusion**

Aviation and shipping both contribute significantly to the global economy, enabling economic development, trade, and business. However, driven to a large part by globalization, they are both growing fast, and their heavy reliance on oil-based fuel means the sectors have important and growing environmental consequences. This is despite both sectors making significant improvements in energy intensity in recent decades. In order to mitigate the climate impact of both aviation and shipping, new technologies, fuels, and operating procedures are required. However, because of the long lifecycles of aircraft and ships, and because of the high costs of new technology, further policy intervention, such as a global emissions trading systems or command and control regulation on emissions intensity, may be required. Such policy intervention would spur technological development as well as slow demand growth, with associated benefits and costs to the global economy. What remains to be seen is whether or not there is sufficient political will across enough global players to make such a policy intervention a reality.

■ REFERENCES

Airbus, 2013. 'Global Market Forecast 2014–2033.' <http://www.airbus.com/company/market/forecast/> (Accessed 13 April 2015).

Airlines for America, 2014. 'Annual Round-Trip Fares and Fees: Domestic', <http://www.airlines.org/Pages/Annual-Round-Trip-Fares-and-Fees-Domestic.aspx?View={e7eeb8b9-cbd2-4345-bbe5-1aac9cac4867}&SortField=AvgRTMiles_x002d_DOM&SortDir=Asc> (Accessed 4 February 2014).

Alcock, C., 2013. <http://www.ainonline.com/aviation-news/ain-air-transport-perspective/2013-10-07/icao-agrees-global-emissions-cap-and-trade-scheme-2020>. (Accessed April 2015).

Anderson, K. and Bows, A. 2012. Executing a Scharnow turn: reconciling shipping emissions with international commitments on climate change. *Carbon Management* 3. 6: 615–28.

ATAG (Air Transport Action Group), 2012. 'Aviation: Benefits beyond borders report,' prepared by Oxford Economics for ATAG, Geneva, Switzerland, <http://aviationbenefitsbeyondborders.org/download-abbb-report> (Accessed 14 February 2014).

Babikian, R., Lukachko, S. P., and Waitz, I. A. 2002. The historical fuel efficiency characteristics of regional aircraft from technological, operational, and cost perspectives. *Journal of Air Transport Management* 8.6: 389–400.

Berghof, R., A. Schmitt, A., Eyers, C., Haag, K., Middel, J., Hepting, M. Grübler, A., and Hancox, R. 2005, CONSAVE 2050—Executive Summary, Competitive and Sustainable Growth, European Community, Growth Programme GROW-2001-4: New Perspectives in Aeronautics, <http://www.dlr.de/consave/CONSAVE%202050%20Executive%20Summary.pdf> (Accessed April 2015).

Boeing, 2012. 'World Air Cargo Forecast 2012–2013.' http://www.aia-aerospace.org/assets/Boeing_World_Air_Cargo_Forecast_2012-2013.pdf (Accessed 13 April 2015).

Boeing, 2013. 'Current Market Outlook 2013–2032.' <http://www.boeing.com/assets/pdf/commercial/cmo/pdf/Boeing_Current_Market_Outlook_2013.pdf> (Accessed 13 April 2015).

Bows, A., Upham, P., and Anderson, K. 'Growth Scenarios for EU & UK Aviation: contradictions with climate policy.' Report for Friends of the Earth Trust Ltd (2005).

Buhaug, Ø., Corbett, J.J., Endresen, Ø., Eyring, V., Faber, J., Hanayama, S., Lee, D.S., Lee, D., Lindstad, H., Markowska, A.Z., Mjelde, A., Nelissen, D., Nilsen, J., Pålsson, C., Winebrake, J.J., Wu, W.-Q., Yoshida, K. (2009). Second IMO GHG study 2009 London: International Maritime Organization (IMO).

CIA (US Central Intelligence Agency), 2013. The world factbook. https://www.cia.gov/library/publications/download/download-2013/index.html. (Accessed 20 March 2014).

Committee on Climate Change, 2009. 'Meeting the UK Aviation target – options for reducing emissions to 2050', <http://archive.theccc.org.uk/aws2/Aviation%20Report%202009/21667B%20CCC%20Aviation%20AW%20COMP%20v8.pdf December 2009> (Accessed 4 February 2014).

Delft CE, 2009, Technical support for European action to reducing Greenhouse Gas emissions from international maritime transport. CE Delft, Delft.

Department of Energy (2011). US Billion-Ton Update: Biomass Supply for a Bioenergy and Bioproducts Industry. R.D. Perlack and B.J. Stokes (Leads), ORNL/TM-2011/224. Oak Ridge National Laboratory, Oak Ridge, TN. 227p.

Dray, L., Evans, A.D., Reynolds, R., Schäfer, A. 2010. Mitigation of aviation emissions of carbon dioxide: an analysis for Europe, *Transportation Research Record*, 2177. 1: 17–26.

Dray, L.M., Evans, A., Reynolds, T., Rogers, H., Schäfer, A., and Vera-Morales, M.. 'Air Transport Within An Emissions Trading Regime: A Network-Based Analysis of the United States and India.' TRB 88th Annual Meeting, Washington DC. 2009.

EC, 2013, Proposal for a regulation of the European parliament and of the council on the monitoring, reporting and verification of carbon dioxide emissions from maritime transport and emending regulation (EU) No. 525/2013. 2013/0224 European Commission. Brussels.

Eide, M., Chryssakis, C., Alvik, S., Endresen, Ø. 2012, Pathways to Low Carbon Shipping—Abatement Potential Towards 2050 DNV Position Paper 14–2012.

Eurocontrol, 2013. <http://www.eurocontrol.int/dossiers/single-european-sky> (Accessed April 2015).

Evans, A.D. and Schäfer, A. 2013, The rebound effect in the aviation sector, *Energy Economics*, 36: 158–65. doi.org/10.1016/j.eneco.2012.12.005.

FAA, 2013, <http://www.faa.gov/nextgen/media/Extended_Executive_Summary_2011.pdf> (Accessed April 2015).

Feenstra, R.C., Inklaar R., and Timmer, M.P. 2013, 'The Next Generation of the Penn World Table' available for download at <http://www.rug.nl/research/ggdc/data/pwt/> (Accessed April 2015).

Geels, F.W. 2002, Technological transitions as evolutionary reconfiguration processes: a multi-level perspective and a case-study. *Research Policy* 31: 1257–74.

Graham, W.R., Hall, C., and Vera Morales, M. 2014, 'The potential of future aircraft technology for noise and pollutant emissions reduction,' *Transport Policy Special Issue on Aviation and the Environment*.

IATA (International Air Transport Association) 2009, 'Halving Emissions by 2050 - Aviation Brings its Targets to Copenhagen', Press Release No.: 54, 8 December 2009.

ICAO (International Civil Aviation Organization) 2006a, 'World Total Traffic 1995 – 2006', <http://www.icaodata.com>.

ICAO 2006b, 'Convention on International Civil Aviation.' Document 7300/9.

IMO 2010a, Marginal abatement costs and cost-effectiveness of energy-efficiency measures, MEPC 61/Inf.18.

IMO 2010b, Reduction of GHG emissions from ships—full report of the work undertaken by the Expert Group on feasibility study and impact assessment of possible Market-Based Measures. MEPC 61/Inf.2.

IMO 2012, Assessment of IMIO mandated energy efficiency measures for international shipping. MEPC 63/Inf.2.

International Energy Agency 2009, 'Transport, Energy and CO2' <http://www.iea.org/publications/freepublications/publication/name,3838,en.html> (Accessed 2 June 2014).

International Energy Agency Data Services 2007, Energy Balances and Energy Statistics for OECD and non-OECD Countries.

Kasarda, J.D., and Green, J. D. 2005, Air cargo as an economic development engine: A note on opportunities and constraints. *Journal of Air Transport Management* 11.6: 459–62.

Kohler, J. and Senger, F. 2012, An Agent-Based Model of Transitions to Sustainability in Deep Sea Shipping. Low Carbon Shipping Conference 2012, Newcastle.

Krammer, P., Dray, L., and Köhler, M. O. 2013, Climate-neutrality versus carbon-neutrality for aviation biofuel policy. *Transportation Research Part D: Transport and Environment* 23: 64–72.

Lee, D.S., Pitari, G., Grewe, V., Gierens, K., Penner, J.E., Petzold, A., Prather, M.J., Schumann, U., Bais, A., Berntsen, T., Iachetti, D., Lim, L.L., and Sausen, R. 2010, Transport impacts on atmosphere and climate: Aviation. *Atmosphere and Environment* 44: 4678–734. doi:10.1016/j.atmosenv.2009.06.005.

Lloyd's Register and UCL Energy Institute 2014, Global Marine Fuel Trends 2030. Published by Lloyd's Register Group Limited (London) and UCL Energy Institute (London).

Mander, S., Walsh, C., Traut, M., Gilbert, P., and Bows, A. Decarbonising the UK energy system and the implications for UK shipping. *Carbon Management* 3.6: 601–14.

Park, Y. and Ha, H-K. 2006, Analysis of the impact of high-speed railroad service on air transport demand. *Transportation Research Part E: Logistics and Transportation Review* 42.2: 95–104.

Penner, Joyce E., (ed.) 1999, *Aviation and the Global Atmosphere: Special Report of the IPCC Working Groups I and III in Collaboration with the Scientific Assessment Panel to the Montreal Protocol on Substances that Deplete the Ozone Layer*. Cambridge University Press.

Rehmatulla, N., Smith, T.W.P., and Wrobel, P. 2013, Implementation Barriers to Low Carbon Shipping. Low Carbon Shipping Conference 2013, London.

Sausen, R., Isaksen, I., Grewe, V., Hauglustaine, D., Lee, D. S., Myhre, G., Kohler, M. O. et al. 2005, Aviation radiative forcing in 2000: An update on IPCC (1999). *Meteorologische Zeitschrift* 14.4: 555–61.

Schäfer A. 2007, 'Long-Term Trends in Global Passenger Mobility,' In *Frontiers of Engineering: Reports on Leading-Edge Engineering from the 2006 Symposium*, p. 85.

Schäfer A, Heywood, J.B., Jacoby, H.D., and Waitz, I.A. 2009, *Transportation in a Climate-Constrained World*. MIT Press.

Sgouridis, S. 2012, Are we on course for a sustainable biofuel-based aviation future? *Biofuels* 3.3: 243–6.

Smith, T.W.P. 2012a, Low carbon ships and shipping. In: O. Inderwildi and Sir David King (eds). *Energy, Transport, & the Environment*, DOI: 10.1007/978-1-4471-2717-8_1. Springer-Verlag: London.

Smith, T.W.P. 2012b, Technical energy efficiency, its interaction with optimal operating speeds and the implications for the management of shipping's carbon emissions. *Carbon Management* 3.6: 589–600.

Stopford, M. 2009, *Maritime Economics*. 3rd Edition, Routledge: London/New York.

theguardian.com. 'UN aviation agency "dragging feet" on efforts to tackle emissions,' BusinessGreen, part of the Guardian Environment Network, theguardian.com, Thursday 5 September 2013. <http://www.theguardian.com/environment/2013/sep/05/un-aviation-agency-emissions-icao> (Accessed April 2015).

UN (United Nations) 2013, *Statistical Yearbook 2011*, Fifty-sixth issue', Department of Economic and Social Affairs, Statistics Division, United Nations, New York.

USA Today 2008. Pilots complain airlines restrict fuel to cut cost, August 8, 2008, <http://www.usatoday.com/money/economy/2008-08-08-519303435_x.htm> (Accessed 4 February 2014).

Vera Morales, M., Graham, W.R., Hall, C.A., and Schäfer, A (2011), 'Techno-economic analysis of aircraft'. TOSCA project work-package 2 final report; <http://www.toscaproject.org/FinalReports/TOSCA_WP2_Aircraft.pdf> (Accessed April 2015).

12 **Carbon capture and storage**

Jim Watson and Cameron Jones

12.1 **Introduction**

Carbon capture and storage (CCS) technologies are often highlighted as a crucial component of future low-carbon energy systems. However, they are still being developed and demonstrated. It is therefore unclear when these technologies will be technically proven at full scale, and whether their costs will be competitive with other low-carbon options. For their supporters, CCS technologies offer a crucial way to square the continued use of fossil fuels with climate change mitigation. According to the International Energy Agency (IEA), fossil fuels will continue to supply the majority of the world's energy to 2035, even if climate change mitigation is taken very seriously (IEA, 2011).

The IEA '450 scenario' considers a global energy system trajectory that provides a technically feasible pathway for limiting average temperature increases to 2 °C. Under this scenario, CCS would be fitted to 32 per cent of the world's coal-fired power plant capacity (410 GW out of 1,270 GW) by 2035, and 10 per cent of global gas-fired capacity (210 GW out of 2,110 GW) by the same date. CCS technologies would therefore account for 22 per cent of the reduction in CO_2 emissions by 2035 when compared to the IEA's alternative 'new policies scenario' in which global greenhouse gas emissions would continue to rise. Modelling conducted by the UK Energy Research Centre for this book (see Chapter 24) includes the deployment of 890 GW of CCS capacity by 2035 in a scenario that limits global temperature rises to 2 °C. This more ambitious scenario for CCS includes further deployment after 2035 so that total CCS deployment reaches 1,280 GW by 2050.

While many governments and companies are now funding and developing CCS technologies, there is a long way to go before we know whether such a role for CCS will be technically and economically feasible. Pilot-scale capture plants are in operation in several countries, CO_2 is routinely transported across large distances in the United States, and CO_2 is being injected successfully at a number of storage sites. But full-scale CCS plants are thin on the ground (Global CCS Institute, 2013). Most demonstrations have focused on gas processing, synthetic fuels, and fertilizer production—applications that are less technically demanding and more economically attractive than CCS in the power sector. To some extent, this makes sense since other low-carbon technologies are available to decarbonize the power sector, while CCS will be essential if emissions from some other industrial sectors, such as steel and cement, are to be reduced significantly.

In the absence of strong climate change mitigation policies, CCS will only increase power production costs when compared to the costs of unabated production. According to the IEA (2011), 'incorporating CCS into a power plant increases the levelised cost of the electricity produced by between 39% and 64%, depending on the technology and fuel source'. This increase is expected for three main reasons. First, the incremental capital costs of adding CCS to a fossil fuel power plant are substantial. Second, the energy penalty of carbon capture is significant. For example, it will reduce power plant efficiency by around ten percentage points. Third, there is a need to pay the investment and operational costs of pipeline and storage infrastructure.

Given these economically unattractive attributes, it is not surprising that there are no full-scale CCS demonstrations in operation at coal- or gas-fired power plants. As discussed later in this chapter, the first plants are currently under construction in the United States and Canada, underpinned by government financial support. There are plans for a number of other CCS power plants, including two full-scale demonstrations in the UK. However, other plans have been cancelled in the last few years, including the Longannet plant in Scotland and the Jänschwalde plant in Germany. While economic and financial factors were significant in the collapse of Longannet, the cancellation of Jänschwalde followed public protests against the planned use of onshore storage. The Global CCS Institute emphasizes the particular importance of economic and policy barriers to such demonstration projects in a recent survey report (Global CCS Institute, 2013).

This chapter builds on UKERC research on CCS in the UK (Watson et al., 2012), and focuses on efforts that are underway in several countries to demonstrate CCS at full-scale on power plants and large industrial facilities, and some of the policy challenges associated with such demonstrations. The chapter provides an important case study of technology demonstration, which is a particularly risky stage of the innovation process (see Chapter 2 of this book).

The chapter comprises three further sections. Section 12.2 provides a survey of the global status of CCS, with a focus on progress with large-scale demonstration projects. Section 12.3 then focuses on demonstrations in the UK and North America, and analyses some of the reasons for the contrasting fortunes of these demonstrations. Finally, Section 12.4 draws some conclusions.

12.2 **Global status of CCS**

This section provides a brief overview of the countries in which large-scale CCS demonstration and deployment projects are progressing. Though the majority of international CCS projects are not yet at this full scale, such 'scaling up' is crucial so that the economic and technical viability of CCS as a route for significant reductions in CO_2 emissions can be tested. Demonstrations at this scale are not only required in the power sector, but also in other industries that include large point sources of CO_2 such as steel, aluminium, and cement.

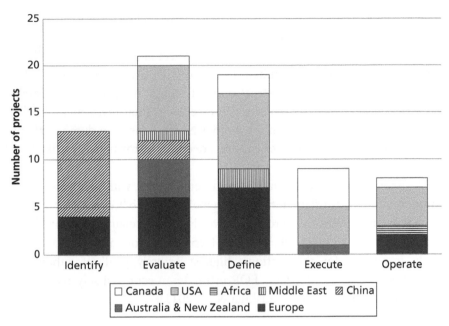

Figure 12.1 The figure provides further details on the status and geographical spread of these projects. As shown, planned projects are divided into three sub-categories to indicate how far advanced their development process is. Confirmed projects are divided into two sub-categories: under construction ('execute') or operational
Source: (Global CCS Institute, 2013).

The Global CCS Institute based in Australia publishes regular overviews of international CCS activity including analysis of what it calls Large-scale Integrated Projects (or LSIPs). These are defined as plants that capture at least 800,000 tonnes of CO_2 per year for coal-fired power plants or 400,000 tonnes of CO_2 per year for gas-fired generation[1] (Global CCS Institute, 2013). Of the seventy-two globally identified LSIPs, only seventeen, representing an anticipated capture capacity of 37 million tonnes of CO_2 per annum (Mtpa), are in some stage of construction or operation. The other fifty-five LSIPs remain in the planning stage. These planned projects represent a potential annual capture rate of 104 million tonnes of CO_2 per year, but for the moment they await a final investment decision.

12.2.1 NORTH AMERICA

Figure 12.1 illustrates the dominance of North American projects in the 'execute' and 'operate' categories. Roughly 75 per cent, or thirteen of the seventeen LSIPs in construction or operation, are situated in North America. This includes the two first full-scale power plant CCS projects: Boundary Dam in Saskatchewan, Canada and Kemper

---

[1] Those projects utilizing carbon dioxide for EOR (Enhanced Oil Recovery) purposes must also adhere to the same capture rates to qualify as an LSIP.

County IGCC in Mississippi, USA. Boundary Dam started operating in 2014, and Kemper County is expected to be online in 2015.

Canada currently has the world's largest CCS project at a syngas production plant in Weyburn, Saskatchewan, where 3 million tonnes of CO_2 are annually injected for EOR purposes. Four LSIPs are in the construction or 'Execute' stage, including Shell's Quest hydrogen production project, which recently received a final investment decision. The Shell project distinguishes itself from most other North American demonstrations in its intent to permanently store 1 Mtpa of CO_2 in a deep saline formation rather than utilize it for EOR.

The USA currently possesses the largest number of LSIPs, with four projects currently in operation, four under construction and fifteen either at the 'Evaluate' or 'Define' stage. The most significant driver of US progress towards commercialized CCS is the domestic demand for CO_2 with EOR intent. The US Department of Energy describes CCS as significantly important for the 'nation's highest priority goal' of enhanced energy security. Current projections for CO_2-EOR estimate an economically recoverable minimum of 67 billion barrels of oil (Kuuskraa et al., 2011).

12.2.2 CHINA

China has focused on CCS R&D rather than full-scale demonstrations so far. An important reason for this is a lack of strong enough policy drivers for CO_2 emissions reduction—and understandable resistance from power companies and many in government to a technology that adds significantly to costs while reducing the efficiency of power generation. For a country where electricity demand is growing very rapidly, these inherent features of CCS technologies are particularly problematic.

China's CCS projects have therefore focused on the testing of technology through pilots rather than on commercialization. The eleven LSIPs in China are all in the very early development stages, with nine in 'Identify' and two in 'Evaluate'. In addition to cost and efficiency barriers, another significant challenge facing potential LSIPs in China is the lack of cross-sectoral collaboration. Many of the power producers keen to develop CCS do not have access to suitable EOR sites and cannot therefore justify a business case (Best and Levina, 2012).

12.2.3 MIDDLE EAST AND NORTH AFRICA

Despite the geographical suitability and high per capita emissions intensity of the Middle East, CCS technology remains marginalized in the region. Only the United Arab Emirates is actively engaged in CCS, with two non-power related industrial projects in the 'Define' phase.

In Northern Africa, Algeria's 'In Salah' project is one of the largest and most extensively studied large-scale CCS projects in the world. Operational since 2004, it is a pre-combustion carbon capture project accounting for the storage of approximately 1 million tonnes of CO_2 annually (Global CCS Institute, 2013). The project developers, British Petroleum (BP), Sonatrach, and Statoil, designed the project with the initial intent of demonstrating CCS as a cost-effective approach to carbon mitigation. It plays a critical role in global knowledge sharing and research networks for CCS, and is studied for both the technical aspects of CO_2 capture, transportation, and storage and the regulatory practices needed for carbon mitigation of this magnitude. Statoil forecasts a storage total of 17 million tonnes over the next twenty years from this project (Statoil, 2009).

12.2.4 EUROPE

In 2007, the European Council agreed to the ambitious target of deploying twelve commercial-scale demonstrations of CCS by 2015 (Global CCS Institute, 2013). Now, six years later, no demonstrations are under construction or in operation. This poor performance to date is due to a combination of European and national factors. European funding was allocated for CCS demonstrations as part of the European Commission's NER300 competition. This funding mechanism was named after the 300 million allowances that were made available from the new entrants' reserve (NER) of the EU emissions trading scheme. New entrants that proposed CCS or renewable energy demonstrations would be made eligible for a number of free allowances. No funding was allocated to carbon capture and storage demonstrations in the first round of NER300 in 2012. This is despite eleven short-listed projects passing technical and financial assessments: €1.2 billion (~£1bn) was allocated to twenty-three renewable energy demonstrations (European Commission, 2012a). When the European Commission announced the scheme it projected that at least £3.8 billion would be raised from the emission allowances, but low carbon prices have resulted in only half the amount expected from the first tranche of sales. Since the first call for applications in 2010, the EC carbon price has declined from £12.65 per tonne to around £4 per tonne at the time of writing. Furthermore, only a fraction of the original NER300 allowances are available for second round funding, and with allowance prices being much lower than anticipated when the NER300 scheme was announced, confidence and progress in CCS deployment has inevitably slowed.

The Don Valley IGCC Power Project in South Yorkshire, England, led the European Commission rankings of CCS projects during the first round of the NER300 scheme (European Commission, 2012b). However, it did not receive funding from that scheme, and has not been shortlisted for UK CCS programme funding. However, it has secured €180 million (~£150m) from the European Energy Program for Recovery (European Commission, 2012b). Advanced design studies indicate this project should be operational by 2016 with a potential capture rate of 5 million tonnes per annum. As discussed in Section 12.4, the UK government's commercialization programme has shortlisted two other CCS projects, White Rose and Peterhead.

The Vattenfall Schwarze Pump facility in Germany has piloted since 2006 a small-scale demonstration of the proprietary CCS technology of Air Products and Chemicals Inc. This is now recognized as the world's pre-eminent demonstration of oxy-fuel based CCS (Air Products, 2011).

Though not a member of the European Union, Norway has made more progress with CCS. Two smaller projects are currently in operation with the injection of CO_2 from natural gas cleaning at the Sleipner and Snøhvit stations. Since 2006, Statoil has stored approximately 8 million tonnes of CO_2 at a rate of 500 Mtpa from Sleipner. Snøhvit, which is a CCS-retrofitted liquefied natural gas (LNG) processing unit, permanently stores 0.7 Mtpa through an offshore sandstone reservoir. No signs of leakage have been observed at either of these operations and both projects were triggered by a carbon tax (Michael et al., 2010).

12.3 Realizing the potential: challenges of technology demonstration

This section will analyse efforts to demonstrate CCS technologies at full scale. It will do so by contrasting the relatively successful demonstration programmes in the USA and Canada with the relatively slow progress in Europe. With respect to European CCS demonstrations, the UK's demonstration programme will be analysed in particular since it has been one of the most ambitious—and has been underpinned by cross-party political support and a lack of public opposition. But despite significant government funding and political will, the UK programme has not yet yielded any firm commitments to construct full-scale demonstration projects.

The section draws heavily on an analysis of CCS uncertainties in the UK (Watson et al., 2012) and on a detailed examination of demonstrations of CCS technologies in Canada and the United States (Jones, 2013). Both of these studies involved new primary data collection, including interviews with representatives from the industry and policy communities in those countries. It includes sub-sections that analyse Canada, the USA, and the UK respectively. For each country, an overview of the country's CCS demonstration programme is provided, along with the political and policy rationales for it. This is followed by a brief comparative analysis of the economic and policy factors that have influenced relative progress to date.

12.3.1 CCS IN CANADA

The Canadian government has ranked CCS as a 'key' mitigation technology and allotted significant public funding to support its demonstration and ultimate commercialization (Global CCS Institute, 2013). While Canada withdrew from the Kyoto Protocol, this focus on the demonstration and deployment of CCS has largely stemmed from mounting international market pressure. As Europe and some American states shift towards standards that make GHG-intensive energy sources less viable, Canada's energy industry faces increasing market uncertainty. Recent policy measures such as the EU Fuel Quality Directive (FQD) and the California Low Carbon Fuel Standard (LCFS) threaten to discourage the economic viability of oil sands by penalizing and effectively limiting imports of oil with high carbon intensity (Krupa and Jones, 2013). While there are proposals to change the FQD so that it does not distinguish between road fuel from oil sands and fuel from other sources of oil (Oliver and Crooks, 2014), these policies led to a growing push within Canada for carbon capture and storage technology to enable the continued competitiveness of its energy sector (Alberta Energy, 2013).

As stated by Jim Prentice, the previous Environment Minister of Canada, 'It is one of the most important technologies that we have to develop if we're going to reduce carbon emissions'. Currently, five CCS demonstration projects have been developed in the neighbouring provinces of Alberta and Saskatchewan. One of which is currently in operation at Weyburn-Midale, where over twenty million tonnes of CO_2 have been stored since 2000 (Global CCS Institute, 2011). The other four demonstrations are in different stages of construction and are expected to be operational by 2015 or 2016.

Four policy incentives are primarily responsible for the development of Canada's five CCS demonstrations. The first, and arguably most important, is that of capital grants. The federal government of Canada has committed over £500 million, and provincial governments over £1.5 billion, to the research development and demonstration of clean energy technologies like CCS (Alberta Energy, 2013). A second important policy measure is the recently instated emission performance standard (EPS) on coal combustion facilities, which effectively limits the CO_2 emissions intensity of electricity to no more than 926 lbs/MWh (~420 g/kWh) (Hallerman, 2012). This should prove particularly conducive to coal-based CCS projects in Alberta and Saskatchewan, where the primary source of GHG emissions is coal-based electricity generation (Saskatchewan, 2012). One feature of the new EPS is a ten-year compliance exemption and a five-year extension to permitted lifetime for operators who choose to install CCS. These extended compliance deadlines have been described by industry stakeholders as being necessary for proper site characterization and for the time needed to select the most appropriate CCS technology.

A third necessary funding mechanism for Canadian CCS is the newly introduced Royalty Credit Program (RCP). Through this policy tool, the government encourages capital investment in CCS by providing royalty deductions to oil and gas companies which invest in the technology. Within Canada, a fee is levied by government on industry for the production of oil, gas, and coal. What the royalty credit programme effectively does is reduce the money owed to government when resources are produced via CCS technology. In which case the project developer can offset their capital investment costs against their royalties payment. The fourth, but a critically important, driver of Canadian CCS has been the availability of enhanced oil recovery (EOR) contracts for power producers. All demonstrations except Shell's Quest project are pursuing CCS with

EOR. The government of Saskatchewan has established a complementary investment tax credit programme that provides all oil producers who opt to procure CO_2 for EOR purposes with a rebate on royalties owed to the government.

12.3.2 CCS IN THE USA

Following the 2008 presidential elections, the US Federal Government has made strong pledges to reinvigorate its climate policy after many years of slow progress and political controversy. The new roadmap for carbon mitigation called for a broad range of climate-related strategies, including improved energy efficiency, cleaner supplies, and a significant reduction of GHGs through increased investment in CCS. While attempts under the Presidential leadership of Barack Obama to pass a climate bill were unsuccessful in 2009 and 2010, US GHG reductions are now primarily driven by technology-specific policies, rather than by Federal climate legislation. Under the Industrial Carbon Capture and Storage (ICCS) programme, the Department of Energy (DoE) helps coordinate cost-sharing arrangements with industry to demonstrate next-phase CCS. There are a multitude of fossil fuel R&D programmes sponsored by the DoE, which prove particularly useful for CCS developments, including the Clean Coal Power Initiative and the Carbon Sequestration Program (van Alphen, 2011). The most important US stimulus package to deliver support to CCS demonstrations, however, has been the American Recovery and Investment Act of 2009, which has appropriated £2.2 billion (Folger, 2014). This recent influx of funding for CCS shifts the federal government's approach from R&D to support for the demonstration and commercialization of carbon capture technology. Those projects eligible for funding must have progressed beyond R&D stages to a scale that can be replicated and deployed by industry. If a company is allocated funding but fails to meet its demonstration target, no federal support will be delivered. A number of CCS demonstrations are in operation for natural gas processing, hydrogen and fertilizer production. Several others are under construction, including the first large-scale power sector project in the USA (at Kemper County, Mississippi), and there are many more in planning (Global CCS Institute, 2013).

The policy incentives and economic factors that have influenced progress with America's CCS demonstration programme are much the same as those in Canada. Many power producers in the USA are reluctant to invest in CCS in the absence of any carbon policies. Consequently, the most effective means of generating investor confidence has been the provision of direct cash subsidies or grants to target projects (van Alphen, 2011). The Department of Energy has set up a £900 billion effort to capture and store carbon from industrial emitters, but so far only a handful of projects have been selected. The second most important driver of American demonstrations has been the availability of Tax Credits under the Energy Improvement and Extension Act of 2008 (National Enhanced Oil Recovery Initiative, 2012). The program, labelled 45Q, makes available credits as a per-tonne payment for CO_2 disposed of through geological storage.

At present, 45Q makes available £6.5 (USD $10) per tonne of CO_2 stored with enhanced oil recovery purposes and £13 (USD $20) per tonne stored in saline formations. As was described in the Canadian context, enhanced oil recovery and enhanced coal bed methane recovery (ECBM) has been crucial to the progress of CCS in the USA. Seven out of the eight projects currently in operation or under construction in the USA have depended on EOR to receive a final investment decision (FiD) (Global CCS Institute, 2011). Some states have explicitly required that companies interested in pursuing CCS must demonstrate an integrated system model that shows both the technical and business arrangements necessary for CO_2-EOR operations, if they hope for political and funding support.

A third driver is more recent, and stems from the use of regulations in the absence of successful Federal climate change legislation. The Environmental Protection Agency (EPA) has proposed an emissions performance standard for new fossil fuel plants that will, if adopted, mean that no new coal plants can be constructed without CCS. The standard would allow new unabated gas plants, however. More recently—in June 2014— the US Administration has announced plans to cut power sector emissions by 30 per cent by 2030 from 2005 levels.[2] This is likely to add further pressure in some States for the deployment of low-carbon technologies including CCS.

12.3.3 CCS IN THE UK

Successive UK governments have emphasized a potential role for CCS technologies in strategies to mitigate greenhouse gas emissions since the early 2000s (Department of Trade and Industry, 2003). The 2005 Carbon Abatement Technology Strategy (Department of Trade and Industry, 2005) suggested a potentially substantial role for CCS in the UK if fossil fuels were retained in the energy mix. This was followed in November 2007 by the launch of a competition to build the UK's first full-scale demonstration plant, which was planned to be operational by 2014. The aim was to 'make the UK a world leader in this globally important technology' (DTI, 2007: 15). Substantial government funding was made available to help meet the capital costs of this demonstration.

However, this competition was not ultimately successful in identifying and funding a full-scale CCS project. This is despite a continuing commitment from the government to allocate up to £1bn to the successful project. It is notable that this commitment survived the 2010 election and significant cuts in public spending that were implemented by the incoming Coalition Government. The competition was narrowly defined, and specified that it would only fund a demonstration of post-combustion capture on a coal-fired power station. Nine competing projects were proposed for funding, and four were

---

[2] See: <http://www2.epa.gov/carbon-pollution-standards/what-epa-doing#overview>.

shortlisted in a pre-qualification stage. The most promising of these projects, at Long-annet in Scotland, was withdrawn by its backers in October 2011. The government cited increased costs and the inability to reach a commercial agreement as the reasons. In a critical report on lessons for the government from the competition, the National Audit Office cited a number of factors that contributed to the collapse of the project (National Audit Office, 2012). These included poor commercial awareness within government, a lack of capacity to procure such large, complex projects, a lack of flexibility with respect to project specifications, and the lack of a business case for the competition.

The UK programme to demonstrate CCS was relaunched in spring 2012 as a new Commercialisation Competition. This renewed the government's commitment to funding demonstrations with up to £1bn of capital support, and expanded the scope of the competition to include all capture technologies and a range of fuels—including gas. These significant changes were in response to the increasing focus on the role of gas in the UK and the continuing uncertainty about which CCS technologies would be the most technically and economically effective (Watson et al., 2012). Once again, a number of projects have been proposed for funding, and the main focus has been on the power sector. Two projects have been selected, and agreements have been signed for an initial tranche of £100m to support Front End Engineering Design (FEED) studies in each case. The projects are White Rose, a coal-fired oxyfuel demonstration at the Drax power plant in Yorkshire, and Peterhead, a gas-fired post-combustion project in Scotland. Final investment decisions for these plants are expected to be taken by the government and developers in late 2015.

The relaunch of the demonstration programme in the UK coincided with the implementation of wide-ranging reforms to the UK electricity market, known as Electricity Market Reform (EMR). In common with support for CCS itself, this broader process of market reform has had cross-party support, and was inherited by the current government from its predecessor. At the heart of the EMR package is a new system of long-term contracts for low-carbon electricity. These contracts for difference (CfDs) are designed to stabilize and top up the revenues of investors in low-carbon technologies, including CCS. During the transition to implementing these contracts, which are eventually intended to be technology neutral, CCS demonstration plant developers can negotiate individual contracts with the government. Through this mechanism, CCS demonstrations will therefore have access to revenue support (linked to the amount of electricity they generate) in addition to capital support (to help reduce up-front costs and risks). In addition, one of the two shortlisted demonstration projects has applied for European 'NER 300' funding in the second round of this mechanism.

12.3.4 COMPARISON OF THE USA, CANADA, AND THE UK

This discussion of progress in Canada, the USA, and the UK shows that CCS demonstrations have been driven by a range of different political factors in these countries

(Meadowcroft and Langhelle, 2009). In Canada, CCS has been supported largely in response to international pressure to mitigate the environmental impacts of its unconventional oil sector in resource rich Provinces—particularly Saskatchewan and Alberta. The decentralization of energy policy and resource ownership in Canada (Jaccard and Sharp, 2009) means that these Provinces have used significant powers to develop policies for CCS technology development and deployment. Support for CCS in the United States stems from a similar focus on technological development, rather than the implementation of Federal climate policies such as emissions targets and/or carbon pricing. It is also linked to the importance of coal, which supplied as much as 50 per cent of US electricity until recently, and to the presence of a well-developed market for CO_2 for enhanced oil recovery (EOR). By contrast, the CCS demonstration programme in the UK was initially developed for international climate diplomacy reasons. The desire to show other countries (particularly China) that continued fossil fuel use could be compatible with climate change mitigation was later joined by domestic policy drivers—and the need to demonstrate and deploy low-carbon technologies to meet legislated climate change targets.

These contrasting political drivers can partly explain the different rates of progress with respect to CCS demonstrations in these three countries. They also provide a context for the development and implementation of policy incentives to support these demonstrations. Table 12.1 summarizes some of these incentives for six demonstration projects in these countries. It distinguishes between upfront capital support, revenue incentives (including carbon pricing, electricity contracts and EOR revenue), and standards (particularly emissions performance standards). At the time of writing in mid 2014, two of these projects are already operating (Weyburn-Midale in Canada and Port Arthur in the USA), whilst the other four are under construction (in the case of Boundary Dam and Kemper County) or undergoing detailed design studies (Peterhead and White Rose).

Taking into account the patterns of incentives shown in Table 12.1, a number of observations can be made about the relative progress of these CCS demonstrations. First, in all cases, support for demonstration involves a complex package of policy incentives. In theory a sufficiently high and stable carbon price should provide enough of an incentive for CCS deployment. However, in practice, comprehensive carbon pricing is largely non-existent or prices are too low to have the desired impact on investment decisions. Furthermore, given that CCS technologies are still at the demonstration stage, additional policies are required to reduce financial and other risks (von Stechow et al., 2011). For most of the projects in Table 12.1, policy support has included both upfront funding for capital costs as well as revenue support through carbon prices or electricity prices. Emissions performance standards for the power sector are also present in all three countries. However, these standards only prevent the construction of new unabated coal plants—and not new unabated gas plants.

This links to a second observation: the CCS power plant demonstrations that are under construction in Canada and the United States benefit from being developed by vertically integrated, monopoly utilities. Unlike CCS power plant developers in the UK, these utilities do not face competition, and can therefore pass some of the costs of CCS

Table 12.1 Incentives for six CCS demonstrations in Canada, the USA, and UK

Country	Project	Capital grant	Carbon price	EPS	Electricity contract	EOR
Canada	Boundary Dam (Saskatchewan)	Yes (plus investment tax credits, ITCs)	No	Yes	Regulated monopoly utility	Yes
	Weyburn-Midale (Saskatchewan)	Yes (R&D grant)	No	n/a	Not applicable (synfuels plant)	Yes
USA	Port Arthur (Texas)	Yes (66% of capital plus investment tax credits)	No. Production subsidy per tonne of stored CO_2	n/a	Not applicable (steam methane reforming)	Yes
	Kemper county IGCC (Mississippi)	Yes ($270m from US DoE & ITCs)	No.	Yes	Regulated monopoly utility	Yes
UK	Peterhead	Yes (UK Government)	Yes	Yes	Competitive market; long-term contract with government	No
	White Rose	Yes (UK Government; NER300 application)	Yes	Yes	Competitive market; long-term contract with government	No

Abbreviations: EOR = Enhanced Oil Recovery; EPS = Emissions Performance Standard; ITC = Investment Tax Credit.

demonstrations through to captive consumers. While in some cases (e.g. Kemper county), the extent of this pass through has been limited by state utility regulators, this ability leads to a significant reduction in risk and cost of capital for developers.

Third, most CCS demonstrations in North America are being partly financed by income from enhanced oil recovery. In both the USA and Canada, there is a well-developed market for CO_2 and an existing onshore CO_2 pipeline network that demonstration projects can connect to. Whilst some CCS projects in the UK, including a previous incarnation of the Peterhead project, have planned to supply CO_2 for enhanced oil recovery, this has not provided such a significant incentive. In the UK, EOR operations would take place offshore. New pipeline infrastructure would need to be constructed and financed. Furthermore, the absence of a large EOR market places more emphasis on the need to identify and characterize CO_2 storage reservoirs.

12.4 **Conclusions**

Modelling by UKERC for this book shows that global efforts to limit temperature increases to 2 °C are likely to require the application of CCS to fossil fuel power plants and other large industrial facilities. The widespread deployment of CCS is the only way for large-scale fossil fuel use to continue while making deep reductions in greenhouse gas emissions. In the absence of CCS technologies, such deep reductions are possible—but may be much more difficult, especially outside the power sector. However, a range of risks including high capital costs and uncertain financial returns have limited the widespread deployment of this technology. The global review of international CCS activities summarized in this chapter indicates that the vast majority of large-scale demonstration projects are in the planning stages. Furthermore, there has been a significant rate of attrition, with many projects around the world being cancelled, delayed, or restructured. Of those projects under construction or in operation, the vast majority are in North America.

This chapter has focused in particular on the contrasting progress of CCS demonstration projects between North America and Europe. Whilst there is significant activity in both the USA and Canada, the UK is the only European Union country that continues to have an ambitious CCS demonstration programme. To best understand the success of North American CCS when compared with the relatively slow progress in the UK, this chapter has focused on both the political drivers for CCS in these three countries and on the package of policy incentives available to support demonstrations.

This comparison shows that it is important to design policy incentives that reduce risks sufficiently for CCS demonstration projects to proceed. However, it also shows that a focus on specific incentives for CCS (e.g. capital grants or electricity feed-in-tariffs) or on generic carbon pricing would miss some important reasons for relative success or failure. It also highlights the importance of revenue from EOR for many successful CCS projects (Global CCS Institute, 2014)—including all four of those projects featured in Table 12.1. In North America, CCS project developers have been able to tap into existing pipeline networks and to benefit from a well-developed market for CO_2 for EOR. While this means that the early deployment of CCS technologies is being used to facilitate the extraction of more oil, this is very unlikely to be sustainable in the medium and longer-term unless its use is constrained by effective climate change legislation.

Neither condition is in place in the UK to the same degree. This makes it more important for the UK demonstration programme to focus on the development of CO_2 pipeline networks and the characterization of storage reservoirs—and not just on providing incentives for CO_2 capture at power plants. Furthermore, the North American experience shows that CCS demonstration programmes should not only focus on the power sector as is the case in the UK. Many early North American projects have been implemented in other industrial sectors where the costs of implementation are lower.

The full-scale power sector CCS demonstrations that are starting to operate in Canada and the United States have been implemented in a regulated utility environment. By contrast, UK CCS developers face competition and do not have access to captive consumers. The contracts for difference that are planned for electricity produced for CCS demonstrations are therefore particularly important, and are required in addition to the substantial capital subsidies that are currently on offer from the UK government. It remains to be seen whether the cross-party support that has been shown for CCS demonstrations in the UK will continue.

■ REFERENCES

AIR PRODUCTS 2011. Air Products Signs Two Agreements to Move Texas Carbon Capture and Sequestration Project Forward. *News Release, 26th Nov 2011*. Air Products.

ALBERTA ENERGY 2013. Carbon Capture & Storage: Summary Report of the Regulatory Framework Assessment. Edmonton, Canada: Electricity and Alternative Energy and Carbon Capture and Storage Division.

BEST, D. & LEVINA, E. 2012. Facing China's Coal Future: Prospects and challenges for carbon capture and storage. *Working Paper*. Paris: International Energy Agency.

DEPARTMENT OF TRADE AND INDUSTRY 2003. Review of the feasibility of carbon dioxide capture and storage in the UK. London: DTI.

DEPARTMENT OF TRADE AND INDUSTRY 2005. *A Strategy for Developing Carbon Abatement Technologies for Fossil Fuel Use*. London: DTI.

DTI 2007. *Meeting the Energy Challenge: A White Paper on Energy*, London: DTI.

EUROPEAN COMMISSION 2012a. Award decision under the first call for proposals of the NER300 funding programme. *Commission Implementing Decision C(2012) 9432 final*. Brussels: European Commission.

EUROPEAN COMMISSION 2012b. NER 300—Moving towards a low carbon economy and boosting innovation, growth and employment across the EU. *Commission Staff Working Document SWD(2012) 224 final*. Brussels: European Commission.

FOLGER, P. 2014. Carbon Capture and Sequestration: Research, Development, and Demonstration at the U.S. Department of Energy. *Congressional Research Service Report*. Washington, DC: CRS.

GLOBAL CCS INSTITUTE 2011. *The Global Status of CCS: 2011*. Canberra, Australia: GCCSI.

GLOBAL CCS INSTITUTE 2013. *The Global Status of CCS: Update January 2013*. Canberra, Australia: GCCSI.

GLOBAL CCS INSTITUTE 2014. *The Global Status of CCS: Update February 2014*. Canberra, Australia: GCCSI.

HALLERMAN, T. 2012. Canada Unveils Softened Final GHG Performance Standard for Coal Units. *GHG Monitor* [Online]. Available: <http://ghgnews.com/index.cfm/canada-unveils-softened-final-ghg-performance-standard-for-coal-units/> (Accessed April 2015).

IEA 2011. *World Energy Outlook 2011*. Paris: OECD/International Energy Agency.

JACCARD, M. & SHARP, J. 2009. CCS in Canada. *In*: MEADOWCROFT, J. & LANGHELLE, O. (eds) *Caching the Carbon. The Politics and Policy of Carbon Capture and Storage*. Cheltenham: Edward Elgar.

JONES, C. 2013. *Policy flexibility in the innovation chain: applications to demonstration stage CCS*. MSc in Energy Policy for Sustainability, University of Sussex.

KRUPA, J. & JONES, C. 2013. Black Swan theory: applications to energy market histories and technologies. *Energy Strategy Reviews*, 1, 286–90.

KUUSKRAA, V., VAN LEEWEN, T., & WALLACE, M. 2011. Improving domestic energy security and lowering CO2 emissions with 'Next Generation' CO2-Enhanced Oil Recovery (CO2-EOR). *Report No. DOE/NETL-2011/1504*. US Department of Energy.

MEADOWCROFT, J. & LANGHELLE, O. 2009. The politics and policy of carbon capture and storage. *In*: MEADOWCROFT, J. & LANGHELLE, O. (eds) *Caching the Carbon. The Politics and Policy of Carbon Capture and Storage*. Cheltenham: Edward Elgar.

MICHAEL, K., GOLAB, A., SHULAKOVA, V., ENNIS-KING, J., ALLINSON, G., SHARMA, S., & AIKEN, T. 2010. Geological storage of CO2 in saline aquifers—a review of the experience from existing storage operations. *International Journal of Greenhouse Gas Control*, 4, 659–67.

NATIONAL AUDIT OFFICE 2012. Carbon capture and storage: lessons from the competition for the first UK demonstration. *Report HC 1829, Session 2011–12*. London: NAO.

NATIONAL ENHANCED OIL RECOVERY INITIATIVE 2012. Recommended modifications to the 45 Q tax credit for carbon dioxide sequestration. Washington DC: NEORI, C2ES and GPI.

OLIVER, C. & CROOKS, E. 2014. Canada poised to dilute EU rules over tar sands oil. *Financial Times*, 6 June 2014.

SASKATCHEWAN, G. O. 2012. Canada and Saskatchewan working together to reduce greenhouse gas emissions. *News Release*. Saskatchewan Geological Survey.

STATOIL 2009. Annual Report 2009. Statoil.

VAN ALPHEN, K. 2011. *Accelerating the development and deployment of carbon capture and storage technologies: An innovation system perspective*. Doctoral Thesis, Universiteit Utrecht.

VON STECHOW, C., WATSON, J., & PRAETORIUS, B. 2011. Policy incentives for CCS technologies in Europe: a qualitative multi-criteria analysis. *Global Environmental Change*, 21, 346–57.

WATSON, J., KERN, F., GROSS, M., GROSS, R., HEPTONSTALL, P., JONES, F., HASZEL-DINE, S., ASCUI, F., CHALMERS, H., GHALEIGH, N., GIBBINS, J., MARKUSSON, N., MARSDEN, W., ROSSATI, D., RUSSELL, S., WINSKEL, M., PEARSON, P., & ARAPOS-TATHIS, S. 2012. Carbon capture and storage: Realising the potential? *Final report*. London: UK Energy Research Centre.

13 Fossil fuels: reserves, cost curves, production, and consumption

Michael Bradshaw, Antony Froggatt, Christophe McGlade, and Jamie Speirs

13.1 Introduction

The purpose of this chapter is threefold: first, to explain the nature of the various reserve definitions that are used to described fossil fuel availability and their associated cost curves; second, to present a range of current reserves estimates for the three fossil fuels—coal, oil, and gas; and third, to present information on current levels of production and consumption. The high level of uncertainty about future energy trends is one of the central themes of this book, thus this chapter does not examine projections about future fossil fuel availability and consumption in any detail (these are discussed in Chapter 1); rather it aims to assess the current reserve base and the level of proven reserves that are thought to be available. Whether or not, in the face of geopolitical concerns and wider concerns about fossil fuels, those reserves are developed at a rate that matches supply with demand is a wider issue that is taken up elsewhere in this volume.

13.2 Reserve definitions

'Reserves' and 'resources' are the most often used terms when discussing the endowment of fossil fuels within a given area, country, or region. Unfortunately they are also often confused and the use, interpretation or comparison of inappropriate or inconsistent terms is a common, but easily avoided, issue when discussing fossil fuel availability. A broad overview and comparison of the different definitions that can be used to report volumes of oil and gas (which are generally similar) is presented Table 13.1; the terms for volumes of coal differ slightly as noted below.

When reporting volumes of oil, the largest figure that can be given is the initial or original oil in place (OOIP). This is the total volume of oil that is estimated to be present in a given field, area or region. Similarly for gas, the term original gas in place (OGIP) is used. Only a proportion of the OOIP or OGIP is ever recoverable and this fraction,

Table 13.1 Brief descriptions of resource and reserves for oil used in this report

Name	Short description	Includes oil in undiscovered formations	Includes oil not economically recoverable with current technology	Includes oil that is not recoverable with current technology	Includes oil that is not expected to become recoverable
Original oil in place	Total volume present	✓	✓	✓	✓
Ultimately recoverable resources	Total volume recoverable over all time	✓	✓	✓	
Technically recoverable resources	Recoverable with current technology	✓	✓		
Economically recoverable resources	Economically recoverable with current technology	✓			
1P/2P/3P reserves	Specific probability of being produced				

Notes: Identical definitions exist for gas

known as the *recovery factor*, can vary substantially depending on the geological conditions, technology used, and prevailing oil prices. The product of the recovery factor and the OOIP or OGIP yields an estimate of the 'resources' that can be recovered. However, a number of different estimates of the 'recoverable resources' can be given.

The ultimately recoverable resource (URR) of a field or region is the sum of all oil or gas that is expected to be recovered from that field or region over all time (McGlade, 2012). This figure includes any oil or gas that is estimated to be undiscovered, is not recoverable with current technology, and/or is not currently economic but which is expected to become so before production ceases. An alternative term for URR is Estimated Ultimate Recovery (EUR), which is more commonly used to refer to single oil or gas wells. A related term is the 'remaining recoverable resources' (alternative names include 'remaining potential', 'future volume' or 'yet-to-produce', but these are essentially identical). Although a precise definition is usually not explicitly stated, the remaining recoverable resources are most commonly defined as the difference between the URR for a region and that region's cumulative production. Two alternative figures sometimes reported are the technically recoverable resources (TRR) and the economically recoverable resources (ERR). The TRR is the oil or gas estimated to be producible with current technology only, that is it excludes any impacts that future technological developments may have. The ERR is a subset of TRR and defines the resource that is estimated to be both technically and economically producible from a field or region under current conditions. Such estimates are sensitive to assumptions about technical and economic conditions and may be expected to change over time.

The final subset of resources is reserves. The exact definition of reserves varies from one source to another but they are generally those portions of the economically recoverable resources that have been discovered and are estimated to have a specified probability of being produced. Generally speaking, for fields from which production has not yet commenced to be considered as reserves there should also be a reasonable timetable for these to be developed. Reserve estimates are frequently given to three levels of confidence, namely: proved reserves (1P); proved and probable reserves (2P); and proved, probable, and possible reserves (3P).

For coal, resources and reserves are again the most frequently used terms. Generally (e.g. Bundesanstalt für Geowissenschaften und Rohstoffe (BGR, 2012)) the definitions used for reserves are very similar to those used for oil and gas, but it is nearly always only the proved (1P) estimates of coal reserves that are reported. Coal resources are different, however, as they are more commonly taken to be the total volume of coal in situ. This definition is therefore more closely associated with the OOIP or OGIP used for oil and gas rather than the URR.

It is therefore not correct to compare resource estimates of coal with resource estimates of oil and gas unless it is explicitly made clear what definitions are being used. It should be clear from the above, however, that the use of resource and reserve terminology is inconsistent, imprecise, and in need of standardization and systematic implementation.

Focusing on reserve estimates, despite using similar language, definitions of the 1P, 2P, and 3P reserves vary widely from one country to another and from one company to another. Some employ a *deterministic* definition (certain qualitative criteria must be satisfied) and others use a *probabilistic* definition (based upon a probability distribution of resource recovery). For example, the Society of Petroleum Engineers Petroleum Resources Management System (SPE/PRMS) (SPE et al., 2011) allows for either deterministic or probabilistic definitions of 1P, 2P, and 3P reserves. The deterministic definition for 1P reserves, for example, is 'those quantities of petroleum which, by analysis of geoscience and engineering data, can be estimated with reasonable certainty to be commercially recoverable'.

Under the SPE/PRMS probabilistic definitions 1P, 2P, and 3P, reserve estimates can be expressed respectively as P90, P50, and P10. P90 (1P) estimates are then interpreted as the volume of oil production that is estimated to have a 90 per cent probability of being exceeded by the time production ceases. Similarly, P50 (2P) and P10 (3P) estimates refer to volumes of oil that are estimated to have a 50 per cent and 10 per cent probability respectively of being exceeded. Under this interpretation, P10 > P50 > P90, and 2P (P50) reserves are equivalent to a median estimate.

In contrast, Russia and members of the Former Soviet Union use the Russian Federation Classification (RFC) system, which has categories A to D3 (Figure 13.1). No probabilistic definition is attached to any of these. Most analysts associate the first three Russian reserve categories (A, B, and C1) with the SPE 2P definition but this is contested. Another reserve and resource classification, the UN Framework Classification

Fundamental class	PRMS			UNFC			RFC	
				E	F	G		
Commercially recoverable	Reserves	1P	Commercial Projects	E1	F1.1	G1	Economic Reserves	A
	Reserves	2P	Commercial Projects	E1	F1.2	G2	Economic Reserves	B & C1
	Reserves	3P			F1.3	G3		C2
Not commercial	Contingent Resources / Sub-marginal	Marginal	Potentially Commercial	E2.1		G1, 2 & 3		
		1C	Non-Commercial	E2.2	F2			
		2C						
		3C						
	Unrecoverable		Additional	E3	F4			
Undiscovered	Prospective Resources	Low	Exploration Projects	E3	F3	G4.1, 4.2 & 4.3	Localised	D1
		Med					Prospective	D2
		High					Undiscovered	D3
	Unrecoverable		Additional	F4		G4		

Figure 13.1 Comparison of three reserve and resource classification schemes: the SPE/PRMS, the UN 2009, and the Russian Federation Classification

Source: ACP (2011); MacDonald (2013); Poroskun et al. (2004); United Nations, (2009).

Notes: Figure 13.1 represents approximation of equivalency between classification schemes and may be inaccurate given the conflicting nature of available literature.

(UNFC) for Fossil Energy and Mineral Reserves and Resources (United Nations, 2009, 2009a), was proposed both as a new system for classification, and as a framework to compare the existing classification systems. Quantities of oil and gas are ranked based on their economic and social viability (E), their project status and feasibility (F), and their geological knowledge (G). There are numerous other examples of reserve definitions in use (e.g. as used by the Federal Institute for Geosciences and Natural Resources in Hannover (BGR) and the UK's Department of Energy and Climate Change).

13. 3 **Reserves and cost curves for oil and gas**

The oil and gas industry tends to focus on the most easy to access, and therefore usually cheapest, resource first, before moving onto more expensive resources. This needs to be considered when extrapolating future production on the basis of past trends. Kaufmann (1991), for example, indicates that as a resource within a region is increasingly depleted production will shift from the large fields developed initially to smaller ones; costs therefore rise as economies of scale are lost. This is now the situation in many mature basins, such as the North Sea and West Siberia. Similarly, as a resource is depleted, production shifts from more easily accessible fields to more complex ones (deeper, offshore etc.) requiring more technical capability and hence greater cost. This problem now challenges the Russian oil industry. Conversely, however, production costs of a given resource can also fall, driven by technical progress and innovation, as exemplified by the drastic reduction in the costs of producing tight oil and shale gas resources in North America. When considering the availability of fossil fuel resources, both the magnitude of the resource (under any of the above definitions) that can be recovered and the costs at which portions of this resource can be recovered are thus important.

The total volume of oil that is available at various costs within a region is usually represented using a 'supply cost curve'. In these figures a measure of the different types of available resource is represented on the *x*-axis while a measure of the different cost associated with producing each of these resource types is represented on the *y*-axis. There are numerous examples that have been produced by a number of organizations: including the IEA (2005, 2008, 2013); Aguilera at al. (2009); Remme et al. (2007), and Farrell and Brandt (2006). One example, for natural gas resources, is presented in Figure 13.2, produced by the IEA (2013).

While this figure provides only a broad representation of the gas resources available and their costs of production, it does suggest that there are 800 trillion cubic metres (Tcm) available at costs of less than $12/million British Thermal Units (MMBTU). This is over 230 times current global production of gas (around 3.4 Tcm/year), and around eight times current global cumulative production over all time. There is a wide range of costs for gas production, however. Some 'conventional gas' is available at costs of around $0.2/MMBTU, while the most expensive volumes in Arctic areas could cost sixty times

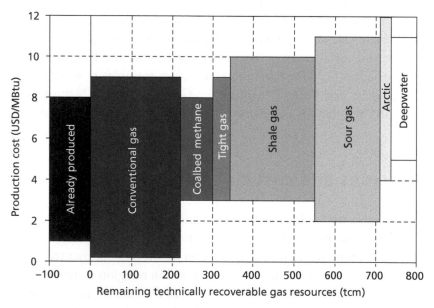

Figure 13.2 Long-term gas supply cost curve

Source: Adapted from IEA (2013).

Notes: MMBtu = million British Thermal Units; Tcm = trillion cubic metres.

more than this. It is noteworthy that despite most interest focusing at present on the potential for shale gas production, this curve suggests that coal bed methane is on average less expensive.

A similar curve is constructed by the IEA (2013) for oil, which if all sources of conventional and unconventional oil were to be included, suggests that there are around 6 trillion barrels of oil, available at costs of less than $100/bbl. This is around five times total cumulative production globally up to 2012, and almost 200 times current production (31 billion barrels/year). A large portion of this resource, around 40 per cent, is considered to be unconventional oil (i.e. has a larger density than water; Sorrell et al., 2009), however this excludes the potential volumes from unconventional liquid sources (such as coal-to-liquids, gas-to-liquids, and biofuels), which could contribute substantially to future production levels. The issue of unconventional oil and gas is discussed in more detail in the next chapter.

To conclude this section, it is important to remember that care is needed with interpreting such supply cost curves and the suggested costs and volumes of oil and gas.

First, there is a huge degree of uncertainty in all the figures stated by these curves; however this is often not displayed or discussed in any detail. The resource figures for all categories of oil and gas presented here are not necessarily well established, especially the volumes currently considered to be unconventional e.g. oil shale (kerogen) and shale gas (see e.g. McGlade et al., 2013; McGlade, 2012). Without knowing the volumes that are present it is also difficult to be particularly confident in the cost figures that are stated. Production costs have varied substantially over the past number of years (see e.g. the capital cost index produced by CERA, 2012), while there is also no commercial recovery technology for some volumes, meaning that the production costs indicated are highly speculative. Second, these curves are usually *long-term* supply cost curves (indeed the examples above are long-term curves). They should not therefore be interpreted as representing the supply costs of oil that can immediately be extracted, or indeed could even come online in the next few years. To prevent such an interpretation, they are sometimes called 'availability cost curves' (e.g. Aguilera et al., 2009).

There are numerous reasons why the oil or gas depicted in a supply (or 'availability') cost curve may not be immediately available at the costs shown. Some examples are presented in Table 13.2.

Third, a number of different costs or prices could be given on the y-axis. The main difference lies between presenting the marginal costs of production or the '*minimum oil price*' necessary to stimulate investment into exploration or production (Figure 13.2 gives the marginal cost of production only). The latter contains a number of additional factors, but two of the most important differences are the inclusion of taxes and the necessary rate of return. The average government tax take (the ratio of revenue taken by a host government to total available revenue for a given project), for example for oil, is around 67 per cent, but this can vary from below 30 per cent to close to 100 per cent. Therefore, whether costs or minimum prices are presented can make a huge difference to the

Table 13.2 Examples of factors limiting the short-term availability of oil and gas resources presented in a supply cost curve

Short-run limiting factors	Example
Political	Production restrictions by national states (such as Saudi Arabia or other members of OPEC) will prevent some new investment and production
Economic	Learning effects are generally included within long-term supply cost curves but do not necessarily represent current costs
Geological/Physical	The initial pressure differential between a reservoir and well (that leads to oil and gas rising to the surface when a new well is drilled) tends falls over time. Production from a well or field is therefore often modelled as exponentially declining over time (Sorrell et al., 2012): all of oil and gas cannot therefore be extracted as quickly as may be desired (without new investment and additional cost)
Technological	The technology for recovery may not yet exist (e.g. large-scale kerogen oil production is not yet possible)

steepness, shape, and intercept of the supply cost curve, as well as shifting the order in which different fields or countries appear on the curve.

Finally, different factors can also be presented on the x-axis. In some cases the volume of current reserves is given, in others all available resources (i.e. including undiscovered volumes, volumes that are not currently economic, large volumes of unconventional oil etc.) are given. An alternative approach is to give the capacities of new fields that could come online at different costs.

An alternative supply cost curve gives snapshots of the reserves or capacities that could come online in various future years. These are potentially more useful for understanding the short and medium-term availability of fossil fuels. However, the volumes or capacities that are available in a given future year depend to a large extent on investments in previous years. Not only will prior investments mean that some reserves or resources will not be available in a future year (if they have already been produced for example), but they will also change the costs of the remaining resource through learning effects (which could have a negative or positive effect). They are much more difficult to construct and are therefore much less common.

The discussion that follows presents reserve estimates for coal, oil, and gas. Three sources are used: the World Energy Council (WEC), German Federal Institute for Geo-Sciences and Mineral Resources (BGR), and BP's Statistical Review of World Energy (BP). Clearly, there are other sources that could be consulted, the IEA and US Energy Information Administration (US EIA) for example, but the aim here is to present a limited sample of estimates to highlight the key characteristics. A further confusion is that the units used to report reserves are different for each class of fossil fuel; often these are reconciled to units of energy equivalent—million tonnes of oil equivalent or mtoe—to enable comparison between energy sources. A final confusion is that even within a particular class of fossil fuels different physical measures are used, for example oil is reported in terms of barrels and in terms of metric tonnes, furthermore metric and imperial units are used, such a cubic feet and cubic metres in relation to natural gas.

All of this leads to a complicated situation and every effort is made below to keep things simple and consistent.

13.4 **Coal**

According to BP (2013: 41) coal accounted for 30 per cent of total primary energy consumption in 2012. Many analysts forecast that by 2020 coal will surpass oil as the major global fuel source. Despite the need to 'decarbonize' the global energy system, coal remains the dominant primary energy source in many national energy systems and coal production and demand continues to outstrip that of oil and gas. This reflects the relative ubiquity of global coal reserves; even though global reserves are concentrated, most countries can call on domestic supplies of coal, even if they are low-grade, and this is often favoured over imported energy supplies. This also means that coal is not subject to the same geopolitical tensions that affect the global oil industry, and latterly international trade in natural gas.

13.4.1 RESERVES

According to the WEC, total world coal reserves, based on 2008 analysis, are 860 Gigatonnes (Gt), of which 405 Gt (47 per cent) are classified as bituminous coal (including anthracite), 260 Gt (30 per cent) as sub-bituminous, and 195 Gt (23 per cent) as lignite (WEC, 2013).[1] There are important differences between the types of coal, the most significant of which is its energy content that can vary from 5 MJ/kg in lignite to 30 MJ/kg for anthracite (GEA, 2012). There are also significant variations between research bodies' estimates as to the extent of these reserves. For example BGR calculates a reserve base of 293 Gt for lignite and 754 Gt for hard coal (in which they include both bituminous and sub-bituminous coal), consequently there are significantly larger overall volumes than the WEC figure. Some of the difference can be attributed to the assessment of the reserves, but also categorization of the different fuels (Table 13.3).[2] According to BP, proven reserves of coal in 2012 were 120 Mt lower than their analysis for 1998, when over the same period reserves for natural gas rose by 28 per cent and oil by 60 per cent (BP, 1999, 2013). This difference is partially due to the large reserves already within the coal sector, acting as a disincentive for further exploration.

---

[1] It is important to note that not all coal is used in the energy sector: steam or thermal coal is mainly used in the power sector, while coking or metallurgical coal is used directly in the manufacturing process for steel. In 2010, approximately 0.7 Gt tonnes of coking coal was used in the production of steel (WCA, 2013).

[2] Other sources, such as the *BP Statistical Review of World Energy*, put lignite and sub-bituminous coal in the same category.

Table 13.3 Categorization of coal

		Low <—Carbon/Energy content of coal—> High			
		High <—Moisture content of coal—> Low			
	Low Rank Coals		Hard Coals		
	Lignite	Sub Bituminous	Bituminous		Anthracite
			Thermal	Metallurgical	
Uses	Power generation	Power generation and cement manufacturing	Power generation and cement manufacturing	Iron and steel manufacturing	Smokeless fuels

Source: WCA (2013).

The resource base of coal is also significant, with the BGR estimating it for both hard coal and lignite to be in the order of 20,000 Gt. This compares to a current annual consumption of 7.6 Gt. According to BP (2013) the current reserve to production ratio for coal reserves is 109 years and therefore the current production to consumption ratio for resources is approximately two millennia, making it the most abundant of the fossil fuels. However, while the reserves and resources are large they are not evenly distributed, with almost 70 per cent of reserves and some 80 per cent of resources located in only four countries: China, India, Russia, and the USA (BGR, 2012).

In coal exporting countries, mining costs range from around \$10/tonne in Indonesia to around \$30/tonne in Australia (Devon, 2010). The cost of mining, however, is only one of many components that form the final price, other important factors include processing, taxation, and inland and sea transportation. Consequently, the prices paid at the power plant for coal also vary both within countries and between regions. Some of the largest in-country difference are in the United States, with coal in the more Central and Eastern States, such as Wyoming and Colorado, selling during 2012 for \$13/ton[3] and \$36/ton respectively, but that produced in Alabama or West Virginia for \$85/ton and \$73/ton respectively, both of which export significant quantities of their production and are therefore are more connected to the international market (US EIA, 2013a). Globally, the Asian market, particularly Japan, has the highest prices, followed by North Western Europe and then North America (BP, 2013).

In the short term, until 2018, there is likely to be increasing convergence of regional prices (US EIA, 2013a). In the longer term, given the extent of known reserves and the technology-related efficiency improvements, the GEA assumes falls in production costs by 2050 ranging from 20–50 per cent (GEA, 2012). However, these production forecasts must be viewed in the context of other costs, particularly those relating to other resources required for production, such as water, that might, in some cases, rise significantly and affect the price.

[3] 1 ton (short ton) = 907 kg: 1 tonne (metric tonne) = 1000 kg.

13. 4.2 PRODUCTION AND CONSUMPTION

The most important country in relation to coal is China, which is now producing and consuming approximately half of the world's total. This is reflected in Figures 13.3 and 13.4. In 1992 China overtook the USA to become the world's largest producer of coal, with 25 per cent of the global total. Since then its total production has risen by 320 per cent, with an even larger increase in coal consumption. Consequently, China has gone from being a net coal exporter to the world's largest importer of coal (IEA, 2012a). In 2012, China imported 289 million tonnes up 57.9 per cent year on year (Platts, 2013). The increase in imports is in part due to the high cost of transporting coal from the production in the West to main consumption centres in the East.

Another significant shift is that the United States, due to the increase in domestic shale gas production, has become a significant exporter of coal. Since 2008 consumption of coal in the USA has fallen by nearly 30 per cent, including by 12 per cent in 2012 alone, while production fell by just 16 per cent over the same five years. As a consequence, it is anticipated that 2013 will be the third year in a row that the USA has exported more than 100 million tons of coal—for both steel and electricity production (US EIA, 2013a). The main destinations of the export are Europe, including the UK—which increased its imports from the USA by over 40 per cent in 2012 (DECC, 2013) and the Netherlands—for subsequent re-export, and Asia, including China and Japan. However, there remain uncertainties over the volume of coal exports in the medium term, as in mid 2013 higher

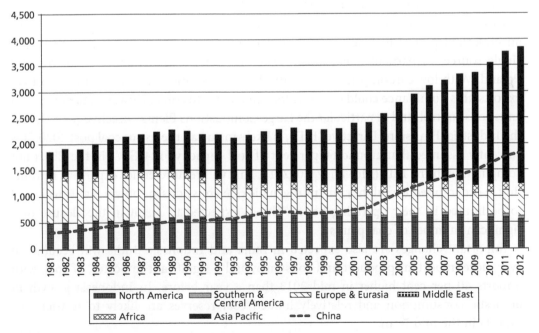

Figure 13.3 Global coal production, 1981–2012 (mtoe)

Source: BP (2013).

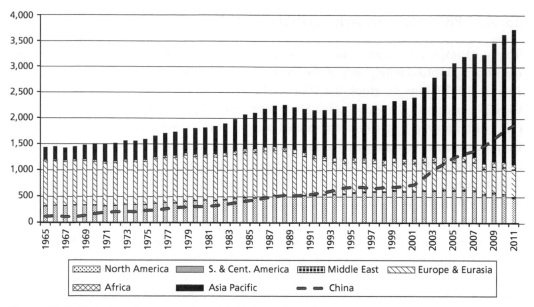

Figure 13.4 Global coal consumption, 1965–2012 (mtoe)

Source: BP (2013).

gas prices in the USA led to resurgence in the domestic use of coal. Therefore the extent of US coal export is significantly influenced by the availability of low-cost shale gas, the sustainability of which is uncertain (US EIA 2013b).

Despite climate objectives, coal—both in the form of hard coal and lignite—remains an important part of Europe's current electricity supply and in 2012 it was the largest single source, in particular in countries such as the Czech Republic, Germany, and Poland. With low carbon prices and relatively low coal prices, in part as a result of US exports, this dominance could continue for some years to come. However, the impact of emissions control, and in particular the Large Combustions Plants Directive (LCPD) and the Industrial Emissions Directive (IED), is likely to curtail the use of coal post 2020. It is currently assumed that around 60 per cent of coal capacity (124 GW) in the EU is not compliant with the IED (McKinley, 2012: 1). This will require either expensive retrofitting in order to reduce sulphur and nitrogen dioxide emissions, or power plants will need to be limited in their operating hours. This is expected to result in the closure of a significant number of coal plants.

Important changes are also expected in India and Indonesia. India will become an ever more important producer and consumer, and a significant importer, of coal, with imports 40 per cent higher in mid 2013 than a year before. In Indonesia growth in domestic consumption and relatively limited coal reserves are likely to restrict coal exports in the mid-term.

However, the future of coal use depends not only on reserves, but also on environmental and resource considerations. According to the International Energy Agency's

(IEA) World Energy Outlook, the use of coal in the 'new policies' scenario,[4] will increase by 21 per cent by 2035 and by 58 per cent under a business as usual scenario. In either case, internationally agreed climate change objectives will not be met and global temperature increases will be in the order of 3–6 °C (IEA, 2012a). With any sort of international agreement to limit GHG emissions, the future of coal, without carbon capture and storage, must be extremely limited.

A report by HSBC suggested that approximately 70 per cent of coal mines are located in water-scarce regions and 40 per cent are estimated to have severe water shortages such that lack of water is already affecting production (HSBC, 2013a). While in India, HSBC also highlighted water as a constraint in the operation of coal stations for the third year running (HSBC, 2013b). With competition for water expected to grow as a result of increased industrialization and urbanization, the coal sector will have to justify and progressively pay for its growing water demand, potentially restricting the sector's ambitious expansion plans. Thus, the future role of coal in the global energy system is uncertain and highly contested, with the level of future consumption constrained not by physical availability, but rather by environmental concerns.

13.5 **Oil**

Oil is the most important of the three fossil fuels, accounting for 33.1 per cent of total primary energy consumption in 2012 (BP, 2013: 41). It is oil that has fuelled the massive improvements in standard of living enjoyed by the 'developed world' during the second half of the twentieth century and it is oil that is essential to increased mobility and the processes of globalization (see Chapter 3). However, because of the complexity of the global oil economy and its increasing 'financialization', the future cost and availability of crude oil has proved impossible to predict with any degree of certainty.

13.5.1 RESERVES

According to the WEC (2013: 42) total proved reserves of oil and natural gas liquids (NGL) are 1,239 billion barrels (169 billion tonnes). In their Statistical Review, BP (2013: 6) gives a figure of 1,668.9 billion barrels (227.6 billion tonnes). The difference between the two estimates is explained by their treatment of unconventional oil (tar sands, extra heavy oil, and shale oil). The WEC treats these as a separate category and their reserve figures are for conventional crude oil and NGL, while BP's figures include

---

[4] The New Policies scenario takes account of broad policy commitments and plans that have been announced by countries, including national pledges to reduce greenhouse gas emissions and plans to phase out fossil energy subsidies, even if the measures to implement these commitments have yet to be identified or announced.

estimates for Canadian oil sands (note that we use the term oil sands here, the term tar sands is also used by environmental groups to describe the same resource) and Venezuelan heavy oil. Thus, about 66 per cent of proved reserves are produced from six countries at a rate of about 1.2 per cent a year with a reserve to production ratio (R/P) of about 85 years. A further 21 per cent is produced at a rate of 3.2 per cent a year with an R/P of 32 years. The remaining 13 per cent is produced by three countries at a rate of 6 per cent a year with an R/P of 17 years. Two issues are raised by these figures. First, there is a high degree of concentration of global reserves in a relatively small number of countries. According to BP (2013: 6), at the end of 2012 the top six reserve holders accounted for 68.6 per cent of global oil reserves and the top ten 85.5 per cent. The top six were: Venezuela (17.8 per cent), Saudi Arabia (15.9 per cent), Canada (10.4 per cent), Iran (9.4 per cent), and Iraq (9.0 per cent). Second, a significant amount of global production is drawn from fields with low R/Ps. Because global demand for crude oil continues to grow, supply and demand cannot be matched by simply relying on existing large fields—which are not being discovered in any number—rather, new production must come from smaller fields in a greater variety of locations (countries). The figures from BP highlight the significance of unconventional oil reserves in future production. Analysis by BGR (2012: 16) suggests that at the end of 2011, 44.1 per cent of remaining potential crude oil (reserves and resources) were in the form of unconventional oil.

When it comes to the remaining conventional crude oil reserves, around 70 per cent are located in OPEC countries and 53 per cent within the Middle East and North Africa (MENA) region. Thus, we can think of two geographies of global oil reserves, an established geography of conventional reserves and production that is concentrated in a relatively small number of countries; and an emergent new geography of unconventional oil reserves and production that at present is equally concentrated—in North America—but has the potential to diffuse to new countries. A further wrinkle is that much of the future incremental conventional production is going to come from regions that are costly and difficult to access—like the Arctic and deep-water offshore—and from countries that present challenges to the international oil companies in terms of access and investment climate and/or that are geopolitically unstable at present (issues illustrated by, for example, the loss of Iraqi production and Iranian exports) (Klare, 2012). Thus, it would seem that when issues of access, cost, and geopolitics are factored in, matching demand with supply will only be possible through the widespread exploitation of unconventional oil reserves (see Chapter 14) and this is the assumption made in most energy forecasts.

Interestingly, the BGR (2012:18) assessment concludes that: 'Crude oil is the only non-renewable energy resource which will no longer be able to keep up with growing demand in future decades.' This would seem to support the 'Peak Oil' argument, which is not about the size of existing reserves, which have been significantly bolstered by unconventional oil, but about the maximum rate that oil can be produced (for a recent discussion of peak oil, see Leggett, 2014). As is discussed in the next chapter, the future availability of unconventional oil is far from certain and it may be that it is only through demand reduction (oil demand has already peaked in the OECD) and efficiency savings and substitution (particularly in the transportation sector) that future demand growth can be constrained.

13.5.2 PRODUCTION AND CONSUMPTION

Oil is a fungible commodity that is traded on a global market. Like coal, there are variations in the quality and content of different crude oils and there are also various benchmark prices. Nonetheless, despite the distinctions between heavy and light crude oil and high and low sulphur content, for example, it is possible to talk of a global market and a global price. Until very recently the geography of global oil production and consumption has been relatively stable, notwithstanding events in the Persian Gulf and North Africa. In the run up to the 2008 financial crisis oil prices rose at a steady rate and then accelerated as production failed to keep pace with the relentless increase in demand, largely in Asia (Mitchell et al., 2012). The 2008 financial crisis and its aftermath dampened demand and price, but as the Asian economies recovered both the price and demand picked up again (see Figure 13.5).

During the 1980s, in response to the actions of OPEC and instability in the Middle East, the OECD economies made a major effort to expand their oil production—during this time production started in Alaska, the Gulf of Mexico, and the North Sea—but during the 1990s, when prices were low, investment stagnated and OECD production began to decline. Thus, when demand increased at the turn of the century more of the burden fell on OPEC (and indirectly Russia) to satisfy the growing demand in Asia. The net result has been an increasing role for OPEC production. Now, the rise of unconventional oil in North America—oil sands and tight oil—threatens to change the balance again; but its most likely effect will be to reduce the size of US imports, which will potentially free up more oil for Asia. According to the US EIA (2013c), US oil import dependency peaked in 2005 and in 2012 net oil import dependency was 40 per cent, a result of both increased domestic production and falling demand. In 2012, over 50 per cent of US crude and petroleum product imports came from the Western Hemisphere and only 29 came from the Persian Gulf countries. Table 13.4 shows the global oil balance in 2012 based on BP's regional groupings, which separate the MENA region between the Middle East and Africa.

As discussed above, global oil production is concentrated in a small number of large reserve holders and this explains the dominance of Eurasia, the Middle East, and Africa as net exporters to the global oil market. In 2012, Saudi Arabia (13.3 per cent) and Russia (12.8 per cent) accounted for more than a quarter of global oil production (BP, 2013: 10). On the consumption side of the equation, the Asia-Pacific is by far the largest net recipient of oil imports (Mitchell, 2010). In 2012, China accounted for 11.7 per cent of total global consumption. China is a major oil producer, accounting for 5 per cent of global production in 2012, but imports accounted for 42.9 per cent of total consumption

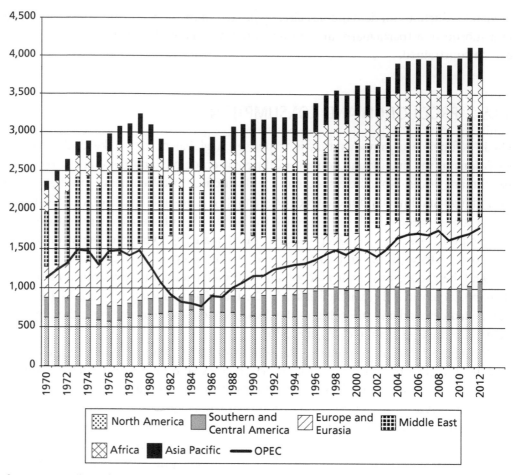

Figure 13.5 Oil production, 1970–2012

Source: BP (2013).

Table 13.4 The global oil balance in 2012

Million Tonnes	Production	% Production	Consumption	% Consumption	Balance
North America	721.4	17.5	1,016.8	24.6	−295.4
S. & Cent. America	378.0	9.2	302.2	7.3	75.8
Europe & Eurasia	836.4	20.3	879.8	21.3	−43.4
Middle East	1,336.8	32.5	375.8	9.1	961.0
Africa	449.0	10.9	166.5	4.0	282.5
Asia Pacific	397.3	9.6	1,389.4	33.6	−992.1
World	4,118.9	100.0	4,130.5	100.0	−11.6

BP (2013: 10–11).

in 2012. By comparison, Japan is the third largest consumer at 5.3 per cent, but this is all imported. The US situation is similar to that of China; the USA is a major producer, 9.6 per cent of total global crude production, but it is also the world's largest oil consumer at 19.8 per cent of global production. The European Union is the third major centre of demand, accounting for 14.8 per cent of global demand in 2012, but only 1.8 per cent of production. Much of that deficit is covered by oil imports from Russia and is thus hidden within the Europe and Eurasian balance, but the EU is also dependent on oil imports from the MENA states.

The reality is that it is too early to tell whether future production from unconventional oil, and from deep-water offshore and Arctic fields, has the potential to change the global oil balance. One other uncertainty is the possibility of increased oil exports from both Iran and Iraq, both of which are major reserve holders. Iraq is struggling to recover from the impact of two Gulf Wars on its oil industry and is still far from politically stable, while Iran is subject to sanctions to protest its nuclear programme that limits the level of oil exports. At the same time, growing domestic demand in the larger oil exporting states of MENA threatens to reduce the size of their exportable surplus. These trends suggest that even if the global oil industry is able to continue to match demand with supply the geographies of both are likely to change quite significantly, with unknown economic and geopolitical consequences.

13.6 **Natural Gas**

Natural gas is not as easily transported as coal or oil and is a relatively low value-to-volume (in both calorific and financial terms) energy source. This means that the majority of natural gas is actually consumed in the countries where it is produced. When natural gas is transported any distance overland, pipelines are the most economic mode. To enable natural gas to cross the oceans in significant volumes it must be turned into liquefied natural gas (LNG), a process that consumes energy, costs money, and requires specialist shipping and regasification terminals. The material distinctiveness of natural gas is important in understanding the current patterns of production, trade, and consumption, which are very different from both coal and oil. However, like oil, the development of technologies to exploit unconventional gas reserves is also changing the prospects for the future. At present, natural gas accounts for 23.9 per cent of total global primary energy consumption (BP, 2013: 41).

13.6.1 RESERVES

Arriving at a reliable estimate of global gas resources and reserves is all but impossible at present because of the large degree of uncertainty surrounding the extent of unconventional—tight gas, shale gas, and coal bed methane—resources (let alone the huge resource of methane that is held in gas hydrates). In their 2012 survey, the WEC (2013: 152) noted that proven natural gas reserves were identified in 103 countries with an aggregate volume of approximately 186 trillion cubic metres (tcm) or 6,550 trillion cubic feet (tcf). This estimate does not include substantial estimates for shale gas beyond proven reserve estimates for the USA at that time. The largest reserves of conventional gas are found in: Russia, Iran, Qatar, Turkmenistan, and Saudi Arabia, with OPEC countries accounting for 50 per cent of those reserves. The WEC reviewed six other surveys of natural gas reserves and concluded that they ranged between 177 and 189 tcm. So an average estimate would be around 183 tcm. The BGR (2012: 19) estimated global gas reserves to be 195 tcm at the end of 2011; again only a small amount proportion of this figure is shale gas reserves. The BGR (2012: 20) estimated global natural gas resources in conventional and non-conventional fields to be around 785 tcm at the end of 2011, with unconventional gas accounting for around 60 per cent of the total. The BGR (2012: 23) concluded that: 'From a geological point of view, natural gas is still available in very large quantities.' At the end of 2012, BP (2013: 20) estimated global proved reserves of natural gas to be 187.3 tcm, with an R/P of 55.7 years. Obviously, the commercial development of unconventional gas has the potential to increase the R/P significantly. According to BP, the major natural gas reserve holders at present are: Iran (18 per cent), Russia (17.6 per cent), Qatar (13.4 per cent), Turkmenistan (9.3 per cent), the USA (4.5 per cent), and Saudi Arabia (4.4 per cent). Together, these six countries account for 67.2 per cent of proved reserves, but not all of them are major exporters of natural gas at present.

The innovative combination of technologies to develop new techniques to extract unconventional natural gas has complicated the assessment of technically recoverable natural gas reserves. McGlade et al. (2012) have conducted a review of regional and global estimates of unconventional gas resources and they reach the following conclusions: 'Taking the best currently available estimates, we find that the globally technically recoverable resource of shale gas may be in the region of 200 tcm, with an additional 70 tcm from tight gas and coal bed methane.' They estimate technically recoverable reserves from unconventional fields to be 425 tcm, within which 190 tcm are currently proved reserves, this is a slightly higher figure than those above. As discussed earlier in this chapter, it is important to stress that the estimates of unconventional reserves are of technically recoverable reserves. However, their main conclusion is a cautionary warning that there is a very high level of uncertainty around these estimates.

As technically recoverable reserves are those that are potentially recoverable with current technology, they do not take account of economics—or other above ground factors such as the costs imposed by regulation and gaining social acceptance—and it is currently not possible to arrive at proved reserve estimates for unconventional gas with any degree of certainty. In 2013 the EIA produced a new set of estimates for technically recoverable reserves of shale gas, based on an assessment of 137 shale formations in 41 countries outside the USA. This US EIA (2013c: 1) assessment estimates totally

Table 13.5 The EIA's top ten countries with technically recoverable shale gas resources

Rank	Country	Tcf	Tcm	%
1	China	1115	31.6	15.3
2	Argentina	802	22.7	11.0
3	Algeria	707	20.0	9.7
4	USA	665	18.8	9.1
5	Canada	573	16.2	7.9
6	Mexico	545	15.4	7.5
7	Australia	437	12.4	6.0
8	South Africa	390	11.0	5.3
9	Russia	285	8.1	3.9
10	Brazil	245	6.9	3.4
	World Total	7,299	206.6	100.0

Source: US EIA (2013d: 10).

technically recoverable reserves of shale gas (including the USA) to be 7,299 tcf or 206.6 tcm of natural gas. They estimate that shale gas resources add 47 per cent to their existing estimate of 15,583 tcf or 440.1 tcm of technically recoverable non-shale gas resources. They also conclude that globally 32 per cent of natural gas resources (technically recoverable) are in shale formations. It is far too early to tell what percentage of those technically recoverable reserves will be translated into proved reserves. The commerciality of a particular gas field is dependent upon: the costs of drilling and completing wells (and those costs cover a wide range of so-called above ground issues), the amount of gas produced over time and the price received for gas production. Clearly, these will vary greatly from field to field and from country to country. Table 13.5 shows the geographical distribution of the EIA's latest estimates.

A number of issues arise from the potential future geography of shale gas production. First, as is explained below, some of the countries in the top ten are also potentially significant importers of natural gas, China being the most obvious. In the US shale gas production has already turned it from planning to import significant amounts of LNG to planning to export LNG. Thus, shale gas development in China would reduce the volume of future gas imports, which would have a knock-on effect in global LNG markets, just as the loss of the US market has done. Second, there are countries here that are not currently significant producers or exporters of natural gas and the development of their shale gas potential could have a significant impact on regional gas balances, Argentina and Brazil are cases in the context of South America. Argentina is already attracting a good deal of interest from the international oil and gas companies. Brazil has other options, such as biofuels and offshore conventional oil and gas. Likewise, Russia has very substantial conventional gas reserves and is unlikely to develop its shale gas potential. In Mexico, shale gas could compensate for declining production in the conventional oil and gas industry. In South Africa, shale gas could provide a viable alternative to its heavily coal-based power generation sector. However, such developments are very much in the realms of potential, and significant new commercial shale gas production outside of the USA is unlikely until the 2020s; but one should not underestimate the impact in the USA, where it accounted for 40 per cent of total gas production in 2012 and, as is made clear in Chapter 15, this is already having an impact on global gas markets.

13.6.2 PRODUCTION AND CONSUMPTION

Unlike oil, the majority of global gas production is not traded across international boundaries. According to the BGR (2012: 20), in 2011 around 1,025 bcm of natural gas, about 31 per cent of global production, was traded internationally (excluding transit countries). Similar figures from BP (2013: 24–28) for 2012 show that total global gas production was 3,336.9 bcm, of which 705.5 bcm (21.1 per cent) was traded via pipelines and 327.9 (9.8 per cent) was traded as LNG (the development of the global LNG industry is discussed in more detail in Chapter 15). Just as a benchmark, the IEA's (2011: 27) 'Golden Age of Gas Scenario' would foresee global gas production increasing to 4,383 bcm by 2025 and 5,132 bcm by 2035, nearly a 54 per cent increase over 2012 levels (but this was only 597 bcm above the level of production in their New Policies Scenario at that time). According to BP's data, in 2012, the major gas-producing countries were: the USA (20.4 per cent), Russia (17.6 per cent), Iran (4.8 per cent), Qatar (4.7 per cent), and Canada (4.6 per cent); thus the top five account for 52.1 per cent of global production. When it comes to pipeline-based internationally traded gas, Russia was the leading exporter in 2012 at 185.9 bcm (all of which was to markets in Europe and the former Soviet Union), followed by Norway at 106.6 bcm (all of which went to markets in Europe) and Canada at 83.8 bcm (all of which went to the USA). These three trade movements accounted for 53.3 per cent of total pipeline trade in 2012. The Gas Exporting Countries Forum (GECF), which started to evolve from 2001 onwards, lacks the formal structure and influence of OPEC, but is beginning to exert itself as a voice in the global gas industry; however, the very nature of the gas industry makes it very unlikely that it will ever have the market power of OPEC (El-Katiri and Honoré, 2012).

The situation with LNG is also concentrated around key producers and key markets, although there is a trend towards diversification. In 2012, Qatar alone exported 105.4 bcm of LNG (32.1 per cent of global exports) and the other major exporters were: Malaysia at 31.8 (9.7 per cent), Australia at 28.1 (8.6 per cent), Nigeria at 27.2bcm (8.3 per cent), and Indonesia 25.0 (7.6 per cent). Thus, the top five LNG exporters account for 66.3 per cent of total exports. The current hierarchy is likely to change for two reasons; first, because some of the Asian producers are facing growing domestic gas demand and are reducing LNG exports (Indonesia now imports LNG); and second, because massive expansion is planned in Australia and there is the prospect of new export capacity in East Africa, North America (based on shale gas), and Russia. During 2012 only one new liquefaction plant came on line (Pluto LNG in Australia) and in 2013 the only major start-up was Angola. However, at the end of 2012 there were 30 LNG trains under construction with a total capacity of 110.1 mtpa or 150 bcm (IGU, 2013).

The gas market globally can be conceived as three different, but interconnected, markets that are each organized on a regional basis—North America, Europe, and Asia (this is discussed in detail in Chapter 15). However, the largest market is the domestic market of gas-producing countries, many of which are not major exporters. A distinction can be made between countries such as Russia who are major producers, consumers, and exporters; countries like the USA who are major producers and consumers, but not major exporters (yet); countries like China that are large producers, and growing consumers, and are importing more and more gas to meet growing domestic demand; and countries that have very small domestic demand and are major exporters—Qatar being the most obvious example. There is also a group of countries with large domestic demand and little or no domestic production, reliant on imports, Japan and South Korea being extreme cases. When it comes to pipeline gas, the markets are divided at a continental scale; here the transport of gas from Russia to Europe is by far the dominant flow. There are also much more modest flows in North and South America and an emerging pipeline network in Asia with Central Asia and Myanmar (and potentially Russia) supplying China.

It is the maritime LNG markets that link producers and consumers across the globe. The global LNG industry is traditionally divided into two basins or hemispheres, the Atlantic basin where the industry first emerged and the Asian basin where the lack of pipelines and the geographically fragmented nature of the market meant that it was the only way to commercialize gas reserves. It is increasingly possible to talk about a globalizing LNG market due to increased interconnection between the two markets, made possible in large part by the dramatic expansion of Qatari LNG export capacity with its ability to serve both markets. However, a key distinction between the two basins is that in the Atlantic basin (particularly the European market) LNG faces competition from both domestic production and large volumes of pipeline imports. In the Asia-Pacific basin, at present, there is relatively little pipeline gas. Because of the costs involved in the liquefaction, transportation, and regasification, pipeline gas is usually less expensive—but much depends on the pricing and contract structures—and that explains the current divergence between prices in Europe and Asia.

It follows from the discussion of the geographical complexity of gas markets, that the pricing of internationally traded gas is also a complicated matter. Allsopp and Stern (2012: 13) note that: 'gas is priced in different ways in different countries (and regions) reflecting different market structures, different national policies and objectives, and barriers, both physical and political, to integration'. The net result is that we have different pricing systems in different export markets and recently there has been a good deal of divergence between them. This is discussed in more detail in Chapter 15. At the moment the highest level of divergence is between the price paid by domestic consumers in the USA and the price paid by LNG importers in the Asian market. The European market, which is supplied by domestic production, pipeline imports, and LNG, occupies a mid-point. There are at least two issues of contention that mean that the future of gas prices is uncertain; first the predominance in Europe and Asia of pricing formulas indexed to oil; and second, the use of long-term contracts often with take or pay clauses. The high price of oil and the low price of gas in the USA are key factors destabilizing the current situation. Gas is indexed to oil for historical reasons, but gas is not presently competing with oil; rather its major source of competition is coal in the power generation sector. How gas is priced is fundamental to its future role in the global energy system, as is the related issue of how carbon emissions are priced. However, the high cost of the LNG supply chain is a complicating factor that makes it very difficult to forecast whether or not there will be a re-convergence of regional gas prices. The net result is that natural gas will play different roles and faces different prospects in different markets; there is unlikely to be a golden age of gas everywhere. In the context of climate change, natural gas can be a positive factor where it replaces or reduces the growth of coal consumption in the power sector. This helps to explain why the global gas industry sees the greatest future demand growth in Asia. In Europe, by comparison, gas may be relegated to providing back to renewable intermittency, rather than generating baseload electricity, and may also suffer demand destruction in the household heating sector.

13.7 Conclusions

The detailed analysis presented in this chapter suggests that physical scarcity is not the key driver of uncertainty in relation to the future role of fossil fuels on the future global energy system. Rather the interplay of geopolitics and climate change policy is likely to exert the greatest influence on the relative demand for the three fossil fuels. Each presents a different challenge within the context of the global energy system. From a climate change perspective, coal presents the greatest challenge. In the absence of widespread utilization of carbon capture and storage (that is discussed in Chapter 16), a global agreement to constrain and then reduce GHG emissions must result in a reduction in coal consumption. The future of global oil production is more likely to be constrained by geopolitical factors. The most accessible and affordable reserves are still held by the member states of OPEC and in the MENA region; but this is a region of continuing geopolitical instability. Consequently, there will continue to be a drive to exploit new conventional reserves in more costly locations, as well as a growing emphasis on unconventional sources of oil production—oil sands, tight oil—as well as gas-to-liquids and biofuels. Whether oil production can keep up with demand remains to be seen; but the OECD has already passed peak demand and the development of relatively abundant natural gas might reduce oil demand in the transportation sector. Of the three fossil fuels, natural gas is potentially the least susceptible to geopolitics—notwithstanding the current situation between Europe and Russia—as abundant conventional and unconventional reserves and the increasingly globalized nature of the industry promise greater security of supply. The limiting factor here is likely to be security of demand and the availability of investment to develop new reserves and expand the global LNG supply chain. The latest IPCC report (2014) sees a role for natural gas as a bridging fuel to a low carbon energy system, but this is likely only to be an opportunity in markets where gas can replace coal. In sum, the future prospects for fossil fuels will be determined by what happens elsewhere in the global system in relation to carbon pricing and demand reduction, and not by fossil fuel scarcity.

■ REFERENCES

Allsopp, Christopher and Stern, Jonathan. 2012. 'The Future of Gas: What are the Analytical Issues Relating to Price?' In *The Pricing of Internationally Traded Gas*, edited by Jonathan Stern, pp. 10–39. Oxford: OUP/OIES.

Aguilera, Roberto, Eggert, Roderick, Lagos, Gustavo, and Tilton, John. 2009. 'Depletion and the future availability of petroleum resources.' *The Energy Journal* 30 (1): pp. 141–74.

BP. 1999. *Statistical Review of World Energy*. London: BP.

BP. 2013. *Statistical Review of World Energy*. London: BP.

Bundesanstalt für Geowissenschaften und Rohstoffe (BGR). 2012. *Energy Study 2012 Reserves, Resources and Availability of Energy Resources*. Berlin: BGR.

Cambridge Energy Research Associates (CERA). 2012. *IHS Upstream Index*. Cambridge, MA: CERA.

Department of Energy and Climate Change (DECC). 2013. *Digest of UK Energy Statistics, Section 2 Solid Fuels and Derived Fuels*. London: TSO, pp. 41–60.

Devon, John. 2010. 'The Competitive Cost of Coal—Analyzing the Major Producing Areas of the World'. Presentation at the Coal Trans Asia Conference, 31 May 2010.

El-Katiri, Laura and Honoré, Anouk. 2012. 'The Gas Exporting Countries Forum: Global or Regional Cartelization?' In *The Pricing of Internationally Traded Gas*, edited by Jonathan Stern, pp. 424–66. Oxford: OUP/OIES.

Farrell, Alex and Brandt, Adam. 2006. 'Risks of the oil transition'. *Environmental Research Letters* 1 (1): pp. 1–6.

Global Energy Assessment (GEA). 2012. *Global Energy Assessment—Towards a Sustainable Future*. Cambridge University Press: IAASA/CUP.

HSBC. 2013a. *China Coal and Power—The water-related challenges of China's coal and power industries*. London: HSBC.

HSBC. 2013b. *India Renewables—Good bye winter, hello spring*. London: HSBC.

IEA. 2005. *Resources to Reserves—Oil and Gas Technologies for the Energy Markets of the Future*. Paris: IEA.

IEA. 2008. *World Energy Outlook 2008*. Paris: IEA.

IEA. 2011. *Are We Entering a Golden Age of Gas?* Paris: IEA.

IEA. 2012. *Coal Medium-Term Market Report 2012, Market Trends and Projections to 2014*. Paris: IEA.

IEA. 2013. *Resources to Reserves. Oil, Gas and Coal Technologies for the Energy Markets of the Future*. Paris: IEA.

International Gas Union (IGU) 2013. *World LNG Report-2013 Edition*. Oslo: IGU.

Klare, Michael. 2012. *The Race for What's Left: The Global Scramble for the World's Last Resources*. New York: Metropolitan Books.

Kaufmann, Richard. 1991. 'Oil production in the lower 48 states: Reconciling curve fitting and econometric models'. *Resources and Energy* 13(1): pp. 111–27.

Leggett, Jeremy. 2014. *The Energy Nations: Risk Blindness and the Road to Renaissance*. London: Earthscan/Routledge.

MacDonald, David. 2013. 'Commodity specifications for UNFC-2009 application of the framework classification using the PRMS'. 75th EAGS Conference & Exhibition incporating SPE EUROPEC 2013, 12 June. London.

McGlade, Christophe, Spiers, Jamie, and Sorrell, Steve. 2013. 'Unconventional gas—A review of regional and global resource estimates'. *Energy* 55: pp. 571–84.

McGlade, Christophe. 2012. 'A review of the uncertainties in estimates of global oil resources'. *Energy* 47(1): pp. 262–70.

McKinley, Roderick. 2012. 'Industrial Emissions Directive: game over for coal?' *Bloomberg New Finance European Power Research Note*. 18 October 2012.

Mitchell, John, Marcel, Valérie, and Mitchell, Beth. 2012. *What's Next for the Oil and Gas Industry*? London: Chatham House.

Mitchell, John. 2010. *More for Asia: Rebalancing World Oil and Gas*. London: Chatham House

Platts. 2013. 'China's coal imports in 2012 rise 583 on year to 289 million tonnes. 24 January.

Poroskun, V.I., Khitrov, A.M., Zykin, M.Y., Heiberg, S. and Sødenå, E. 2004. 'Reserves and resource classification schemes used in Russia and Western Countries: a review and comparison'. *Journal of Petroleum Geology* 27 (1): pp. 85–94. [Full names not available in the original]

Remme, Uwe, Blesel, Markus, and Fahl, Ulrich. 2007. *Global Resources and Energy Trade: An Overview for Coal, Natural Gas, Oil and Uranium*. Stuttgard: Institute für Energiewirtschaft und Rationelle Energieanwendung.

Sorrell, Steve, Speirs, Jamie, Bentley, Roger, Brandt, Adam, and Miller, Richard. 2009. *Global Oil Depletion: An Assessment of the evidence for a near-term peak in global oil production*. London: UKERC.

Sorrell, Steve, Spiers, Jamie, Bentley, Roger, Miller, Richard, and Thompson, Erica. 2012. 'Shaping the global oil peak: A review of the evidence on field sizes, reserve growth, decline rates and depletion rates'. *Energy* 31(1): pp. 709–24.

SPE, AAPG, et al. 2008 (2011). Petroleum resources management system. Available at: <http://www.aapg.org/geoDC/PRMS_Guidelines_Nov2011.pdf> (Accessed April 2015).

United Nations (UN) 2009. Mapping of United Nations framework classification for fossil fuel energy and mineral resources. *Energy Series No. 33*. New York: UN Economic Commission for Europe.

UN 2009a. *United Nations framework classification for fossil energy and mineral reserves and resources*. New York: UN Economic Commission for Europe.

US EIA. 2013a. *NYMEX Coal Future Near-Month Contract Final Settlement Price 2013*. 26 July. Washington DC: US EIA.

US EIA. 2013b. *Annual Energy Outlook 2013*. Washington DC: US EIA.

US EIA 2013c. *Technically Recoverable Shale Oil and Shale Gas Resources: An Assessment of 137 Shale Formations in 41 Countries Outside the United States*. Washington DC: US EIA.

World Coal Association (WCA). 2013. Coal and Steel. WCA website: <http://www.worldcoal.org> (Accessed April 2015).

World Energy Council (WEC). 2013. *2010 Survey of Energy Resources*. London: WEC.

14 Unconventional fossil fuels and technological change

Michael Bradshaw, Murtala Chindo, Joseph Dutton, and Kärg Kama

14.1 Introduction

Building on the reserve classification as set out in the previous chapter on conventional fossil fuels, this chapter explains the 'conditional' nature of much of the fossil fuel resource base with the result that 'proven reserves' are dynamic and change as technology, in combination with economic and operating conditions, changes. The category of 'unconventional fossils fuels' includes a group of resources that are particularly conditional, whereby the interactions between technological change, economic viability, and environmental consequences both create and destroy proven reserves. Often the existence of a potential resource base is sufficient to trigger the processes of innovation that develop new technologies and new applications of existing technologies to enable commercial exploitation of the resource. The tight oil and shale gas revolution is a case in point. Shale has been long known as the source rock for conventional oil and gas, but only recently have technologies been deployed to enable commercial extraction of liquid hydrocarbons from shale formations. Equally, the prospect of falling oil prices and environmental constraints (the prospect of carbon taxes or clean fuel requirements) drives innovation to reduce costs and socio-ecological impacts. The development of the Canadian oil sands is a case in point. As the previous chapter revealed, the category of unconventional oil and gas reserves has become increasingly important to matching supply with demand, but the analysis here warns that realizing this production will require sustained high oil and gas prices and trade-offs between energy security and environmental security (in terms of both local impacts and global climate change). Some of these new extractive economies may remain localized in scope, rather than acquiring global significance.

Figure 14.1 places unconventional energy resources within the wider context of conventional resources. This is a complex landscape and there is considerable confusion as how best to define and categorize each type of resource (see the Society of Petroleum Engineers et al. 2011for detailed discussion of this matter). In this chapter the distinction between conventional and unconventional resources is drawn between heavy oil that is considered conventional and oil sands (bitumen) that is considered unconventional. Thus, tight oil (sometimes also referred to as shale oil and much lighter than heavy crude and oil sands) and oil shales are also considered unconventional. In this chapter, shale oil

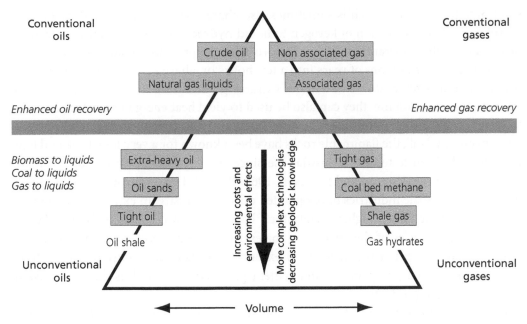

Figure 14.1 The resource triangle
Source: Arkonsuo (2013: 36).

refers to unrefined crude oil products obtained from oil shales; whereas tight oil refers to oil obtained from shale rock using the same technologies as in shale gas extraction (this reflects established terminology in the geosciences). The gas side of the equation is less complicated, tight gas is considered as conventional and coal bed methane and shale gas are considered unconventional. Gas hydrates are also an unconventional gas resource, but they are not discussed here. What follows is three detailed case studies of different unconventional resources: oil shales (written by Kärg Kama), oil sands (written by Murtala Chindo), and shale gas (written by Michael Bradshaw and Joseph Dutton). Tight oil is not the subject of a case study, but its production in the United States has developed in tandem with shale gas. Similarly, coal-bed-methane (CBM) is not considered for much the same reason in that it uses similar technologies as shale gas, although CBM is actually more widely established than shale gas. Each case study illustrates, in different ways, the 'conditionality' of unconventional fossil fuels and their specific historical and geographical contexts that explain the key technological advances and challenges they face and that influence their future prospects.

14.2 **Oil shales**

The term 'oil shale' denotes a wide range of fine-grained sedimentary rocks that are found all over the globe. Unlike liquid-bearing shales and bituminous sands, oil shales contain no oil and gas as such, but to a smaller or lesser degree kerogen, a mixture of solid organic material, which is sometimes also characterized as an 'immature form of petroleum'.[1] The conversion of kerogen to liquid hydrocarbons requires the rocks to be heated at a high temperature in airtight retorts that compress the geological processes of oil formation over millions of years into a few hours or, alternatively, three to four years if the mineral is retorted in situ without extracting and lifting it to the ground. When such stones are set on fire, they can also be used to yield heat energy in a manner similar to brown coal. Although oil shales have more recently become regarded as a source of 'unconventional oil', the flammable rocks have been known for a very long time and have continuously been tested for industrial use since the late nineteenth century in many parts of the world (Duncan and Swanson, 1965; Russell, 1990). Throughout the 'age of oil', smaller oil shale industries have operated outside mainstream extractive economies, especially at times when higher quality fuels have been scarce, such as the two World Wars or the oil crises of the 1970s. Oil shales are not considered solely as a potential liquid fuel resource, but the majority of extraction to date has occurred due to direct burning in thermal power plants and cement production, as well as due to its use in various chemical products.

Oil shales primarily captivate those nation-states that lack conventional hydrocarbon reserves. Being abundant, indigenous fossil fuel resources, their development is expected to solve structural energy dependence problems, lower foreign debt, and provide for future economic and social stability (e.g. Bartis et al., 2005; Dammer, 2007; Ogunsola, et al. 2010). While such promise has repeatedly failed to be realized, significant industries continue to exist in Estonia, China, and Brazil. The former Soviet republic of Estonia has been the leading producer of oil shale under both socialism and capitalism, accounting for nearly half of the global output to date, although it is estimated to possess only a fraction of the world's potential resources (Laherrère, 2005; Francu et al., 2007). In conjunction with rising oil prices over the past decade, however, several other states are either resurrecting earlier development projects or considering the exploitation of their deposits for the first time, most notably in the USA, the Levant region, and the former Communist bloc. The aspirations of local government officials are backed up by longstanding hopes in the geosciences and industry for a technological breakthrough similar to shale gas and oil sands development (in this chapter we use the term oil sands, but note that others use the term tar sands to stress the bituminous origins of this type of oil). The USA is currently in the midst of a heated public debate over the economic and social acceptability of another surge in developing its vast resources, which are mostly located on federal lands, after having already experienced two boom and bust cycles in

---

[1] Recent reports have introduced the term 'kerogen oil' for oil shales in order to avoid confusion with tight oil and other liquid crude reservoirs found in shale formations (e.g. McGlade, 2012; IEA, 2013b). However, although kerogen is often taken to be a common characteristic of oil shales, it has different origins and is not an entirely homogeneous substance, while occurring also in other carbonaceous fuels (Yen and Chilingar, 1976; Hutton et al., 1994). To avoid further confusion, this chapter reserves 'shale oil' for the heavy oil extracted from oil shales, adhering to long-term terminology in the geosciences and technological literature.

previous decades (Gulliford, 1989; Hanson and Limerick, 2009). Meanwhile, several projects in the MENA countries are now suspended or delayed as a result of the regional instability that followed the Arab Spring.

As a rule, oil shale development occurs currently outside the Kyoto Annex I countries in order to avoid regulatory constraints, public opposition, and the prospect of carbon pricing that is expected to become the major driver for production costs. By contrast, Estonia has persisted in maintaining its industry in the face of the stringent environmental and market regulations of the European Union due to energy security concerns, and is increasingly used as a reference point for the promotion of oil shale development in other locations.

Given the sheer quantity of organic-rich sedimentary rocks in the earth's crust, the proponents of oil shale development have invariably claimed that such resources exceed conventional crude oil availability by many times. However, although several experts have attempted to specify the global resource, such estimates remain highly speculative as they are based on inconsistent national reports and evaluation methodologies (e.g. Duncan and Swanson, 1965; Russell, 1990; Dyni, 2006; WEC, 2007). The uncertainties of resource assessment are not just bound to the fact that countries use different cut-off grades to delimit their resource base (McGlade, 2012), but derive from more substantial problems related to the lack of geological studies, the high variability of host rocks, and the use of different measurement practices and analytical units that reflect the disparate legacies of technology development (Dyni, 2006; Speight, 2012), as epitomized in the 'shale-to-liquids' and 'shale-to-power' discourses (Table 14.1). These difficulties are aggravated by the fact that calculations of the mineral mass need to be translated into a liquid fuel, which varies across deposits, and that there is very little information available on technical recoverability and the economics of production (Kama, 2013). As a result, existing estimates of 'in-place' resources of shale oil cannot be compared with the proven reserves of crude oil.

While Estonia represents a unique case of successful oil shale production largely due to utilizing its relatively high-grade resources for power supply, there is very little evidence of commercially viable production of *shale oil*, not to mention refined synthetic crude usable in transport. The few existing industries are based on various methods of retorting oil shale once the mineral has been mined through traditional methods, and closely integrated into the economics of other extractive or energy supply operations. Since above-ground retorting yields nearly the same quantity of spent shale as extracted, its main costs are related to extensive material handling and environmental pollution. Below-ground conversion is expected to be less expensive in the long term, but involves much more knowledge-intensive and similarly polluting technologies. These technologies are yet to be proven technologically and economically feasible, despite repeated in situ experiments conducted since the early 1970s. These experiments are led by major oil companies in Colorado, most notably by Shell, who has completed the first pilot tests, but also Chevron, ExxonMobil, and Total[2] under the RD&D leases issued by the federal

---

[2] Total has formed a 50/50 joint venture with Genie Energy, called American Shale Oil (AMSO).

Table 14.1 'Shale-to-power' and 'shale-to-liquids' discourses of oil shale development

	Shale to power	Shale to liquids
Producing countries	Estonia, China, Israel	Estonia, China, Brazil
Emerging developers	Jordan, Morocco	USA (Colorado, Utah), Australia (Queensland), Jordan, Morocco, Israel, Egypt, Turkey, Syria, Russia, Serbia, Ukraine, and others
Utilization	Shale is burnt directly as solid fuel in thermal power stations similarly to the combustion of coal for electricity generation.	Shale is retorted into oil, for which there is no commercially proven technology, but many are under development, including:
		1) in situ: Shell's ICP, ExxonMobil's Electrofrac, Chevron's CRUSH, and AMSO's CCR processes
		2) ex situ: Alberta-Taciuk, Fushun, Galoter (Enefit and Petroter), Kiviter, Paraho, Petrosix, Unocal, and a multitude of other processing methods.
Technology	Several combustion techniques have been adapted to oil shales, but require innovation for the reduction of emissions, improved energy efficiency, and readiness to adjust the process to feedstock of lower quality.	Except for some operational experience in surface retorting, all technologies are still in need of major innovation, especially concerning net energy gain (EROEI), environmental impacts and adaptability to various host rocks.
Resource calculations	Based on mineral mass (tonnage) or heating value measured with a calorimeter.	Based on *oil yield* measured with the Fischer assay, or alternative methods of assessment.
Environmental problems	Very high greenhouse gas emissions, wastes/tailings, water use, landscape degradation etc.	Water use, wastes/tailings, but fewer emissions compared to burning, since carbon is released by end-consumers.

Source: adjusted from Kama (2013).

government. Chevron's and Shell's abrupt decisions to halt experiments and abandon their leases in 2012 and 2013 respectively are a serious drawback to the industry's reputation. By contrast, smaller private and state companies are currently upgrading ex situ technologies, often by capitalizing on the legacies of earlier projects (for recent overviews, see Bartis et al., 2005; Dammer, 2007; Speight, 2012). There is little competition between these approaches, not least because they can be applied to different deposits, depending on whether the resource is accessible to mining. So far all available technologies have been developed in accordance with the characteristics of particular deposits; however, the industry is aiming to invent more universal solutions that could be adapted to deposits of varied composition and quality.

The major obstacle for both in situ and ex situ developers is therefore raised by the material and spatial heterogeneity of world resources, which has precluded economies of scale and technology transfer to other locations. Although several businesses in surface retorting produce various shale oil fragments that increasingly have market value, mostly as heating oil or bunker fuel, their technologies are adapted to the local resource base and are of limited use elsewhere. In this context, the Kingdom of Jordan has turned out the main testing ground for commercial production, having leased a quarter of its entire territory to Shell, while also cooperating with several other technology holders in retorting and power generation.[3] However, even if commercial-scale production will be demonstrated eventually, a further question is whether the refining business is willing to adapt their processes to such low grade crudes that are contaminated with various mineral compounds and particulate matter from host rocks. Despite successful tests with upgrading small samples of shale oil in the US refineries, the prospects of supplying synthetic crude for transport fuel remains highly uncertain.

Nevertheless, the industry's future viability should not be regarded as unilaterally defined by technological progress, but rather the result of both capital investment and local regulatory and market arrangements that will ultimately judge whether a technology is deemed to 'work' sufficiently well. First, while larger companies are increasingly wary about allocating money and people to prolonged research, as Chevron's and Shell's recent decisions show, the expansion of surface retorting depends on the industry's ability to attract both direct funding and speculative capital in the shadows of the US 'shale boom' (cf. Mitchell 2013). Second, as indicated by the first commercial leases signed in Jordan, the competitiveness of shale oil and power production hinges on novel fiscal regimes set up by state governments (regarding royalties, taxes, environmental liabilities, and subsidies), as well as international developments in carbon pricing and trading. The major forthcoming task for the industry is to influence the regulation of energy systems and markets by stressing the strategic importance of oil shale to local economies, as demonstrated, for example, by on-going disputes over the carbon footprint-based standards of the EU Fuel Quality Directive.

Oil shale development has thus proven to be a historically persistent and highly specific issue based on national energy security dilemmas, which is consequently less related to global oil availability and prices, nor ruled out by international demands that countries undertake a 'low-carbon transition'. Rather, the future likelihood of oil shale becoming a global energy resource and, more specifically, an oil resource, depends on the extent to which developers will succeed in transforming both regional energy markets and the downstream sector to accommodate the resource's specific material qualities and ensure its competitiveness, and how this process will be judged against wider social and political processes in reconfiguring the carbon economy.

14.3 **Oil sands**

The basic process for the recovery and processing of oil sands was developed following the pioneering work of Karl Clark in the 1920s separating bitumen from sand. The first commercial quantity of recovered bitumen occurred in 1930 with the production of 300 barrels of bitumen from the Athabasca oil sands of Canada (Demaison, 1977; Millington

---

[3] In addition to concession agreements signed with Shell and two ex situ developers associated with the Alberta-Taciuk and Enefit technologies, the Jordanian government is in negotiations with Total, ExxonMobil, Petrobras, and several other companies.

et al., 2012). Research, sponsored by the Alberta Research Council, continued on the economic and technical feasibility of oil sands mining through the 1960s. The Great Canadian Oil Sands Company, now Suncor Energy, began large-scale commercial production of 12,000 barrels per day using open-pit mining (Meyer et al., 2007). The development of a commercial in situ project was championed by the industry in 1985. By 1989, Imperial Oil had pioneered the production of 140,000 bpd of bitumen in Cold Lake. Since the 1990s, numerous projects have been developed using these inventions to exploit oil sands worldwide (Bowman, 2008). Later, continuous research and technological advances and innovations focused on lowering production costs to compete with conventional oil, enhanced recovery, and more recently to lower energy requirement and comply with environmental regulations.

Global oil sands[4] resources are large: identified in-place resources are estimated to be between 2.2–3.7 trillion barrels of oil in place, far exceeding conventional oil reserves (WEC, 2010: 124–25; McManus, 2011; Millington and Mei, 2011). Large deposits have been found in limited parts of the world (Figure 14.2), but none of the known deposits come close to the supergiant oil sands trapped on the flanks of the Canadian province of Alberta. According to Alberta Energy (2014), Alberta's oil sands underlie 54,132 square miles (140,200 square kilometres) of land in the Cold Lake, Peace River, and Athabasca areas in northern Alberta. Together, these oil sands areas sit on an estimated 1.7 trillion barrels of crude bitumen in-place. At the beginning of 2012, about 179 billion barrels of this volume is considered to be a proven reserve under current technology (or about 13 per cent of global oil reserves), representing a potential supply larger than the conventional crude oil reserves of Iran, Iraq, or Kuwait, and are third only to those of Saudi Arabia and Venezuela (McManus, 2011). The largest volumes of oil sands outside Canada are in Kazakhstan and Russia, but both also have huge deposits of conventional oil and less incentive to develop oil sands. Significant resources are also found in Angola, Azerbaijan, Colombia, Congo Brazzaville, Democratic Republic of Congo, Georgia, Germany, Indonesia, Israel, Italy, Switzerland, Syria, Tajikistan, Tonga, and Uzbekistan (WEC, 2010). These resources are not currently exploited due to isolation, low grade and/or quantity, unknown reserve estimates, political risks, lack of technological solutions, limited economic viability, and environmental trade-offs. The United States, Romania, Trinidad and Tobago, Nigeria, Albania, and Madagascar have key sites of significant reserves for which development is proposed, under construction, or in production stages. The Orinoco Petroleum Belt, Venezuela, contains 90 per cent, or nearly 1.9 trillion barrels of the world's deposit of a similar substance, extra-heavy crude in-place (extra-heavy crude is much denser oil, similar to oil sands, much harder to

---

[4] The term 'oil sands' includes the crude bitumen, minerals, water, and rocks that are found together but excludes any related natural gas. Crude bitumen is the thick, black, sticky, extra-heavy oil within the oil sands areas. Oil sands contain about 83 per cent sand, 10 per cent bitumen, 3 per cent clay, and 4 per cent water.

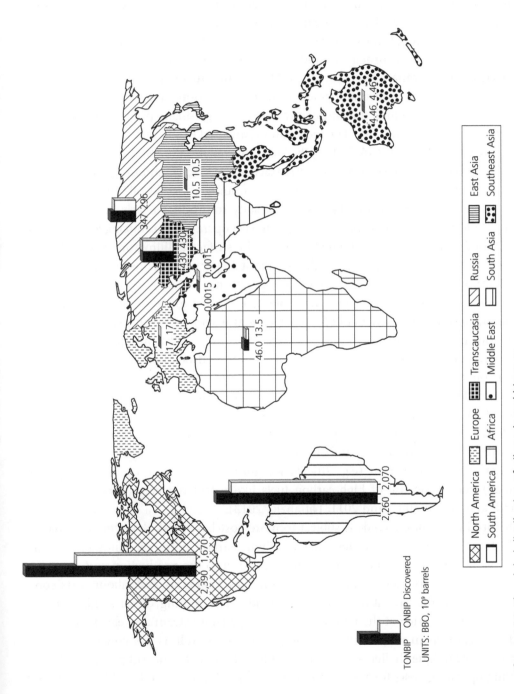

Figure 14.2 The global distribution of oil sands and bitumen

Source: Meyer et al. (2007).

extract with an API of <10°), and has already been in commercial production for a number of years (Huc, 2010; Pascal and Azpurua, 2008).

The mining, extraction, and upgrading process in oil sands is unique and complex (Figure 14.3). The reserve depth is the determining factor for the choice of mining or in situ technology. Mining is most efficient on large-scale oil sands deposits covered by less than 75 metres of overburden; beyond that depth, the in situ method becomes the preferred option (Atanasi and Meyer, 2007). The in situ method uses thermal techniques where wells are drilled and cased in the oil formation, and steam, water, and/or solvents are injected to melt the bitumen reservoir and gradually pump it to the surface (Madden and Morawski, 2010). The most commonly used in situ technologies are Steam Assisted Gravity Drainage (SAGD), Cyclic Steam Stimulation (CSS), Cold Flow, Toe-to-Heel Air Injection (THAI), and Vapour Extraction (VAPEX), choosing between them depends on reservoir characteristics. To achieve economies of scale, oil sands projects are large. While Syncrude Canada Ltd has the capacity to produce nearly 375,000 bpd by mining, Cold Lake–Imperial Oil is the largest in situ operation, producing an average of 160,000 bpd from more than 400 metres below the surface (Millington et al., 2012). Of the total proven oil sands reserves of Alberta, about 20 per cent is considered recoverable by surface mining and 80 per cent by in situ operations.

The oil that is part of the bitumen obtained from mining or in situ extraction (including the diluent which is added to enable the transport of highly viscous petroleum and is usually natural gas condensate) is separated (Figure 14.3b) and upgraded to create synthetic crude oil (SCO or Syncrude) or lighter hydrocarbon, either by thermal conversion to remove carbon (coking) or by applying heat, hydrogen, and pressure (hydro-cracking) to break the large carbon chains that make up the bitumen molecule (Figure 14.3c). Adding hydrogen reduces carbon waste and delivers 18 per cent more barrels for every 100 barrels of processed bitumen. Upgrading is the most energy-intensive part of the operation and emits 15 to 20 per cent more carbon than conventional crude oil production (e.g. 20–200 kilograms of CO_2 per barrel of upgraded bitumen) due to the use of high-carbon fuels like petroleum coke during extraction (Brandt and Farrell 2007, and 2011; Lattanzio, 2013).

The production of oil sands is subject to more cost-based risk than conventional oil production. Figure 14.4 illustrates where, without carbon costs, oil sands (both extra heavy oil and bitumen in Figure 14.4) are in relation to the production cost curve for different energy resources. Crude prices, natural gas,[5] material, and labour costs comprise about half the operating costs of oil sands. The mining of oil sands is partly dependent on the price of the natural gas used to generate steam and electricity and to produce hydrogen in associated upgrading facilities; as well as gas condensate costs. Because of the unique challenges associated with each project, different projects will have differing operating costs for the production of a barrel. For example, SAGD requires a

---

[5] For mining, the natural gas demand is around 250 cubic feet per barrel; for in situ, the gas requirement is ca. 1000 cubic feet per barrel. Upgrading to SCO requires an additional 330–730 cubic feet per barrel.

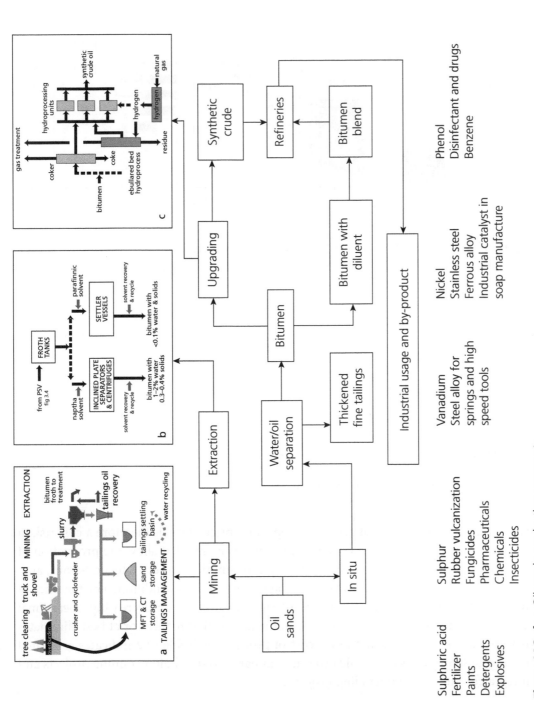

Figure 14.3a,b,c Oil sands production processes

Source: Alberta Chamber of Resources (2004); Humphries (2008).

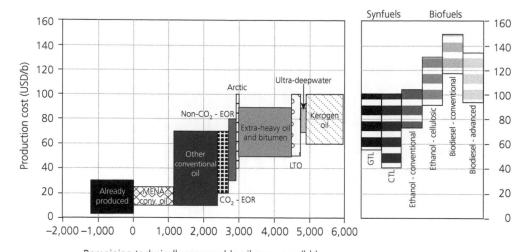

Figure 14.4 Comparing production costs to other energy sources
Source: International Energy Agency (2013).

Table 14.2 Costs of production using major oil sands recovery methods

Method of extraction		Oil sands Dollars per barrel at Plant Gate	
	Crude type	Operating cost $	*Supply cost $
Cold production—Wabasca Seal	Bitumen	6–8	13–17
Cold Heavy Oil Production with Sand (CHOPS)	Bitumen	7–9	15–18
Cyclic Steam Stimulation (CSS)	Bitumen	10–14	19–22
Steam Assisted Gravity Drainage (SAGD)	Bitumen	9–13	17–20
Mining/extraction	Bitumen	8–12	17–19
Integrated mining/Upgrading	Synthetic crude	17–22	33–40

* The cost price includes capital and operating costs, maintenance, taxes, and royalties.
Source: National Energy Board (2006).

significant amount of natural gas—the supply and price of gas is therefore a cost risk in production and profit. Continuous process improvements that allow the production of oil faster, more easily, and with lower energy consumption, lower GHG emissions, and, therefore, lower supply costs have been—and will always be—conditioned by the price of natural gas. The aspect that has the most significant effect on decreasing operational costs is finding the right technology—as in the past[6]—that reduces or at best eliminates natural gas use, resulting in a lower cost of production. Table 14.2 indicates estimated plant gate supply costs for SAGD projects, compared to stand-alone mining projects and integrated mining and upgrading projects.

---

[6] Replacing the draglines and bucket-wheel reclaimers with trucks and power shovels and the hydro-transport system that replaces the conveyor belts were the major innovations in the 1990s that cut production costs.

History indicates that technological innovation has been a key driving force behind the oil sands industry since its inception. All aspects of operations have gone through substantial change, sometimes step changes, in their evolution into today's industry. Advances in technology have made it possible to increase reserves, reduce costs, and significantly improve environmental performance. Although research and development efforts related to oil sands are currently concentrated in Canada, they are being carried out in many parts of the world, covering a wide array of scientific disciplines and technologies. In addition to today's processing techniques, new technologies are in various stages of field, pilot, and experimental testing, including LASER, Steam Flooding, CSP, NAE, and hybrid thermal/solvent processes, all aimed at further expanding the size of the recoverable reserves volume, reducing energy requirements, decreasing GHG intensity, decreasing water use and costs (Table 14.3).[7] Some of these new techniques use alternatives to steam for mobilizing the bitumen, including warm solvents, electricity, and even creating a fire within the reservoir (Spirov et al., 2013).

A good example of technological advancement is the Cycling Solvent Process. In this process, instead of injecting steam into the bitumen reservoir to recover bitumen, propane is injected instead. It mixes with the bitumen, reduces its viscosity, and allows it to flow—achieving the same effect as steam but without the addition of heat or water. The process significantly improves energy efficiency and reduces carbon emission intensity by about 90 per cent. Similarly, Excelsior Energy advanced a method called Combustion Overhead Gravity Drainage (COGD) that doubles the recovery rate of SAGD using virtually no water and will utilize only 20 per cent of the energy needed by SAGD operations (Cooper, 2010). Many similar technologies are currently the subject of innovation efforts, particularly with the aim of reducing environmental impacts related to the use of water, tailings disposal, and other environmental footprints. More examples are in the Alberta Chamber of Resources (2004) Oil Sands Technology Roadmap (see also Chhina, 2013). Similarly, various carbon capture and storage projects to capture and inject 2.76 megatons of CO_2 by late 2015 and 139 megatons by 2050 are at an experimental stage (Orcutt, 2013).

The first boom in oil sands development was just before the 2008 economic recession, when oil prices were at US$70/barrel and higher. The industry experienced unprecedented growth due to rising oil prices, the availability of capital, and emerging new projects in new frontiers. With this unforeseen growth came cost inflation—project costs began to escalate and there was a shortage of skilled labour, a weaker US currency, competition for the same materials (particularly steel), and, most noticeably, increased

---

[7] **IRR** (Internal Rate of Return) shows how fast profits are being made and the attractiveness of the cash flow relative to the initial investment. **NPV** (Net Present Value) normally discounted at 10 per cent per annum shows the appreciating (↑) or depreciating (↓) value of the project. It also indicates the total amount of profit that is being made with the new technology. **Opex** (Operating Expenditure) is the ongoing cost for running and maintaining production well. Any increase (↑) or decrease (↓) in Opex affects the supply cost of bitumen or synthetic crude.

Table 14.3 Some technologies at various phases of development

Development phase	Technology	Expected impact	IRR	NPV	Opex	Environment
Pilot/ Experimental	Thin pay wells	Increased resource base	↑	↑	–	No environmental impact
	Farmer wells	Increased production	↑	↑	↓	Reduce GHG emissions Reduce water usage intensity
	Steam flooding	Increase recovery	↑	↑	↓	Reduce GHG by 30%
	Chemically induced micro-agglomeration	Tailings Dewatering Speed up tailings rock formation Recover water from rock	↑	↑	↓	Aid land reclamation Reduce water wastages
	Paraffinic froth flotation	Produce diluted bitumen onsite Eliminate cost of pipeline	↑	↑	↓	• Reduce footprint • Low land disturbance
Commercial	Liquid addition to steam to enhance recovery (LASER)	Reduce viscosity	↑	↑	↓	Reduce GHG intensity by 25%
	Cyclic solvent process	Reduce viscosity Increased recovery	↑	↑	↓	Reduce emission intensity by 95%
	Non- Aqueous Extraction (NAE)	Eliminate need for water Increase dry tailings Less water consumption	↑	↑	↓	Eliminate need for wet tailings
Improvement	Solvent Assisted SAGD	More recovery Less energy	↑	↑	↓	15% reduction in GHG intensity
	Skystrat drilling rig	Efficiency Environmental benefits	↑	↑	–	• Reduced land disturbance • Smaller well pad

Source: modified from Chhina (2013) and Millington et al. (2012).

global attention towards oil sands as investment havens, as well as targets of environmental protests. Then, in 2008, the economic recession hit and brought the industry to a halt: projects were postponed indefinitely or cancelled. However, the recession also brought down the capital costs of oil sands, creating opportunities for developers and investors (similar impacts can be seen in the LNG industry). As some, especially emerging, economies experienced renewed growth, the price of oil remained in a range of US$70–100 and made oil sands projects profitable again, a situation that is projected to continue into the future. Managing today's cycle more efficiently and in a more sustainable manner than in the first boom poses the greatest challenge to the industry; requiring costs to be managed better and environmental performance and the impacts on the indigenous communities to be continuously improved.

Debates (often polarized and emotionally charged) about the long-term sustainable development of the industry dominate the media. The postponement of a decision by President Obama on the Keystone XL pipeline in the face of environmental protests is clearly a source of uncertainty, but it is likely that similar protests will become the new norm in the future. Environmental criticism such as this could potentially create a barrier to investment and limit the future prospects for oil sands projects. However, a substantial reduction in production and new investment is unlikely, as oil sands investments have been shown to be extremely profitable for their operators, as well as their host governments. At the same time, the growth of both shale gas and tight oil in North America poses both a threat and an opportunity to Albertan oil sands production. A threat in that increased tight oil production in the USA has reduced the market for oil sands and also reduced the price, thus, projects are now being proposed to expand the pipeline capacity to move Albertan oil to the Pacific coast to access lucrative Asian markets. However, it should be noted that tight oil produces light crude oil and the US refineries on the Gulf of Mexico are currently designed to handle heavy crude—previously from Mexico and Venezuela—so there is still a market for the heavier crude oil produced by oil sands. An opportunity, in that development of tight gas and shale gas in Western Canada provides a new source of natural gas to reduce the cost of oil sands production. However, there are also plans to move that gas to the Pacific coast and export it as LNG.

In sum, innovation has opened new opportunities and reduced the environmental footprint of oil sands projects by making operations more efficient. An enduring commitment to innovation and technology can continue to enable substantial progress, expanding opportunities for the economic development of oil sands to support growing energy needs. Investing in research, technology, and policies with constancy of purpose is critical to finding cleaner, cost-effective, and more efficient ways of developing the oil sands, taking into account decreasing reservoir. However, at present there seems only a limited prospect of the Albertan oil sands experience resulting in worldwide development of this unconventional fossil fuel. In part, this is because of the high levels of capital investment required and the fact that many reserve holders also have significant conventional oil resources and/or are high-risk locations due to political instability; but it is also because the carbon intensity of oil sands production puts it at odds with climate change policy.

14.4 Shale gas

The wider impact of shale gas production is discussed in the next chapter, here the focus of the discussion is on the origins and nature of the key technologies that have enabled the so-called 'shale gas revolution'. Shale gas was first produced in the USA as far back as the 1820s, with street lighting in Fredonia (New York State) powered by shale gas; production remained at a cottage-industry scale into the mid twentieth century. Though fracturing was developed and deployed as a well stimulation method as far back as the 1940s, development of shale and tight gas reserves began in earnest in the 1980s. During a period of falling US domestic gas production and fear of oil supply disruption following the Arab oil embargo, the US government began funding a series of research programmes into fossil fuel technologies in an attempt to increase domestic output. In 1976 the Bureau of Mines and the Morgantown Energy Research Centre (now NETL, National Energy Technology Laboratory) initiated the Eastern Gas Shales Project—a public, private, and university collaboration to determine the extent, thickness, and structural complexity of a number of Devonian shale basins, and to develop new drilling, stimulation, and recovery technologies to increase production potential (Pennsylvania Department of Conversation & Natural Resources , 2014). Over this period there was also government intervention on the cost and market side of development. Following a period of gas supply shortages in the USA, the 1978 Natural Gas Policy Act (NGPA) introduced a series of measures aimed at increasing financial incentives for developing natural gas. Section 107 provided incentive pricing for 'high cost' shales (such as Devonian age shales) and re-deregulated (removed price regulation) wellhead prices in the USA, leading to a competitive gas-to-gas (as opposed to oil indexed) priced national market (see Michot Foss (2012) for a discussion of natural gas pricing in North America). Section 29 of the Crude Oil Windfall Profits Tax Act (1980) provided gas production tax credits (up until 2002) of $0.50 per Mcf for tight gas and $1 per Mcf for shale gas in an attempt to provide financial incentives for new developments and the maintenance of existing production. Consequently activity boomed in the years following their introduction, with shale gas production growing 428 per cent from 1978–99 and reserves increasing by 428 per cent (Kuuskraa and Guthrie, 2001). During this time, technological innovation in drilling and well stimulation greatly increased the productivity and commercial viability of shale gas. Rising wholesale gas prices accelerated the growth in shale development in the 2000s, with the wellhead price rising from $1.55/Mcf in 1995 to $7.33/Mcf by 2005 before reaching a peak of $7.97/Mcf in 2008. This upward price trend greatly increased profitability of US shale operators and led to large-scale drilling programmes; subsequently a number of high profile acquisitions of independent operators by larger companies took place (this is discussed further in Chapter 15).

Key upstream technologies that facilitated the shale gas sector were a combination of those developed specifically for shale and others that were innovations for the oil and gas industry in general and subsequently applied to shale plays. A 2001 National Research Council (2001) study listed the most significant technologies as horizontal drilling and hydraulic fracturing, with 3D seismic imaging making a general contribution to the upstream sector. Research at the Morgantown Energy Research Centre (MERC) led to rudimentary horizontal drilling in 1976 (Trembath et al., 2012: 5–6), while in the early 1980s operators at BP and Total achieved successful commercialization of horizontal oil wells (EIA, 2013a). Mitchell Energy first drilled horizontal wells in the Barnett Shale in 1991,[8] and following a period of drilling processes improvement, falling upstream costs, and rising domestic gas prices, the number of horizontally drilled production wells in the Barnett grew from just 1 in 2000 to 2901 in 2008.

Although hydraulic fracking was first utilized early in the twentieth century, it was not until the 1980s that it became commercially viable and widespread. Gustafson (2012) provides a fascinating account of how Western engineers used fracking to boost oil production in West Siberia in the 1990s and the technique is commonplace in the global oil industry. During the late 1970s and early 1980s Mitchell Energy began developing hydraulic fracking methods in the Barnett Shale with 'frack zones' of up to 1,500 ft. The process was greatly refined, with the content and volume of flack fluid and proppants (small grains, usually sand, that hold open the fractures in the shale allowing the gas to flow more freely) altered to reduce costs but increase fracturing. Fracking fluids changed from CO_2 and NO_2 treatments to gelled water, before eventually in 1996 Mitchell Energy applied slick water frack treatments to the Barnett; a method that used large volumes of water (now known as high volume hydraulic fracturing or HVHF) with sand as the proppant, and this has been used in more than one million—mainly vertical—oil and gas wells in the USA (Givens and Zhao, 2004: 5–7). Like hydraulic fracturing, extended reach horizontal drilling also developed in the conventional oil and gas industry and is a widely used technology to enhance recovery and also develop hard to reach fields (for example, accessing offshore fields from onshore as in the case of Wytch Farm in Dorset in the UK). The combination of horizontal drilling and HVHF has enabled operators to overcome the low porosity and flow rates of shale gas formations, and increase the ultimate rate of recovery and gas produced (see Figure 14.5). In more recent years the advent of pad drilling has further lowered costs and increased efficiency in field development. A single surface pad holds the wellheads for multiple directionally and horizontally drilled wellbores, meaning a greater area can be exploited with shorter lead times, and reduced costs and surface footprint. The use of 3D seismic imaging for reservoir evaluation and fracture monitoring has also provided operators with improved data before carrying out drilling and exploration (Wang and Krupnick, 2013: 13–14), while the use of geosteering (or 'logging while drilling') allows the direction and characteristics of a wellbore to be altered during drilling, with geological data being captured simultaneous to the drilling itself. Thus, the direction of travel is for increasingly sophisticated geological appraisal work and complex drilling pads to both reduce the cost and the environmental impact of shale gas development. There is also research underway to reduce the amount of water used in the fracking process and even the possibility of alternatives to HVHF that would significantly reduce the water demand of shale gas production. However, this increasing technological sophistication also raises the costs of entry and is likely to change the types of company that get involved in shale

---

[8] Mitchell Energy drilled only 4 horizontal wells in the Barnett compared with more than 800 vertical up until its acquisition by Devon Energy in 2002.

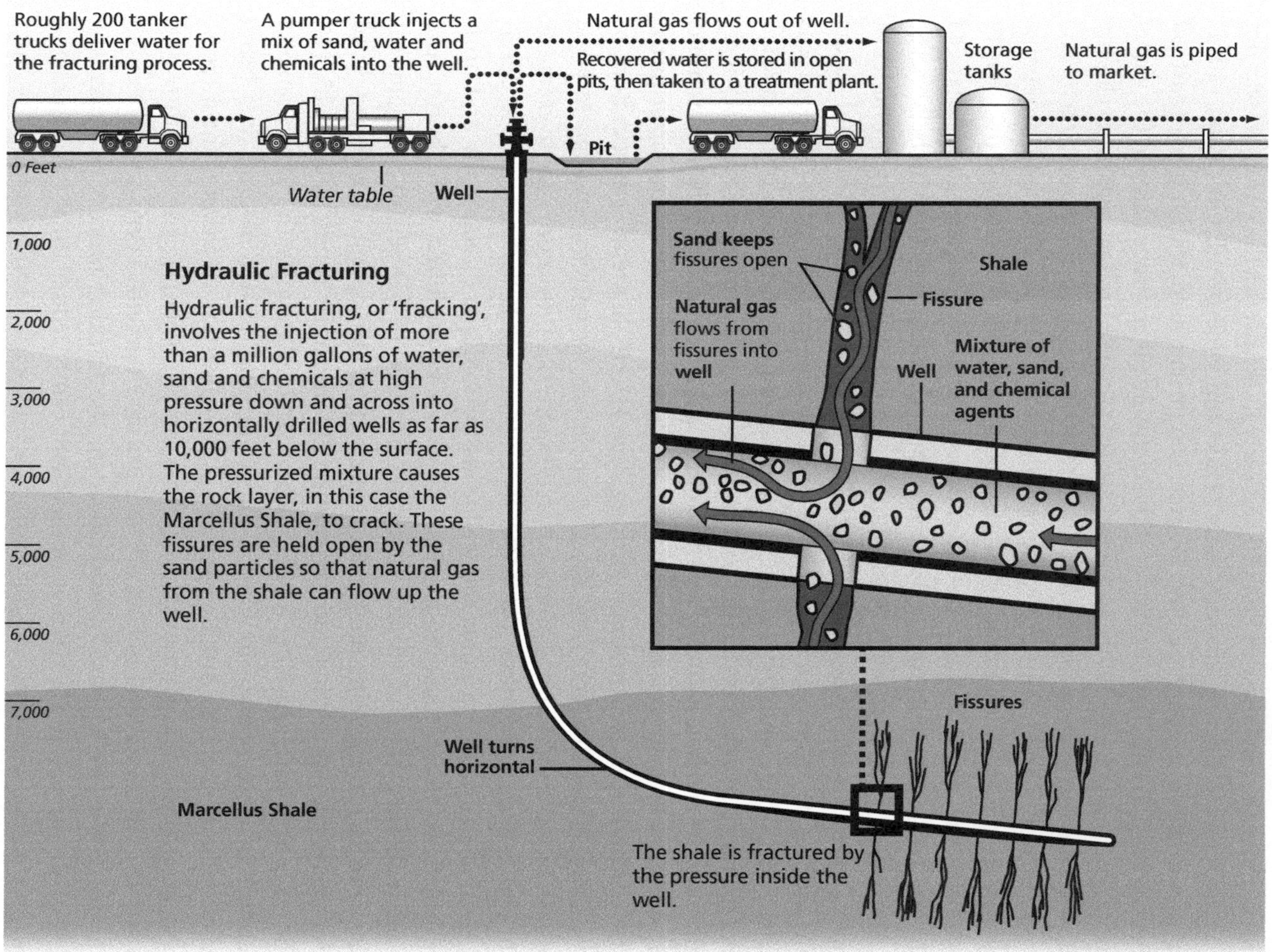

Roughly 200 tanker trucks deliver water for the fracturing process.
A pumper truck injects a mix of sand, water and chemicals into the well.
Natural gas flows out of well.
Recovered water is stored in open pits, then taken to a treatment plant.
Storage tanks
Natural gas is piped to market.
Pit
0 Feet
Water table
Well
1,000
2,000
3,000
4,000
5,000
6,000
7,000
Hydraulic Fracturing
Hydraulic fracturing, or 'fracking', involves the injection of more than a million gallons of water, sand and chemicals at high pressure down and across into horizontally drilled wells as far as 10,000 feet below the surface. The pressurized mixture causes the rock layer, in this case the Marcellus Shale, to crack. These fissures are held open by the sand particles so that natural gas from the shale can flow up the well.
Sand keeps fissures open
Natural gas flows from fissures into well
Shale
Fissure
Well
Mixture of water, sand, and chemical agents
Fissures
Well turns horizontal
Marcellus Shale
The shale is fractured by the pressure inside the well.

Figure 14.5 The shale gas production process

Source: Al Granberg/ProPublica <http://www.propublica.org/special/hydraulic-fracturing-national>.

gas development. This is especially so in places like Europe with tight regulatory requirements and vocal opposition to shale gas development.

This discussion of shale gas in the USA has focused on the development of the key technologies; however, it is well understood that there are a range of 'above ground' factors that have been critical to the rapid expansion of shale gas (and tight oil) production in the USA, these include: the extensive knowledge base about shale geology, the existence of a highly developed onshore oil and gas service industry, initial regulatory exemptions and tax incentives, private ownership of the sub-soil rights, the availability of finance—less of an advantage post the 2008 financial crisis— and greater public acceptance of oil and gas drilling—though this varies regionally (Stevens, 2012: 9).

Public opposition to shale gas in Europe and the perceived dangers of fracking are the primary barriers to development, due to strong environmental and anti-fossil fuel and fracking movements both nationally and internationally. However, there are also other issues that may hamper or prevent development. The greater geological complexity of European shales compared with the USA, tighter environmental regulation, and current upstream supply chain deficiencies mean production costs are likely to higher than in the USA. European shale would be entering an increasingly gas-to-gas priced market place supplied with existing conventional gas, pipeline imports from North Africa, Norway, and Russia, as well as LNG imports. With production costs as yet unknown the long-term commercial viability of the sector in Europe is unclear (Dutton, 2012). Table 14.4 shows the top ten countries in terms of technically recoverable shale gas resources (see Chapter 13 for a discussion of the various types of resource and reserve estimates).

The shale sector outside of the North America is not commercialized, with exploration ongoing or only beginning in a number of potential markets. Europe has been the focus so far, but activity is increasing in China, Argentina, Australia, South Africa, and across

Table 14.4 Top ten countries with technically recoverable shale gas resources

Rank	Country	Shale Gas Resources (Trillion Cubic Feet)
1	China	1,115
2	Argentina	802
3	Algeria	707
4	USA	665
5	Canada	573
6	Mexico	545
7	Australia	437
8	South Africa	390
9	Russia	285
10	Brazil	245
	World Total	7,299

Source: EIA (2013b: 10).

the MENA region. Poland was widely viewed as the next major market for shale gas production outside of the USA; with the EIA estimating risked (this a judgement based on a play success probability factor—the chance of attractive flow rates—and prospective area success factor—the wider status of the shale play on the basis of existing knowledge) technically recoverable resources of 148 tcf (EIA, 2013b: 1–9). The Polish government has issued over 100 exploration licenses with more than 40 test wells drilled across the country, but drilling results have as yet been indifferent. In 2012 Chevron withdrew from the country, while in 2013 ExxonMobil, Talisman Energy, Marathon, and ENI ended their operations and Total left in early 2014. Despite this, independent operators such as PG and San Leon are still active in Poland. In August 2013 Lane Energy reported they had achieved a flow rate of 8,000 cu.m/day, representing the largest production volume of shale gas in Europe to date. There is a very limited knowledge base in Poland and the regulatory regime has also been a problem; however, the potential reserves are significant and the Polish government is developing incentives in the hope of attracting new investment.

In 2012 the EIA estimated risked, technically recoverable shale resources of 26 Tcf in the UK, with a subsequent British Geological Survey report estimating resources of 822 Tcf[9] in the Bowland Shale basin (Lancashire) alone (BGS, 2013). Numerous companies have planning permission for exploration wells, but no fracking programmes or development wells have been completed. Fracking was carried out at Cuadrilla Resource's Preese Hall-1 well in the Bowland basin in 2010 but operations were suspended before completion after two seismic events were attributed to the fracking (Green et al., 2012). A previously drilled oil appraisal well[10] was re-entered by Cuadrilla in Balcombe in August 2013, but work was halted due to a security threats posed by protestors outside the site. The UK Department of Energy and Climate Change is now consulting on the 14th Onshore oil and Gas Licensing round that could open up 60 per cent of the UK to unconventional oil and gas exploration (tight oil, shale gas, and coal bed methane). The UK government hopes to see thirty to forty exploratory shale gas wells drilling over the next three to four years to enable a proper assessment of the resource base. However, this is fostering considerable public concern about the environmental impacts of such activity and the compatibility of shale gas development with the UK climate change policy.

14.5 Conclusions

The purpose of this chapter has been to examine the nature of unconventional fossil fuel resources and to identify the key technologies and challenges that influence their current

[9] This figure is a low range gas-in-place resource estimate.
[10] Conoco Phillips originally drilled the exploration well in March 1987. The well was drilled in an oil formation, but the licence area in the Weald basin held by Cuadrilla includes shale gas prospects.

and future levels of development. It is clear that oil shales and oil sands represent very different prospects from shale gas (and tight oil). Oil shales have been developed in just a few locations and in very specific historical circumstances, with industries currently operating in Estonia, Brazil, and China. Although there are estimates of substantial resources elsewhere, the prospects for worldwide commercial development seem rather limited at present, especially considering the recent pull-out of oil majors in Colorado. Likewise, large-scale development of oil sands (bitumen) is highly concentrated in Alberta in Western Canada, where it is of great significance. The recent developments in Alberta illustrate how cost and environmental considerations are driving innovation and technological change, but the industry remains vulnerable to a rise in natural gas prices and requires a high oil price to remain commercially viable. The industry also illustrates the diminishing return on energy invested that is associated with unconventional oil and gas production. Simply put, it takes more energy to produce a barrel of synthetic crude oil than it does to produce conventional crude. This is one of the reasons for the high GHG emissions associated with both oil sands and oil shales exploitation besides the carbon content of the resources. Again, despite substantial reserves globally, it seems unlikely, and many would argue undesirable, that there will be a significant expansion of oil sands development beyond Canada. This is where shale gas and tight oil are different. Although it has taken twenty years to develop the industry in North America (there is also significant production in Canada), in less than ten years it has had a dramatic impact on US oil and gas production (the latter is discussed in Chapter 15). The key question is how well and how far will the shale gas and tight oil revolution travel beyond North America? There are significant resources in place that are considered technically recoverable, many of which are in oil and gas importing states, and the entry costs are much lower when compared to oil sands, for example. However, it will take time to develop understanding of the shale geology, to develop the required infrastructure and to put in place the necessary regulatory and fiscal regimes to enable commercial production without endangering the environment. Thus, we are unlikely to see significant international shale gas and tight oil production until the early 2020s, but the potential is there to have a major impact on the global oil and gas industry.

■ REFERENCES

Alberta Chamber of Resources. 2004. *Oil Sands Technology Roadmap: Unlocking the Potential.* Edmonton: Alberta Chamber of Resources.

Alberta Energy. 2014. *Oil Sands: Facts and Statistics.* Accessed 5 February 2014. <http://www.energy.alberta.ca/oilsands/791.asp>.

Arkonsuo, Hannu. 2013. *Risks, strengths and weaknesses of Russian oil and gas.* Helsinki: Arewcon Development Oy.

Attanasi, E., and R. Meyer. 2007. 'Natural bitumen and heavy oil'. In *2007 Survey of World Energy Resources.* London: The World Energy Council. Accessed 19 July 2009. <http://bit.ly/axjUlu>.

Bartis, James, Tom LaTourrete, Lloyd Dixon, D. J. Peterson, and Gary Cecchine. (2005). *Oil Shale Development in the United States: Prospects and Policy Issues*, Santa Monica, California: RAND Corporation.

British Geological Survey (BGS) 2013. 'The Carboniferous Bowland Shale gas study: geology and resource estimation.' Accessed 4 March 2014. <https://www.gov.uk/government/uploads/sys tem/uploads/attachment_data/file/226874/BGS_DECC_BowlandShaleGasReport_MAIN_ REPORT.pdf>.

Bowman, Clement. 2008. *The Canadian oil sands: past, present and future*. Global Energy International Foundation Prize 2008. Accessed 4 March 2014. <http://www.assembly.ab.ca/ lao/library/egovdocs/2008/ca6/gei/173121.pdf>.

Brandt, Adam. 2011. *Upstream greenhouse gas (GHG) emissions from Canadian oil sands as a feedstock for European refineries*. Prepared for European Commission. Accessed 5 February 2014. <https://circabc.europa.eu/d/d/workspace/SpacesStore/db806977-6418-44db-a464- 20267139b34d/Brandt_Oil_Sands_GHGs_Final.pdf>.

Brandt, Adam, and Alexander Farrell. 2007. 'Scraping the bottom of the barrel: CO_2 emission consequences of a transition to low-quality and synthetic petroleum resources.' *Climatic Change*, 84(3–4): 241–63.

Chhina, Harbir. 2013. *Building on our track record of innovation*. Investor Day, Calgary, June 18 2013. Accessed 3 March 2014. <http://www.cenovus.com/invest/docs/2013/investor-day-06- chhina.pdf>.

Cooper, Dave. 2010. New technique could double recovery rate. *Edmonton Journal*. January 27. Accessed 20 August 2013. <http://tinyurl.com/l9rjbxy>.

Dammer, Anton. 2007. *Secure Fuels from Domestic Resources: The Continuing Evolution of America's Oil Shale and Tar Sands Industries*. Washington, DC: U.S. Department of Energy, Office of Naval Petroleum and Oil Shale Reserves.

Demaison, G. J., 1977. Tar sands and supergiant oil fields. *Bulletin of American Association of Petroleum Geologists*, 61: 1950–61.

Dutton, Joseph. 2012. 'Public policy and public opinion on shale gas development.' In *Shale Gas: A Practitioners Guide to Shale Gas & Other Unconventional Resources*, edited by Vivek Bakshi and Fraser Milner Casgrain, pp. 143–54. London: Globe Law and Business.

Duncan, Dondal, and Vernon Swanson. 1965. *Organic-rich Shale of the United States and World Land Areas*. Washington, DC: Department of the Interior.

Dyni, John. 2006. *Geology and Resources of Some World Oil-Shale Deposits*. Reston, Virginia: U.S. Geological Survey.

EIA 2013a. 'Drilling Sideways – A Review of Horizontal Well Technology and Its Domestic Application', *Natural Gas Monthly*, DOE/EIA-TR-0565, April 2013.

EIA 2013b. *Technically Recoverable Shale Oil and Shale Gas Resources: An Assessment of 137 Shale Formations in 41 Countries Outside the United States*. Washington DC: EIA.

Francu, Juraj, Barbara Harvie, Ben Laenen, Andres Siirde, and Mihkel Veiderma. 2007. *A study on the EU oil shale industry—viewed in the light of the Estonian experience*. A report by EASAC to the Committee on Industry, Research and Energy of the European Parliament. Brussels, European Academies Science Advisory Council. Brussels: European Parliament.

Givens, Natalie, and Hanks Zhao. 2004. 'The Barnett Shale: Not So Simple After All' American Association of Petroleum Geologists Annual Meeting Program. Accessed 2 March 2014. <http://www.beg.utexas.edu/pttc/archive/barnettshalesym/notsosimple.pdf>.

Green, Christopher, Peter Styles, and Brian Baptie. 2012. 'Preese Hall Shale Gas Fracturing. Review and Recommendations for Induced Seismicity.' Accessed 2 March 2014. <https://www. gov.uk/government/uploads/system/uploads/attachment_data/file/15745/5075-preese-hall-shale-gas-fracturing-review.pdf>.

Gulliford, Andrew. 1989. *Boomtown Blues: Colorado Oil Shale, 1885–1985*. Niwot, Colorado: University Press of Colorado.

Gustafson, Thane. 2012. *Wheel of Fortune: The Battle for Oil and Power in Russia*. Cambridge, MA: The Belknap Press of Harvard University Press.

Hanson, John, and Patty Limerick. 2009. *What Every Westerner Should Know About Oil Shale: A Guide to Shale Country*. Boulder, CO: Center of the American West, University of Colorado at Boulder. Accessed 2 March 2014. <http://www.centerwest.org/publications/oilshale/print/oilshale.pdf>.

Holditch, Stephen, and Walter Ayers. 2010. How Technology Transfer Will Expand the development of Unconventional Gas Worldwide. Unpublished paper, Department of Petroleum Engineering, Texas A&M University.

Huc, Alaine-Yves. 2010. *Heavy Crude Oils: From Geology to Upgrading-An Overview*. Paris: Editions Technip.

Humphries, Marc. 2008. *North American Oil Sands: History of development, prospects for the future*. Washington DC: Congressional Research Service Report for Congress.

Hutton, Adrian, Sunil Bharati, and Thomas Robl. 1994. Chemical and petrographic classification of kerogen/macerals. *Energy & Fuels* 8(6): 1478–88.

IEA (International Energy Agency). 2013. *Resources to reserve 2013: Oil, Gas and Coal Technologies for the Energy Markets of the Future*. Paris: International Energy Agency.

Kama, Kärg. 2013. *Unconventional Futures: Anticipation, Materiality, and the Market in Oil Shale Development*. DPhil thesis in the School of Geography and the Environment, University of Oxford.

Kuuskraa, Vello, and Hugh Guthrie 2001. Translating Lessons Learned From Unconventional Natural Gas R&D To Geologic Sequestration Technology. NETL. Accessed 4 March 2014. <http://www.netl.doe.gov/publications/proceedings/01/carbon_seq/1a3.pdf>.

Lattanzio, Richard. 2013. *Canadian Oil Sands: Life-Cycle Assessments of Greenhouse Gas Emissions*. Washington DC: Congressional Research Service Report for Congress.

Madden, Peter, and Jacek Morawski. 2010. The future of the Canadian oil sands: engineering and project management advances. *AMEC Oil Sands 2010-001*. Accessed 4 March 2014. <http://www.worldenergy.org/documents/congresspapers/402.pdf>.

McManus, Bob. 2011. *Canada's Oil Sands*. Canadian Centre for Energy Information. Accessed 4 March 2014. <https://www.centreforenergy.com/Shopping/uploads/12.pdf>.

Meyer, Richard, Emil Attanasi, and Philip Freeman. 2007. Heavy oil and natural bitumen resources in geological basins of the World. *US Geological Survey Open-File Report 2007–1084*, p. 3–36. Accessed 10 July 2009. <http://pubs.usgs.gov/of/2007/1084/>.

Michot Foss, Michelle. 2012. 'Natural Gas Pricing in North America.' In *The Pricing of Internationally Traded Gas*, edited by Jonathan P. Stern, pp. 40–84. Oxford: OUP/OIES.

Millington, D. and Mei, M., 2011. Canadian Oil Sands Supply Costs and Development Projects (2010–2044). Calgary, Alberta: Canadian Energy Research Institute, Study no. 122.

Millington, D., Murillo, C. A., Walden, Z., and Rozhon, J. 2012. Canadian Oil Sands Supply Costs and Development Projects (2011–2045). Calgary, Alberta: Canadian Energy Research Institute, Study no 128.

Laherrère, Jean. 2005. 'Review on oil shale data.' Accessed 3 November 2011. <http://www.hubbertpeak.com/laherrere/OilShaleReview200509.pdf>.

McGlade, Christophe. 2012. A review of the uncertainties in estimates of global oil resources. *Energy* 47(1): 262–70.

Mitchell, Timothy. 2013. 'Peak Oil and the New Carbon Boom.' Accessed 24 July 2013. <http://www.dissentmagazine.org/online_articles/peak-oil-and-the-new-carbon-boom>.

National Energy Board. 2006. *Canada's Oil Sands. Opportunities and Challenges to 2015: an update*. Ottawa: National Energy Board.

National Research Council. 2001. 'Energy Research at DOE: Was it Worth it: Energy Efficiency and Fossil Energy Research 1978–2000.' Washington, DC: National Academy Press.

Ogunsola, Olayinka, Arthur Hartstein, and Ogunsola Olubunmi. 2010. *Oil Shale: A Solution to the Liquid Fuel Dilemma*. Washington DC, ACS Symposium Series 1032. American Chemical Society.

Orcutt, Michael. 2013. 'Can Carbon Capture Clean Up Canada's Oil Sands?' *MIT Technology Review*, 9 May. Accessed 30 August 2013. <http://www.technologyreview.com/news/514221/can-carbon-capture-clean-up-canadas-oil-sands/>.

Pascal, Larry, and Ramon Azpurua. 2008. The Venezuelan oil and gas sector–are there still opportunities in the era of petronationalism? *Latin American Law & Business Report*, 16(7): 1–5.

Pennsylvania Department of Conversation & Natural Resources. 2014. Eastern Gas Shales Project. Accessed 4 March 2014. <http://www.dcnr.state.pa.us/topogeo/econresource/oilandgas/marcellus/marcellus_egsp/index.htm>.

Russell, Paul. 1990. *Oil Shales of the World, Their Origin, Occurrence and Exploitation*. Oxford: Pergamon.

Society of Petroleum Engineers (SPE). 2011. *Guidelines for Application of Petroleum Resources Management System*. Richardson, Texas: SPE.

Speight, James. 2012. *Shale Oil Production Processes*. Oxford: Elsevier.

Spirov, Pavel, Svetlana Rudyk, Anastasios Tyrovolas, and Ismaila Jimoh. 2013. 'The Bitumen extraction from Nigerian tar sand using dense carbon dioxide.' *Chemical Engineering Transactions*, 32: 283–8.

Stevens, Paul. 2012. *The 'Shale Gas Revolution': Developments and Change*. London: Chatham House.

Trembath, Alex, Jessa Jenkins, Ted Nordhaus, and Michael Shellenberger. 2012. *Where the Shale Gas Revolution Came From*. Oakland, California: Breakthrough Institute for Energy & Climate Program.

Wang, Zhingmin, and Alan Krupnick. 2013. *A Retrospective Review of Shale Gas Development in the United States*. Washington DC: Resources for the Future Discussion Paper 13–14.

WEC (World Energy Council) 2007. *Survey of Energy Resources 2007*. <London: World Energy Council>.

WEC (World Energy Council). 2010. *2010 Survey of Energy Resources*. London: World Energy Council. Accessed 5 February 2014. <http://www.worldenergy.org/documents/ser_2010_report_1.pdf>.

Yen, Teh Fu, and George Chilingar, Eds. 1976. *Oil Shale*. Developments in Petroleum Science 5. Amsterdam: Elsevier.

15 The geopolitical economy of a globalizing gas market

Michael Bradshaw, Joseph Dutton, and Gavin Bridge

15.1 Introduction

In 2011 the International Energy Agency (IEA, 2011) published a report entitled: *Are We Entering a Golden Age for Gas?* Some, such as the International oil companies (IOCs), preferred to see the report as prophetic, but it actually considered the circumstance under which the world might see an increased role for gas and considered the implications for the energy system and climate change (in 2012 the IEA 2012 subsequently published a report listing the 'Golden Rules for a Golden Age of Gas'). The initial report was prompted, in part, by the possibility that a shale gas revolution on a global scale would make natural gas abundant and affordable. The IOCs argue that natural gas is also acceptable from an environmental point of view because is the 'cleanest fossil fuel' from a carbon dioxide emissions perspective, though methane is itself a greenhouse gas. However, the IEA (2011) points out that its 'Golden Age for Gas' scenario puts the world on a trajectory that would see 3.5 °C global warming, well above the internationally agreed target of 2 °C. The detail of the report suggested that there would be significant regional differences in the future role of natural gas and that the non-OECD countries would account for 80 per cent of total demand increase between 2010 and 2035.

This chapter focuses on recent developments in the global gas market and in particular on developments in North America, Europe, and Japan, which still today account for the vast majority of global gas consumption. The chapters examine the various ways in which the shale gas revolution in the United States has been changing the global game through indirect impacts on the LNG market and on the way in which gas is traded and priced. The analysis also considers the current situation in Europe, which is very far from a golden age for gas and which is now further complicated by a political desire to reduce dependence on Russian gas imports. Finally, the chapter considers the impact of the Fukushima disaster in March 2011 on the global LNG market. The conclusion seeks to identify the key issues that have emerged during the recent turmoil that are likely to influence the prospects for a golden age for gas.

15.2 **The shale gas revolution in the United States**

As explained in the previous chapter, the commercial development of shale gas took decades to achieve (Stevens, 2012), but once proven actual production grew very rapidly in the mid 2000s, with production growing from 11 billion cubic metres (bcm) in 2005 to 226 bcm by 2011. In 2012 shale gas output reached 265 bcm–39 per cent of all US gas production (EIA, 2013a). According to the 2014 Annual Energy Outlook Reference Case, shale gas will account for 53 per cent of total gas production in 2040 with production more than double its current level at 554 bcm (EIA, 2014). In adding greatly to natural gas production volumes in the USA, shale gas has transformed the domestic energy market. The boom in gas output over such a short period of time led to a steep fall in the Henry Hub wholesale price (the benchmark gas price in the US) from $12.69 in June 2008 to a ten-year low of $1.82 in April 2012—as a consequence utilities and power generators switched from more expensive coal to now cheaper natural gas, while the petrochemicals industry also moved to natural gas as its chemical feedstock, shifting away from the more expensive naphtha sourced from crude oil. Although the whole-sale gas price rebounded to over $3.72/mmbtu in 2013, the Henry Hub price is still at a level substantially lower than before the shale gas boom (see Figure 14.1). As a consequence of the huge growth in domestic output, US gas imports fell from 130 bcm in 2007 to 87 bcm in 2012—a trend that will soon see the USA become a net exporter of natural gas. The speed with which the boom in shale gas production changed the US gas market and the extent to which it was largely unforeseen is demonstrated by its impact on the US LNG import industry. The country had previously been forecast to become the world's largest LNG importer with a 23 per cent LNG import market share by 2010, and LNG imports in excess of 169 bcm (EIA, 2005). In 2011 installed US LNG regasification capacity (the process by which cooled liquefied natural gas is warmed to return it the gaseous state) stood at a quarter of the country's natural gas demand, but only 13 per cent of the regasification capacity was utilized, falling to 3 per cent in 2012 (Krijgsman, 2011).

This rapid decline in import demand resulting from the shale boom means the USA is set to become a net exporter of natural gas by 2016. The US government has given LNG export permits to four terminals, while there are a further twenty-one applications pending at eighteen sites, predominantly on the Gulf coast. There are also terminals planned on both the east coast and west coast of Canada. In the USA, the export of natural gas to countries that the USA does not have a free trade agreement with (FTA) requires presidential permission and export licences under the 1938 Natural Gas Act. The export of LNG is a hotly debated issue in the USA, with some believing it may cause the price of gas to rise again and therefore hurt domestic, petrochemical, and industrial customers that have benefited greatly from cheap and plentiful indigenous gas supplies. However, an increase in the domestic price of gas would bring 'dry gas' production (that is gas that does not have natural gas liquids and/or is not associated with oil production) back into play, much of which is shut in at present, as the current low prices cannot cover production costs. Thus, the development of export capacity could act as a safety valve as the increased price would stimulate greater domestic production and with it economic benefits. Even at US$6–7 mmbtu, US gas prices would still be historically low and would remain considerably lower than those prevalent in Europe and Asia, thus giving the USA a competitive advantage in energy intensive sectors. However, higher gas prices would also allow coal to return to the US power generation mix at the expense of gas (something that happened in 2014 following a very cold winter and gas shortages). Thus, there is a fine balance to be struck between encouraging gas production and encouraging decarbonization. Recent events in the Ukraine have added geopolitical impetus to the development of LNG export capacity as some politicians, both in the USA and in Europe, see it as a means of reducing Europe's reliance on Russian gas. However, the economic reality is that the majority of US LNG exports will be destined for Asia markets where they can attract a higher price.

15.3 Impact of US shale gas production on global LNG markets

While the USA is not expected to export any substantial volumes of natural gas until later this decade, already the growth in US domestic shale gas production and the subsequent loss of the country as the largest importer of LNG has impacted upon global LNG markets in terms of supply, demand, and pricing. With domestic production increasingly meeting US natural gas demand, the country's LNG imports fell by 77 per cent from a decade-peak of 21.8 bcm in 2007 to 2.8 bcm in 2013 (EIA, 2013a). The price impact of shale production also influenced North American exports, with pipeline deliveries to Mexico growing 99 per cent from 2008 to 2012. This growth was due to the falling price of gas in the USA also lowering the export price. Consequently, Mexico made a strong shift from LNG to US pipeline gas due to the import price falling from $8.25/Mcf in 2008 to $2.94/Mcf in 2012 (EIA, 2013b).

With the loss of US gas imports, LNG cargoes initially destined for US terminals were delivered to other markets. The displacement of LNG onto international markets at a time of reduced European gas demand led to an oversupply of LNG in the Atlantic basin and relatively cheap LNG was able to compete against pipeline gas in European markets. In the UK, for example, this availability of surplus LNG coincided with commissioning of new LNG import terminals—South Hook and Dragon LNG, both at Milford Haven. This was exacerbated by the start up of a large volume of new liquefaction projects, making the global capacity 52 per cent higher in 2011 than 2006. Liquefaction growth was led by Qatar and the new 'mega-trains', with the country seeing an 80 per cent increase in capacity between 2006 and 2011, accounting for 27 per cent of global liquefaction capacity and 31 per cent of global exports in 2011 (IGU, 2011). Due to the oversupply of LNG and a drop in demand (primarily in North America and Europe) in 2012 the global import terminal utilization rate fell to 37 per cent—down from 45 per cent in 2007 (GIIGNL, 2013).

In anticipation of the oversupply in Atlantic basin LNG remaining, European importers increased their LNG import capacity between 2006 and 2011. Over this period following the boom in US shale production, six European countries brought online a combined total of 75 bcm/y in new regasification capacity, with the UK alone accounting for 31.6 bcm of this figure; although the investment decisions were made well before the loss of the US LNG market and in anticipation of falling UK domestic gas production (GIIGNL, 2013). The large increase is reflective of how LNG demand and supply in Europe were expected to at least maintain their level and possibly increase. However, the loss of the USA as an LNG import market has altered the pricing dynamics of LNG globally. Although gas had been predominantly priced regionally, the US–Henry Hub, UK–National Balancing Point (NBP), Japanese Crude Cocktail (JCC), and continental European hubs had remained on parity (see Figure 15.1) in the period before the US shale boom. However the fall in the Henry Hub price began a significant divergence from other hubs that has remained. The Henry Hub is a gas-to-gas priced hub (the price is discovered on the basis of a competitive market and the laws of supply and demand between gas suppliers and gas consumers) and experienced a huge drop as a result of shale production, whereas the other regional hubs are partially (the NBP and European

Figure 15.1 Trends in global gas prices 1995–2012

Source: BP Statistical Review of World Energy 2013.

hubs) or fully (the JCC) indexed against the global crude oil price (for an extensive discussion see Stern, 2012). In the wake of this hub price divergence, large importers of LNG and consumers of pipeline gas are clamouring for a move away from oil-indexed pricing (which is also suffering because of the high price of oil) to gas-to-gas competition such as that used in the USA (Rogers and Stern, 2014). Equally, there is growing demand from Europe and Asia for the USA to export natural gas that would be priced at the lower Henry Hub cost rather than the higher prices many importers currently pay. As noted earlier, events in Ukraine have added a geopolitical dimension to Europe's desire to import US LNG, although the reality is that future volumes are unlikely to make much impact on European reliance on Russian gas imports.

The transformation of the USA from potentially the largest importer of natural gas to a net exporter has also had an impact upon the upstream gas industry. Development of Gazprom's Shtokman field in the Barents Sea was delayed indefinitely in 2012 due to a combination of project economics and the loss of the USA as a target market for the field's gas, with disagreements between project partners Total and Statoil. In 2013 the Gazprom deputy chairman was quoted as saying the project should be left for future generations, with the company cancelling a call for bids to prepare project documentation for the LNG facility (Khrennikova and Bowles, 2013). Expensive new LNG liquefaction projects in Australia—such as Browse, Gorgon, Queensland Curtis, and Wheatstone—may also struggle to compete against cheaper US LNG exports into Asian markets, despite US LNG having higher shipping costs. Thus, the development of new LNG capacity in North America is adding to an already complex and potentially crowded situation on the supply side, but there have also been further shocks on the demand side.

15.4 **The impact of Fukushima on global LNG demand**

The earthquake and subsequent tsunami in Japan on 11 March 2011 caused widespread damage, and most notably led to a nuclear meltdown at the Fukushima nuclear power plant on 12 March. Following the accident the Japanese government shut down all fifty of the country's reactors, which were providing 30 per cent of the country's power at the time. As a result the power generation sector lurched quickly towards gas to fill the power generation gap, leading to an 11 per cent increase in LNG imports in 2011 compared with 2010. Japan's 2011 LNG imports of 78.8 mmtpa (110 bcm) were 33 per cent of global LNG imports, and more than double the next highest importer, South Korea. LNG imports to the country grew 24 per cent from 2010–2012, reaching 87.3 mmpta (122 bcm) (IGU, 2013a). They increased further in 2013, reaching a record level of 87.49 mmtpa, and trade figures for 2012 show that Japan spent $68.98 billion on LNG imports (Reuters, 2014).

The principal impact of Japan's import growth following the Fukushima accident has been a significant shift in the LNG demand from the Atlantic basin to the Asia-Pacific.

The previous oversupply of LNG in the Atlantic basin was short lived as increased demand from Asia coupled with a drop in European and North American demand resulted in arbitrage and redelivery of cargoes from the Atlantic basin, and an overall tightening of the global market. The market is tight because very little new LNG production capacity has come online since 2011, but substantial expansion is expected after 2015 with Australia expected to add 62 mmtpa of new capacity (IGU, 2014). While Japan's growth was a result of the nuclear shutdown, other Asian countries' economic growth also contributed to the region's overall surge in demand, with Japan, South Korea, and China accounting for 58.4 per cent of global LNG imports in 2012, rising to 61 per cent in 2013. South Korea and Taiwan are also experiencing their own nuclear power problems and LNG imports are expected to increase as a result. Much of this new growth was satisfied by redeliveries of cargoes and LNG purchased through spot sales. Accordingly, spot market sales grew from 10 per cent of the global LNG market in 2004 to 25 per cent in 2011 and reached 33 per cent in 2013; Asian customers accounted for 72 per cent of total spot purchases in 2012 (up from 61 per cent in 2011), while European spot sales fell from 21 per cent of the total to 12 per cent (GIIGNL, 2013; IGU, 2014). Figures for inter-regional LNG trade reflect the shifting geographies of demand and subsequent redelivery, with Atlantic–Pacific shipments growing to 15.6 mmtpa in 2012 from none in 2000, while Middle East–Pacific LNG trade increased from 15.4 mmtpa to 69.7 mmtpa over the same period (IGU, 2013a). Spot sales remained flat in 2012 at 25 per cent of global trade, but there was a large increase in the amount of reloading and redelivery of cargoes from 44 in 2011 to 75 in 2012. This highlights the falling gas demand in Europe, as fifty of the reported cargoes totaling 2.66 mmt (3.73 bcm) were from Belgium, France, Portugal, and Spain. The primary destinations for the cargoes were Japan (eleven), Argentina (nine), and Brazil (seven) (GIIGNL, 2013). Amid the tightening in supplies of LNG and stronger demand from Asian markets, Japanese LNG prices rose from $10.91 in 2010 to $16.75 in 2012 (BP, 2013). While the higher price of the JCC compared with the European and North American markets hubs underlay this increase, it was compounded by Asian importers' willingness to pay a premium for securing supplies of LNG during periods of high demand. Overall European gas imports fell 3 per cent from 2011–12, but year-on-year LNG imports fell by 23 per cent, with imported LNG's market share in Europe falling from 19.7 per cent to 15.5 per cent (BP, 2013).

All of this detail supports a pattern of growing global market share in Asia and falling market share in Europe. Evidence suggests that while the LNG market remains tight Asian buyers will attract any surplus LNG. This is reflected in the continued expansion of spot market transactions. Thus, the LNG market is becoming more globally inter-connected with Atlantic basin supplies serving Asia buyers. The question remains: what happens post 2015 when new capacity comes online? Equally, what happens when Japan re-starts some of its nuclear power capacity? There is the possibility of an LNG supply glut, with surplus spot cargoes seeking buyers. If this happens, then perhaps LNG will return to Europe to displace Russian pipeline gas.

15.5 **The impact of US shale gas on European gas markets**

Although the US shale gas revolution was perceived by some as a blueprint for development of a shale sector in Europe, the more immediate impact is that US shale has indirectly impacted European gas markets via changes to the US domestic energy market. Following the steep decline in the wholesale price of gas in the US as a result of shale development, US utilities—who purchase 92 per cent of US coal output—made a shift from coal to natural gas for thermal generation. The share of gas-fired electricity generation rose from 18 per cent in 2002 to 30 per cent in September 2012, with the share of coal falling from 50 per cent to 37 per cent over the same period. Consequently, coal prices for 2014 delivery to the Netherlands—regarded as the European benchmark—fell to $81.80 per ton in October 2013, the lowest for a next-year contract since 2009, and with the price falling 21 per cent in 2013 (Strzelecki, 2013). As a result of domestic changes in price and demand, US coal producers sought to increase their exports to both Europe and Asia. Consumption of coal fell in the USA by 18 per cent from 2010 to 2012, but in Europe imports of US coal in 2012 were 29 per cent higher than in 2011. The increased coal burn in Europe was also facilitated by very low carbon prices on the EU Emissions Trading System (Garman and Kahya, 2012).

The increase in coal imports from the USA coincided with both reduced industrial gas and electricity demand as a result of ongoing economic problems (therefore affecting overall gas consumption) and pressure on wholesale gas prices in Europe. Strong LNG demand growth in Asia from post-Fukushima Japan and industrializing economies saw increased sales at higher-priced Asian market hubs, impacting both the available supply and price of LNG cargoes in Europe. The high relative cost of gas in Europe compared to coal imports has seen the power sector make the switch opposite to that of the USA— going from gas to coal. Compounded by the prevalence of oil-indexed gas sales in Europe, continental European gas prices fell by only a 5 per cent from 2012–13, compared with a 19 per cent drop in coal prices on European markets (Hrdomadko, 2013). The switch from gas to coal underlies the perceived current poor market conditions for natural gas in Europe, with a number of power generators delaying investment in new gas power plants or mothballing existing facilities. For example, in March 2013 UK utilities firm SSE reported it was withholding new-build investment in over 1600 MW of gas-fired capacity until 2015 and mothballing over 1460 MW of capacity, citing poor market conditions for electricity generation from gas (SSE, 2013). GDF Suez has closed 12 GW of gas plant capacity in Europe, and E.ON has shut 7 GW and is reviewing a further 5–7 GW of capacity for potential closure. In Germany 7.3 GW of new hard-coal fired generation capacity from ten new units is set to be brought online by 2015, boosting coal capacity by 33 per cent. The share of coal in German power generation has increased from 16.6 per cent in 2010 to 23.9 per cent in 2013 (World Nuclear News, 2014). The financial rationale driving the current dash for new coal can be seen from the estimated power generation profit margins in Germany, where in 2013 coal had an current operating profit of €9.16 per MWh of power produced, while gas fired generation is making a loss of €19.31 per MWh (Hrdomadko, 2013).

The displacement of coal by gas from the US domestic market has altered baseload production for the European power generation sector, but the redirection of US-bound LNG also placed pressure on the existing European wholesale gas pricing mechanisms. Continental European gas sales are dominated by oil-indexed supply contracts for Russian and North African pipeline gas, but the redelivery of (US-bound) Atlantic basin LNG cargoes to Europe and growth in the LNG imports from spot-market sales means there has been a growing volume of gas-to-gas priced supply entering a predominantly oil-indexed market. Europe as a whole saw the proportion of gas-to-gas priced trade grow from 15 per cent in 2005 to 45 per cent in 2012, with oil-indexed trade falling from 78 per cent to 50 per cent over the same period. This change has been led by Northwest Europe (Belgium, Denmark, France, Germany, Ireland, Netherlands, and the UK) where oil-indexed and gas-to-gas trade have switched from 72 per cent and 25 per cent of the total in 2005 to 28 per cent and 72 per cent in 2012 respectively, driven primarily by the Netherlands and the UK (with the latter's declining domestic oil-indexed production replaced by import gas-to-gas supplies). But also by Statoil decision to supply its gas into these markets on the basis of gas-to-gas competition on trading hubs (Makan, 2013). In Central Europe (Austria, Czech Republic, Hungary, Poland, Slovakia, and Switzerland) oil-indexation has fallen from 85 per cent to 50 per cent from 2005 to 2012 with gas-to-gas growing from almost zero to 35 per cent, while in the Mediterranean oil-indexed gas has fallen from 100 per cent to 90 per cent over the period. The situation is very different in Southeast Europe (Bosnia, Bulgaria, Croatia, FYR Macedonia, Romania, Serbia, Slovenia), however, where there are still no gas-to-gas priced trades, with the market instead a combination of oil-indexation—41 per cent— and government regulated prices—59 per cent (IGU, 2013b).

Gazprom currently sells some gas to Europe on hub-based pricing; in 2010 16 per cent of E.ON's 20 bcm annual gas supply from Russia was linked to the TTF hub, while 15 per cent of the 10 bcm per year GDF Suez purchased from Russia was also linked to hub prices.[1] There is growing pressure from European importers on Gazprom and other exporters to change contract arrangements to reflect the evolving nature of gas markets and pricing. In 2012 German utility RWE successfully re-negotiated long-term procurement contracts with Statoil, following financial losses in its midstream division from exposure to oil-indexed imports (Hromadko, 2012). In 2013 RWE also won an arbitration case against Gazprom that saw the awarding of €1bn in retroactive payments to RWE due to overcharging and the introduction of spot indexation into pricing formulae (RIA Novosti, 2013). Similarly in 2012 E.ON and Gazprom reached a negotiated settlement affecting all long-term oil-indexed gas supply contracts between the two

---

[1] Argus European Natural Gas (15 November 2011) p. 8.

companies (E.ON, 2012), and Italian company Edison is currently seeking arbitration in relation to Russian and Libyan gas supplies to ensure long-term contract terms reflect wholesale gas prices. As a result of both reduced gas demand in recession-hit Europe and increased flexibility in gas supply contract structure and pricing, Gazprom's gas market sales in Europe fell by 6 per cent in 2012, while Statoil's increased by 12 per cent. Over this period Qatar greatly increased its market share in Europe, increasing exports to the region from 5 bcm in 2006 to 44 bcm in 2011, satisfying 11 per cent of gas demand (Schultz and Bidder, 2013).

However, Gazprom's exports into Europe began to recover in 2013 with year-on-year growth of 14.5 per cent, and deliveries to Europe in mid 2013 at a five-year-high of 470mn m³/day, highlighting growing European demand amid falling indigenous production (Chazan, 2013). At the end of 2013 Gazprom reported record market share in its sales to Europe of 162.7 bcm and 30 per cent of the European and Turkish gas market (Moscow Times, 2014). By comparison, in 2012 Gazprom's gas sales to Europe totalled 151 bcm, of which 92.6 per cent (139.9 bcm) was sold on long-term contracts (Khodyakova, 2013). Although pricing mechanisms are changing with customers seeking hub-based prices instead of oil-indexation, the prevalence of long-term contracts demonstrates that the nature of supply agreements (specifically in the case of Gazprom customers) largely remains the same. The long-term nature of Gazprom's supply contracts also makes it very difficult for Europe to redirect gas imports away from Russia in the face of events in Ukraine. However, all this may change with the outcome of the current EU anti-trust case against Gazprom, much of which is concerned with the contracting and pricing practices of the company (Riley, 2012).

The growth of subsidized renewable energy in Europe as EU member states strive to reach energy sector decarbonization targets has also reduced the competitiveness of gas-powered generation across the region. For example, in Germany 22 per cent of electricity was generated from renewable energy in 2012, while in the UK renewables had an 11.3 per cent share of electricity generation. The low operational and marginal costs of renewables generation mean a lower wholesale electricity price than gas or coal, thus squeezing the share of thermal generation on the grid. This trend is set to continue due to the ambitious targets for installed renewable capacity across the EU. In Germany the ownership structure of renewables is also having a financial impact on large utilities. By 2012 only 7 per cent of the 71 GW of installed renewables capacity in Germany was owned by the traditional utilities, with the market instead dominated by individuals, agriculture, and industry (De Clercq, 2013). With a substantial share of renewables now in the German power market and a falling wholesale price, the fossil fuel-based utilities are coming under more pressure to increase their shareholding in renewables, especially as the role of coal-fired power generation will be reduced in the future by EU air pollution legislation.

The European market conditions for gas may remain poor in the short to medium term, but coal consumption in the EU will decline as a result of the European Commission's 2001 'Large Combustion Plants Directive' (LCPD, which has now been rolled into the Industrial Emissions Directive). The legislation seeks to reduce emissions of sulphur dioxide and nitrogen oxides from combustion plants (and industrial facilities) with a thermal output of 50 MW or higher. Specifically all plants built after 1987 must comply with pollutant limits prescribed by the LCPD, while those that entered operation before 1987 can install emission abatement equipment or opt-out of the LCPD. Importantly, those plants that have been opted out of the LCPD are required to cease production by 2015 or after 20,000 operational hours since 2007. As of 2012, 221 thermal combustion plants across Europe had opted-out and are set to close, with Romania (41) Poland (40), and France (36) having the highest number of closures. The total thermal input loss across the EU is set to be 99,161 MWth, of which 38,194 MWth will be from the UK in sixteen plant closures and 12,669 MWth from France on thirty-six plant closures (EEA, 2013). Despite the aim of the LCPD being a reduction in air pollution across Europe, it has inadvertently contributed to the short-term rise in coal consumption in Europe as operators of plants set to close are using up their remaining operational hours and burning more coal, while it is relatively cheap. In the UK, for example, the share of gas in electricity generation fell from 40 per cent in 2011 to 28 per cent in 2012 (147 TWh to 100 TWh), while the share of coal grew from 30 per cent to 39 per cent (109 TWh to 143 TWh) (DECC, 2013).

Despite the current adverse market conditions for gas relative to coal in Europe, gas demand is highly likely to return in the medium term following the enforcement of the LCPD and closure of certain thermal generation plants across Europe. The lost thermal generation resulting from plant closures will be replaced through greater use of gas for electricity generation, alongside continuing growth in installed renewables capacity. Eventually, gas will migrate to a role primarily as back up for renewable intermittency. This shift in power generation, combined with a return to economic growth and subsequent industrial output towards the end of this decade, is likely to lead to an increase in European gas imports. However, it is not clear from where this rebound in European demand will be satisfied. A situation that is likely to become more complex given Europe's growing commitment to reduce its reliance on Russian gas imports in light of events in the Ukraine in 2014. However, future demand growth is likely to be located outside of Europe, which means the European consumers will have to compete on global markets. The IEA forecasts global gas demand to grow by 16 per cent from 2012 to 2018, with Asia and South America the primary demand centres along with rapidly developing domestic demand in gas exporting Middle Eastern countries (IEA, 2013).

15.6 **The future pricing of LNG**

Although spot and short-term LNG sales have been increasing as a share of overall sales, medium to long contract agreements (contracts for five years or more) have increased in number and volume of gas to be delivered. From 2010 to 2011 the number of concluded

LNG contracts rose from four totalling 8.4 mmtpa to twenty contracts totalling 33.5 mmtpa. This upward trend continued in 2012 with twenty-three concluded contracts totalling 33.5 mmtpa (GIIGNL various years). The increase was driven by Asia, with Japan and South Korea concluding thirty of the forty-three contracts in 2011 and 2012. Following the post-Fukushima nuclear shutdown and subsequent shift to gas, Japan concluded twelve contracts totalling 10.9 mmtpa in 2011, with eleven of these contracts sourcing LNG from Australian fields. Similarly, in 2012 Japan concluded eleven contracts totalling 6.3 mmtpa, with eight of these for Qatari and Australian LNG. The growth in medium–long term contracts in Asia is emblematic of how countries with high natural gas import dependency are securing supplies through longer term contracts, and in Asia these importers are paying a higher price than their European counterparts due to regional price divergence. However, the level they are willing to pay is a key political and economic issue. In 2013 there are signs that the number of concluded medium–long contracts will fall due to Asian importers' insistence on reviewing pricing structures. The continuing price gap of over $15/mmbtu between the JCC and US Henry Hub and increasing volumes of LNG imports since 2011 have impacted the Japanese national economy as well as the finances of utilities, with the lowering of prices now a priority for importers. Asian buyers are looking to the USA for future LNG supply but also as a pricing benchmark for the Pacific LNG market (Rogers and Stern, 2014). Qatar cut prices in 2013 to its Asian LNG customers to ensure trade in anticipation of the USA, Australian, and East African LNG entering the global market place in the coming years. Tension is developing between the buyers and sellers in this respect, as a lowering of LNG prices and a wider shift away from oil-indexation may impact the future development of expensive large LNG projects. Chevron, the developer of the Gorgon LNG project in Australia, has affirmed that the price of crude oil should dictate LNG prices due to the inextricable nature of the oil and gas supply chain and upstream construction processes. Furthermore, LNG contracts, according to Chevron, should only be priced relative to the market from where the feedstock is obtained; therefore LNG in the Asia-Pacific market should not be benchmarked against the Henry Hub arbitrarily. However, it is just as difficult to see the logic of pricing natural gas relative to another energy commodity with very different market fundamentals. The increasing volumes of spot-priced gas may also undermine LNG exporters' attempts to use high-priced bilateral medium-to-long-term contracts with importers to secure the finances of future LNG projects (Ernst and Young, 2013, 2014).

Despite the move towards short term and spot market trades with gas-to-gas pricing in Europe, blended pricing with oil-indexation (and therefore a higher price) found in Asia is more attractive for those selling LNG. Therefore although European gas pricing is shifting towards hub-based indices this does not mean the price of gas will fall; gas-to-gas pricing merely provides a more accurate representation of actual supply and demand in a market. Equally, should the oil price fall significantly, then oil-indexed prices could also fall. Nonetheless, should Asian demand remain strong through into the next decade, it is likely that European LNG importers will be forced into paying more to ensure security of supply.

15.7 **Conclusions: prospects for a globalizing gas market?**

Given that most analysts agree that future natural gas demand growth is likely to be concentrated in non-OECD Asia, it might seem strange that this chapter has focused on developments in the OECD. The reason for this is quite straightforward: at the moment this is where the key changes are taking place that will determine the future pricing and contract structures within the global gas industry, and thus the terms on which it might satisfy future energy demand in the non-OECD. The game changing nature of the US shale gas revolution, at present at least, lies not in the prospects of LNG exports; rather it is the impact that it has already had in destabilizing the way in which internationally traded gas is priced and contracted. The high price of oil, the impact of the global economic crisis, and the growing role of renewables (supported by subsidies) are also significant factors that have combined to dampen gas demand in Europe; while in OECD Asia the current high cost of oil-indexed LNG is leading to concerted efforts to revise the price down, if not find a new pricing mechanism. A direction of travel away from oil indexation and long-term contracts does not necessarily mean cheaper gas. First, the oil price may fall and could result in oil-indexation discovering a lower gas price than that produce by gas-to-gas competition, such as Henry Hub. Second, the LNG supply chain is capital intensive and thus there is a limit to how cheap LNG can be and still stimulate investment in new projects. Third, Asian customers may continue to be willing to pay a premium for the 'security of supply' that is provided by long-term contracts.

At present we have a regionally structured globalizing gas market. This may sound like an oxymoron: we have an almost self-contained North American market based on gas-to-gas competition that is on the brink of delivering LNG exports, we have a European market that is evolving away from oil indexation and long-term contracts towards an integrated hub-based system and gas-to-gas competition—a process that is likely to be accelerated in response to events in Ukraine, and we have an Asian LNG market that is dominated by long-term contracts and oil indexation. The direction of travel is towards greater reliance on gas-to-gas competition (Stern, 2012); but convergence in relation to price formation is not the same as converging gas prices. Here geography matters and the market fundamentals mean that there will continue to be different prices set in different gas markets. The European gas market is very different from the Asia LNG market and will remain so in the future. Here we return to the prospects for non-OECD gas demand. The reality is that natural gas can only grow new markets in the emerging Asian economies if it is cost competitive against coal. Of course, there are very good reasons for reducing future coal demand that relate to climate change and air pollution; there are also mechanisms that can be used to push up the price of coal relative to gas, such as a carbon tax. But the reality remains that many countries see coal as cheap and secure; which means that gas does need to be abundant and affordable if it is to create new markets. The hope is that the current changes underway in OECD gas markets will enable gas to compete against coal in emerging markets. However, a final caveat: as gas itself is a fossil fuel, any golden age of gas must be relatively short-lived if global warming is to be limited to 2 °C.

■ REFERENCES

BP. 2013. BP *Statistical Review of World Energy June 2013*. London: BP.

Chazan, Guy. 2013. 'Gazprom profits up 5% as exports to Europe recover.' *Financial Times*. 3 September. Accessed 11 May 2014. <http://www.ft.com/cms/s/0/09cced56-1497-11e3-a2df-00144feabdc0.html?siteedition=uk#axzz31P73U9Dk>.

De Clercq, Geert. 2013. 'Renewables turn utilities into dinosaurs of the energy world.' *Reuters*. 8 March. Accessed 11 May 2014. <http://www.reuters.com/article/2013/03/08/us-utilities-threat-idUSBRE92709E20130308>.

De Clercq, Geert, and Steitz, Christoph. 2013. 'E.ON-GD Suez Earnings hit by European Energy Crisis.' *Reuters*, 13 November. Accessed 11 May 2014. <http://uk.reuters.com/article/2013/11/13/utilities-europe-idUKL5N0IY0SX20131113>.

DECC. 2013. *Digest of UK Energy Statistics 2013*. London: DEC. Accessed April 2015. <https://www.gov.uk/government/uploads/system/uploads/attachment_data/file/65818/DUKES_2013_Chapter_5.pdf>.

EIA. 2005. *Annual Energy Outlook 2005*. Washington DC: EIA. Accessed 27 April 2014. <http://www.eia.gov/forecasts/archive/aeo05/pdf/0383(2005).pdf>.

EIA. 2013a. 'North America leads the world in production of shale gas,' *Today in Energy*, 23 October. Washington DC: EIA. Accessed April 2015. <http://www.eia.gov/todayinenergy/detail.cfm?id=13491>.

EIA. 2013b. U.S. *Natural Gas Imports*. Washington DC: EIA. Accessed 9 May 2014. <http://www.eia.gov/dnav/ng/ng_move_impc_s1_m.htm>.

EIA. 2014. *Annual Energy Outlook 2014*. Washington DC: EIA.

Ernst and Young. 2013. *Global LNG: will new demand and supply mean new pricing?* Accessed April 2015. <http://www.ey.com/Publication/vwLUAssets/Global_LNG_New_pricing_ahead/$FILE/Global_LNG_New_pricing_ahead_DW0240.pdf>.

Ernst and Young. 2014. *Competing for LNG demand: the pricing structure debate*. Accessed 14 May 2014. <http://www.ey.com/Publication/vwLUAssets/Competing-for-LNG-demand/$FILE/Competing-for-LNG-demand-pricing-structure-debate.pdf>.

European Environment Agency (2013): *Large Combustion Plants (LCP) opted out under Article 4 (4) of Directive 2001/80/EC*. Accessed 11 May 2014. <http://www.eea.europa.eu/data-and-maps/data/large-combustion-plants-lcp-opted-out-under-article-4-4-of-directive-2001-80-ec-3>.

E.ON. 2012. 'E.ON Reaches Settlement with Gazprom and long-term gas supply contracts and raises outlooks for 2012.' 7 March. Accessed 11 May 2014. <http://www.eon.com/en/media/news/press-releases/2012/7/3/eon-reaches-settlement-and-raises-group-outlook-for-2010.html>.

Garman, Joss and Kahya, Damian. 2012. 'Why is Europe burning more coal?' *Greenpeace UK*, 25 January. Accessed 16 May 2014. <http://www.greenpeace.org.uk/newsdesk/energy/analysis/why-europe-burning-more-coal>.

GIIGNL. 2013. *The LNG Industry 2012*. Paris: GIIGNL. Accessed 9 May 2014. <http://www.giignl.org/sites/default/files/PUBLIC_AREA/Publications/giignl_the_lng_industry_2012.pdf>.

Hromadko, Jan. 2012. 'RWE Upgrades Outlook for Gas Midstream on Statoil's Accord.' *Wall Street Journal*, 14 August. Accessed 11 May 2014. <http://www.4-traders.com/RWE-AG-436529/news/RWE-Upgrades-Outlook-for-Gas-Midstream-Business-on-Statoil-Accord-14459256/>.

Hromadko, Jan. 2013. 'Shale Boom Is a Bust for Europe's Gas Plants.' *Wall Street Journal*, 8 May. Accessed 11 May 2014. <http://online.wsj.com/news/articles/SB10001424127887323744604578470841012284404>.

IEA. 2011. *Are We Entering a Golden Age of Gas?* Paris: IEA.

IEA. 2012. *Golden Rules for a Golden Age of Gas*. Paris: IEA.

IEA. 2013. 'IEA sees growth of natural gas in power generation slowing over next 5 years.' 20 June 2013. Accessed 14 May 2014. <http://www.iea.org/newsroomandevents/pressreleases/2013/june/name,39014,en.html>.

IGU. 2011. *LNG Report 2011*. Oslo: IGU. Accessed 9 May 2014. <http://www.igu.org/gas-knowhow/publications/igu-publications/LNG per cent20Report per cent202011.pdf>.

IGU. 2013a. *LNG Report 2013*. Oslo: IGU. Accessed 9 May 2014 <http://www.igu.org/gas-knowhow/publications/igu-publications/IGU_world_LNG_report_2013.pdf>.

IGU. 2013b. *Wholesale Gas Price Survey 2013*. Oslo: IGU. Accessed 11 May 2014. <http://www.igu.org/news/gas-knowhow/publications/igu-publications/Wholesale per cent20Gas per cent20Price per cent20Survey per cent20- per cent202013 per cent20Edition.pdf>.

IGU. 2014. *LNG Industry 2014*. Oslo: IGU. Accessed 9 May 2014. <http://www.igu.org/gas-knowhow/publications/igu-publications/igu-world-lng-report-2014-edition.pdf>.

Khodyakova, Yelena. 2013. 'Gazprom ups 2013 Europe gas export forecast.' *Vedomosti*, 3 July. Accessed 11 May 2014. <http://rbth.co.uk/business/2013/07/03/gazprom_ups_2013_europe_gas_export_forecast_27717.html>.

Khrennikova, Dina, and Bowles, Alisdair. 2013. 'Russia's Gazprom Cancels call to prepare project for the Shtokhman LNG Plant," *Platts*, 23 August. Accessed 9 May 2013. <http://www.platts.com/latest-news/natural-gas/moscow/russias-gazprom-cancels-call-to-prepare-project-26211338>.

Krijgsman, Richard. 2011. 'Too much gas? *Energy Evaluate*. Accessed 17 February. <evaluateenergy.com/by-sector/lng/gas>.

Makan, Ajay. 2013. 'Statoil breaks oil-linked gas pricing.' *Financial Times*. 19 November. Accessed 11 May 2014. <http://www.ft.com/cms/s/0/aad942d6-4e25-11e3-b15d-00144feabdc0.html#axzz31P73U9Dk>.

Moscow Times. 2014. 'Gazprom Captures Largest-Ever Share of European Gas Market.' *The Moscow Times*, 19 February. Accessed 11 May 2014. <http://www.themoscowtimes.com/news/article/gazprom-captures-largest-ever-share-of-european-gas-market/494834.html>.

Reuters. 2014. 'Japan's 2013 LNG imports hit record high on nuclear woes.' 27 January. Accessed 9 May 2014. <http://www.reuters.com/article/2014/01/27/energy-japan-mof-idUSL3N0L103N20140127>.

Riley, Alan. 2012. *Commission v. Gazprom: The antitrust clash of the decade?* CEPS Policy Brief. London: Centre for European Policy Studies.

Rogers, Howard V, and Stern, Jonathan. 2014. *Challenge to JCC Pricing in Asian LNG Markets.* OIES Paper: NG 81. Oxford: OIES.

Ria Novosti. 2013. 'Gazprom to cut gas price for RWE after Arbitration.' 7 July. Accessed 11 May 2014. <http://en.ria.ru/business/20130709/182144457.html>.

Schultz, Stefan, and Bidder, Benjamin. 2013. 'Under Pressure: Once Mighty Gazprom Loses its Clout.' *Spiegel Online International*, 1 February. Accessed 11 May. <http://www.spiegel.de/international/world/gazprom-gas-giant-is-running-into-trouble-a-881024.html>.

SSE. 2013. 'Review of thermal generation operations.' 21 March 2013. Accessed 11 May 2014. <http://sse.com/newsandviews/allarticles/2013/03/review-of-thermal-generation-operations/>.

Stern, Jonathan (ed.). 2012. *The Pricing of Internationally Traded Gas.* Oxford: OIES/OUP.

Stevens, Paul. 2012. *The Shale Gas Revolution: Development and Changes.* London: Chatham House.

Strzelecki, Marek. 2013. 'European Coal Falls to Record Low as Imports from Columbia to Resume.' *Bloomberg News*, 9 October. Accessed 11 May 2014. <http://www.bloomberg.com/news/2013-10-09/european-coal-falls-to-record-as-imports-from-colombia-to-resume.html>.

World Nuclear News. 2014. 'Coal Taints Germany's Energy Mix.' *World Nuclear News.* 12 March. Accessed 11 May. <http://www.world-nuclear-news.org/EE-Coal-taints-Germanys-energy-mix-1203141.html>.

16 Nuclear power after Fukushima: prospects and implications

Markku Lehtonen and Mari Martiskainen

16.1 Introduction: the ups and downs of civilian nuclear power

The civilian nuclear industry has gone through alternating phases of development and decline. The rise of the industry in the immediate aftermath of the Second World War was largely based on the idealistic and often utopian notions of converting a destructive military technology to the service of the well-being of the entire humankind—an ethos manifested in the US 'Atoms for Peace' initiative. Until the early 1970s, the atmosphere was one of optimistic belief in nuclear providing clean, abundant electricity 'too cheap to meter' (e.g. Herring, 2003: 143–8). The mastery of nuclear technology also offered a promise for former world powers (e.g. France and the UK) to regain some of their lost 'greatness' (e.g. Gowing and Arnold, 1974; Hecht, 2009). Policy was largely in the hands of a small elite of technical experts with virtually complete autonomy to develop the technology without public contestation, government control, or parliamentary oversight (e.g. Winskel, 2002; Hecht, 2009).

The first oil crisis promised another golden opportunity for the nuclear industry, with many Western countries planning to expand their nuclear power production capacity. However, rising public scepticism, stagnating electricity demand, and the Three Mile Island accident—TMI (1979)—put an end to the euphoria, and brought into the limelight the dangers of nuclear power. Falling gas prices and increasing investment costs caused by technical problems and record-high interest rates led to significant downscaling of the planned nuclear programmes. Nuclear technology was criticized for being incompatible with democratic policymaking, secrecy and centralization of knowledge were perceived as its inherent qualities, and nuclear institutions were portrayed as remote and powerful bureaucracies that threatened civil liberties (Herring, 2003: 158). The hitherto nuclear-enthusiastic media accounts gradually gave way to more critical reporting, echoing the arguments of the emerging environmental and anti-nuclear movement (e.g. Herring, 2003).

Nuclear safety—in particular risks of a major accident on one hand and proliferation of nuclear weapons on the other—was among the central concerns in debates during the first wave of anti-nuclear contestation in the mid 1970s. Concerns over a possible yet unlikely catastrophic accident and over the unresolved waste management and plant decommissioning challenges attracted increasing attention. Deep geological disposal emerged as the preferred solution for high-level waste management, as dumping of radioactive waste in the ocean was banned. The development of fast breeder reactors, seen until then as the logical culmination point of nuclear programmes, stalled as concerns over proliferation, nuclear waste, and safety accumulated (e.g. Cochran et al., 2010). In the USA, these concerns added to the economic troubles that had led to a slowdown of the country's nuclear programme in the mid 1970s (Hultman and Koomey, 2013).

By the mid 1980s, new nuclear plans were underway in many countries, as economic growth had picked up again, the memory of TMI had faded, oil prices were low, and the anti-nuclear movement focused on campaigning against nuclear weapons. However, the Chernobyl accident in 1986 again put an end to the renewed hopes of the nuclear industry, although France, for example, accelerated its nuclear programme. In Germany, the accident left a lasting impact on attitudes and policies, paving the way for subsequent phase-out decisions (e.g. Mez, 2012; Schreurs, 2012, 2013).

The TMI and Chernobyl accidents prompted improvements in reactor technologies, safety control, and transparency. In the 1990s, attention gradually moved towards the presumed benefits of nuclear power, which became increasingly perceived as an affordable low-carbon solution to the problems of climate change and energy security. Public opposition against nuclear declined in Europe and the USA (Kessides, 2012: 186), while many NGOs now prioritized climate change over anti-nuclear campaigning (e.g. Twena, 2006; Lammi, 2009). The strengthened communication efforts undertaken by the nuclear industry, which had lost a lot of its earlier autonomy and direct state support (e.g. Winskel, 2002; Twena, 2006) further spurred what was to be called a 'nuclear renaissance'.

The Fukushima accident, triggered by a tsunami and an earthquake on 11 March 2011, seriously threatened the 'nuclear renaissance', and brought safety back to the forefront of the debate. Beyond the immediate radiation effects, the issues on the agenda included concerns over reactor safety (leading to the subsequent national and EU-level 'stress tests' and safety audits); the ability of current probabilistic risk calculation methods to adequately consider 'beyond-design' accidents (e.g. Goldberg and Rosner, 2011); the importance of the 'human factor'; institutional issues (especially independence of the safety authority); and natural hazards as safety risks (e.g. Elliott, 2013).

The rest of this chapter will explore the potential of and obstacles to the development of nuclear power in the post-Fukushima world. After a brief survey of the current situation of civilian nuclear power, the chapter reviews the motivations for the construction of nuclear power, examines the potential of new nuclear technologies, and then looks at the current obstacles and challenges faced by the nuclear industry. The concluding section discusses the future prospects of nuclear energy.

16.2 **Nuclear energy after Fukushima**

Whether a 'true' nuclear renaissance in fact was underway in the early 2000s has been contested (e.g. Thomas, 2012; Hultman and Koomey, 2013; Szarka, 2013). The French-led nuclear renaissance had stalled by 2011 for financial and organizational reasons (Szarka, 2013), the unanticipated surge of natural gas in some countries (e.g. in the USA) had reduced the appeal of nuclear (Kessides, 2012), and technical problems, increasing costs, and construction delays, especially in the EPR plant sites at Flamanville and Olkiluoto, added to the problems. Even in China the pace of ordering showed signs of a slowdown already prior to Fukushima. The Fukushima accident prompted several countries to halt or scale down their nuclear programmes.

16.2.1 WORLDWIDE SITUATION AND TRENDS

According to the World Nuclear Association, at the end of 2013 there were 434 reactors in thirty-one countries operating in the world, with a combined installed capacity of 372 GWe.[1] A further sixty-four were under construction (sixty-seven according to Schneider et al., 2014: 7), of which twenty-one were in China and ten in Russia. However, many of these reactors had been under construction for a long time, and their completion schedule was unclear. The share of nuclear in the energy mix and the total amount of nuclear electricity generated in the world have declined over time. The share of nuclear in electricity generation dropped from its peak of 17 per cent in 1993 to about 11 per cent in 2013, and the share of nuclear in global commercial primary energy production dropped to 4.4 per cent, a level last seen in 1984 (Schneider et al., 2014: 155; see Figure 16.1). The industry faces significant challenges of decommissioning and reactor replacement in the upcoming years, given that the average age of operating reactors in mid 2013 was twenty-eight and a half years and growing, with almost half of the reactors having operated for thirty years or more (ibid: 4). For the total number of reactors, operating capacity, and nuclear generation in the world, see Figure 16.2.

Since Fukushima, international organizations have scaled down their predictions for nuclear generation. In 2010, the IAEA (2010) projected 546–803 GW nuclear generation capacity by 2030, while two years later, its lower projection had dropped to 456 GW, and the upper projection down to 740 GW (IAEA, 2012). The IEA, in its flagship publication, World Energy Outlook, also scaled down its predictions after Fukushima. Production was forecast to grow by almost 60 per cent from 2756 TWh in 2010 to about 4370 TWh in 2035, but other generation sources were expected to grow even more rapidly, and

---

[1] <http://www.world-nuclear.org/info/Nuclear-Fuel-Cycle/Power-Reactors/Nuclear-Power-Reactors/>. However, according to Schneider et al. (2014: 6), the true number of operating reactors in mid 2014 was 388, with 43 nominally operating reactors being in long-term outage.

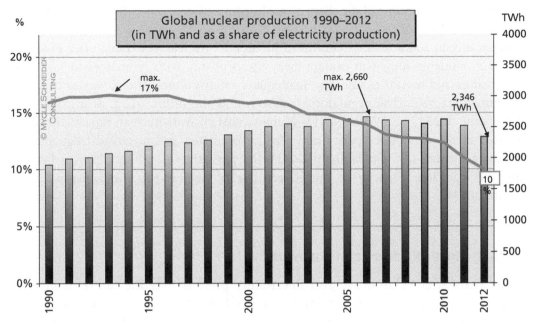

Figure 16.1 Number of reactors, operating capacity (1954–2013), and nuclear generation (1990–2012)

Source: Schneider et al. (2013: 17).

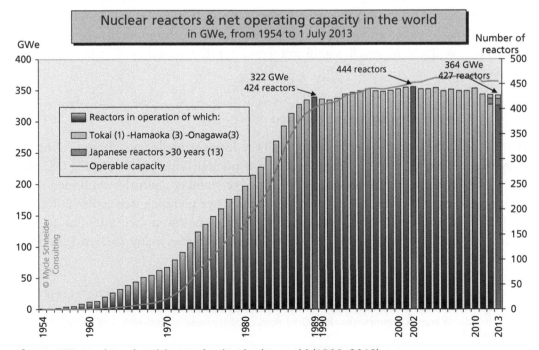

Figure 16.2 Nuclear electricity production in the world (1990–2012)

Source: Schneider et al. (2013: 11).

thereby prompt a decline of the share of nuclear in total generation from 13 per cent to 12 per cent (IEA, 2012).[2]

Especially in countries that continued to support nuclear new-build after Fukushima, governments sought to explain away the problem, with references to Japanese specificities: old reactor technology, inadequate safety culture, and close ties between the bureaucracy, nuclear regulators, industry, and the utilities (e.g. Srinivasan and Rethinaraj, 2013: 732; Walker, 2013; Ramana, 2013). Anti-nuclear experts, in turn, pointed at the great discrepancy between the predicted and observed frequency of serious accidents (e.g. Goldemberg, 2011), the pervasiveness of the 'human factor', and the traditional opacity of the governance of nuclear (e.g. Greenpeace, 2012; Elliott, 2013).

16.2.1.1 Japan: Nuclear at a Crossroads?

Three years after Fukushima, great uncertainty prevails over the future of nuclear power in Japan. In contrast with the previous administration's plan to mothball all reactors, the new conservative government that took power after the December 2012 elections has called for the restart of reactors deemed safe by regulators.[3] By contrast, Schneider et al. (2013: 6) estimated that many of the country's remaining forty-four reactors might never restart, and public opposition against restarting the reactors, buttressed by the still incipient anti-nuclear movement, remained strong in early 2014.[4] Schreurs (2013: 108) deemed it highly unlikely that Japan would increase its dependence on nuclear, yet replacing nuclear supply with renewables or imported fossil fuels is likely to entail significant costs for the economy (Nesheiwata and Crossa, 2013).

16.2.1.2 Europe: Divided and Saturated

The EU landscape remains polarized between the fundamentally anti-nuclear countries (Austria and Ireland)[5] and the almost equally pro-nuclear (Finland, France, the UK, and many Eastern European countries). In 2013, twelve member countries continued their nuclear programmes, while eleven[6] were constructing or planning new units (Euratom, 2013: 36). Germany, Italy, Switzerland, and Belgium announced their intention to halt nuclear new-build or phase out the existing plants, whereas many Eastern European

---

[2] <http://www.world-nuclear-news.org/ee-iea_cuts_nuclear_power_growth_forecast-1211124.html>.

[3] <http://www.reuters.com/article/2014/03/13/us-japan-nuclear-restarts-idUSBREA2C0F620140313>.

[4] <http://digitaljournal.com/news/world/poll-80-percent-of-japanese-against-restart-of-nuclear-plants/article/376563>.

[5] The Irish Electricity Regulation Act 1999 forbids nuclear fission-based power stations (Nuttall, 2009).

[6] France, the UK, Sweden, the Czech Republic, Finland, Hungary, the Slovak Republic, Romania, Bulgaria, the Netherlands, and Slovenia.

countries, with highly pro-nuclear public opinion,[7] seek to reduce their dependence on Russian gas through building nuclear. Somewhat paradoxically, the Russian state-owned nuclear company, Rosatom, is likely to play an increasing role as a supplier of nuclear technology in these countries.[8]

The prospects for significant expansion of nuclear power in Europe seem limited, not least because of public opposition in greenfield sites. Of the five scenarios in the EU Roadmap towards an 80 per cent reduction of GHG emissions by 2050,[9] three foresee a 14–20 per cent contribution of nuclear to the EU electricity generation mix (Euratom, 2013: 40). EU climate policy and the potential special treatment of nuclear in decision-making as a 'low-carbon' energy source are likely to be crucial for the future of nuclear in Europe. The replacement of ageing plants, decommissioning, and high-level radioactive waste management appear as the main challenges and opportunities for the nuclear industry.

16.2.1.3 USA: Phase-out for Economic Reasons?

In the USA, despite the government's continued commitment to new-build, the nuclear renaissance seems to have halted, as eight of the twenty-eight proposed reactors were suspended indefinitely and sixteen delayed (Schneider et al., 2013: 56). Five reactors are under construction (Schneider et al., 2014: 100).[10] For the first time in fifteen years, operating reactors are being closed as uneconomic (ibid: 4).[11] The US nuclear industry has a long history of permitting problems, cost overruns, and construction delays that have dissuaded financial markets from backing new nuclear construction (Lévêque, 2013; Mecklin, 2013: 9). The drop in gas prices provoked in part by the shale gas 'revolution' has dented the economic viability even of existing nuclear plants, while making new reactor investments unattractive (Bradford, 2013: 12).

16.2.1.4 India, China, and Russia: The New Eldorado?

The greatest potential for nuclear expansion seems to lie in Asia, the Middle East, and to some extent in Eastern Europe. Two thirds of global growth in nuclear power is based in just three countries: China, India, and Russia (Schneider et al., 2014: 7).

---

[7] According to a Eurobarometer study published in March 2010, 86 per cent of Czechs, 76 per cent of Slovenians and Hungarians, and 70 per cent of Poles were favourable to maintaining or increasing the share of nuclear in their countries' electricity supply. <http://www.monde-diplomatique.fr/2013/07/BIENVENU/49358>.

[8] <http://www.monde-diplomatique.fr/2013/07/BIENVENU/49358>.

[9] The European Commission issued its 2050 Energy Roadmap at the end of 2011. It was then discussed at Council level and led to Presidency conclusions in 2012 (Euratom, 2013: 40).

[10] <http://world-nuclear.org/nucleardatabase/rdResults.aspx?id=27569>.

[11] See also <http://www.nytimes.com/2013/06/15/business/energy-environment/aging-nuclear-plants-are-closing-but-for-economic-reasons.html?_r=0>.

In 2013, *Russia* had thirty-three reactors totalling 24,253 MWe in operation and ten new reactors under construction (WNA, 2013c). The majority of the existing reactors have been granted a life extension, while the construction of a further forty-six reactors is planned for 2020–30 (ibid.). After Fukushima, further safety checks were made in all reactors, and Rosenergoatom, the country's only nuclear operator, announced a $530 million safety programme, still a modest sum compared to the country's announced $55 billion nuclear development programme in 2006 (ibid.).

The IEA (2013: 44) foresees nuclear power to have a limited role in *Southeast Asia* over the next couple of decades, due to 'complexities of developing a nuclear power programme' and the slow progress of nuclear plans thus far. Most new-build plans were postponed after Fukushima, with the exception of Vietnam (Srinivasan and Rethinaraj, 2013: 733). Malaysia and Thailand foresee new-build soon after 2020 (IEA, 2013).[12]

India continues with its ambitious objectives of 60 GWe of new nuclear capacity by 2030 (Srinivasan and Rethinaraj, 2013: 733),[13] and a 25 per cent share of nuclear in electricity generation by 2050—up from the 3.7 per cent in 2011 (WNA, 2013b). It seeks to become a world leader in nuclear technology, especially in fast reactors and the thorium fuel cycle (WNA, 2013b). India is not part of the Nuclear Non-Proliferation Treaty and has developed its nuclear programme largely without external assistance (ibid.). The government has thus far suppressed local opposition at construction sites, but conflicts may well resurface as new projects advance (Srinivasan and Rethinaraj, 2013: 733; IEA, 2013).

China was by far the world's largest market for nuclear plants in the past decade and expected until recently to add up to more than 80 GWe of new capacity by 2020 (Srinivasan and Rethinaraj, 2013: 733; WNA, 2013a). In mid 2014, twenty-one reactors were in operation and twenty-eight under construction (Schneider et al., 2014: 105). China also may become a proving ground for generation III+ designs (Schneider et al., 2013: 34), and small modular reactors (WNA, 2013a). Following Fukushima, the government called a temporary moratorium on new-build, initiated safety checks of the country's fourteen existing plants, developed a new nuclear safety plan (Schreurs, 2013: 103), and reduced the construction target in October 2012 to 60 GWe by 2020. China counts on fast reactors as an increasing proportion of new-build from 2020 onwards, with a capacity of at least 200 GWe by 2050 and 1400 GWe by 2100 (WNA, 2013a). In 2012, nuclear accounted for just under 2 per cent of the rapidly increasing total electricity

[12] In Taiwan, street protests and parliamentary disputes over safety have postponed new-build projects for years, and nine of South Korea's reactors have been closed (Reuters, 9 August 2013: <http://www.reuters.com/article/2013/08/09/us-asia-nuclear-idUSBRE9780BD20130809>).

[13] Announced after the successful negotiation of the civilian nuclear deal with the United States in 2008 and its subsequent clearance by the Nuclear Suppliers Group (NSG).

BOX 16.1. 'GENERATIONS' OF NUCLEAR REACTOR DESIGNS (SOURCES: GOLDBERG AND ROSNER, 2011; THOMAS, 2012)

Generation I designs include the prototype and demonstration plants of the 1960s, whereas **generation II** designs include the vast majority of the currently operating pressurized and boiling-water reactors (PWRs and BWRs), including those built in the 1970s and 1980s. These reactors were designed to be economical and reliable for a typical operational lifetime of forty years. The safety features are mostly 'active', involving electrical or mechanical operations that are initiated automatically and, in many cases, can be initiated by the operators of the nuclear reactors.

Generation III and III+ nuclear reactors, ordered from the mid 1980s onwards, differ from gen II in that they were designed with the lessons of TMI accident fully incorporated, hence integrating evolutionary, state-of-the-art design improvements in the areas of fuel technology, thermal efficiency, modularized construction, safety systems (greater use of passive systems),[14] and standardized design. These features aim at extending the operational lifetime, potentially to more than sixty years, to help expedite the reactor certification review process and thus shorten construction schedules, as well as to reduce fuel consumption and waste production thanks to an increased fuel burn-up. No clear criteria exist for distinguishing between gen III and III+. Goldberg and Rosner (2011) site as examples of gen III designs the Westinghouse 600 MW advanced PWR (AP-600), the GE Nuclear Energy Advanced Boiling Water Reactor (ABWR), the Enhanced CANDU 6, and System 80+, a Combustion Engineering design. Gen III+, in turn, would include VVER-1200/392M Reactor, Advanced CANDU Reactor (ACR-1000), AP1000, European Pressurized Reactor (EPR), Economic Simplified Boiling-Water Reactor (ESBWR), APR-1400, EU-ABWR, Advanced PWR (APWR), and ATMEA I. Only four gen III reactors, all ABWRs, are in operation today.

generation in China. For a description of the different 'generations' of nuclear reactors, see Box 16.1.

Open questions persist concerning the expansion of nuclear in China. The likely strengthening of safety requirements may force a greatly accelerated introduction of generation III reactors (only two are under construction at present, and the near-term policies are based on generation II plants).[15] This could create the kind of technical and economic challenges already encountered in the construction of generation III reactors in Europe and the USA. The priority given to domestic manufacturing and fuel assembly (WNA, 2013a), growing public opposition, the increasing importance of economics in Chinese policymaking, and the controversies over the quality and independence of safety regulation may further slow down nuclear expansion (Schneider et al., 2013).

---

[14] Passive systems would not require active controls or operator intervention (Kessides, 2012: 198).

[15] Indeed, China is the only country currently building gen-II units in large numbers, with fifty-seven (53.14 GWe) reactors currently being planned (WNA, 2013a).

16.3 **Motivations for building nuclear**

16.3.1 SECURITY OF ELECTRICITY SUPPLY

In the face of a projected increase of electricity demand worldwide (see e.g. IEA, 2013), nuclear power is frequently portrayed as a low-carbon, secure, and affordable source of stable and reliable baseload power, capable of enhancing diversity and independence of supply. Whether nuclear indeed strengthens energy security depends on context-specific factors, such as the share of electricity-intensive industry in the economy. As long as uranium comes from abroad, nuclear is not a fully indigenous energy source. While dependence on imports—whether of electricity or fuel—is not necessarily a bad thing in itself, dependence on unstable and unreliable supply sources is. A high proportion of nuclear may also increase vulnerability to supply disruptions, as demonstrated most dramatically by the post-Fukushima situation of forced demand control in Japan (Froggatt et al., 2012). Watson and Scott (2009) argue that nuclear power may help UK energy security by hedging against rapid fossil fuel price increases, but would seem powerless in the face of other security risks such as inadequate investment in gas infrastructure, while aggravating the vulnerability to domestic activism and terrorism.

16.3.2 CLIMATE CHANGE AND THE ROLE OF NUCLEAR IN A LOW-CARBON WORLD

Especially in Europe and China, climate change has been a major argument in favour of nuclear (e.g. Euratom, 2013: 4). Currently, nuclear energy does not benefit from the advantages accorded to energy sources classified as 'renewable' in EU legislation, while for instance the UK government maintains its official position of not offering nuclear power subsidies more favourable than to other generation options.[16] Most observers agree that nuclear is far superior to fossil generation and comparable to renewables in terms of lifecycle GHG emissions (MacKerron, 2012). The respective GHG benefits of an *electricity system* with and without nuclear is a far more contested issue. Again, the answer depends on the context, including the technological, political, and institutional compatibility of nuclear with other low-carbon, more decentralized generating options (e.g. Broderick et al., 2013: 56–8); the likely back-up sources of electricity; and the costs and speed of building the systems in the face of urgency of climate policies. Even the advocates of nuclear acknowledge that a several-fold increase in generation capacity would be needed if nuclear were to replace a significant share of fossil electricity: for

---

[16] The government's policy is one 'of not providing a public subsidy for new nuclear unless similar support is provided to other types of low-carbon generation' (<https://www.gov.uk/government/news/initial-agreement-reached-on-new-nuclear-power-station-at-hinkley>).

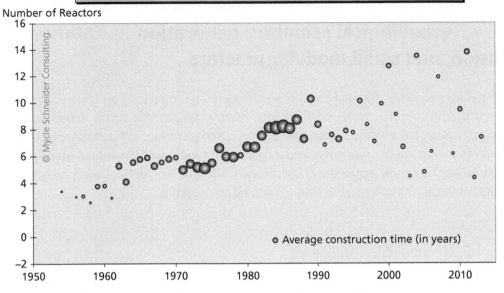

Figure 16.3 Average annual nuclear construction times (1954–2013)

Source: Schneider et al. (2013).

example Kim and Edmonds (2007) suggest that only a 5–10-fold increase over current capacity would constitute a true contribution to climate mitigation.

Those arguing in favour of nuclear as a solution to climate change refer to 'the enormous technical and economic hurdles' for quick and large-scale deployment of renewables (Srinivasan and Rethinaraj, 2013: 734), whereas critics point to recent experience as a demonstration of the superiority of renewables in reducing GHG emissions (Elliott, 2013; Schneider et al., 2013: 73–83), and the ever-longer average construction time of nuclear reactors (see Figure 16.3). Historic data on highly-nuclearized countries like France is used by some to demonstrate the effectiveness of nuclear in keeping down GHG emissions, and by others to show precisely the opposite.[17] Many claim that the German phase-out decision inevitably led to an increase in GHG emissions, yet the evidence is indecisive: German CO_2 emissions fell by 2.4 per cent in 2011, despite the closure of eight of the country's seventeen nuclear reactors within six months of the Fukushima accident (Froggatt et al., 2012). However, continued German imports of electricity from other countries would increase emissions in these electricity-

---

[17] For example, according to Schneider (2013: 21) French annual per capita consumption of 10 barrels of oil in 2011 was greater than that of Italy (which abandoned nuclear power in the 1980s), and above the EU average of 9.8 barrels per capita per year.

exporting countries (ibid.). As for Japan, the IEA (2013: 1) concludes that GHG emissions increased by 70 Mt following the reactor shutdown after Fukushima.

16.4 **Technological promises: generation IV, thorium, fusion, and small modular reactors**

Ever since the 1950s, thermal reactors were widely seen merely as an intermediate step towards advanced and technically superior reactor designs. Producing more fuel than they consume, fast breeder reactors were expected to introduce a 'sustainable fuel cycle', whereas fusion energy was advocated as a virtually inexhaustible source of electricity. Today, R&D on these options still continues, while two new options are under serious consideration: thorium reactors and small modular reactors.

16.4.1 GENERATION III/III+

Generation III/III+ reactors are designed to correct a number of the imperfections that plague generation II reactors, in particular those that stem from the origin of these reactors in military applications. Only four generation III reactors in the world were in operation in October 2013—all of one type, advanced boiling water reactor (ABWR), and located in Japan.[18] A further three units were near completion in Taiwan and Japan, while four more units were being planned in Japan. In China, EDF and the China General Nuclear Power Company (CGN) were jointly constructing four EPRs, and a Westinghouse AP 1000 reactor was soon to be completed. Several reactor types have been approved for construction and are under consideration in Europe and the USA.

The claims of improved safety and shorter construction times of gen III+ reactors have been challenged. The unusual containment structure of the Westing-house AP 1000 design and the Hitachi ESBWR (Piore, 2011) has drawn criticism, as have the safety arrangements in the EPR reactors.[19] The generation III+ reactors have in the West not proven quicker to construct than their predecessors, as most dramatically shown by the construction delays and budget overruns at the EPR reactors at Olkiluoto 3 and Flamanville,[20] whereas the construction of EPRs in

---

[18] See, for example, IOP (2013) and WNA website: <http://www.world-nuclear.org/info/Nuclear-Fuel-Cycle/Power-Reactors/Advanced-Nuclear-Power-Reactors/>.

[19] A joint statement by the British, Finnish, and French safety regulators called into question the adequacy of the EPR's safety arrangements, notably the presumed independence of the back-up safety systems from the control systems used to operate the plant under normal conditions. <http://www.hse.gov.uk/newreactors/joint-regulatory-statement.pdf>.

[20] In September 2014, Areva announced that the Olkiluoto plant was expected to start operating in 2018, that is nine years behind the original schedule. <http://www.areva.com/EN/news-10288/updated-schedule-for-olkiluoto-3.html>. In November 2014, EDF in turn informed about a new delay at Flamanville,

Taishan, China, has advanced without major delays. Potential reasons for the apparent Chinese success include a highly qualified and cheap labour force; well-organized nuclear industry; the vast size of the nuclear programme, which allows efficiency gains and learning; and the around-the-clock operation of the construction works (Lévêque, 2013: 44). The EDF reactors planned for the UK are of the EPR type.

16.4.2 GENERATION IV: REVIVAL OF THE FAST BREEDERS?

As opposed to the incremental improvements entailed in the generation III/III+ reactors, the diverse set of designs classified under the heading 'generation IV' promises a more fundamental leap in technology, similar to the one envisaged by the fast breeder reactors in the past (e.g. Cochran et al., 2010). Six design concepts were chosen as the basis of the Generation IV International Forum (GIF),[21] set up in 2001 to develop these systems (Kessides, 2012: 199). Expected benefits of generation IV include improved safety and reliability (low likelihood and degree of reactor core damage, and reduced need for offsite emergency response); sustainability (more efficient resource utilization and minimal generation of radioactive waste); economic performance (lifecycle cost advantage over other sources of energy); and resistance to proliferation (unattractive route for diversion or theft of weapons-usable materials) (ibid.: 199).

GIF estimates the new designs to be deployable between 2020 and 2030. In light of the considerable technical and economic problems encountered in earlier attempts to develop fast breeder reactors (from the 1950s to the 1990s), these estimates appear optimistic (e.g. Cochran et al., 2010; Thomas, 2012; Le Renard et al., 2013). Grimes and Nuttall (2010), in turn, recognize the benefits of these reactors, but argue that these entail greater safety, security, and proliferation risks than the current reactors.

Low uranium prices and the slowdown of nuclear programmes may further reduce the appeal of a technically immature option whose basic rationale relies on more efficient raw material use. Even authors otherwise favourable to nuclear consider fast breeder prospects—and thereby the 'solution' to the uranium problem—still to be a long time off (e.g. Srinivasan and Rethinaraj, 2013: 734).

foreseeing the plant to become operational in 2017, five years behind schedule, while the construction costs for both EPRs have nearly tripled over initial estimates. <http://www.lesechos.fr/18/11/2014/lesechos.fr/0203947861905_le-demarrage-de-l-epr-de-flamanville-est-a-nouveau-reporte.htm>.

[21] In April 2014, the membership of the Generation IV International Forum (GIF) comprised twelve countries and Euratom.

16.4.3 FUSION

Nuclear fusion continues to absorb a major share of R&D funding devoted to the energy sector, the bulk of the expenditure being dedicated to the joint international ITER project in France. Industrial-scale deployment is not foreseen until 2050, which makes fusion irrelevant for the current efforts to curb CO_2 emissions. Doubts prevail concerning reliability and sustainability—for instance because of the use of helium coolants (Grimes and Nuttall, 2010). Yet the most optimistic observers expect fusion to generate 'viable electricity' in the 2020s, while providing 'half of all mankind's energy needs' by 2050.[22]

16.4.4 THORIUM REACTORS

Thorium reactors are increasingly evoked as a promising new option, notably given their allegedly superior safety compared to the sodium-cooled fast breeder reactors. Tickell (2012) estimates that around thirty reactors in India, Germany, the Netherlands, and the USA already use thorium fuel in conjunction with fissile uranium or plutonium. Thorium-only reactors operate in India, while Canada, China,[23] Israel, Japan, Norway, the UK, and the USA conduct R&D in the area (Tickell, 2012).

Thorium is estimated to be 3–4 times more abundant in the world than uranium (Tickell, 2012), and thorium reactors would produce less radioactive waste than current reactors (Grimes and Nuttall, 2010: 803). However, since thorium is not fissile, the fuel cycle hence needs to be complemented by another existing fissile material, such as uranium or plutonium (Tickell, 2012). Fuel performance and the safety of thorium extraction may become critical questions should this option become widespread (Grimes and Nuttall, 2010). Accelerator-driven subcritical reactors (ADSRs) could avoid the need for fissile materials, while being potentially safer, and producing less long-lived radioactive waste (ibid.: 803).

Elliott (2013: 109) estimates that commercial thorium reactors would not become available earlier than in a decade, while Grimes and Nuttall (2010: 803) argue that thorium could become 'an important nuclear fuel' after 2030.

16.4.5 SMALL MODULAR REACTORS

Small modular reactors (SMRs)—with an output up to 300–400 MWe, compatible with assembly-line manufacture, less complex to build, and therefore more affordable than

[22] Professor Steven Cowley, chief executive of the UK Atomic Energy Authority (<http://peakoil.com/alternative-energy/nuclear-fusion-now-a-viable-energy-source-claims-scientist>).

[23] <http://www.theaustralian.com.au/business/opinion/china-moving-to-thorium-as-safe-nuclear-fuel/story-fnciihm9-1226550688296#>.

the current reactors—are seen as among the most promising alternatives for the near-term future (Lyman, 2011: 52). SMRs could reduce investment and accident risks; enable greater standardization and thereby learning; have a diverse set of useful applications including those suitable to developing countries;[24] enable a greater range of potential siting options; and reduce proliferation hazards (Kessides, 2012).[25] Risk of catastrophic accidents would be reduced thanks to the substantially smaller radioactive inventory of SMRs as compared to current reactors (ibid.).

Several countries[26] are exploring the various concepts and designs of advanced SMRs (Kuznetsov, 2008). The designs include light-water reactors, high-temperature gas-cooled reactors, and liquid metal and gas-cooled fast reactors (Kessides, 2012: 203–5). The expected competitiveness benefits of SMRs, resulting from lower investment cost and shorter construction times, could be outweighed by the absence of economies of scale, that is, substantially higher specific capital costs as compared to large-scale reactors (Kessides, 2012: 201–2). Sceptics doubt the possibility of scaling down of nuclear reactor technology, and the ability of SMRs to catch up the 'decades of head start enjoyed by small modular renewables' (Lovins, 2013: 61). They point to the scant interest expressed by buyers towards most new LWR options, and note that the designs are still far from mature enough to be submitted for review by safety regulators (Thomas, 2012: 16).

16.5 **Constraints**

A number of factors can limit the rate and extent of deployment of nuclear energy in the near future. Five specific challenges will be highlighted in the following.

16.5.1 RISKS AND SAFETY

Fukushima brought safety back to the centre of the nuclear debate, at varying intensity in different countries. The European experience shows the wide cross-country variation in the priority given to safety in nuclear debates, depending on factors such as the importance of the nuclear industry and number of plants in the country (which explains the intensive debate in France); the history of nuclear power (e.g. the long-term public

---

[24] For example low-carbon electricity generation in remote locations with little or no access to the grid, industrial process heat, desalination or water purification, and co-generation applications (e.g. in the petrochemical industry) (Kessides, 2012: 201).

[25] 'Compared to large-scale reactors, SMRs have a larger surface-to-volume ratio (easier decay heat removal), lower core power density (more effective use of passive safety features), smaller core inventory relative to traditional large-scale reactors, and multi-year refuelling so that new fuel loading is needed very infrequently' (Kessides, 2012: 201).

[26] For example Argentina, Brazil, China, Croatia, India, Indonesia, Italy, Japan, the Republic of Korea, Lithuania, Morocco, Russia, South Africa, Turkey, the USA, and Vietnam.

opposition to nuclear in Germany); and 'cultural' factors such as trust in state institutions, authorities, and experts (e.g. high trust in the Nordic countries), the role and nature of experts in society (e.g. Stankiewicz, 2013), and risk perceptions.

Beyond the uncertainties and challenges concerning radiation leaks from Fukushima, issues on the agenda have included reactor safety (leading to the subsequent national and EU-level 'stress tests' and safety audits); the importance of the 'human factor'; institutional issues (especially independence of the safety authority); and natural hazards as safety risks. The reassessment of nuclear risk has been particularly salient in Germany (Schwägerl, 2011). Fukushima also fuelled discussion on the relative safety benefits of generation III/III+ designs, with some arguing that these reactors would have avoided problems of the kind encountered in Fukushima. Sceptics have retorted that the empirical evidence shows a far greater frequency of major nuclear accidents than earlier predicted (Goldemberg, 2011). In particular, they have pointed out that gen III/III+ reactor development is underpinned by the same probabilistic risk assessment methods (Goldberg and Rosner, 2011) that seem unable to address 'beyond-design' accidents resulting from a cascading series of failures—which have allegedly been at the root of all major nuclear accidents thus far (e.g. Dorfman et al., 2013).

It is still too early to judge what the accident will imply in terms of nuclear safety (Thomas, 2012: 13). Past major accidents (including events such as the 9/11 terrorist attack) have prompted significant improvements in safety control institutions and reactor technology, measures to ensure the plants can withstand the impact of a civil aircraft (Thomas, 2012: 13), and, after Fukushima, debate on the prospects of moving beyond probabilistic risk assessment towards 'imagining the unimaginable'. What such an approach would mean in practice remains unclear, and integrating the lessons from Fukushima into new reactor designs will take years (Thomas, 2012: 13).

16.5.2 RESOURCES: URANIUM, SKILLS, AND INSTITUTIONAL CAPACITIES

Concern over the sufficiency of uranium supplies has diminished since the considerable downscaling of nuclear programmes in the mid 1970s. The IAEA and OECD-NEA deem current resources more than adequate for projected needs, even in a high-demand scenario, which foresees 782 GWe nuclear and would require less than half of the identified resources documented in 2009.[27] Furthermore, uranium currently represents as little as 2 per cent of the current electricity price (MacKerron, 2012: 2), while the cost of capital represents by far the highest single line of expense (e.g. Boccard, 2014; see also Figure 16.4). Grimes and Nuttall (2010) predict that today's uranium fuel cycle will become unsustainable by around 2060, which would prompt price increases and development of alternative fuel cycles. Abundant resources of unconventional uranium exist (e.g. uranium phosphates

---

[27] <http://www.iea.org/aboutus/faqs/nuclear/>.

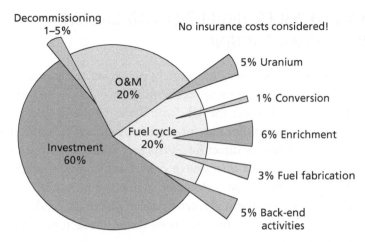

Figure 16.4 Typical cost breakdown of nuclear electricity (O&M: Operating and Maintenance.)
Source: IAEA & Haas and Hiesl (2013).

and uranium in sea water), but according to WNA about a 10-fold increase in uranium price is estimated as necessary to make these options economically viable.[28]

Large-scale worldwide expansion of current thermal nuclear reactors would ultimately be constrained by world uranium resources, yet a much more immediate resource bottleneck is the 'skills gap' triggered by the decline of reactor orders since the 1970s. Many of the still active and experienced nuclear engineers are near retirement age, whereas the economically struggling sector has until recently found it hard to attract young engineers. The phase-out decisions may further erode the skills base (Euratom, 2013: 46), affecting not only new-build programmes, but also regulatory safety inspection (Grimes and Nuttall, 2010), decommissioning, and waste management. The expansion of nuclear in developing countries would entail particularly significant challenges, including the considerable demands that the construction and safe management of the reactors would place on the institutional, human, and financial capacities of these countries (Jewell, 2011). On the other hand, the need to fill the 'skills gap' by training a new generation of nuclear specialists is often highlighted as a formidable opportunity to create high-skilled jobs (Watson and Falcan, 2013: 45–8).[29]

Uranium mining presents a host of problems concerning human rights, development, and security, not least in the context of the recent 'uranium rush' (Conde and Kallis, 2012) fuelled by the prospects of a 'nuclear renaissance'. Many of the major resources are located in developing countries, especially in Africa, and in remote, politically unstable, or underdeveloped regions in the Global North.[30]

---

[28] <http://www.world-nuclear.org/info/Nuclear-Fuel-Cycle/Uranium-Resources/Supply-of-Uranium/>.

[29] The planned EU programme for nuclear reactor lifetime extension and safety upgrade is estimated to add, over the period 2015–35, another 50,000 jobs into the existing 900,000 currently associated with the nuclear sector in Europe, for a total investment of 4.5 billion/year (Euratom, 2013: 40).

[30] Kazakhstan is presently the leading producer of mined uranium (33 per cent), followed by Canada (18 per cent), Australia (11 per cent), and Namibia (8 per cent) (Conde and Kallis, 2012).

16.5.3 BACK-END ISSUES: HIGH-LEVEL RADIOACTIVE WASTE AND DECOMMISSIONING

Since the 1980s, the nuclear industry has acutely perceived the unresolved question of high-level radioactive waste management as the Achilles' heel of the sector (e.g. Grimes and Nuttall, 2010). While France, the UK, and India, for example, reprocess spent nuclear fuel, initially to extract plutonium for producing atomic weapons, most countries have chosen temporary storage or direct final disposal. Today, continued reprocessing is typically justified on the grounds of resource conservation and reduction of the most hazardous types of radioactive waste. However, while the plutonium from the reprocessed spent fuel can be used in thermal reactors as MOX (mixed oxide) fuel, most of the reprocessing products consist of uranium, whose reuse in reactors is not economically viable at current uranium prices.

Despite constant efforts since the mid 1980s, and the 'participatory turn' that radioactive waste management policy took in many countries in the 1990s (Sundqvist and Elam, 2010), only Finland, Sweden, and France have identified a willing host community for a waste repository, and have near-to-implementation plans for geological disposal—the solution preferred by international organisations for managing high-level radioactive waste.[31] At present each country is responsible for the management of the radioactive waste produced in its territory, yet the potential success of the projects in Finland, Sweden, and France may help legitimize industry calls for an international repository.

The decommissioning of the ageing reactor fleet will in the near future absorb increasing amounts of resources. In the UK, the official cost estimates for the decommissioning of the country's nuclear fleet have been creeping up, standing at £52 bn in 2012, largely because of the complexity of operations and escalating costs at the Sellafield site.[32] These estimates are contested, and some experts fear the bill might exceed £100 bn.[33] Cost estimates for other countries are generally lower, but vary greatly according to the method and underlying assumptions (e.g. Cour des Comptes, 2012: 84–115). These uncertainties underline the importance of adequate institutional arrangements for ensuring the availability of funds for decommissioning and waste management long into the future.[34] However, decommissioning might also constitute

---

[31] The Finnish and Swedish sites are to become operational in the early 2020s. In France, a repository site has been identified and is planned to become operational in 2025, yet the schedule seems likely to be postponed, as further 'real-life' experimentation will be needed. Many European countries have made progress, but are still a long way from implementation, whereas in the USA, the government withdrew the long-planned Yucca Mountain repository project in 2009, and left the waste problem unresolved.

[32] UK's nuclear clean-up programme to cost billions more than expected. Terry Macalister, *The Guardian*, 23 June 2013.

[33] UK's nuclear clean-up programme to cost billions more than expected. Terry Macalister, *The Guardian*, 23 June 2013.

[34] The sufficiency and adequacy of financing arrangements have been subject to intense debate in a number of countries, including Sweden (Daoud and Elam, 2012: 20–8) and France (e.g. Cour des Comptes, 2012; Lévêque, 2013: 14–24).

a potential topic of future joint R&D among European countries with starkly contrasting nuclear policies, while the provision of decommissioning expertise could help the nuclear industry survive even if phase-out decisions generalize across the world.

16.5.4 PUBLIC OPINION, THE ROLE OF EXPERTS, AND DECISION-MAKING: TRUST AND TRANSPARENCY

The nuclear sector still suffers from a reputation of a particularly closed and non-transparent expertise and decision-making, inherited from the times when the military and civilian nuclear technologies were developed as part of the same 'complex'. Persistent public opposition and declining direct government support have forced the nuclear industry to learn more participatory and dialogical ways of dealing with the public and the media. In practice, public engagement has often been motivated by the desire to obtain acceptance for specific construction projects, and hence failed to generate trust (e.g. Sundqvist and Elam, 2010; Lehtonen, 2010; Strauss, 2010).

Attitudes and opinions concerning nuclear power have proven relatively stable over time (e.g. Siegrist and Visschers, 2013). The post-Fukushima debate in most countries seems to confirm a pattern whereby the public opinion relatively quickly regains its pre-accident levels after a temporary dip in support to nuclear. Japan may well turn out to be an exception, as the share of people in favour of phasing out, or at least reducing the dependence on, nuclear continued to increase during the months following the accident (Froggatt et al., 2012) and opposition remains strong. While Bickerstaff et al. (2008) detected a 'reluctant acceptance' among the British, that is, relatively high acceptance of nuclear provided that it can help combat climate change, Siegrist and Visschers (2013; see also Corner et al., 2011) argue that climate concerns have only modestly increased the acceptance of nuclear power plants. In France, public support for nuclear had been declining ever since the Chernobyl accident (OCDE-AEN, 2010: 44–5), partly because of the alleged government attempts to cover up the consequences from Chernobyl, and the subsequent numerous public health and safety scandals that eroded public trust in the governance of science and technology (e.g. Chateauraynaud and Torny, 1999). More generally, public support for nuclear appears to be higher in countries with nuclear power plants than in those without any, and higher in Eastern than in Western Europe. The interpretation of public opinion surveys is complicated by the scarcity of longitudinal studies applying consistent methods, as well as by the fact that most respondents are often relatively neutral towards nuclear power (Ipsos Mori, 2011), hence constituting a large stock of 'moving voters', susceptible to change opinion in response to accidents and other policy events.

16.5.5 ECONOMICS: FROM 'CHEAP' ELECTRICITY TO JOB CREATION

After Fukushima, the economic situation of the nuclear sector is as difficult as ever. The income of most major nuclear energy utilities has declined and their debts increased, and the firms have seen their credit ratings and share values decline (Schneider et al., 2013: 8).

In contrast with other electricity generation technologies, nuclear power has seen its construction costs increase steadily over time (e.g. Grubler, 2010; Cour des Comptes, 2012; Lévêque, 2013; Boccard, 2014). Construction delays and budget overruns have become a systematic feature, especially in Europe and the USA (e.g. Haas and Hiesl, 2013). Throughout its history, nuclear energy has benefited from a range of subsidies, for instance in the form of government-funded R&D, price advantage given to nuclear by the energy industry, and low-interest bank loans (Thomas, 2005).

Estimates of the rate of increase in construction costs ($/kW installed) of nuclear new-build vary across studies. Schneider et al. (2013: 4) argue that the increase over the past decade was seven-fold (from $1000 to $7000 per kW), while Grubler (2010) identifies 4.4-times increase between the four first and the four last French reactors. Relying on Cour des Comptes (2012), Lévêque (2013: 32) refutes the figures for France as exaggerated, contrasting the five-fold increase in the USA between the beginning of the 1970s and early 1990s with the 'mere' doubling of the costs in France (see also Boccard, 2014). Since technological learning from accumulating experience has been absent in the construction of nuclear reactors, some have suggested that nuclear would suffer from diseconomies of scale (Cantor and Hewlett, 1988; Grubler, 2010). The complexity of the construction operations and the site-specific character of reactors are cited as factors limiting learning (Grubler, 2010; Kessides, 2012), while plant upgrades required by the safety authorities tend to push investment and operating costs further upwards.[35] Recent debates on nuclear safety have indeed paid increasing attention to cost and economics (e.g. costs of a major accident, of plant safety upgrades).

The low price of natural gas has reduced the attractiveness of nuclear investments (e.g. Ramana, 2013: 73), while policy uncertainty especially after Fukushima has added a further risk premium (Srinivasan and Rethinaraj, 2013: 735). For Boccard (2014), nuclear electricity is affordable to 'the vast majority of advanced country citizens' in view of the current retail price of electricity, even though the price of new nuclear compares unfavourably with alternative sources. Srinivasan and Rethinaraj (2013: 735) conclude that after Fukushima, 'building new nuclear power plants will not be feasible without state subsidies',[36] while Lévêque (2013: 58) identifies a high carbon price as

[35] According to the estimates of the EDF, the extra cost from investments in improved safety required after Fukushima would reach EUR 10 billion on top of the already planned investments of EUR 40–50 billion (Le Monde, 05/01/2012).

[36] Some would argue that nuclear never has been built without subsidies, either direct or indirect.

essential if nuclear is to compete with gas, and Watson and Falcan (2013: 24) predict that other low-carbon options are likely to be cheaper than nuclear in the UK. Lévêque (2013: 59) highlights the contextual factors (e.g. the role of the safety authority, benefits of series construction, and the cost of capital) and assumptions concerning the future (technological progress, costs of alternative generation options, discount rate, etc.) as crucial determinants of cost estimates.

Several governments are considering options for the support needed to attract investors. The construction of the Finnish Olkiluoto 3 reactor was made possible through an arrangement whereby the owners of the plant—primarily large energy-intensive industrial companies—were able to buy electricity at cost price. The USA is considering federal loan guarantees (Thomas, 2012: 15), while many call for the introduction of a carbon tax to increase the competitiveness of nuclear (Hultman and Koomey 2013: 69; see also Lévêque 2013: 58). In the UK, the government in practice abandoned in early 2014 its longstanding absolute 'no-subsidies' policy, by granting EDF a thirty-five-year price guarantee—at about double the current wholesale market price—for the electricity generated by the foreseen EDF-built nuclear plants. By contrast, in France—the country of 'dirigisme'—economic arguments seem to occupy an increasingly central position in nuclear debates. Since the first comprehensive independent analysis of nuclear economics in France just over a decade ago (Charpin et al., 2000), discussion on the costs of nuclear has been increasingly lively (e.g. Cour des Comptes, 2012; Lévêque, 2013).

16.6 Conclusions: prospects for nuclear energy in a post-Fukushima world

The 'renaissance' of nuclear power had in most Western countries either stalled or slowed down prior to the Fukushima accident, largely for economic and financial reasons, which were further aggravated by the onset of the financial crisis in 2008. The new generation III/III+ reactor designs have thus far largely failed to fulfil their promises, and EPR plants under construction in Finland and France have instead tarnished the image of nuclear as a low-cost power generation option.

The most promising markets are clearly in Asia—notably China and India—and Russia.[37] These countries, together with Korea, are also key potential vendors (Thomas, 2012), while the future of the Western nuclear industries seems more uncertain. In many ways, the situation in the Western world appears similar to that of the immediate aftermath of the first oil crisis in the mid 1970s, 'in a world of fast-changing competition and fickle demand growth', when the rapid expansion of US reactor orders in the first half of the 1970s was followed by a rapid decline of orders after 1974 (Hultman and Koomey, 2013: 64–5). A lot will depend on the climate policies, and the

---

[37] Including to a lesser extent the former Soviet bloc countries.

policy treatment of nuclear as a potential low-carbon generation option. Outside of Europe, energy security and independence have had greater traction as an argument in favour of nuclear, notwithstanding its inherent limitations in securing security of supply.

Thomas (2012) and Grimes and Nuttall (2010: 803) concur with the idea that nuclear has better prospects in the long than short or medium term. Grimes and Nuttall (2010: 803) predict that a second, larger phase of nuclear development would be required beyond 2030, in order to find resource-efficient solutions to combat climate change. Key future challenges include the need for new reactor designs, extension of the lifetime of gen III/III+ plants (provided that these will be constructed on a large scale in the upcoming years), improvements in regulation and inspection, high-level radioactive waste management, fuel availability, and local public opinion (Grimes and Nuttall, 2010). Moreover, in light of the repeated past failures of new technologies to fulfil their often unrealistic promises, the potential of new technologies such as fusion, generation IV, and small modular reactors should be viewed with caution. None of these technologies will be ready for large-scale deployment soon enough to help solve the current climate problems.

The history of the ups and downs of nuclear energy has shown the inventiveness of the industry and the broader nuclear 'establishment' in campaigning in favour of the technology. Thomas (2012: 15) hence envisages a potential revival and large-scale attempt to relaunch nuclear technology perhaps in a decade. The strategies of 'othering' (Walker, 2013), frequently employed by the industry to dispel safety concerns following major accidents,[38] are likely to be employed in these discursive battles, and will certainly intensify should the current situation in Fukushima worsen.

Finally, success in other areas of energy policy, in particular renewables and radioactive waste management, will be of vital importance for the future of nuclear energy. If policies in support of renewables and energy efficiency fail to advance, the appeal of nuclear may increase. The progress in radioactive waste policies in Finland, Sweden, and France will be followed closely by the international nuclear community. More generally, the future of the powerful French nuclear industry will influence policies, especially in Europe. Fukushima did not revolutionize French nuclear policy overnight, yet President Hollande's commitment to reducing the share of nuclear from 75 per cent to 50 per cent of electricity supply, and discussions of an eventual nuclear phase-out may signal potential changes to come (see e.g. Schneider, 2013).

---

[38] The Chernobyl accident was hence explained away by blaming not only Soviet technology (which was indeed also used in some Western countries, including Finland), but in particular the Soviet management style and serious negligence (Schreurs, 2013; Walker, 2013). The anti-nuclear movement in turn used Chernobyl as an example of the fatal risks of the technology, and pointed at the pervasive opacity of the governance of the 'nucleocracy' also in the Western countries.

▓ REFERENCES

Bickerstaff, K., Lorenzoni, I., Pidgeon, N.F., Poortinga, W., and Simmons, P. 2008. Reframing nuclear power in the UK energy debate: nuclear power, climate change mitigation and radioactive waste. *Public Understanding of Science* 17(2): 145–69.

Boccard, N. 2014. The cost of nuclear electricity: France after Fukushima. *Energy Policy* 66 (March): 450–61.

Bradford, P.A. 2013. How to close the US nuclear industry: Do nothing. *Bulletin of the Atomic Scientists* 69(2): 12–21.

Broderick, J., Anderson, K., Jones, C., and Watson, J. Eds. 2013. *A Review of Research Relevant to New Build Nuclear Power Plants in the UK*, London, UK: Friends of the Earth, England, Wales and Northern Ireland.

Cantor, R. and Hewlett, J. 1988. The economics of nuclear power: Further evidence on learning, economies of scale, and regulatory effects. *Resources and Energy* 10(4): 315–35.

Charpin J-M., Dessus, B., and Pellat, R. 2000. Etude Economique Prospective de la Filière Electrique Nucléaire, Rapport au Premier Ministre, La Documentation Française, Paris. <http://www.ladocumentationfrancaise.fr/rapports-publics/004001472/index.shtml>.

Chateauraynaud, F. and Torny, D. 1999. *Les sombres précurseurs. Une sociologie pragmatique de l'alerte et du risqué*. Paris: Editions de l'EHESS.

Cochran, T.B., Feiveson, H.A., Patterson, W., Pshakin, G., Ramana, M.V., Schneider, M., Suzuki, T. and von Hippel, F. Eds. Fast Breeder Reactor Programs: History and Status. Research Report 8, International Panel on Fissile Materials. February.

Conde, M. and Kallis, G. 2012. The global uranium rush and its African frontier. Effects, reactions and social movements in Namibia. *Global Environmental Change* 22(3): 596–610.

Corner, A., Venables, D., Spence, A., Poortinga, W., Demski, C., and Pidgeon, N. 2011. Nuclear power, climate change and energy security: exploring British public attitudes. *Energy Policy* 39 (9): 4823–33.

Cour des comptes. 2012. Les coûts de la filière électronucléaire. Rapport public thématique. Paris: Cour des comptes. Janvier 2012.

Daoud, A. and Elam, M. 2012. Identifying remaining socio-technical challenges at the national level: Sweden. InSOTEC Working Paper WP 1—MS 11, 1 May 2012.

Dorfman, P., Fucic, A. and Thomas, S. 2013. Late lessons from Chernobyl, early warnings from Fukushima. In: *Late Lssons from Early Warnings: Science, Precaution, Innovation*. EEA Report 1/2003. Summary. European Environment Agency: Copenhagen. ISSN 1725–9177. p. 28.

Elliott, D. 2013. *Fukushima: Impacts and Implications*. Houndmills, Basingstoke: Palgrave Macmillan.

Euratom. 2013. Benefits and limitations of nuclear fission for a low-carbon economy. 2012 Interdisciplinary Study. Defining priorities for Euratom fission research & training (Horizon 2020). Synthesis report. February. Luxembourg: European Union.

Froggatt, A., Mitchell, C., and Shunsuke, M. 2012. Reset or Restart? The Impact of Fukushima on the Japanese and German Energy Sectors. Chatham House (the Royal Institute of International Affairs).

Goldberg, S. and Rosner, R. 2011. *Nuclear Reactors: Generation to Generation*. American Academy of Arts and Sciences: Cambridge, MA.

Goldemberg, J. 2011. Have Rising Costs and Increased Risks Made Nuclear Energy a Poor Choice? Science and Development Network, 4 October 2011. <http://oilprice.com/Alternative-Energy/Nuclear-Power/Have-Rising-Costs-and-Increased-Risks-Made-Nuclear-Energy-a-Poor-Choice.html>. (Accessed April 2015).

Gowing, M. and Arnold, L. 1974. *Independence and Deterrence: Britain and Atomic Energy, 1945–1952*. Vol.1 (*Policy making*) and Vol. 2 (*Policy Execution*). London: Macmillan.

Greenpeace. 2012. The lessons from Fukushima. February <http://www.greenpeace.org/international/Global/international/publications/nuclear/2012/Fukushima/Lessons-from-Fukushima.pdf>. (Accessed April 2015).

Grimes, R.W. and Nuttall, W.J. 2010. *Generating the Option of a Two-Stage Nuclear Renaissance*. Science. Washington: American Association for the Advancement of Science.

Grubler, A. 2010. The costs of the French nuclear scale-up: a case of negative learning by doing *Energy Policy* 38(9): 5174–88.

Haas, R. and Hiesl, A. 2013. The economics of nuclear power. Paper presented at the 18th REFORM group meeting, Salzburg, 26–30 August 2013.

Hecht, G. 2009. *The Radiance of France: Nuclear Power and National Identity after World War II*. Cambridge, MA: MIT Press.

Herring, H. 2003. Energy utopianism and the rise of the anti-nuclear power movement in the UK. PhD thesis, Open University, London.

Hultman, N. and Koomey, J. 2013. Three Mile Island: The driver of US nuclear power's decline? *Bulletin of the Atomic Scientists* 69(3): 63–70.

IAEA. 2010. Energy, Electricity, and Nuclear Power Estimates for the Period up to 2050. Vienna: International Atomic Energy Agency. <http://www-pub.iaea.org/books/iaeabooks/8580/Energy-Electricity-and-Nuclear-Power-Estimates-for-the-Period-up-to-2050-2010-Edition>. (Accessed April 2015).

IAEA. 2012. Energy, Electricity, and Nuclear Power Estimates for the Period up to 2050. Vienna: International Atomic Energy Agency. <http://www-pub.iaea.org/books/iaeabooks/10358/Energy-Electricity-and-Nuclear-Power-Estimates-for-the-Period-up-to-2050-2012-Edition>. (Accessed April 2015).

IEA. 2012. *World Energy Outlook 2012*. International Energy Agency: Paris.

IEA. 2013. *World Energy Outlook Special Report 2013: Southeast Asia Energy Outlook*. September. International Energy Agency: Paris.

IOP. 2013. Institute of Physics, Nuclear Industry Group newsletter no. 3, September 2013.

Ipsos. 2011. Global Citizen Reaction to the Fukushima Nuclear Plant Disaster. <http://www.ipsos-mori.com/Assets/Docs/Polls/ipsos-global-advisor-nuclear-power-june-2011.pdf>. (Accessed April 2015).

Jewell, J. 2011. Ready for nuclear energy?: An assessment of capacities and motivations for launching new national nuclear power programs. *Energy Policy* 39(3): 1041–55.

Kessides, J.N. 2012. The future of the nuclear industry reconsidered: Risks, uncertainties, and continued promise. *Energy Policy* 48(February): 185–208.

Kim, S.H. and J.A. Edmonds. 2007. The Challenges and Potential of Nuclear Energy for Addressing Climate Change. Available from: <http://www.pnl.gov/main/publications/external/technical_reports/PNNL-17037.pdf>. (Accesed April 2015).

Kuznetsov, V. 2008. Progress Toward Deployment and Opportunities for Advanced Small and Medium Sized Reactors. International Conference "Today's Safe Nuclear

Power-Management of the Plants and Nuclear Waste" 9–10 October, Tallinn, Estonia. <http://www.ben.ee/public/Tuumakonverentsi%20ettekanded%202008/V.Kuznetsov_091008. pdf>. (Accessed April 2015).

Lammi, H. 2009. Social dynamics behind the changes in the NGO anti-nuclear campaign during 1993–2002. In: Kojo, M. and Litmanen, T. (Eds) *The Renewal of Nuclear Power in Finland.* Basingstoke: Palgrave Macmillan: pp. 69–87.

Le Renard, C., Jobert, A., and Lehtonen, M. 2013. The diverging trajectories of Fast Breeder Reactor development in France and the UK (1950–1990): a tentative comparison. Paper presented at the 6th Tensions of Europe plenary conference, 'Democracy and Technology: Europe in Tension from the 19th to the 21st century'. Paris, 19–21 September 2013.

Lehtonen, M. 2010. Deliberative decision-making on radioactive waste management in Finland, France and the UK: Influence of mixed forms of deliberation in the macro discursive context. *Journal of Integrative Environmental Sciences* 7(3): 175–96.

Lévêque, F. 2013. *Nucléaire On/Off. Analyse économique d'un pari.* Paris: Dunod.

Lovins, A. B. 2013. The economics of a US civilian nuclear phase-out. Bulletin of the Atomic Scientists 69(2): 44–65.

Lyman, E. S. 2011. Surviving the one-two nuclear punch: Assessing risk and policy in a post-Fukushima world. *Bulletin of the Atomic Scientists* 67(5): 47–54.

MacKerron, G. 2012. A nuclear renaissance for Europe? Paper presented at the British Institute for Energy Economics (BIEE) Conference, Oxford, 19 September 2012.

Mecklin, J. 2013. Introduction: US nuclear exit? *Bulletin of the Atomic Scientists* 69(2): 9–11.

Mez, L. 2012. Germany's merger of climate change and energy policy. *Bulletin of the Atomic Scientists* 68(6): 22–9.

Nesheiwata, J. and Crossa, J.S. 2013. Japan's post-Fukushima reconstruction: A case study for implementation of sustainable energy technologies. *Energy Policy* 60(September): 509–19.

Nuttall, W.J. 2009. Nuclear Energy in the Enlarged European Union. University of Cambridge Electricity Policy Research Group (EPRG) Working Paper 0904.

OCDE-AEN. 2010. L'opinion publique et l'énergie nucléaire. AEN 6860. Paris, OCDE-AEN. L'Agence de l'OCDE pour l'énergie nucléaire.

Piore, A. 2011. Planning for the black swan. *Scientific American* 304(6): 48–53.

Ramana, M. V. 2013. Nuclear policy responses to Fukushima: Exit, voice, and loyalty. *Bulletin of the Atomic Scientists* 69(2): 66–76.

Schneider, M. 2013. Nuclear Power and the French energy transition: It's the economics, stupid! *Bulletin of the Atomic Scientists* 69(1): 18–26.

Schneider, M. and Froggatt, A., with Hosokawa, K., Thomas, S., Yamaguchi, Y., and Hazemann, J. 2013. World Nuclear Industry Status Report. Paris and London: Mycle Schneider Consulting. <http://www.worldnuclearreport.org/World-Nuclear-Report-2013.html>. (Accessed April 2015).

Schneider, M. and Froggatt, A., with Ayukawa, Y., Burnie, S., Piria, R., Thomas, S., and Hazemann, J. 2014. World Nuclear Industry Status Report. Paris, London and Washington, D.C.: Mycle Schneider Consulting. <http://www.worldnuclearreport.org/-2014-.html>. (Accessed April 2015).

Schreurs, M. A. 2012. The politics of phase-out. *Bulletin of the Atomic Scientists* 68(6): 30–41.

Schreurs, M. A. 2013. Orchestrating a low-carbon energy revolution without nuclear: Germany's response to the Fukushima nuclear crisis. *Theoretical Inquiries in Law* 14(1): 83–108.

Schwägerl, C. 2011. Germany's Unlikely Champion of a Radical Green Energy Path, Yale Environment 360, University of Yale, School of Forestry and Environmental Studies. <http://e360.yale.edu/feature/germanys_unlikely_champion_of_a_radical_green_energy_path/2401/>. (Accessed April 2015).

Siegrist, M. and Visschers, M. 2013. Acceptance of nuclear power: The Fukushima effect. *Energy Policy* 59(August): 112–19.

Srinivasan, T. S. and Rethinaraj, G. 2013. Fukushima and thereafter: Reassessment of risks of nuclear power. *Energy Policy* 52 (January): 726–36.

Stankiewicz, P. 2013. The role of experts in opening controversial energy projects to stakeholders' participation. Presentation at the 8th International Interpretive Policy Analysis Conference (IPA), 'Societies in Conflict: Experts, Publics and Democracy', Vienna, 3–5 July 2013.

Strauss, H. 2010. Involving the Finnish public in nuclear facility licensing: participatory democracy and industrial bias. *Journal of Integrative Environmental Sciences* 7(3): 211–28.

Sundqvist, G. and Elam, M. 2010. Public involvement designed to circumvent public concern? The 'participatory turn' in European nuclear activities. *Risk, Hazards & Crisis in Public Policy* 1(4): 198–224.

Szarka, J. 2013. From exception to norm—and back again? France, the nuclear revival and the post-Fukushima landscape. *Environmental Politics* 22(4): 646–63.

Thomas, S. 2005. The economics of nuclear power: analysis of recent Studies. Public Services International Research Unit (PSIRU), University of Greenwich. July.

Thomas, S. 2012. What will the Fukushima disaster change? *Energy Policy* 45(June): 12–17.

Tickell, O. 2012. Thorium: Not 'green', not 'viable', and not likely. Policy Briefing, April/May 2012. <http://www.nuclearpledge.com/reports/thorium_briefing_2012.pdf>. (Accessed April 2015).

Twena, M. 2006. Nuclear energy: Rise, fall and resurrection. CICERO Working Paper 2006:01. CICERO—Center for International Climate and Environmental Research, Oslo. <http://www.cicero.uio.no>.

Walker, G. 2013. Turning point or one way street? The reaction to Fukushima in the UK. Presentation at the 8th International Interpretive Policy Analysis Conference (IPA), 'Societies in Conflict: Experts, Publics and Democracy', Vienna, 3–5 July 2013.

Watson, J. and Falcan, J. 2013. Economics of nuclear power. In: John Broderick, Kevin Anderson, Christopher Jones, Jim Watson (Eds) *A Review of Research Relevant to New Build Nuclear Power Plants in the UK*. London, UK: Friends of the Earth, England, Wales and Northern Ireland: pp. 23–40.

Watson, J. and Scott, A. 2009. New nuclear power in the UK: a strategy for energy security? *Energy Policy* 37(12): 5094–104.

Winskel, M. 2002. Autonomy's end: Nuclear power and the privatization of the British electricity supply industry. *Social Studies of Science* 32(3): 439–67.

WNA. 2013a. *Nuclear Power in China (October 2013)* [Online]. World Nuclear Association. Available: <http://world-nuclear.org/info/Country-Profiles/Countries-A-F/China–Nuclear-Power/ - .UmZNDxYwaLI> (Accessed 22 October 2013).

WNA. 2013b. *Nuclear Power in India (October 2013)* [Online]. World Nuclear Association. Available: <http://world-nuclear.org/info/Country-Profiles/Countries-G-N/India/ - .Um7JNha6GLI> (Accessed 28 October 2013).

WNA. 2013c. *Nuclear Power in Russia (September 2013)* [Online]. World Nuclear Association. Available: <http://world-nuclear.org/info/Country-Profiles/Countries-O-S/Russia–Nuclear-Power/ - .Um6vvRa6GLI> (Accessed 28 October 2013).

17 Bioenergy resources

Raphael Slade and Ausilio Bauen

17.1 Introduction

Biomass is the oldest fuel used by humankind and was the main source of energy for cooking and keeping warm from the dawn of civilization to the industrial revolution. Biomass is defined to include any non-fossilized organic material of plant and animal origin, and most types of biomass can, at least in principle, be used to provide energy services. The most important sources, however, are materials derived from forestry and agriculture, along with industrial and municipal residues and wastes. Specially cultivated energy crops such as coppiced wood and perennial grasses may also play an important role in the future.

Until the eighteenth century, humans were almost completely reliant on wood and charcoal for all of their energy needs. When coal use began in earnest in the early 1800s (and later, oil and gas) the use of biomass declined. Fossil fuels were cheaper, higher energy density, easier to handle, and better able to support rapid industrialization and the demands of a growing population. Yet despite the considerable advantages of fossil fuels, biomass continued to be an important energy resource. Currently biomass accounts for around 50 EJ (~10 per cent) of global primary energy supply. The majority of which (~8 per cent, ~39 EJ) is used by the world's poorest people to provide rudimentary energy services such as cooking and heating[1] (IEA, 2010; IPCC, 2011). The remaining ~2 per cent (~11–12 EJ) includes the provision of high-quality energy services—heat, power, and transport fuels—enjoyed by affluent countries and delivered using modern and efficient conversion technologies.

Over the last three decades there has been resurgent interest in these modern applications of bioenergy. This interest has been driven by concerns about energy security, increasing prices of fossil fuels, and climate change. All of which can be addressed, at least in part, by increasing the proportion of bioenergy in the global energy

---

[1] Globally, it is estimated that around 2.6 billion people are still reliant on traditional uses of biomass and burn wood, straw, charcoal, and dung to provide basic energy services such as cooking and heating (REN21, 2010). This use is predominantly restricted to rural areas of developing countries, and it is associated with poverty and deforestation (Ludwig, et al., 2003; Hall, et al., 1983). The quantity of traditional biomass consumption is known with far less certainty than commercially traded energy sources and may be systematically underestimated in government statistics because production and use is largely informal (IPCC, 2011: 9). It is estimated that the least developed countries still rely on biomass for over 90 per cent of their energy needs (IEA, 2009a, 2009b).

mix. Energy scenarios, such as those developed by the International Energy Agency (IEA) and the Intergovernmental Panel on Climate Change (IPCC), also indicate that bioenergy could make a major contribution to a future low-carbon energy system (IPCC, 2007; IEA, 2010; IEA, 2012b). Many governments (including the G8 plus five[2] and all European member states) have responded by giving bioenergy a role in their energy strategies and introducing policies to increase deployment (Faaij, 2006; GBEP, 2007).

Sustained interest has also helped biomass conversion processes develop and improve. Technologies now on the market can more easily accommodate the varied physical and chemical composition of biomass feedstocks and deliver automated and reliable service. Technologies at the research, development, and demonstration stage also promise more efficient and cleaner conversion into an ever broader range of products. The aspiration is that modern bioenergy technologies should be able to provide energy services at comparable levels of convenience and cost as fossil fuels, and with greatly improved environmental performance.

Yet as efforts to accelerate the introduction of bioenergy have gathered pace, the prospect of mobilizing the large quantities of biomass required has become increasingly controversial. Biomass availability tends to be intertwined with activity in other major economic sectors—agriculture, forestry, food processing, paper and pulp, building materials, and so on—and as feedstocks are diverted from established markets some impact on these sectors is almost inevitable (Faaij, 2006). The way in which land resources are used may also be changed, and many commentators foresee growing land and resource conflicts between bioenergy and food supply, water use, and biodiversity conservation. Their fear is that the benefits offered by increased bioenergy production could be rapidly outweighed by the costs, and that increased production could exacerbate existing environmental problems. Sources of concern include both direct impacts, such as the effect of domestic stoves on urban air quality, and indirect impacts, such as land use change mediated through changing market prices (Searchinger, et al., 2008; Eide, 2008; Creutzig, et al., 2012; Agostini, et al., 2013).

Ultimately, the contribution that bioenergy makes to the global energy mix will depend not only on the efficacy of the conversion technology, but also the availability of biomass and the social acceptability of large-scale adoption. In this chapter we explore each of these aspects of modern bioenergy deployment. We start by providing an overview of conversion pathways, and examine how biomass is currently being used to provide heat, power, and transport fuels. We then explore the range of global biomass availability estimates and consider some of their merits and limitations in helping us form a view on what the resource might be. Finally we address the challenge of ensuring biomass supply is sustainable and consider how this might constrain future expansion.

---

[2] The G8 countries are Canada, France, Germany, Italy, Japan, Russia, the United Kingdom, and the United States. The plus five are the five leading emerging economies: Brazil, China, India, Mexico, and South Africa.

17.2 Competing options for providing energy services from biomass

Biomass resources include an incredibly diverse range of feedstocks including both wet and dry waste materials (e.g. sewage sludge and municipal solid waste). Generally, drier and un-contaminated feedstocks are easier and cheaper to convert into energy carriers than wet or contaminated ones. This difference is reflected in their relative price and consequently a balance must be struck between the cost of the conversion process and the quality and price of the feedstock. It is important to note that no single conversion technology can use biomass indiscriminately in all its forms. The main biomass energy conversion pathways are shown in Figure 17.1.

Thermo-chemical pathways preferentially use dry feedstocks and include combustion, gasification, and pyrolysis. Combustion involves the complete oxidation of biomass to provide heat. This may be used directly, or may be used to raise steam and produce electricity. Gasification involves the partial oxidation of the biomass at high temperatures (>500 °C) and yields a mixture of carbon monoxide and hydrogen (syngas), along with some methane, carbon dioxide, water, and small amounts of nitrogen and heavier hydrocarbons (Hamelinck et al., 2004). The quality of the gas

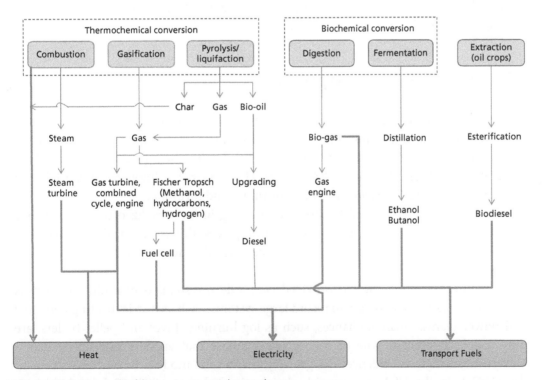

Figure 17.1 The major bioenergy conversion pathways
Source: Adapted from Turkenburg (2000).

depends on the temperature of the gasification process: a higher temperature process will yield more syngas with fewer heavy hydrocarbons. Syngas may be converted into a wide range of fuels and chemicals; alternatively, it can be used to produce electricity, or cleaned and injected into the gas distribution network. Pyrolysis involves heating biomass in the absence of oxygen at temperatures up to 500 °C and produces an energy-dense bio-oil along with some gas and char. This bio-oil is corrosive and acidic, but could in principle be upgraded for use as a transport fuel. Bio-oil from pyrolysis most often receives attention as a pre-treatment and densification step that could make the long distance transport of biomass more economic (Faaij, 2006).

Biochemical conversion pathways use microorganisms to convert biomass into methane or simple alcohols, usually in combination with some mechanical or chemical pre-treatment step. Anaerobic digestion is a well-established technology and is suited to the conversion of homogenous wet wastes that contain a high proportion of starches and fats—for example food waste—to methane. Fermenting sugars and starches to alcohols using yeast is also a fully mature technology. Woody biomass can potentially be used as a feedstock for both anaerobic digestion and fermentation processes, but requires an additional pre-treatment step in order to release the sugars that these feedstocks contain; technologies adopting this approach are being demonstrated but are not yet fully mature.

Lastly, plant oils may be extracted mechanically, reacted with alcohols or treated with hydrogen and used as substitute for diesel and other fuels.

17.3 **Biomass for heat, power**

Biomass can be used to generate heat at all scales, ranging from a single household to a large industrial complex, and using all types of biomass. Systems are fully commercial, and in many cases they are also cost competitive with their fossil fuel alternatives, particularly in off-grid locations (IEA, 2009a). The principal technologies used to deliver modern biomass heat are *combustion, gasification,* and *digestion.*

Biomass is extensively used for domestic heating in industrialized countries. While some biomass continues to be used in open fireplaces, this is principally for aesthetic reasons, and where biomass is the main source of heating automated boilers for logs or pellets are widely available.

Estimating how much heat is provided via modern systems is difficult, however, as government statistics tend only to record large systems, such as district heating networks and power plants. Small appliances, such as log burning stoves and pellet boilers, are only visible at point of sale or if they use biomass that appears in retail statistics. Nevertheless, there is evidence of a growing market for modern boilers and stoves in the OECD. In the USA, for example, it is estimated that ~800,000 households use wood as their primary heat source, whereas in the top nine bioenergy-using European countries[3] the number of domestic stoves and boilers is estimated to exceed 1.3 million and accounts for the majority of solid biomass sold in Europe (AEBIOM, 2010; REN21, 2010). Bioenergy demand in the residential sector is expected to also double in OECD countries from ~3 EJ in 2009 to 6 EJ in 2050, driven predominantly by space heating demand (IEA, 2012a).

The northern European countries—Sweden, Finland, Denmark, and Germany—lead in the deployment of large-scale biomass heating systems. Much of this heat is generated in large combined heat and power (CHP) facilities and delivered via extensive district heating networks. Sweden provides a good example of a country that has successfully increased bioenergy provision, and in 2013, over 70 per cent of total district heating fuel demand was provided from biomass (Bayar, 2013). The Danish government has similar ambitions, and to encourage greater biomass use and connection to district heating networks has banned fossil-fuel fired boilers in new buildings from 2013 (Leidreiter, 2013).

Large combustion plants offer a number of advantages. They can deliver greater thermal efficiency than domestic boilers and this may result in a reduced capital cost per unit of heat delivered. They may also be able to use lower quality or contaminated biomass, such as waste derived fuels. These fuels are cheaper than alternatives but usually require flue gas cleaning technologies that are only economically viable at a larger scale (Dornburg and Faaij, 2001). CHP systems based on biomass combustion can be very efficient (60–90 per cent) (see Box 17.1), although for maximum efficiency they require large and stable heat loads and they are therefore most economical in colder climates where district heating is installed, or where there is an industrial heat demand.

Biomass is also used to provide process heat to industry, most frequently in the agricultural and forestry product processing industries where biomass is a byproduct of the main process and can be used as fuel. On larger sites CHP is widely adopted, for example co-production of steam and electricity from sugar cane residue (bagasse) is common practice in Brazilian sugar and ethanol mills. Industrial bioenergy demand appears set to increase, and the IEA envisages that it will be one of the fastest growing sectors, potentially increasing from ~ 8 EJ in 2009 to ~22 EJ in 2050 (around 15 per cent of industrial final energy demand) (IEA, 2012a).

Whereas statistics for biomass heat provision are somewhat sketchy, far better data exists for global power generation. In 2011 it is estimated that global power production from biomass and wastes was in the region of ~ 1.3 EJ (355 TWh), roughly equivalent to the total annual electricity generation of the UK (EIA, 2014). Although this represents only a small fraction of global electricity supply, generation capacity has grown rapidly over the last ten years, particularly in Europe and Asia where policy incentives have been favourable and fossil fuel prices comparatively high, see Figure 17.2.

---

[3] Austria, Belgium, Denmark, Finland, France, Germany, Italy, Spain, Sweden.

BOX 17.1. CHP IN SWEDEN—THE IGELSTA CO-GENERATION PLANT

One of the world's largest and most efficient biomass CHP plants is the Igelsta plant in Sodertalje, near Stockholm, Sweden (Söderenergi, 2010). Commissioned in March 2010, this plant produces 200 MW of heat and 85 MW of electricity, sufficient to heat ~50,000 private houses and provide power for 100,000 homes.

The combustion technology used at Igelsta is a sophisticated circulating fluidized bed and incorporates state-of-the-art flue gas cleaning. The heat produced is used to raise steam: used first for electricity production (85 MW) and then to deliver low grade heat to the local district heating network (140 MW). By condensing the steam in the flue gas the plant is able to deliver an additional 60 MW heat to the heating network, and this gives the combined system an overall efficiency in excess of 90 per cent.

The fuel used in the Igelsta plant is a combination of wood chip from forest residues (75 per cent) and recovered fuels from waste (25 per cent). When running at full capacity the plant uses ~17,000 tonnes of biomass per week. Wood chips from forestry operations are transported by road, rail, and boat from all over Sweden and neighbouring Baltic countries. The recovered fuels include scrap paper, wood, and plastic that is not suitable for recycling and is sourced from offices, shops, and industries in the Stockholm region. Pelletized waste from similar sources is imported by boat from Germany and the Netherlands, around 100,000 tonnes of waste wood is also imported from Norway, Belgium, and the UK. The municipality that owns the plant considers that such fuel flexibility will be critical as competition for bio-fuels increases.

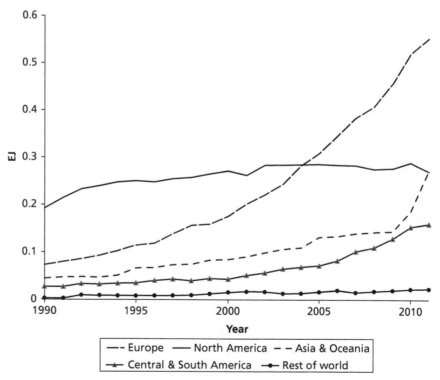

Figure 17.2 Global net electricity generation from biomass and waste
Source: Data from (EIA, 2014).

The vast majority (~90 per cent) (REN 21, 2013) of biomass electricity is generated from burning solid biomass and includes the following applications:

- co-firing wood pellets in coal power stations;
- combustion-based CHP plants—for countries that possess district heating systems and industries with available process residues (e.g. pulp and paper);
- municipal solid waste (MSW) incineration; and,
- stand-alone power plants where large amounts of agricultural residues are available (e.g. the UK has a 38 MW dedicated straw-burning power station at Ely in East Anglia).

All these applications depend on locally available biomass, with the exception of internationally sourced wood pellets.[4] Wood pellets have emerged over the last ten years as a commodity energy vector and are now traded internationally, albeit in far smaller quantities than coal, oil, and gas. Global pellet production has grown rapidly from ~10 Mt in 2007 to ~18 Mt in 2011, and Europe is the largest consumer (estimated at 12.3 million tonnes in 2011). The majority of demand for pellets comes from the Netherlands, Belgium, Denmark, Sweden, and the UK and is a direct result of aggressive biomass co-firing targets (Verhoest and Ryckmans, 2012; Joudrey et al., 2012).

The remaining ~10 per cent of global bio-electricity is generated from biogas that is either captured from landfill sites or produced by anaerobic digestion. Europe is the leading producer of biogas and the market growth has been driven by both renewable energy targets and increasingly strict waste handling legislation (see Box 17.2). In 2011, total European biogas production was ~0.42 EJ, around just over half of which (56 per cent) was produced by anaerobic digestion from agricultural residues, putrescible waste, and dedicated crops such as maize silage. The remainder was produced from landfill sites and water treatment works (DENA, 2013). Biogas systems are also increasingly common in China where large numbers of small-scale systems have been installed for rural electrification (REN 21, 2013).

17.4 **Transport fuels from biomass**

Biofuels make a modest contribution to global transport fuel supply (~3 per cent, 3 EJ) and production is dominated by two liquid fuels: ethanol, which can substitute for gasoline, and biodiesel (produced from vegetable oil), which can substitute for diesel. Many other biofuel options exist, for example biogas is a viable fuel for fleet vehicles such as busses and bio-kerosene is attracting increasing interest for aviation, but the use of these alternatives is negligible on a global scale.

---

[4] Some very large CHP facilities, such as Igelsta in Sweden, also source biomass internationally.

BOX 17.2. BIOGAS IN GERMANY

Germany is a technology leader in anaerobic digestion and generates over half of all the biogas produced in the EU. The German market has grown exponentially over the last eighteen years supported by favourable policy incentives and generous feed in tariffs (see Figure 17.3). By the end of 2012, there were over 7500 biogas plants (mostly biogas CHP facilities) operating in Germany and supplying around 83 PJ (23 TWh) of electricity. Gas clean-up and injection into the national gas grid started in 2009 and has become an increasingly attractive option for large plants; by July 2013, 116 biogas plants had adopted this technology (DENA, 2013).

Germany also hosts the world's largest biogas plant at Güstrow, Western Pomerania. This plant came online in 2009 and can digest 450,000 tons of biomass per year including maize silage, grain, and crop silage cultivated on an area of ~10,000 ha. The plant's output is equivalent to roughly 0.58 PJ (160 GWh) of electricity and 0.65 PJ (180 GWh) of heat. This is sufficient to cover the energy needs of a small town of ~50,000 households (Nawaro, 2013; EnviTech, 2008–9).

The use of maize silage as an energy crop has become increasingly controversial as it is perceived to compete with food production. Legislative changes introduced in 2012 sought to address this concern by limiting the use of maize to a maximum of 60 per cent of input biomass, and mandating for a minimum level of heat recovery. This change in the subsidy regime has caused a sharp decline in the domestic industry and many technology developers are now focusing their attention on the export market (RENI, 2013; DENA, 2013).

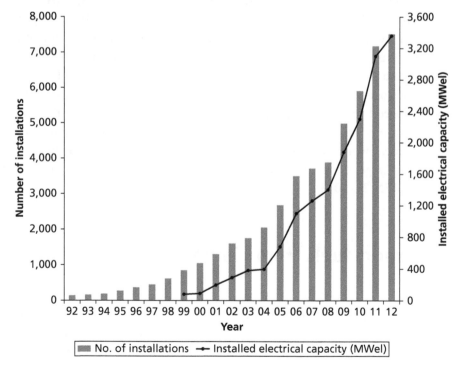

Figure 17.3 Biogas production in Germany 1992–2013
Source: Data from (DENA, 2013).

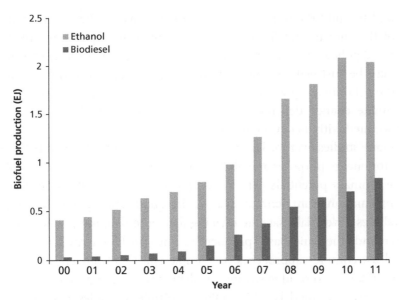

Figure 17.4 Global production of bioethanol and biodiesel 2000–11
Source: Data from (EIA, 2014).

In 2011, bioethanol accounted for over two thirds (71 per cent, 2 EJ) of global biofuel supply, the majority of which was produced from maize in the USA (1.2 EJ) and sugar-cane in Brazil (0.5 EJ). The global production of biodiesel is smaller but still significant (29 per cent, 0.8 EJ). The most important global producers were Argentina, Brazil, and the USA (0.32 EJ from soy oil), Germany and France (0.18 EJ from rape seed oil), and Indonesia and Thailand (0.06 EJ from palm oil)[5] (EIA, 2014).

Biofuel production has grown rapidly in the last ten years (see Figure 17.4), and this has resulted in the diversion of large quantities of commodity food crops to energy production. The scale of the change has triggered a backlash against biofuel policies and mandates and this may limit future production growth from these feedstocks. At the time of writing, global use of cereals for biofuels had decreased by ~2 per cent and virgin vegetable oils by ~10 per cent over 2012 levels, and this decline has been largely attributed to policy changes brought about by sustainability concerns (F. O. Lichts, 2013).

17.5 **The future availability of biomass**

Expanding the use of biomass to make a major contribution to the global energy mix would require significant and sustained investment, both to develop sustainable sources

---

[5] Malaysia is one of the largest producers and exporters of palm oil; however, due to high palm oil prices, subsidized petroleum, and little domestic demand the majority of Malaysia's biodiesel plants were idle in 2011 (Wahab, 2012).

of supply and to build the infrastructure required to use it effectively. In this context, estimates of the current and future biomass resource underpin many of the strategic investment and policy decisions that must be made. Investments in new technology, for example, may be justified on the basis that a large and accessible resource exists. Similarly, the prominence given to biomass in international negotiations as a means to mitigate climate change depends on both a quantification of the resource and the impacts associated with its development.

A great many studies over the last twenty years have sought to quantify the availability of biomass for energy purposes at global, regional, and sub-regional scales. Models used to calculate biomass potentials vary in complexity and sophistication, but all aim to integrate information from sources such as the Food and Agriculture Organization's (FAO) databases, field trials, satellite imaging data, and demand predictions for energy, food, timber, and other land-based products, to elucidate bioenergy's future role (see Box 17.3). The least complex approaches use simple rules and judgement to estimate the future share of land and residue streams available for bioenergy. The most complex use integrated assessment models which allow multiple variables and trade-offs to be analysed. There is good agreement, however, about the modelling parameters that are most important. These are: the availability (and productivity) of land for energy crops and food, and the accessibility of residues and wastes from existing and anticipated economic activity. Land availability estimates are strongly influenced by assumptions about how much land should be set aside for nature conservation, and how much will be needed to feed a growing population. Anticipated dietary changes are also important as a meat-rich diet requires far more land per person than is needed to support a vegetarian diet. Land productivity estimates are strongly influenced by technology scenarios and assumptions about how fast crop yields might be increased. Particularly important is the potential to increase crop yields and close the gap between optimal yields and those achieved by farmers when faced with environmental constraints such as water and nutrient scarcity, soil degradation, and climate change (Berndes et al., 2003; Lysen et al., 2008; Thrän et al., 2010).

Modelling results are most often discussed in terms of a hierarchy of potentials: theoretical > technical/geographic > economic > realistic/implementable, although these terms are not always used consistently. A theoretical potential estimate, for example, might be made by assuming that all net primary productivity (NPP) not needed for food could be available for bioenergy purposes. This assumption would lead to a very large and abstract number because it would ignore all competing land uses and socio-economic constraints. At the other end of the spectrum, an economic potential would constrain the useable quantity of biomass to the amount that could be produced at a specific price. This would lead to an inherently more subjective and smaller number, but one that may be far more appropriate for informing policy decisions.

The most important potential sources of biomass and the range of technical potentials found in the academic literature are energy crops (22–1272 EJ), agricultural residues (10–66 EJ), forestry residues (3–35 EJ), wastes (12–120 EJ), and forestry (60–230 EJ),

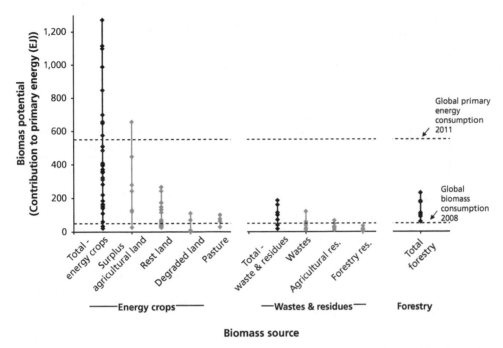

Figure 17.5 Estimates for the contribution of energy crops, wastes, and forest biomass to future energy supply
Source: adapted from (Slade et al., 2011).

summarized in Figure 17.5. Not all studies include all these categories in their analysis, and the broad range of estimates reflects the fact that some of the studies aim to test the boundaries of what might be physically possible rather than explore the boundaries of what might be socially acceptable or environmentally responsible. In particular, biomass extraction from forests is not considered by many authors because of concern about the potential impacts on biodiversity and carbon stocks. By way of comparison, the total human appropriation of net terrestrial primary production (including the entirety of global agriculture and commercial forestry) is around 320 EJ, of which 220 EJ is consumed and 100 EJ discarded as residues or otherwise destroyed during harvest. This is considerably less than current global primary energy supply (~500 EJ).

As the proportion of energy supplied from biomass in future global energy scenarios increases, the modelling assumptions required to make sufficient biomass available become increasingly demanding. The most important combinations of assumptions used to calculate estimates of the future technical biomass potential are summarized in Figure 17.6.

Estimates up to ~100 EJ (around 1/5th of current global primary energy supply) assume that there is limited land available for energy crops. This assumption is driven by scenarios in which there is a high demand for food, limited productivity gains in food production, and limited expansion of land under agriculture. Diets are assumed to evolve along the existing trend of increasing meat consumption. The contribution

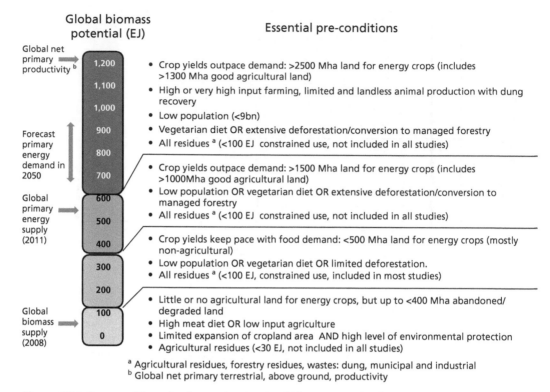

Figure 17.6 illustration content:

Global biomass potential (EJ)

Essential pre-conditions

Global net primary productivity[b] — 1,200

- Crop yields outpace demand: >2500 Mha land for energy crops (includes >1300 Mha good agricultural land)
- High or very high input farming, limited and landless animal production with dung recovery
- Low population (<9bn)
- Vegetarian diet OR extensive deforestation/conversion to managed forestry
- All residues [a] (<100 EJ constrained use, not included in all studies)

1,100

1,000

Forecast primary energy demand in 2050 — 900

800

700

- Crop yields outpace demand: >1500 Mha land for energy crops (includes >1000Mha good agricultural land)
- Low population OR vegetarian diet OR extensive deforestation/conversion to managed forestry
- All residues [a] (<100 EJ constrained use, not included in all studies)

Global primary energy supply (2011) — 600

500

400

- Crop yields keep pace with food demand: <500 Mha land for energy crops (mostly non-agricultural)
- Low population OR vegetarian diet OR limited deforestation.
- All residues [a] (<100 EJ, constrained use, included in most studies)

300

200

Global biomass supply (2008) — 100

0

- Little or no agricultural land for energy crops, but up to <400 Mha abandoned/degraded land
- High meat diet OR low input agriculture
- Limited expansion of cropland area AND high level of environmental protection
- Agricultural residues (<30 EJ, not included in all studies)

[a] Agricultural residues, forestry residues, wastes: dung, municipal and industrial
[b] Global net primary terrestrial, above ground, productivity

Figure 17.6 Pre-conditions for increasing levels of biomass production
Source: adapted from (Slade et al., 2014).

from energy crops (8–71 EJ, ~140-400 Mha) predominantly comes from agricultural land identified as abandoned, degraded or deforested, and from limited expansion of energy crops onto pasture. The contribution from wastes and residues is considered in only a few studies, but where included the net contribution is in the range 17–30 EJ. Most studies in this range exclude biomass extraction from non-commercial forestry.

Estimates falling within the range 100–300 EJ (roughly half current global primary energy supply at the top end), all assume that increasing food crop yields keep pace with population growth and the trend of increasing meat consumption. Limited good quality agricultural land is made available for energy crop production, but these studies identify areas of natural grassland, marginal, degraded, and deforested land ranging from twice to ten times the size of France (100–500 Mha) yielding 10–20 odt. ha^{-1}. In scenarios where demand for food and materials is high, achieving biomass potentials in this range implies a decrease in the global forested area (up to 25 per cent), or replacing mature forest with younger, more rapidly growing, forest. The majority of estimates in this range also rely on a larger contribution from residues and wastes (60–120 EJ). This is partly achieved by including a greater number of waste and residue categories in the analysis, and partly by adopting more ambitious assumptions on recoverability.

BOX 17.3. CALCULATING BIOMASS POTENTIALS

Estimating the potential of energy crops

The future availability of energy crops depends on the availability and productivity of land. Two broad approaches to modelling future land availability can be distinguished: *availability factors* and *land balance models*. The *availability factor* approach simply identifies different categories of land and multiplies the area in each category by the fraction deemed suitable for energy crops. This fraction may be informed by information about agricultural surpluses, or may be purely hypothetical. An influential study from 1993, for example, assumes that 100 per cent of all areas of logged forest in Africa may be suitable for re-afforestation (Johansson et al., 1993). This approach has the advantage of a high level of transparency, but is simplistic and cannot capture the dynamics of competing demands for land or spatial variation in yields. *Land balance models* in contrast identify land areas on which crops may be cultivated (depending on soil, climate, and terrain);[6] they then exclude areas required for food production and other land uses such as urbanization and nature conservation. The area that remains is allocated to energy crops (see for example Hoogwijk et al. (2005) and Erb et al. (2009)). This more sophisticated approach can investigate the interactions between changing food demand, climate change, and land availability over time. Yet it may also overestimate the land available because uncultivable areas that only show up at high resolution may not be excluded. Also, land in cultivation may be underestimated in national statistics, and some land uses such as human settlements, forest, and conservation areas may not be recognized.

If food crop yields can be increased then agricultural land may become available for energy crops. Similarly, if energy crop yields can be increased then more energy can be produced for any given amount of land. Crop yields are a function of the amount of sunlight, the proportion of that light intercepted by the crop, the efficiency with which it is converted to biomass by photosynthesis, and the proportion of that biomass partitioned to the harvested product (Monteith, 1977; Hay and Walker, 1989). At any given location, the yield achieved will be determined by complex interactions between plant physiology, local ecology and climate, and management practices. Yields that can be achieved on poor quality soil, or in areas where water is scarce, may be far less than those achieved under optimum conditions. There are two approaches to estimating the productive yield: extrapolation from case-studies and sample plots, and model-based predictions. Uncertainty about how model parameters will change with location and over time, and limitations in the number of sample plots available, mean that both these methods are highly uncertain (Berndes et al., 2003).

Estimating the potential of agricultural residues

In contrast to the uncertainties that beset energy crop estimates, comparatively good data about the production of major food crops is collated and published by the FAO. From this data it is possible to estimate the quantity of residues produced by applying availability factors. The basic calculation for each crop is as follows:

*Resource = Total crop * Harvest index * Recoverability − Residues dedicated to other uses*

The harvest index is the fraction of the above ground biomass that is the primary crop. In the case of wheat and barley in the UK this is ~51 per cent, and for rapeseed it is about 30 per cent (Kilpatrick, 2008). Because past improvements in the major food crop species such as wheat have largely resulted from increases in the harvest index rather than increases in the total biomass produced by each plant (Hay, 1995), residue production may decrease as cereal yields increase. This effect may, however, be offset by increases in total crop production. It

(continued)

---

[6] Most assessments use the FAO Agro-ecological Zoning (AEZ) methodology to match crop and land types.

BOX 17.3. (CONTINUED)

---

should also be noted that not all biomass residues will be recoverable: some may be left in the field to maintain soil fertility or may already be dedicated to existing uses—such as animal bedding.

Estimating the potential of wastes and residues

Robust data on waste production is seldom available. Consequently, most attempts to quantify the resource are limited to top-down estimates of the amount of waste likely to be produced per unit of economic activity in different industrial sectors, per head of population, or per head of livestock.[7] The basic calculation for each waste sub-category is:

*Resource = Level of economic activity * Waste generation fraction* Recoverability*

Estimates may also be projected into the future, moderated by judgements about the effect of economic growth or other anticipated changes such as increased recycling rates. The principal source of variation between estimates is the inclusion/exclusion of waste categories in the resource inventory. The main source of data is the FAO.

Estimating the potential of forestry

Forestry residues may be estimated in the same way as other wastes: as a fraction of the unused biomass produced by existing forest industries—again relying on FAO data. Harvesting biomass from mature forests, however, is a more controversial area. Many recent studies exclude mature forestry directly from biomass-for-energy estimates considering it better to retain the carbon stored in mature forest. The rationale for this is twofold: firstly, the impact on biodiversity would be unacceptable; and secondly, the carbon emitted as a result of changing the land use could be significant. Nevertheless a number of studies do include estimates of wood production from natural forests in their calculations (Smeets et al., 2007; Fischer and Schrattenholzer, 2001; Yamamoto et al., 1999, 2000, 2001). There is very limited data on the harvest intensity of mature forests and so the approach taken is to estimate the gross annual forest growth increment (a measure of net primary production, or NPP) as a proxy for the technical potential, and limit this by the fractions deemed available and accessible. Implicit in this approach is that a proportion of mature forest would become managed 're-growth' forest. This category of biomass would also overlap with traditional firewood gathering.

Estimates in excess of 300 EJ and up to 600 EJ (600 EJ is slightly more than current global primary energy supply) are all predicated on the assumption that increases in food crop yields could significantly outpace demand for food, with the result that an area of high yielding agricultural land the size of China (>1000 Mha) could be made available for energy crops. In addition, these estimates assume that an area of grassland and marginal land larger than India (>500 Mha) could be converted to energy crops. The area of land

---

[7] For example, Johansson et al. (1993), assume that Municipal Solid Waste (MSW) in OECD countries will be generated at a constant rate of 300 kg per capita per year, and that 75 per cent of this will be recoverable for energy purposes. In another example, Yamamoto et al. (1999) estimates that 20 per cent of food supply will end up as kitchen refuse and that 75 per cent of this could be used for energy purposes. These authors also estimate that 20 per cent of food supply will end up as human faeces and that 25 per cent of this could be recovered.

allocated to energy crops could thus occupy over 10 per cent of the world's land mass, equivalent to the existing global area used to grow arable crops. For most of the estimates in this range a high meat diet could only be accommodated with extensive deforestation. It is also implicit that most animal production would have to be landless to achieve the level of agricultural intensification and residue recovery required.

Estimates in excess of 600 EJ are extreme. The primary purpose of scenarios in this range is to provide a theoretical maximum upper bound and to illustrate the sensitivity of the models to key variables such as population, diet, and technological change. Estimates in this range are not intended to represent *socially acceptable* or *environmentally responsible* scenarios and none of the studies analysed here suggests that they are plausible.

Global biomass potential studies do not try to describe what is likely to happen. Rather, they describe scenarios in which biomass makes an increasing contribution to primary energy supply while attempting to minimize the negative impacts by imposing environmental constraints on development. They are optimistic in the sense that they try to describe sustainable paths as opposed to unsustainable ones. What they are *not* is forecasts extrapolated from empirical observations or any practical experience of trying to achieve these sorts of transitions at a global scale. They therefore provide limited insight into how biomass supply would actually develop if demand was to increase.

In a special report on renewable energy sources for climate change mitigation published in 2011, the view of the Intergovernmental Panel on Climate Change (IPCC) was that the global biomass technical potential could reach 100–300 EJ by 2050. However, the authors of this report also concluded that the technical potential cannot be determined precisely because it depends on 'factors that are inherently uncertain' while societal preferences are unclear. Moreover, increased biomass consumption could evolve in a sustainable or unsustainable way and this could present a challenge for effective governance (IPCC, 2011).

17.6 **The sustainability and governance challenge**

In the 1990s bioenergy was generally regarded as an uncontroversial and environmentally benign alternative to fossil fuels. At this time food and energy prices were comparatively low. Resource scarcity was not high on the agenda, and agricultural land had been taken out of production in Europe and North America to limit food surpluses. Bioenergy was seen by policymakers and politicians as an attractive and low-risk option, and one that could help meet a wide range of policy goals. This favourable view, however, was largely untempered by experience. Outside of a small number of industrial sectors such as pulp and paper, and countries such as Brazil which had been an early adopter of ethanol for transport, there was very limited practical understanding of deploying bioenergy technologies at scale.

By 2015, this situation had to a large extent reversed. Companies and policymakers had accumulated a wealth of knowledge from real projects, but the sustainability of biomass supply had become highly controversial as an increasingly complex and contested picture of the potential impacts and benefits emerged.

By far the most heated debate has been around the production of transport biofuels from commodity food crops. The principal argument against using food crops in this way is that it will increase competition for land, thereby driving up the price of food and setting in motion a cascade of undesirable indirect effects. For example, it is argued that increased demand will not only cause the poor to suffer but will lead to increased conversion of pasture and forested land to arable production. This land conversion may be associated with greenhouse gas emissions if, for instance, newly exposed carbon-rich soils begin to oxidize, and these emissions could negate many of the environmental benefits that provided the rationale for supporting biofuels in the first place. Similarly, agricultural intensification could lead to greater fertilizer use and emissions of nitrous oxides which are also a potent greenhouse gas. Some of the more vociferous opponents claim that biofuels will lead to famine, deplete water resources, destroy biodiversity and soils, as well as being primarily responsible for the food price spikes that occurred in 2008 (Eide, 2008; Mitchell, 2008).

Those seeking to counter these arguments acknowledge the potential for competition but question both the *scale of the effect* and the *direction of travel*. In 2012/13 roughly 137 Mt of cereals (~14 per cent of global production) was used to produce bioethanol, but because one of the co-products of ethanol production is a protein-rich animal feed, the net additional demand for cereals would have been somewhat less (around 9 per cent of global production)[8] (F. O. Lichts, 2013). It is also argued that the 2008 price spikes could better be attributed to a multitude of factors *in addition to biofuels*. These include: the depreciation of the US dollar, increased oil prices, export restrictions on rice, weather shocks leading to poor harvests in some regions, commodity speculation, and increased meat consumption in China and India (Headey and Fan, 2008). The direction of travel is also important because bioenergy proponents do not advocate that an ever larger proportion of good quality farmland should be used to produce biofuels from sugar, starch, and vegetable oil. Rather, they envisage that technological advances will lead to a new generation of conversion technologies able to convert residues and waste products into fuels. A number of researchers have also suggested that there may be beneficial synergies from co-producing food and energy crops. Perennial energy crops, for instance, may also be used to mitigate some of the environmental impacts of intensive agriculture—such as nitrate run-off and soil erosion (Wicke et al., 2011; Berndes, 2008). Using biomass to provide energy services in developing countries might even help prevent wastage in food supply-chains and provide a route for the introduction of sorely needed agricultural infrastructure and knowhow (Lynd and Woods, 2011).

---

[8] Net additional demand for cereals is ~2/3rds gross input to biofuels (Keller, 2010).

Arguments about sustainability, however, are not restricted to the use of food crops and agricultural land. The rapid increase in wood pellet imports from North America to Europe has attracted opprobrium from non-governmental organizations including Greenpeace and Friends of the Earth (RSPB, 2012). In this case the principal objection is that producing wood pellets could directly (or indirectly through market-mediated changes in demand) reduce the standing stock of forest biomass. Their fear is that this could result in carbon emissions increasing in the short term generating a *carbon debt* that would only be repaid over a much longer timescale as the trees regrow and reabsorb the carbon. Whether a debt arises has been reasonably well examined in the academic literature and depends on how a forest is managed and the balance of impacts between natural disturbance (wind, fire, pests) and human disturbance (harvesting). A debt is most likely to occur when unmanaged forest is brought into management, but this also depends on assumptions about what, if anything, is done to increase the productivity of the forest, such as accelerating the rate of re-establishment after harvesting. If, alternatively, you consider a forest with a population of different age trees that is already under management, and each year only the annual growth increment is harvested, no debt arises (Matthews et al., 2014). Carbon debt has only recently entered the public consciousness and has gained salience because of the rapid growth of the wood pellet market and because the timeframes for policy and forest management decisions are so dramatically different. It has yet to be seen how this heightened concern will affect political support for co-firing projects in the UK and Europe.

These examples of on-going debates surrounding sustainability serve to illustrate the complexity of sourcing large quantities of biomass for energy production almost no matter what the ultimate source of the biomass might be. At European level the policy response has been twofold. Firstly, proposals (expected to be adopted in 2014) have been put forward to amend the legislation that mandates increasing quantities of bioenergy in member states (Renewable Energy Directive). These proposals are intended to favour sources of biomass such as residues that are considered less likely to raise sustainability concerns. Other proposed changes include increasing the minimum greenhouse gas saving threshold for new installations, accounting for land use change impacts, and setting limits on the quantities of food-crop-based biofuels. The second element of the policy response has been to place increased reliance on biomass certification schemes that attempt to ensure that biomass entering the energy supply chain meets minimum environmental criteria.

Certification schemes aim to translate broad sustainability principles into decision making criteria that can be evaluated on the basis of detailed and specific indicators. Although the concept of certification is not new—familiar examples include fairtrade coffee and the Forestry Stewardship Council (FSC) standard for wood products—the introduction and design of certification standards for bioenergy as a means to ensure sustainability is not straightforward. Identifying appropriate criteria for certification schemes presents a trade-off between efficacy and ease of adoption.

If criteria are overly detailed and too stringent, compliance may be difficult to demonstrate or they may act as a barrier to trade as reporting costs may become excessive. Conversely, if criteria are too general, they may become meaningless. There is also a risk of leakage if measures are applied to bioenergy production in isolation from the rest of the agricultural and forestry system. For instance, if food and feed crops do not face limitations in land use conversion then areas currently used for food could be diverted to bioenergy and be replaced with newly deforested land elsewhere. Adoption of standards is also essentially a voluntary approach as the implementation of binding requirements is limited by World Trade Organization rules.[9]

A review of bioenergy standards in 2010 identified sixty-seven ongoing certification initiatives (van Dam et al., 2010a). The way these initiatives had approached developing criteria, prescribing calculation methodologies and adopting default values for indicators, was found to be very diverse, reducing transparency and making comparison between schemes difficult (van Dam et al., 2010b). The majority of schemes also focused only on the environmental sustainability of liquid biofuels (and to a lesser extent wood pellets) ignoring social criteria such as peasant farmers' access to land, water, and other natural resources. This proliferation of schemes presents a risk for confusion and potential for a race to the bottom if companies shop around for a standard that can demonstrate regulatory compliance with minimum effort and no change in production methods.

One of the most comprehensive sets of sustainability criteria and indicators has been developed by the Global Bioenergy Partnership—summarized in Table 17.1. These indicators aspire to be value-neutral and do not provide thresholds or limits. Nonetheless they show the breadth of impacts that policymakers might wish to consider in a national or international certification scheme.

The greater the role that bioenergy has in meeting future energy demand, the more important it will become to ensure that biomass production delivers sustainability benefits over the fossil fuel alternatives. Because biomass supply is intertwined with so many different production systems these benefits may be hard to identify. For many biomass resources the economic, social, and environmental impacts are diverse, difficult to quantify, and often contested. Biomass certification is the principal approach to ensuring biomass supply meets public expectations for sustainability, and initiatives have developed and evolved as supply has increased. Sustainability concerns nonetheless present a constraint on the uptake of modern bioenergy technologies and new approaches may be required to reconcile competing demands for food, energy, and environmental protection.

---

[9] Although the use of standards is voluntary, conformity to European standards may constitute a presumption of conformity to the legal requirements of European Directives.

Table 17.1 The Global Bioenergy Partnership's sustainability indicators for bioenergy

Themes		
Environmental	Social	Economic
Greenhouse gas emissions. Productive capacity of the land and ecosystems. Air quality. Water availability, use efficiency and quality. Biological diversity. Land use change, including indirect effects.	Price and supply of a national food basket. Access to land, water, and other natural resources. Labour conditions. Rural and social development. Access to energy. Human health and safety.	Resource availability and use efficiencies in bioenergy production, conversion, distribution and end use. Economic development. Economic viability and competitiveness of bioenergy. Access to technology and technological capabilities. Energy security/diversification of sources and supply. Energy security/ infrastructure and logistics for production and use.
Indicators		
• Lifecycle GHG emissions • Soil quality • Harvest levels of wood resources • Emission of non-GHG air pollutants, including air toxics • Water use and efficiency • Water quality • Biological diversity in the landscape • Land use and land-use-change related to bioenergy feedstock production	• Allocation and tenure of land for new bioenergy production • Price and supply of a national food basket • Jobs in the bioenergy sector • Change in unpaid time spent by women and children collecting biomass • Bioenergy used to expand access to modern energy services • Change in mortality and burden of disease attributable to indoor smoke • Incidence of occupational injury, illness and fatalities	• Productivity • Net energy balance • Change in consumption of fossil fuels and traditional use of biomass • Training and requalification of the workforce • Energy diversity • Infrastructure and logistics for distribution of bioenergy • Capacity and flexibility of use of bioenergy

Source: adapted from GBEP (2011).

17.7 Summary and conclusions

Biomass can be used to provide the full range of modern energy services—heat, power, and transport fuels—using established and proven technology. Advanced conversion technologies that offer improved performance and the ability to use a broader range of feedstock are also beginning to enter the market. Bioenergy production has grown rapidly over the last ten years and estimates of global biomass potential suggest that there is scope for its role in the energy mix to increase further. Yet as bioenergy production has increased, concerns about the environmental and social impacts of biomass supply have emerged that have resonated with public sentiment and gained significant political traction. These concerns appear increasingly likely to constrain future growth, and it has yet to be seen how rapidly the combination of more advanced conversion technologies together with approaches such as biomass certification will provide a way forward. Thus the future development of the bioenergy sector is uncertain. The efficacy of the conversion technologies is not in doubt, but the sheer complexity of biomass supply chains and the acceptability of their environmental and social impacts could limit its contribution to global energy supply.

■ REFERENCES

AEBIOM (2010) AEBIOM European biomass statistics [Internet], European Biomass Association (AEBIOM), available from: <http://www.aebiom.org> [Accessed: December 2010].

AGOSTINI, A., GIUNTOLI, J., and BOULAMANTI, A. (2013) *Carbon Accounting of Forest Bioenergy—Conclusions and Recommendations from a Critical Literature Review*. European Commission Joint Research Centre (JRC), Luxembourg.

BAYAR, T. (2013) Sweden's bioenergy success story [Internet], available from: <http://www.renewableenergyworld.com/rea/news/article/2013/03/swedens-bioenergy-success-story> [Accessed: November 2013].

BERNDES, G. (2008) *Water Demand for Global Bioenergy Production: Trends, Risks and Opportunities*. WBGU, Berlin.

BERNDES, G., HOOGWIJN, M., and VAN DEN BROEK, R. (2003) The contribution of biomass in the future global energy supply: a review of 17 studies. *Biomass and Bioenergy*, 25, 1–28.

CREUTZIG, F., POPP, A., PLEVIN, R., LUDERER, G., MINX, J., and EDENHOFER, O. (2012) Reconciling top down and bottom-up modelling on future bioenergy deployment. *Nature Climate Change*, 2, 320–7.

DENA (2013) Renewables made in Germany, information about German renewable energy industries, companies and products [Internet], German Energy Agency (DENA), available from: <http://www.renewables-made-in-germany.com/en/renewables-made-in-germany-start/biogas/biogas.html> [Accessed: May 2014].

DORNBURG, V. and FAAIJ, A. (2001) Efficiency and economy of wood-fired biomass energy systems in relation to scale regarding heat and power generation using combustion and gasification technologies. *Biomass and Bioenergy*, 21, 91–108.

EIA (2014) [Internet], US Energy Information Administration (EIA)—International Energy Statistics, available from: <http://www.eia.gov/countries/data.cfm> [Accessed: January 2014].

EIDE, A. (2008) *The right to food and the impact of liquid biofuels (agrofuels)*. Rome, Food and Agriculture Organization of the United Nations, Rome.

ENVITECH (2008–9) *Biogas projects—operators and their plants*. EnviTech Biogas, (Corporate brochure). Available from: <http://www.envitec-biogas.com/fileadmin/Brochures/Reference_Brochure/EnviTec_Biogas_Referenzen1en.pdf> [Accessed: April 2015].

ERB, K.-H., HABERL, H., KRAUSMANN, F., LAUK, C., PLUTZAR, C., STEINBERGER, J. K., MÜLLER, C., BONDEAU, A., WAHA, K., and POLLACK, G. (2009) *Eating the Planet: Feeding and Fuelling the World Sustainably, Fairly and Humanely—A Scoping Study*. Institute of Social Ecology and PIK Potsdam, Vienna.

F. O. LICHTS (2013) *World Ethanol and Biofuels Report—May 2013*. Agra-Net, London.

FAAIJ, A. P. C. (2006) Bio-energy in Europe: changing technology choices. *Energy Policy*, 34, 322–42.

FISCHER, G. and SCHRATTENHOLZER, L. (2001) Global bioenergy potentials through 2050. *Biomass and Bioenergy*, 20, 151–9.

GBEP (2007) *A Review of the Current State of Bioenergy Development in G8 + 5 countries*. Rome, Global Bioenergy Partnership (GBEP)/Food and Agriculture Organisation of the United Nations (FAO).

GBEP (2011) *The Global Bioenergy Partnership Sustainability Indicators for Bioenergy—First Edition*. Food and Agricultural Organization of the United Nations (FAO), Rome.

HALL, D. O. and MOSS, P. A. (1983) Biomass for energy in developing countries. *GeoJournal*, 7.1, 5–14.

HAMELINCK, C. N., FAAIJ, A. P. C., DEN UIL, H., and BOERREGHTER, H. (2004) Production of FT transportation fuels from biomass; technical options, process analysis and optimisation, and development potential. *Energy*, 29, 1743–71.

HAY, R. K. M. (1995) Harvest index: a review of its use in plant breeding and crop physiology. *Annals of Applied Biology*, 126, 197.

HAY, R. K. M. and WALKER, A. J. (1989) *An Introduction to the Physiology of Crop Yield*. Longman Scientific and Technical, Harlow.

HEADEY, D. and FAN, S. (2008) Anatomy of a crisis: the causes and consequences of surging food prices. *Agricultural Economics*, 39, 375–91.

HOOGWIJK, M., FAAIJ, A., and EICKHOUT, B. (2005) Potential of biomass energy out to 2100, for four IPCC SRES land-use scenarios. *Biomass and Bioenergy*, 29, 225–57.

IEA (2009a) *Bioenergy—a sustainable and reliable energy source a review of status and prospects*. IEA Bioenergy. Available from: <http://www.ieabioenergy.com/> [Accessed: April 2015].

IEA (2009b) *World Energy Outlook*. International Energy Agency (IEA), Paris.

IEA (2010) *Energy Technology Perspectives 2010: Scenarios and Strategies to 2050*. International Energy Agency, Paris.

IEA (2012a) *Technology Roadmap Bioenergy for Heat and Power*. International Energy Agency, Paris.

IEA (2012b) *World Energy Outlook*. International Energy Agency, Paris.

IPCC (2007) *Contribution of Working Group III to the Fourth Assessment Report of the Intergovernmental Panel on Climate Change*. Cambridge University Press, Cambridge, United Kingdom and New York, NY, USA.

IPCC (2011) *Bioenergy in IPCC Special Report on Renewable Energy Sources and Climate Change Mitigation*. Cambridge University Press, Cambridge, United Kingdom and New York, NY, USA.

JOHANSSON, T. B., KELLY, H., REDDY, A. K. N., and WILLIAMS, R. H. (1993) A renewables-intensive global energy scenario (RIDGES) (appendix to Chapter-1). IN T. B. JOHANSSON ET AL. (Ed.) *Renewable Energy: Sources for Fuels and Electricity*. Island Press, Washington, D.C.

JOUDREY, J., MCDOW, W., SMITH, T., and LARSON, B. (2012) *European Power from U.S. Forests: How Evolving EU Policy Is Shaping the Translatlantic Trade in Wood Biomass* [Internet], Environmental Defense Fund, available from: <http://www.edf.org/bioenergy> [Accessed: July 2012].

KELLER, C. (2010) Personal communication to R. Slade (February 2010). Deputy editor F.O. Lichts.

KILPATRICK, J. (2008) *Addressing the land use issues for non-food crops, in response to increasing fuel and energy generation opportunities*. ADAS Rosemaund Ltd. Hereford. Available from: <http://www.nnfcc.co.uk/tools/addressing-the-land-use-issues-for-non-food-crops-in-response-to-increasing-fuel-and-energy-generation-opportunities-nnfcc-08-004> [Accessed: April 2015].

LEIDREITER, A. (2013) Denmark puts the brakes on heating costs with new legislation [Internet], available from: <http://www.renewableenergyworld.com/rea/blog/post/print/2013/02/denmark-puts-the-brakes-on-heating-costs-with-new-legislation> [Accessed: November 2013].

LUDWIG, J., MARUFU, L. T., HUBER, B., ANDREAE, M. O., and HELAS, G. (2003) Domestic combustion of biomass fuels in developing countries: a major source of atmospheric pollutants. *Journal of Atmospheric Chemistry*, 44, 23–37.

LYND, L. R. and WOODS, J. (2011) A new hope for Africa. *Nature*, 474, S20–S21.

LYSEN, E., VAN EGMOND, S., DORNBURG, V., FAAIJ, A., VERWEIJ, P., LANGEVELD, H., VAN DE VEN, G., WESTER, F., VAN KEULEN, H., VAN DIEPEN, K., MEEUSEN, M., BANSE, M., ROS, J., VAN VUUREN, D., VAN DEN BORN, G. J., VAN OORSCHOT, M., SMOUT, F., VAN VLIET, J., AIKING, H., LONDO, M., and MOZAFFARIAN, H. (2008) *Biomass Assessment: Assessment of Global Biomass Potentials and their Links to Food, Water, Biodiversity, Energy Demand and Economy*. Netherlands Environmental Assessment Agency MNP, Bilthoven, Netherlands.

MATTHEWS, R., MORTIMER, N., MACKIE, E., HATTO, C., EVANS, A., MWABONJE, O., RANDLE, T., ROLLS, W., SAYCE, M., and TUBBY, I. (2014) *Carbon impacts of using biomass in bioenergy and other sectors: forests (DECC project TRN 242/08/2011—Final report: Parts a and b)*. Forest Research, UK.

MITCHELL, D. (2008) *A Note on Rising Food Prices*. The World Bank, Washington, DC.

MONTEITH, J. L. (1977) Climate and the efficiency of crop production in Britain [and discussion]. *Philosophical Transactions of the Royal Society of London*, 281, 277–94.

NAWARO (2013) *Güstrow BioEnergie Park* [Internet], Nawaro Bioenergy AG, available from: <http://www.nawaro.ag/en/company/projects/guestrow-bioenergy-park/> [Accessed: November 2013].

REN21 (2010) *Renewables 2010: Global Status Report*. Renewable Energy Policy Network for the 21st century (REN21), Milan.

REN 21 (2013) *Renewables 2013 Global Status Report*. Renewable Energy Policy Network for the 21st Century, Milan.

RENI (2013) *Biogas an all-rounder—new opportunities for farming, industry and the environment (3rd edition)*. Renewables Insight—Energy Industry Guides (RENI). [Available from: <http://www.german-biogas-industry.com/> [Accessed: April 2015].

RSPB (2012) *Dirtier than coal? Why government plans to subsidise burning trees are bad news for the planet*. The Royal Society for the Protection of Birds (RSPB). [Available from: <http://www.rspb.org.uk/Images/biomass_report_tcm9-326672.pdf> [Accessed: April 2015].

SEARCHINGER, T., HEIMLICH, R., HOUGHTON, R. A., DONG, F., ELOBEID, A., FABIOSA, J., TOKGOZ, S., HAYES, D., and YU, T.-H. (2008) Use of U.S. croplands for biofuels increases greenhouse gases through emissions from land-use change. *Science*, 319, 1238–40.

SLADE, R., SAUNDERS, R., GROSS, R., and BAUEN, A. (2011) *Energy from Biomass: The Size of the Global Resource*. Imperial College Centre for Energy Policy and Technology and UK Energy Research Centre, London.

SLADE, R., BAUEN, A., GROSS, R. (2014) Global bioenergy resources. *Nature Climate Change*, 4, 99–105.

SMEETS, E., FAAIJ, A., LEWANDOWSKI, I., and TURKENBURG, W. (2007) A bottom-up assessment and review of global bio-energy potentials to 2050. *Progress in Energy and Combustion Science*, 33, 56–106.

SÖDERENERGI (2010) *The Igelsta CPH plant—Sweden's largest bio-fuelled cogeneration facility* [Internet], available from: <http://www.soderenergi.se> [Accessed: December, 2010].

THRÄN, D., SEIDENBERGER, T., ZEDDIES, J., and OFFERMANN, R. (2010) Global biomass potentials—resources, drivers and scenario results. *Energy for Sustainable Development*, 14 200–5.

TURKENBURG, W. C. (2000) Renewable Energy Technologies, Chapter 7 of the *World Energy Assessment of the United Nations, UNDP, UNDESA/WEC*. Published by UNDP, New York.

VAN DAM, J., JUNGINGER, M., and FAAIJ, A. P. C. (2010a) From the global efforts on certification of bioenergy towards an integrated approach based on sustainable land use planning. *Renewable and Sustainable Energy Reviews*, 14, 2445–72.

VAN DAM, J., JUNGINGER, M., and FAAIJ, A. P. C. (2010b) From the global efforts on certification of bioenergy towards an integrated approach based on sustainable land use planning. *Renewable and Sustainable Energy Reviews*, 14, 2445–72.

VERHOEST, C. and RYCKMANS, Y. (2012) *Industrial Wood Pellets Report*. PellCert, Laborelec-GDF Suez and Intelligent Energy Europe.

WAHAB, G. (2012) *Malaysia biofuels annual*. Global Agricultural Information Network (GAIN). Report number MY2006. Washington, DC. Available from: <http://gain.fas.usda.gov/Recent%20GAIN%20Publications/Biofuels%20Annual_Kuala%20Lumpur_Malaysia_8-3-2012.pdf> [Accessed: April 2015].

WICKE, B., SMEETS, E., WATSON, H., and FAAIJ, A. (2011) The current bioenergy production potential of semi-arid and arid regions in sub-Saharan Africa. *Biomass and Bioenergy*, 35, 2773–86.

YAMAMOTO, H., FUJINO, J., and YAMAJI, K. (2001) Evaluation of bioenergy potential with a multi-regional global-land-use-and-energy model. *Biomass and Bioenergy*, 21, 185–203.

YAMAMOTO, H., YAMAJI, K., and FUJINO, J. (1999) Evaluation of bioenergy resources with a global land use and energy model formulated with SD technique. *Applied Energy*, 63, 101–13.

YAMAMOTO, H., YAMAJI, K., and FUJINO, J. (2000) Scenario analysis of bioenergy resources and CO_2 emissions with a global land use and energy model. *Applied Energy*, 66, 325–37.

18 Solar energy: an untapped growing potential?

Chiara Candelise

18.1 Introduction

Solar energy is the most abundant of all energy resources, about 885 million TWh reach the earth's surface in a year, that is 6,200 times the worldwide primary energy consumed in 2008 (IEA, 2011). If the yearly energy received by the sun could be entirely captured and stored it would represent more than 6,000 years of total worldwide energy consumption at 2010 energy rates (IEA, 2010b), which compares with 46 years for oil, 58 years for natural gas, and almost 150 years for coal based on proven energy reserves (IEA, 2011). The technical potential of solar energy, that is how much solar energy can actually be tapped for human uses by deployment of available technologies or practices, varies across technology deployed, over different regions of the earth (depending on local factors such as solar irradiance, land availability and uses, demands for energy services), according to conversion efficiencies as well as the assessment methodologies used. A study commissioned by the German Environment Agency on renewable energy potential found that the technical potential of photovoltaics (PV) and concentrating solar power (CSP) ranges significantly across studies (from 1,338 to 14,778 EJ/yr for PV and 248 to 10,791 EJ/yr for CSP), mainly due to the different assumptions used on land area availability and conversion efficiency (Krewitt et al., 2009). Moreover, such technical potential can change over time, as technologies and solar energy harvesting improve. A recent study from NREL (National Renewable Energy Laboratory) estimating land-use for various solar technologies in the USA points to how total land-use requirements for solar power plants range across technologies and can also reduce over time thanks to technological improvements (Ong et al., 2013).

Despite this assessment variability, solar energy technical potential is estimated to be greater not only than current global primary energy consumption but also than other renewable energies' potential. An Intergovernmental Panel on Climate Change report comparing estimates of global technical potential of renewable energy sources from a wide number of studies shows how the lowest estimate of the technical potential for direct solar energy is greater than the highest estimate of any other renewable energy potential (Figure 18.1).

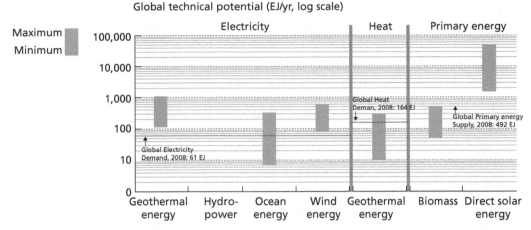

Figure 18.1 Technical potential of renewable energy sources global

Source: (IEA, 2011).

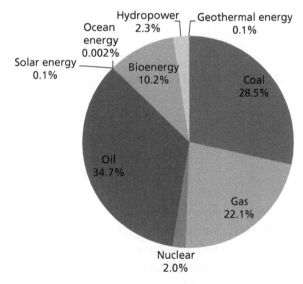

Figure 18.2 Shares of energy sources in total global primary energy supply in 2008

Source: (IPCC, 2011).

In spite of such huge potential, solar energy still accounts for a very small share of primary energy consumed (Figure 18.2). However, since the climate change mitigation and renewable deployment commitments started playing a role in European and global governments' agendas in the late 1990s, solar energy has been the fastest-growing energy sector (mainly driven by PV), experiencing dramatic market expansion (Section 18.3) and achieving previously unexpected reduction in prices (Section 18.4). Indeed, in only the last ten years, and due to PV market expansion, solar has quickly begun to make a noticeable contribution to the national energy mix in countries that have consistently supported its deployment in the recent years. For example in Germany, Italy, and Spain solar PV alone managed to increase its share of the total national electricity generation from 0 per cent in 2005 to respectively 3 per cent, 3.6 per cent, and 2.8 per cent in 2011 (IEA, 2012). Not only does solar energy have the potential to contribute to the energy mixes of developed energy systems and economies, but solar powered devices can also provide easily implementable and cost-effective solutions for million-sof people without access to grid electricity in developing countries (see Section 18.2).

This seems to show that, once the right technical, economic, and regulatory conditions are in place, solar technologies can be successfully deployed, account for a reasonable share of the energy mix, and contribute to the climate change mitigation challenge. This chapter will address these conditions by looking into the recent technological and market developments (Sections 18.2 and-18. 3), analysing drivers for solar technologies' competitiveness, and discussing potential and challenges for future deployment. After an introduction to solar heat technologies (Section 18.2) the chapter will focus the discussion on solar electricity technologies, in particular on photovoltaics (Sections 18.3–18.5).

18.2 Solar technologies and applications

There are two fundamental ways to capture solar energy—heat and photoreaction—delivering two energy vectors: heat and electricity. These translate into several techniques/technologies, at different levels of maturity, for tapping solar energy. They can be grouped into those that produce (or use) *heat*, that is passive solar and active heating (cooling), and those producing *electricity*, that is photovoltaic, where solar energy is directly converted into electricity through photoreaction, and concentrating solar power (CSP), where high-temperature heat is firstly produced then converted into electricity through a heat engine and generator. An alternative way of harnessing solar energy is its conversion into hydrogen, and thence perhaps into other chemical *fuels*, which constitute an interesting option to store and transport solar energy. Direct conversion of solar energy into fuel is not yet widely demonstrated or commercialized and solar fuel production technologies are in an earlier stage of development. However, the solar hybrid fuel production system and generation of hydrogen from water electrolysis using electricity produced through PV or CSP appear to be the most commercially feasible options in the near to medium term (IPCC, 2011; IEA, 2011).

Harnessing solar *heat* is relatively easy as any material object absorbs thermal energy from sunlight. *Passive solar heating (and cooling)* techniques exploit the solar irradiance incident on buildings through the use of glazing (windows, sun spaces, conservatories) and other transparent materials, managing heating energy consumption based on the natural energy flows of radiation, conduction, and convection and without the dominant use of pumps or fans. Passive solar can provide an important contribution to building energy efficiency and net-zero-energy homes (along with onsite renewable generation), in particular in Europe where the Energy Performance of Buildings Directive (Directive 2010/31/EC) provides that all new buildings must be nearly zero-energy buildings by 31 December 2020. It is estimated that well-designed passive solar systems decrease the need for additional comfort heating requirements by about 15 per cent for existing buildings and about 40 per cent for new buildings (IPCC, 2011).

Active solar heating can be used in a wide range of applications, including domestic hot water, solar cookers and ovens (mainly used in developing countries), space heating and cooling, industrial process heat as well as electricity generation (see Section 18.2.1 on CSP) and manufacturing fuels. Solar cooling for buildings can also be achieved, for example, by using solar-derived heat to drive thermodynamic refrigeration absorption or adsorption cycles. Solar heat can be produced with a variety of devices and technologies (all essentially offering a receptive surface heated by direct or diffuse sunlight) which generally vary depending on the temperature levels required for end-uses, for example ranging from 25 °C (e.g. for swimming pool heating) to 1,000 °C (e.g. for dish/Stirling concentrating solar power), and even up to 3000 °C in solar furnaces. Among these, solar water heating for domestic and commercial buildings is the most mature solar thermal technology and one with a growing market. The world installed capacity of solar thermal systems increased from 180 GWth in 2009 to 268.1 GWth in 2012, with Europe and China accounting for about 80 per cent of the total installed capacity (Mauthner and Weiss, 2013). Indeed, solar thermal is already second only to wind energy among renewables (excluding traditional hydro and biomass) in contributing to world energy demand (Figure 18.3).

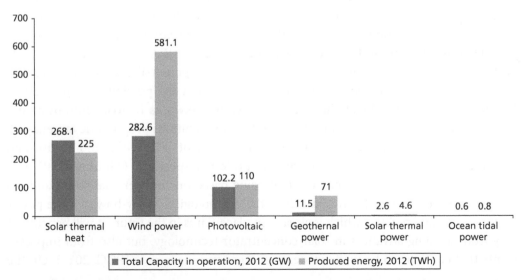

Figure 18.3 Renewable technologies total capacity in operation and produced energy, 2012

Source: (Mauthner and Weiss, 2013).

The remainder of this section will focus on the two main solar technologies for the production of *electricity*. Both approaches are currently in use, but differ quite substantially particularly in terms of: (1) type of applications, with PV being a very modular technology deployed through a wide diversity of uses and plant sizes and CSP mainly deployed through medium- to large-scale plants and complex/capital-intensive investments; (2) market expansion, with PV experiencing stronger market expansion than CSP; (3) positioning along the innovation chain, with CSP still bridging between demonstration stage and full market deployment versus an already mature global PV industry (Section 18.3); (4) dispatchability, that is non dispatchable PV versus dispatchable CSP with storage.

18.2.1 CONCENTRATING SOLAR POWER

In Concentrating Solar Power (CSP) solar radiation heats up a fluid which is then used to drive a thermodynamic cycle. There are four main CSP technology types, classified according to the way the sun is concentrated. Concentration is either to a line (linear focus) as in parabolic trough or linear Fresnel systems, or to a point (point focus) as in tower or dish systems (IEA, 2011). Despite such variety (even within each family there exist various options for the heat transfer fluid and thermodynamic cycle), the majority of installed plants to date use parabolic trough technology. The first commercial CSP plants were deployed in California between 1985 and 1999 and continue to operate commercially today. A new wave of CSP plant construction was initiated in the mid 2000s, and by 2011 around 1.8 GW have been deployed worldwide, although mainly concentrated in the USA and Spain with smaller investments in a few other countries (Algeria, Egypt, and Morocco) (IEA, 2012). China is likely to become the third market in the years to come, as the 12th Five Year Plan for Energy Technology provides for several CSP demonstration projects to be implemented by 2017 (IEA, 2012). MENA (Middle East and North Africa) region countries also have a high potential, also considering both their solar resource and their proximity to Europe, and the possibility of developing HVDC (High Voltage Direct Current) transmission networks between European and MENA countries.[1] Saudi Arabia is implementing an ambitious programme targeting 25 GW of CSP by 2032, Morocco is currently building a plant of 500 MW, and Tunisia and Algeria have announced 2 GW and 500 MW projects respectively (Muirhead, 2013).

An advantage of CSP technology is that being based on conventional steam and gas turbine cycles it can benefit from much of the technological know-how of large power station design and practice which is already in place. Thus in the future it will benefit not only from ongoing advances in solar concentrator technology, but also from improvements that continue to be made in steam and gas turbine cycles (IPCC, 2011). On the

---

[1] See for example the Desertec initiative <http://www.desertec.org/concept>.

other hand the various technological options have advantages and disadvantages which increase complexity, and CSP plants need to be optimized to meet local and regional conditions.

The possibility of incorporating thermal storage is another key advantage of CSP technology as it provides dispatchability, helping the grid operator and grid stability as well as making the technology more cost-effective in markets where peak demand and high tariffs occur in the evening. Hybridization with non-renewable fuels is another way in which CSP can be designed to be dispatchable. Hybrid solar/fossil plants have received increasing attention in recent years, and several integrated solar combined-cycle (ISCC) projects have been either commissioned or are under construction in the Mediterranean region (Morocco, Egypt) and the USA (IPCC, 2011). Therefore, CSP with storage could potentially provide a cost-effective low-carbon means of incorporating intermittent renewables sources into electricity systems.

However, there still are some challenges to be overcome for full mass deployment of CSP. CSP only uses direct solar irradiation thus requiring large amounts of direct sunlight. Hence they are generally situated in arid and semi-arid regions where water cooling needs can be an issue. Wider implementation of CSP would require such water needs to be met or technologies with lower water use to be implemented.

CSP plants are rather complex projects, they involve several actors (engineering, procurement, and construction—EPC; contractor; technology suppliers; owners; financing institutions), and imply high capital investments and complex contractual arrangements to manage investment risks. These complexities affect costs as well as cost assessment, that is there is no single figure for cost of electricity from CSP. In 2010 levelized costs of electricity (LCOE)[2] from CSP were reported to range between 15–22 €cents/KWh depending on technology, size, and location. To reach cost parity with fossil fuel generation cost reductions still need to be achieved. Provided that commercial deployment of CSP continues to grow cost reductions in the 50–60 per cent range are deemed achievable, coming half from more efficient technologies and processes and half from economies of scale and volume production (EASAC, 2011). Operation and maintenance costs are also expected to decrease with CSP technology development and implementation. CSP is estimated to reach cost competitiveness with fossil-based generation between 2025 and 2030 (EASAC, 2011; IPCC, 2011).

Despite the fact that a large scale allows for economies of scale, the high capital intensity of CSP plants has tended to inhibit investment and the perceived risk of the novel technology (with respect to conventional power plants) has increased the cost of capital. In addition, larger power stations require strong infrastructural support, and new or augmented transmission capacity may be needed. These elements are a clear disadvantage in particular when compared to very modular, less technologically complex, and (more recently) less expensive PV technology, a direct competitor among renewable

---

[2] LCOE is generally defined as the discounted lifetime PV system CAPEX divided by the discounted lifetime generation of the PV system.

technologies. The steep cost reductions achieved by PV in recent years (Section 18.4) have, for example, led some planned CSP projects in the USA to turn to PV in 2011 (IEA, 2011). Moreover, PV market and research capacity is currently bigger than for CSP and PV systems can be implemented much more quickly. On the other hand, it is true that a large share of the installed grid connected PV systems (i.e. residential and commercial systems) compete with retail prices, which are higher than the wholesale prices that CSP has to face. In addition, PV cannot provide dispatchable electricity, so that its overall value is lower than fully dispatchable CSP electricity. A full assessment of CSP potential should also take into account the value of CSP with storage in electricity markets both as grid balancing and back-up capacity to cover other variable renewable sources (EASAC, 2011; Denholm and Mehos, 2011). The value of CSP technologies in the future will depend on the price difference between dispatchable and non dispatchable electricity as well as on the share in the countries' energy mixes of intermittent renewable sources as wind and PV (and the relative need for grid balancing).

18.2.2 PHOTOVOLTAICS

A photovoltaic system is an integrated assembly of modules and other components designed to convert solar energy into electricity. The main component of a PV system is the module, being the device responsible for the conversion of sunlight into electricity and accounting for the largest share of the PV system cost (see Section 18.4). All the other components needed to build up a PV system are, by convention, called Balance of System (BOS). Usually, BOS refers to all PV system components and cost elements except for the modules, thus including technical components such as inverter, mounting structures, cables and wiring, battery (for off-grid systems), metering (for grid-connected applications) as well as other costs such as installation, design, and commissioning costs.

There is a wide variety of PV module technologies at different levels of maturity. Commercial PV are mainly flat plate modules based on either crystalline silicon—c-Si— (also often called *1st generation*) or thin film (*2nd generation*), more specifically Cadmium Telluride (CdTe), Amorphous Silicon (a-Si), and Copper Indium (Gallium) (di) Selenide (CIGS) technologies.

C-Si module technologies are based on abundant silicon material, and wafer-based c-Si are a mature technology dominating the PV market since the 1950s and still accounting for the majority of the market share. Wafer-based c-Si modules are currently mass produced with production capacity in the GW range (Section 18.3). They are the technologies responsible for the dramatic market expansion and price reduction experienced by the PV sector in the last decade (see Sections 18.3 and 18.4). The silicon feedstock bottleneck experienced by the PV industry in the mid 2000s fostered interest and investments in thin film technologies, as their lower use of semiconductor material and production processes seemed to guarantee the achievement of lower production costs (Chopra et al., 2004; Zweibel and National Renewable Energy, 2005; EU PV

Technology Platform, 2011) (see Section 18.4). Thin film cells are made by depositing a thin layer of semiconductor material on rigid substrates, such as glass in flat plate modules, but also on flexible substrates such as polymer or metal, producing flexible and lightweight solutions which allow for product differentiation and diversity of uses (e.g. integration into buildings, BIPV). Despite achieving low production costs (see Section 18.4) thin film still account for a smaller share of the market, mainly as their lower efficiencies partially offset their cost advantage. Recent studies have also raised concerns over the future potential growth of these technologies due to scarcity of some critical semiconductor materials, such as indium and tellurium used in CdTe and CIGS thin film (Angerer et al., 2009; EC, 2010; Wadia et al., 2009). However, further analysis seems to show that it is more the impact on production costs of materials' price increases that has potential negative implications for the technologies relying on them, rather than the absolute availability of the materials per se (Candelise et al., 2011, 2012).

A range of other PV technologies (often defined as *3rd generation*) are also emerging, varying from technologies under demonstration (e.g. high efficiency, multi-junction concentrating PV) to concepts still at the R&D stage (e.g. quantum-structured PV cells) (EU PV Technology Platform, 2011). Among these, dye-sensitized and organic solar cells are an emerging niche, due to their potential to achieve very low costs through the use of cheap and abundant materials and low-cost processes, and their potential of being deposited on different substrates (Azzopardi et al., 2011; Carlé and Krebs, 2013; Krebs and Jørgensen, 2013). Their initial target market is consumer devices, but their low efficiencies and short lifetimes are still challenges to be overcome, which make their future success and use in the power sector still unclear.

PV systems can be grouped into two main categories: off-grid, which operates independently from the grid network, and grid-connected. Although off-grid applications still play a relevant role in global PV markets (in particular in developing countries), grid-connected PV currently accounts for the major market share and is responsible for the dramatic expansion of the last decade (Section 18.3). *Grid-connected PV systems* can be further sub-divided into grid-connected *distributed*, where the electricity generated satisfies local loads and only the excess is fed into the grid, and grid-connected *centralized*, where all electricity generated is fed into the grid. Grid-connected PV systems can also be divided according to the market segment they satisfy and the system type/size. Typical PV market segments are: residential (systems of small size—below 10 kW), commercial (systems of medium size—in the hundreds of kW range), and utility (large ground-mounted systems—in the MW range). Residential and commercial systems are generally utilized as distributed systems, and are on top of buildings or building integrated (BIPV) when they displace conventional building materials. This wide variation of grid-connected system types and sizes implies variation in system costs (as discussed in Section 18.4). *Off-grid PV systems* for rural electrification include solar home systems (SHS), solar lanterns, and centralized installations such as solar PV mini-grids, solar DC micro-grids, and solar charging stations (Palit, 2013). SHS are the most commonly implemented, although centralized installations outperform

SHS in terms of enhanced electrical performance and reduction of storage needs (Chaurey and Kandpal, 2010; Thirumurthy et al., 2012). An emerging solar application for rural electrification is the PicoPV system, a small PV-system with a power output of 1 to 10 W, mainly used for lighting and thus able to replace unhealthy and inefficient sources such as kerosene lamps and candles. PicoPV systems are often defined as a 'pre-electrification' option, as they provide a cheap solution where SHS costs would not be affordable or the energy demand is too low to be powered by SHS. They are over-the-counter consumer products and do not need specific know-how for installation or O&M (Reiche et al., 2010).

The next sections of this chapter will focus on and discuss in more detail recent developments in PV markets, costs, and policies. The impressive market growth experienced by the PV sector in the last decade makes it a relevant technology not only for its potential to contribute to climate mitigation and the future energy mix (see Sections 18.3 and 18.5), but also as an interesting case of energy technology innovation in particular in relation to its technology costs trend and successful learning (as further discussed in Section 18.4).

18.3 **PV sector growth and dynamics**

The growth of the global PV market in the last decade has been impressive. Worldwide cumulative installed capacity increased from 1.4 GW in 2000 to over 100 GW in 2012, of which more than 60 per cent was installed only in 2011 and 2012 (IEA, 2011; EPIA, 2013a). Global annual PV installations (which are a measure of the size of the PV market) have increased dramatically (Figure 18.4), in particular in the most recent years, despite the difficult economic environment. Although up to the mid 1990s off-grid applications accounted for more than 70 per cent of the PV market share (IEA-PVPS, 2005), grid-connected PV systems are responsible for this dramatic PV market growth and currently account for the majority of the market. Indeed, PV sector expansion has been mainly driven to date by demand-pull policy incentives for grid-connected PV systems implemented since the late 1990s, with the first PV rooftop programmes introduced in countries such as Japan and Germany. However, it has been the implementation of feed-in-tariff schemes[3] in key countries such as Japan and Germany in the early 2000s and more recently in other countries (both in Europe such as Spain, Italy, Czech Republic, the UK, and elsewhere such as China and India) which has triggered the exponential growth of the last decade.

Driven by the high PV market demand the solar industry has also been steadily growing in the last decade. Worldwide PV production has increased from only 280 MW

---

[3] Under feed-in-tariff schemes electricity utilities are obliged to enable renewable generators to connect to the electricity grid and they must purchase electricity generated by renewables at a set rate, typically based on the cost of generation of each technology.

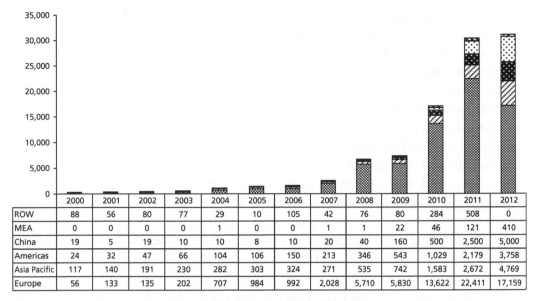

	2000	2001	2002	2003	2004	2005	2006	2007	2008	2009	2010	2011	2012
ROW	88	56	80	77	29	10	105	42	76	80	284	508	0
MEA	0	0	0	0	1	0	0	1	1	22	46	121	410
China	19	5	19	10	10	8	10	20	40	160	500	2,500	5,000
Americas	24	32	47	66	104	106	150	213	346	543	1,029	2,179	3,758
Asia Pacific	117	140	191	230	282	303	324	271	535	742	1,583	2,672	4,769
Europe	56	133	135	202	707	984	992	2,028	5,710	5,830	13,622	22,411	17,159

⊠ROW ⊠MEA ⊡China ▓Americas ▨Asia Pacific ⊠Europe

Figure 18.4 Evolution of global PV annual installations, 2000–12 (MW)

Note—MENA: Middle East and North Africa; ROW: Rest of the World; Americas: USA and Canada; Asia Pacific: Japan, Korea, Australia, Taiwan, and Thailand.

Source: (EPIA, 2013a).

in 2000 to about 50 GW in 2012 (Photon International, 2009; Q-Cell, 2012), with annual growth rates above 40 per cent since the late 1990s and 100 per cent in 2010, when worldwide production jumped to 27 GW from 12 GW in only one year (Photon International, 2011). Indeed, the PV sector has been an extremely dynamic sector, not just because of its exponential growth, but also in terms of markets and industry dynamics, as key players both in demand and supply have been changing very quickly over the recent years.

In terms of demand, Europe has been the driving force for most of the last decade (accounting in 2012 for 70 per cent of the global cumulative installations (EPIA, 2013a)), but this situation has been rapidly changing in the last couple of years, with China quickly switching from being mainly an exporter of PV modules to a leading PV market (from only 40 MW of annual installations in 2008 to 5 GW in 2012) and other countries such as the USA, Australia, India (with a cumulative capacity of respectively 7.8 GW, 2.4 GW, and 1.2 GW in 2012), and several countries from North Africa and the MENA region ramping up installed capacity. Even within Europe, apart from Germany which has experienced a steady growth since the mid 2000s reaching a cumulative installed capacity above 30 GW in 2012 (EPIA, 2013a), several national markets have been booming from one year to the next (and in some unsuccessful cases collapsing): for example Italy reached a market size of 9.3 GW in 2011 from only a few hundred MWs in 2008 (EPIA, 2012b); similarly the UK achieved an annual market of 1.3 GW in 2012, from few hundreds of MWs in 2011 (EPIA, 2012b; DECC, 2013); Spain and the Czech Republic experienced overheated market development (respectively in 2008 with a 2.5 GW market (EPIA, 2009) and in 2010 with around 2 GW (EPIA, 2012b)) to then bust in the next year because of a sudden reduction in policy incentives.

The supply side of the PV sector has been even more dynamic, with China and Taiwan entering the worldwide market and managing to outpace the leading manufacturers (Japan and Germany) in less than five years. In 2011 China alone accounted from more than 65 per cent of worldwide capacity, quickly ramping up from only 8 per cent in 2005 (Jager-Waldau, 2012). This aggressive entry of Chinese manufacturers into the worldwide PV industry has had important implications both for module prices and in terms of PV industry consolidation (as discussed in Section 18.4). Moreover, it illustrates the strong global dimension of the PV sector and industry, characterized by distance between the production and the consumption centres (made possible by a modular and easily shipped technology) and where new players could easily enter the market thanks to the availability of mature and quickly scalable technology and manufacturing processes.

The importance of demand-pull policy incentives implemented in key countries in triggering the PV sector expansion of the last decade has already been pointed out. However, supply-push support and funding to PV R&D and product development, both from the private and the public sector, has also played a role in the development and deployment of PV. Cumulative public R&D investments in PV over the 1974–2008 period are reported to have been around €8bn (which compares to €3.5bn for wind power, €12.7bn all other RES, and €173.9bn for nuclear energy), of which more than 80 per cent have been historically funded by Japan, Germany, and the USA (IEA, 2013). In 2009 public R&D spending on solar energy has been estimated to be €663.8m (OECD countries, plus China, India, and Russia) (IEA, 2009). However, this figure compares with an estimate of corporate PV R&D total investments of €3100m in 2009 (Breyer et al., 2010). In other words, in the last decade companies have been earning financial resources thanks to policy-driven market expansion (demand-pull) and have been investing significant resources in R&D, outpacing public funding and driving technological (often incremental) innovation. Analysis of trends in PV patent filing (historically mainly in the USA, Germany, the UK, and France; more recently the UK and France have lost ground and South Korea, Taiwan, and China have been gaining a growing share of international PV patent families) shows a boom in PV innovation in the last decade resulting from the combined effect of public investments in R&D (historical) and a fast rate of growth in markets for these technologies (Bettencourt et al., 2013; Breyer et al., 2010).

18.4 **PV cost competitiveness and drivers behind it**

PV is an interesting case of renewable technology innovation as it has experienced decades-long learning thanks to sustained policy support and massive investment in

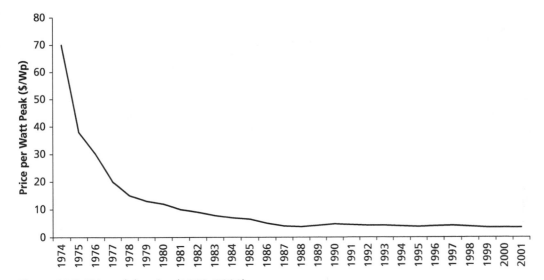

Figure 18.5 PV module price (1968–2000)

Source: (Maycock, 2011; Solarbuzz, 2012).

manufacturing capacity (Section 18.3). Indeed, PV rates of deployment and learning, particularly of the last decade, constitute an exemplary case of policy driven 'innovation outcome' and the analysis of PV cost/price reduction dynamics provides interesting insights into the challenges for technology forecasting tools, as further discussed below.

PV costs can be measured in terms of factory gate cost of individual PV system components (in €/Wp), in terms of cost of investing in a PV system (i.e. capital expenditure, CAPEX in €/Wp), and in terms of generation costs (i.e. levelized cost of electricity, LCOE[4] in €/kWh). Most of the cost reduction and PV roadmapping efforts have been concentrating on the PV module, being the major component and contributor to the total PV system costs (ranging from 35 per cent to 55 per cent depending on the PV system type and application (EPIA Greenpeace, 2011; EU PV Technology Platform, 2011)).

PV module prices have experienced sustained reductions over time (Figure 18.5), dropping from about 70 $/Wp in the 1970s to about 4 $/Wp in early 2000s (Maycock, 2011; Solarbuzz, 2012). The last decade's impressive PV market growth (Section 18.3) coupled with continuous industrial and R&D developments over time have allowed further price reductions (Nemet, 2006; van Sark et al., 2010; IEA, 2010a; EU PV Technology Platform, 2011; Candelise et al., 2013). Such historical reductions reflect the development and deployment of c-Si technologies, which still account for the majority of the PV market (about 87 per cent in 2011

---

[4] LCOE is generally defined as the discounted lifetime PV system CAPEX divided by the discounted lifetime generation of the PV system.

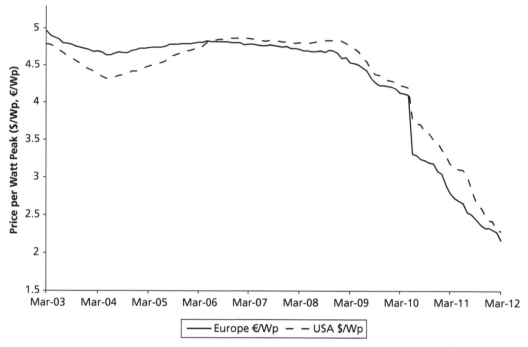

Figure 18.6 PV module retail price index (2003–12, €2012 and $2012)

Note: average retail prices in Europe and the USA based on a monthly online survey, encompassing a wide range of module prices, varying according to the module technology (with thin film modules generally cheaper than c-Si), the module model and manufacturer, its quality, as well as the country in which the product is purchased.

Source: (Solarbuzz, 2012).

(Photon International, 2012)). Indeed, as estimated by the experience curves literature, c-Si module technologies have shown an historical learning rate in the order of 20 per cent (ranging from 18 per cent to 22 per cent depending on studies and the reference dataset used (Candelise et al., 2013)). In other words, this means a reduction of about 20 per cent in c-Si module prices for every doubling of production capacity.

The only inversion in the historical module price trend occurred in the mid 2000s (Figure 18.6) as a consequence of a serious silicon feedstock shortage, which caused silicon (and module) prices to rise. This bottleneck triggered innovation to improve material utilization (Flynn, 2009) and new investments in feedstock production, which allowed cheap silicon feedstock to feed through into a dramatic fall in c-Si module manufacturing costs from the late 2000s and early 2010s. Indeed, PV module prices have dropped dramatically in the last few years, falling about 45 per cent between mid 2010 and March 2012 (Figure 18.6) (Solarbuzz, 2012). Among thin film technologies, CdTe thin film has been particularly successful in quickly expanding production capacity (mainly driven by a single US company, First Solar) and in achieving very low production costs. Indeed, First Solar has been the first PV manufacturer to reduce production cost below the 1$/Wp threshold, in 2009 (First Solar, 2011). However the recent dramatic drop in c-Si module prices has severely affected the market positioning of thin film technologies, which have lower efficiencies and are currently struggling to keep the pace with the incumbent/conventional c-Si technology.

This recent dramatic drop in c-Si prices is the outcome of a complex mix of factors affecting underlying production costs as well as resulting from market forces (see (Candelise et al., 2013; Marigo and Candelise, 2013) for more details). In particular, production costs have been reducing thanks to technological development and optimization of production processes; a reduction in the cost of input materials, in particular an unprecedented drop in silicon feedstock prices; and a massive increase in the scale of production (in China in particular), which have allowed economies of scale and cost optimization along the whole value chain (Marigo and Candelise, 2013). Prices have also been heavily affected by a sustained oversupply situation in the global PV market (annual PV production capacity has been higher than worldwide annual installation since 2010, e.g. in 2012 50 GW versus 30 GW (Q-Cell, 2012)) which has put strong downward pressure on c-Si module prices.

Thus, while historical trajectories describe a sustained long-term cost and price reduction along a prescribed curve, a closer look shows price volatility in the most recent years strongly linked to market dynamics and industry expansion, in China in particular. Thanks to a combination of policy support (mainly through finance and availability of credit), economies of scale, process optimization, and industry restructuring Chinese manufacturers have managed to massively increase production capacity (see Section 18.3) and have been able to supply the global market with module prices between 32 and 34 per cent cheaper than their EU or Japanese competitors (Marigo and Candelise, 2013). Average c-Si module prices are reported to be sold at 0.77 €/Wp from Germany and at 0.56 €/Wp from China in June 2013 (which compares with an average market price in Europe of 1.95 €/Wp only in March 2010) (PvXchange, 2013). This has triggered a strong worldwide industry consolidation, with several companies going out of business since late 2011, and has led both the USA and the European Commission to launch antidumping investigations and to impose antidumping tariffs in an attempt to support their domestic PV industries.

Such rapidly changing dynamics of PV prices pose a challenge for technology forecasting tools and for policymaking relying on energy technology cost assessment in order to develop coherent energy strategies and support for emerging energy technologies. None of the available technology forecasting tools (i.e. experience curves and engineering assessment) have been able to anticipate such unprecedented reduction in costs and prices (Candelise et al., 2013), suggesting the need to form expert judgements across a wide canvas of expertise in order to anticipate more immediate, volatile, and often country-specific forces affecting PV costs and prices (Candelise et al., 2013).

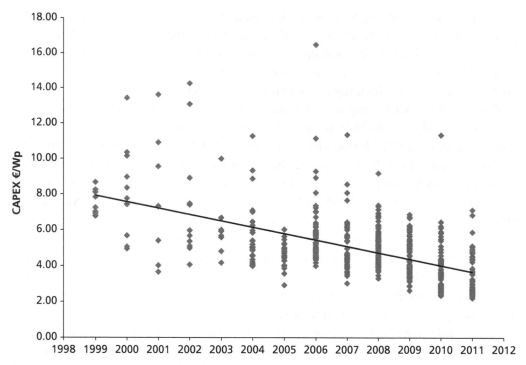

Figure 18.7 PV system price across European countries

Source: (ARUP/DECC, 2011; Candelise et al., 2010; Castello and De Lillo, 2010; Castello et al., 2003, 2004, 2007, 2008, 2009, 2010; Energy Saving Trust, 2008; IEA-PVPS, 2005; Mott Mcdonald, 2011; Rudkin et al., 2007; Sonnenertrag.eu, 2011).

Nonetheless, these reductions in module prices have considerably improved the cost competitiveness of PV, feeding into the costs of investing in PV systems. The CAPEX of grid-connected PV systems varies across market segment (and system size), system types, and countries.[5] Despite this variability, PV system CAPEX have been decreasing over time across segments and countries (see Figure 18.7 for the trend in Europe). However, this has resulted from a combination of global module price reductions discussed above and BOS learning effects driven by more local/national driving forces. In particular, evidence shows the strong correlation between system CAPEX cost reductions and national market expansion (Barbose et al., 2011, 2012; Candelise et al., 2013; Werner et al., 2011; Wiser et al., 2009), emphasizing the importance of developing

---

[5] They do not scale linearly with system size, and thus tend to be higher in residential markets compared to medium-size commercial systems and large utility scale systems; they also differ across countries (as affected by national market size and implementation conditions) and across PV system types (with e.g. building integrated PV systems being more expensive than standard roof top applications).

domestic markets to guarantee price reductions at system level (beyond international module price dynamics).

Price reductions as well as the price variability in PV system prices are also reflected in the estimates of grid-connected PV LCOE, which varies considerably according to the type of PV system assumed (being a function of PV system CAPEX) and is very location and country specific (as PV electricity generation is strongly dependent on climatic conditions and irradiation levels). Nonetheless, the LCOE of grid-connected PV has also reduced over time, moving in Europe from 0.20–0.60 €/kWh in 2005 to a 0.16–0.35 €/kWh in 2011 (EU Commission, 2005; EPIA, 2011).

Cost and prices of PV are expected to further decrease, although estimates for future cost reductions of PV systems vary according to source, market segment, and country of reference. EPIA (European Photovoltaic Industry Association) predicted European grid connected PV system prices will reduce by 36–51 per cent over the next ten years, with the LCOE of PV generation declining by around 20 per cent by 2020 (EPIA, 2011). Recent US estimates see utility-scale PV system prices reducing to 1.71–1.91 $/Wp and residential systems to 2.29 $/Wp by 2020 (Goodrich et al., 2012). The average annual reduction of balance of system (BOS) costs (see Section 18.2) has been estimated to be in the range of 8–9.5 per cent (Ringbeck and Sutterlueti, 2013). Grid parity (i.e. retail price competitiveness) is expected to be achieved in southern Europe (and in high electricity price countries such as Italy) by the end of 2013 and spread across all Europe by 2020 (EPIA, 2011).[6] Similarly, PV LCOE is estimated to compete with residential electricity prices in a wide range of US regions by 2020 and grid parity achieved in high-cost regions by 2015 (IRENA, 2012).

18.5 PV sector future scenarios and challenges

Further reduction in PV prices is going to guarantee a sustained growth of the PV sector in the future.

Grid-connected PV uptake is estimated to continue growing worldwide. The European Photovoltaic Industry Association (EPIA) estimates PV could supply 8 per cent of European electricity demand by 2020, which represents about 200 GW of installed capacity. At the same time Europe's share of the global market is reducing (it dropped from 74 per cent in 2011 to 55 per cent in 2012 and is expected to further decrease), as PV is being taken up by other areas of the world, including China, India, South Asia, Latin America, and MENA countries. The Asia Pacific region is expected to be the

---

[6] In Italy utility and commercial PV systems have already been installed under grid parity, that is they are cost competitive with grid electricity without government incentives. For example, the first grid parity 200 kWp commercial roof-top plant was installed in October 2013 in North Italy, <http://www.green energyjournal.it/index.php/42-notizie-green/1108-in-veneto-il-primo-impianto-aziendale-grid-parity-in-italia#.UoSmIni-a8w>.

leading regional market, with China, for example, recently raising an already aggressive PV deployment target by 20 per cent (now aiming at 12 GW of PV capacity nationwide by 2014, up from a previous target of 10 GW) of which 8 GW is of distributed PV (Young, 2013). Worldwide cumulative installed capacity is estimated on current trends to double by 2020 (i.e. 200 GW) and could go above 400 GW under policy-driven scenarios (EPIA, 2013a).

A recent analyst report predicts global annual installations to double by 2020 (up to 73.4 GW) and such growth as mainly driven by small/medium size, distributed PV systems below 1 MW (Navigant Consulting Inc., 2013). Indeed, recent price declines are opening up new markets for distributed PV, as grid parity is going to be quickly achieved in high-cost retail electricity markets. Moreover, low prices (not just for solar panels but for low-energy DC-powered light-emitting diode (LED) lamps) are also opening up new opportunities for off-grid applications, such as PicoPV systems (see Section 18.2), in particular in developing countries where upfront cost of PV systems is still often a barrier for their wider deployment.

The impressive reduction in module prices could also open the possibility of new PV applications and business opportunities previously not economical for PV manufacturing, which has been to date mainly dominated by flat plate modules. For example, thanks to their flexibility of uses, thin film PV as well as novel organic and dye-sensitized solar cells (DSSC) PV (see Section 18.2) have the potential for product differentiation, gaining market share in new niche applications, including integration into consumer products, portable rechargers, vehicles or into building components (BIPV). Emerging thin film PV in particular can also be produced in different colours and have a performance advantage over c-Si when operating at low irradiance conditions, making them particularly suited for indoor applications, for example. Moving novel technologies toward commercialization as well as developing product differentiation are potentially new business strategies that the European PV industry could adapt to leverage domestic R&D capabilities and counteract the Chinese PV industry, which is currently mainly focused on the volume c-Si market (Marigo and Candelise, 2013).

Overall PV is starting to be a noticeable contribution to energy mixes (grid connected PV in particular), for example in 2012 it contributed 2.6 per cent of European electricity demand and 5.2 per cent of peak electricity. However, it is a variable, non dispatchable, and distributed renewable energy source and it is already starting to pose challenges to the management and structure of electricity systems. The need for a more flexible, interconnected electricity system capable of integrating variable sources of generation is discussed in Chapter 21. However, it is here relevant to note that at the distribution grid level (where PV is likely to play an increasing role in the years to come) several solutions have already been identified to increase hosting capacity and overcome voltage and congestion limitations. These span from measures that need to be implemented by distribution grid operators—DSO—(such as network reinforcement, advanced voltage control, DSO storage) to prosumer solutions including DMS (demand side management), prosumer/small scale storage, active power control by PV inverters (Sonvilla et al., 2013). Policy support and regulatory changes are needed for the adoption of some of these measures, for example regulating the need for an inverter to be equipped with voltage control and/or curtailment options, setting up a regulatory framework to incentivize self-consumption (EPIA, 2013b) or providing support for storage (Germany has already implemented in May 2013 the first public support scheme for small-scale storage associated with domestic PV systems (Barth, 2013)). On the other hand, the costs associated with increased PV penetration into the grid might have an impact on PV competitiveness, as PV system owners might be required to pay grid costs and taxes even when self-consuming electricity (EPIA, 2012a).

■ REFERENCES

ANGERER, G., MARSCHEIDER-WEIDEMANN, F., LULLMANN, A., ERDMANN, L., SCHARP, M., HANDKE, V., & MARWEDE, M. 2009. Raw materials for emerging technologies. English summary. *Fraunhofer Institute for Systems and Innovation Research (ISI) and Institute for Future Studies and Technology Assessment (IZT). Commissioned by the German Federal Ministry of Economics and Technology Division IIIA5—Mineral Resources.* Available at: <http://www.isi.fraunhofer.de/isi-en/service/presseinfos/2009/pri09-02.php?WSESSIONID= 1656a06b68afc5f050c3cfee1c3c259e&WSESSIONID=1656a06b68afc5f050c3cfee1c3c259e>. Accessed: April 2015.

ARUP/DECC 2011. Review of the generation costs and deployment potential of renewable electricity technologies in the UK. *Report commissioned by UK Department of Energy and Climate Change (DECC) to Arup. June 2011.*

AZZOPARDI, B., EMMOTT, C. J. M., URBINA, A., KREBS, F. C., MUTALE, J., & NELSON, J. 2011. Economic assessment of solar electricity production from organic-based photovoltaic modules in a domestic environment. *Energy & Environmental Science,* 4, 3741–53.

BARBOSE, G., DARGHOUTH, N., & WISER, R. 2012. Tracking the Sun V. An Historical Summary of the Installed Price of Photovoltaics in the United States from 1998 to 2011. *Lawrence Berkeley National Laboratory. November 2012.*

BARBOSE, G., DARGHOUTH, N., WISER, R., & SEEL, J. 2011. Tracking the Sun IV. An Historical Summary of the Installed Cost of Photovoltaics in the United States from 1998 to 2010. *Lawrence Berkeley National Laboratory. September 2011.*

BARTH, B. Results from PV GRID research for Germany. PV curtailment and prosumer storage. First European PV GRID Forum 2013 London, October 2013.

BETTENCOURT, L. M. A., TRANCIK, J. E., & KAUR, J. 2013. Determinants of the Pace of Global Innovation in Energy Technologies. *PLoS ONE* 8(10), e67864. October 2013.

BREYER, C., BIRKNER, C., KERSTEN, F., GERLACH, A., GOLDSCHMIDT, J., STRYI-HIPP, G., MONTORO, D. F., & RIEDE, M. Research and development investments in PV—a limiting factor for a fast PV diffusion. 25th European Photovoltaic Solar Energy Conference 2010 Valencia, Spain, September 2010.

CANDELISE, C., GROSS, R., & LEACH, M. 2010. Conditions for photovoltaics deployment in the UK: the role of policy and technical developments. *Proceedings of the Institution of*

Mechanical Engineers, Part A: J. Power and Energy, 224 (A2), 153–66. DOI 10.1243/09576509JPE768.

CANDELISE, C., SPEIRS, J. F., & GROSS, R. J. K. 2011. Materials availability for thin film (TF) PV technologies development: A real concern? *Renewable and Sustainable Energy Reviews*, 15, 4972–81.

CANDELISE, C., WINSKEL, M., & GROSS, R. 2012. Implications for CdTe and CIGS technologies production costs of indium and tellurium scarcity. *Progress in Photovoltaics: Research and Applications*, 20, 816–31.

CANDELISE, C., WINSKEL, M., & GROSS, R. 2013. The dynamics of solar PV costs and prices as a challenge for technology forecasting. *Renewable and Sustainable Energy Reviews*, 26, 96–107.

CARLÉ, J. E. & KREBS, F. C. 2013. Technological status of organic photovoltaics (OPV). *Solar Energy Materials and Solar Cells*, 119, 309–10.

CASTELLO, S. & DE LILLO, A. 2010. National Survey Report of PV Power Applications in Italy, 2009. *International Energy Agency, Co-operative Programme on Photovoltaic Power Systems, Task 1. Prepared by ENEA, May 2010.*

CASTELLO, S., DE LILLO, A., & GUASTELLA, S. 2003. National Survey Report of PV Power Applications in Italy, 2002. *International Energy Agency, Co-operative Programme on Photovoltaic Power Systems, Task 1. Prepared by ENEA, May 2003.*

CASTELLO, S., DE LILLO, A., & GUASTELLA, S. 2004. National Survey Report of PV Power Applications in Italy, 2003. *International Energy Agency, Co-operative Programme on Photovoltaic Power Systems, Task 1. Prepared by ENEA, May 2004.*

CASTELLO, S., DE LILLO, A., & GUASTELLA, S. 2007. National Survey Report of PV Power Applications in Italy, 2006. *International Energy Agency, Co-operative Programme on Photovoltaic Power Systems, Task 1. Prepared by ENEA, May 2007.*

CASTELLO, S., DE LILLO, A., GUASTELLA, S., & PALETTA, F. 2008. National Survey Report of PV Power Applications in Italy, 2007. *International Energy Agency, Co-operative Programme on Photovoltaic Power Systems, Task 1. Prepared by ENEA, May 2008.*

CASTELLO, S., DE LILLO, A., GUASTELLA, S., & PALETTA, F. 2009. National Survey Report of PV Power Applications in Italy, 2008. *International Energy Agency, Co-operative Programme on Photovoltaic Power Systems, Task 1. Prepared by ENEA, May 2009.*

CASTELLO, S., DE LILLO, A., GUASTELLA, S., & PALETTA, F. 2010. National Survey Report of PV Power Applications in Italy, 2009. *International Energy Agency, Co-operative Programme on Photovoltaic Power Systems, Task 1. Prepared by ENEA, May 2010.*

CHAUREY, A. & KANDPAL, T. C. 2010. A techno-economic comparison of rural electrification based on solar home systems and PV microgrids. *Energy Policy*, 38, 3118–29.

CHOPRA, K. L., PAULSON, P. D., & DUTTA, V. 2004. Thin film solar cells: an overview. *Progress in Photovoltaics: Research and Applications*, 12, 69–92.

DECC 2013. Feed-in Tariff statistics. *Department of Energy & Climate Change Statistics.* Available at: <https://www.gov.uk/government/collections/feed-in-tariff-statistics>. Accessed: November 2013.

DENHOLM, P. & MEHOS, M. 2011. Enabling greater penetration of solar power via the use of CSP with thermal energy storage. *National Renewable Energy Laboratory. Technical Report. NREL/TP-6A20-52978. November 2011.*

EASAC 2011. Concentrating solar power: its potential contribution to a sustainable energy future. *European Academics Science Advisory Council Policy Report 16. November 2011.* Available at: <http://www.easac.eu>. Accessed: April 2015.

EC 2010. Critical raw materials for the EU: Report of the Ad-hoc Working Group on defining critical raw materials. Brussels: European Commission.

ENERGY SAVING TRUST 2008. Statistics on PV installation funded through Low Carbon Building Programme. *Excel spreadsheet personally gathered from Energy Saving Trust representative. March 2008.*

EPIA 2009. 2013. Global market outlook for photovoltaics until 2013. *European Photovoltaic Industry Association Report. April 2009.* Available at: <http://www.epia.org/index.php?id=18>. Accessed: April 2015.

EPIA 2011. Solar photovoltaics competing in the energy sector, on the road of competitiveness. *European Photovoltaic Energy Association Report.* Available at: <http://www.epia.org/fil eadmin/user_upload/Publications/Competing_Full_Report.pdf >. Accessed: April 2015.

EPIA 2012a. Connecting the sun. Solar photovoltaics on the road to large-scale grid integration. *European Photovoltaic Industry Association Report. September 2012.* Available at: <http://www. epia.org/uploads/tx_epiapublications/Connecting_the_Sun_Shorter_version.pdf>. Accessed: April 2015.

EPIA 2012b. EPIA Market Report 2011. *European Photovoltaic Industry Association Report. January 2012.* Available at: <http://www.epia.org/index.php?id=18>. Accessed: April 2015.

EPIA 2013a. Global market outlook for photovoltaics 2013–2017. *European Photovoltaic Industry Association Report. May 2013.* Available at: <http://www.epia.org/?id=22>. Accessed: April 2015.

EPIA 2013b. Self consumption of PV electricity. *European Photovoltaic Industry Association, Position Paper. July 2013.* Available at: <http://www.epia.org/fileadmin/user_upload/Position_ Papers/Self_and_direct_consumption_-_position_paper_-_final_version.pdf>. Accessed: April 2015.

EPIA GREENPEACE 2011. Solar generation 6. Solar photovoltaic electricity empowering the world. *European Photovoltaic Industry Association, Greenpeace Report.* Available at: <http:// www.greenpeace.org/international/Global/international/publications/climate/2011/Final% 20SolarGeneration%20VI%20full%20report%20lr.pdf>. Accessed: April 2015.

EU COMMISSION 2005. A vision for photovoltaic technology. *Report by the Photovoltaic Technology Research Advisory Council (PV-TRAC).*

EU PV TECHNOLOGY PLATFORM 2011. A strategic research agenda for photovoltaic solar energy technology, Edition 2 *Report prepared by Working Group 3 'Science, Technology and Applications' of the EU PV Technology Platform.* Office for Official Publications of the European Union, Luxemburg. Available at: <http://www.eupvplatform.org/publications/strategic-research- agenda-implementation-plan.html#c2783>. Accessed: April 2015.

FIRST SOLAR. 2011. *First Solar Overview* [Online]. Available: <http://www.firstsolar.com/Down loads/pdf/FastFacts_PHX_NA.pdf>. Accessed: March 2012.

FLYNN, H. 2009. Photon Consulting's monthly silicon update. *Photon International, the photo- voltaic magazine*, January 2009.

GOODRICH, A., JAMES, T., & WOODHOUSE, M. 2012. Residential, Commercial, and Utility- Scale Photovoltaic (PV) System Prices in the United States: Current Drivers and Cost- Reduction Opportunities. *NREL Technical Report. NREL/TP-6A20-53347. February 2012.*

IEA 2009. Global Gaps in Clean Energy Research, Development and Demonstration. *International Energy Agency Report, prepared in support of the Major Economies Forum (MEF) Global Partnership. December 2009.* Available at: <http://www.iea.org/publications/freepublications/publication/Global_gaps_in_Clean_Energy.pdf>. Accessed: April 2015.

IEA 2010a. Technology Roadmap. Solar photovoltaic energy. *International Energy Agency Report.* Available at: <http://www.iea.org/publications/freepublications/publication/pv_roadmap.pdf>. Accessed: April 2015.

IEA 2010b. World Energy Outlook 2010. *OECD/IEA, Paris, November.*

IEA 2011. Solar energy perspectives. *International Energy Agency Report, November 2011.*

IEA 2012. Renewable energy. Medium term market report 2012. Market trends and projections to 2017. *OECD/IEA Report. 2012.*

IEA 2013. International Energy Agency RD&D Statistics. Available at: <http://www.iea.org/statistics/topics/rdd/>. Accessed: April 2015.

IEA-PVPS 2005. Trends in photovoltaic applications. Survey report of selected IEA countries between 1992 and 2004. *Report IEA—PVPS T1—14:2005.*

IPCC 2011. Special Report on Renewable Energy Sources and Climate Change Mitigation. *Prepared by Working Group III of the Intergovernmental Panel on Climate Change* (O. Edenhofer, R. Pichs-Madruga, Y. Sokona, K. Seyboth, P. Matschoss, S. Kadner, T. Zwickel, P. Eickemeier, G. Hansen, S. Schlömer, C. von Stechow (eds)). Cambridge University Press, Cambridge, UK.

IRENA 2012. Photovoltaics. *IRENA Working Paper. Renewable Energy Technologies: Cost Analysis Series. Volume 1: Power Sector. Issue 4/5. June 2012.*

JAGER-WALDAU, A. 2012. PV Status Report 2012. *European Commission, DG Joint Research Centre, Institute for Energy, Renewable Energy Unit Report. EUR 25749-2012.*

KREBS, F. C. & JØRGENSEN, M. 2013. Polymer and organic solar cells viewed as thin film technologies: What it will take for them to become a success outside academia. *Solar Energy Materials and Solar Cells,* 119, 73–6.

KREWITT, W., NIENHAUS, K., KLEßMANN, C., CAPONE, C., STRICKER, E., GRAUS, W., HOOGWIJK, M., SUPERSBERGER, N., VON WINTERFELD, U., & SAMADI, S. 2009. Role and Potential of Renewable Energy and Energy Efficiency for Global Energy Supply. *Climate Change 18/2009, ISSN 1862-4359, Federal Environment Agency, Dessau-Roßlau, Germany.*

MARIGO, N. & CANDELISE, C. 2013. What is behind the recent dramatic reductions in photovoltaic prices? The role of China. *Economia e Politica Industriale—Journal of Industrial and Business Economics,* 40, 5–41.

MAUTHNER, F. & WEISS, W. 2013. Solar Heat Worldwide. Markets and Contribution to the Energy Supply 2011. *AEE—Institute for Sustainable Technologies Report, International Energy Agency Solar Heating and Cooling Programme. Edition 2013.* Available at: <http://www.iea-shc.org/Data/Sites/1/publications/Solar-Heat-Worldwide-2013.pdf>. Accessed: April 2015.

MAYCOCK, P. D. 2011. PV module price data. *Directly from correspondence from Paul Maycock, 2011.*

MOTT MCDONALD 2011. Costs of low-carbon generation technologies. *Report commissioned by Committee on Climate Change to Mott McDonald, May 2011.*

MUIRHEAD, J. 2013. MENA CSP: what lies beyond Saudi? *CSP Today, 21st June 2013.* Available at: <http://social.csptoday.com/markets/mena-csp-what-lies-beyond-saudi>. Accessed: April 2015.

NAVIGANT CONSULTING INC. 2013. Solar PV Market Forecasts. Installed Capacity, System Prices, and Revenue for Distributed and Non-Distributed Solar PV. *Navigant Consulting*

Report. November 2013. Available at: <http://www.navigantresearch.com/research/solar-pv-market-forecasts>. Accessed: April 2015.

NEMET, G. F. 2006. Beyond the learning curve: factors influencing cost reductions in photovoltaics. *Energy Policy*, 34, 3218–32.

ONG, S., CAMPBELL, C., DENHOLM, P., MARGOLIS, R., & HEATH, G. 2013. Land-Use Requirements for Solar Power Plants in the United States. *NREL Technical Report. NREL/TP-6A20-56290. June 2013.* Available at: <http://www.nrel.gov/docs/fy13osti/56290.pdf>. Accessed: April 2015.

PALIT, D. 2013. Solar energy programs for rural electrification: Experiences and lessons from South Asia. *Energy for Sustainable Development*, 17, 270–9.

PHOTON INTERNATIONAL 2009. Photon Market survey on modules 2009. *Photon International, the photovoltaic magazine*, January 2009.

PHOTON INTERNATIONAL 2011. Market survey. Cell and module production 2010. *Photon International, the photovoltaic magazine*, March 2011.

PHOTON INTERNATIONAL 2012. Market survey. Cell and module production 2011. *Photon International, the photovoltaic magazine*, March 2012, 132–61.

PVXCHANGE. 2013. *PV price index* [Online]. <http://pvXchange.com>. Accessed: March 2013.

Q-CELL 2012. Q-Cell Extraordinary General Meeting. *Q-Cell Power Point Presentation, 9 March 2012, Leipzig.*

REICHE, K., GRÜNER, R., ATTIGAH, B., HELLPAP, C., & BRÜDERLE, A. 2010. What difference can a PicoPV system make? Early findings on small Photovoltaic systems—an emerging lowcost energy technology for developing countries. *Report by Deutsche Gesellschaft für Technische Zusammenarbeit GmbH (GTZ). Eschborn, May 2010.*

RINGBECK, S. & SUTTERLUETI, J. 2013. BoS costs: status and optimization to reach industrial grid parity. *Progress in Photovoltaics: Research and Applications*, 21, 1411–28.

RUDKIN, E., THORNYCROFT, J., NJOKU, C., & COGZELL, J. 2007. PV Large Scale building integrated field trial. Third technical report—Case studies. *Halcrow Group report for BERR (Department for Business Enterprise & Regulatory Reform). 2007.*

SOLARBUZZ. 2012. *Solarbuzz module price survey* [Online]. <http://solarbuzz.com/facts-and-figures/retail-price-environment/module-prices>. Accessed: April 2015.

SONNENERTRAG.EU. 2011. *Large international photovoltaic database* [Online]. <http://www.solar-yield.eu/home/main>. Accessed: April 2015.

SONVILLA, P. M., ZANE, E. B., POBLOCKA, A., & BRUCKMANN, R. 2013. PV Grid. Initial project report. October 2013. Available at: <http://www.pvgrid.eu/fileadmin/PV_GRID_Initial_Report_April_2013.pdf>. Accessed: April 2015.

THIRUMURTHY, N., HARRINGTON, L., MARTIN, D., THOMAS, L., TAKPA, J., & GERGAN, R. 2012. Opportunities and Challenges for Solar Minigrid Development in Rural India. *National Renewable Energy Laboratory Technical Report, NREL/TP-7A40-55562. September 2012.*

VAN SARK, W., NEMET, G. F., SCHAEFFER, G. J., & ALSEMA, E. A. 2010. Photovoltaic solar energy. *In:* JUNGINGER, M. (ed.) *Technological Learning in the Energy Sector: Lessons for Policy, Industry and Science.* Edward Elgar, Cheltenham.

WADIA, C., ALIVISATOS, A. P., & KAMMEN, D. M. 2009. Materials availability expands the opportunity for large-scale photovoltaics deployment. *Environmental Science & Technology*, 43, 2072–7.

WERNER, C., GERLACH, A., ADELMANN, P., & BREYER, C. Global cumulative installed photovoltaic capacity and respective international trade flows. *26th European Photovoltaic Solar Energy Conference and Exhibition, 2011 Hamburg September 2011.*

WISER, R., BARBOSE, G., & PETERMAN, C. 2009. Traking the sun. The installed cost of photovoltaics in the U.S. from 1998–2007. *Lawrence Berkeley National Laboratory. February 2009.*

YOUNG, D. 2013. Beijing Ups Solar Energy Targets as Suntech Liquidates. *Renewable Energy World, 8 November 2013.* Available at: <http://www.renewableenergyworld.com/rea/blog/post/2013/11/beijing-ups-solar-energy-targets-as-suntech-liquidates?cmpid=SolarNL-Tuesday-November12-2013>. Accessed: April 2015.

ZWEIBEL, K. & NATIONAL RENEWABLE ENERGY. 2005. *The Terawatt Challenge for Thin-Film PV,* National Renewable Energy Laboratory, Golden CO.

19 Water: ocean energy and hydro

Laura Finlay, Henry Jeffrey, Andy MacGillivray, and George Aggidis

It has long been acknowledged that the power of water can make a considerable contribution to energy generation around the world. There are a number of different forms of energy generation from both fresh and salt water, and this chapter aims to provide an overview of the energy sectors powered by both hydropower and the oceans. The chapter is split into two distinct sections; the ocean energy section provides an overview of ocean energy, including the associated resource, technology, technology and research challenges, costs, and policies relating to the sector. The hydropower section provides a review of the installed capacity and generation, technology and operations, future developments, and the social and environmental constraints.

19.1 Ocean energy

19.1.1 GENERAL INTRODUCTION

Government commitments targeted at reducing carbon emissions and other polluting gases, a declining carbon based energy resource, a likely future increase in energy demand, and uncertainty in the security of supply from more traditional energy sources continue to ensure that less traditional sources of energy, such as ocean energy, are perceived as feasible future alternatives. Ocean energy is receiving increasing attention as a viable energy source with significant potential due to the reliability, predictability, and the overall size of the global resource.

Approximately 71 per cent of the surface of the earth is covered by seawater, and there is potential to harness abundant energy for those countries where a suitable resource and coastline exists. The ocean energy resource is made up of:

- Waves: Surface and upper water column motion of water particles resulting from a transfer of energy from the wind to the surface of the ocean.
- Ocean currents: Continuous motion of ocean water mainly driven by wind, thermohaline circulation, gravity, and the Coriolis effect due to the rotation of the earth.
- Tidal range: Energy derived from tidal rise and fall of the sea level, caused by gravitational forces of the moon, sun, and earth.
- Tidal currents: The flow of ocean water, often accelerated through constrictions between or around land masses, driven by the gravitational forces of the moon, sun, and earth.

- Salinity gradient (osmotic power): Energy derived from the difference in salinity between ocean water and fresh water at the mouth of rivers.
- Ocean thermal energy: Energy derived from the difference between the warm surface temperatures and cold deep water layers of the ocean.

These resources are not distributed evenly throughout the world, and each type of ocean energy resource requires very different technologies for energy conversion. This is due to the nature of the resource, that is ocean thermal energy is generally distributed in areas where the surface temperature is relatively high, such as in equatorial regions; however, wave energy potential is increased in the higher latitudes and in the vicinity of large oceans where there is a significant fetch (the distance across an open water surface over which the wind blows in order to generate the resulting waves). With each type of ocean energy, different characteristics prevail in different regions; however, there is, in general, at least some form of ocean energy available for significant energy extraction in every continent.

For the purpose of this chapter, the term ocean energy will be used to describe wave and tidal stream energy technologies only. While tidal impoundment or barrage technology could be considered a more mature tidal energy technology, there are considerable differences in the scale of construction, resulting in very different financial and environmental requirements using mature hydro turbines and civil technology. The other identified sources of ocean energy, while potentially offering opportunities for future energy extraction, are still within the very early stages of research. Wave and tidal stream energy technologies are perceived to be on the cusp of commercialization, and as a result will be the mainstay and focus of this chapter.

Globally, ocean energy (specifically wave and tidal) has the potential to make a significant contribution to the security of the world's energy supply, while also playing a role in carbon savings and job creation. Over the next decade it is anticipated that the contribution from ocean energy will be relatively small. Nonetheless, the Ocean Energy Systems Implementing Agreement (OES), an intergovernmental collaboration between countries that operates under a framework established by the International Energy Agency (IEA), estimate that there is the potential to develop 337 GW of wave and tidal energy worldwide by 2050, contributing to the creation of almost 300,000 jobs, and the saving of 0.8 billion tonnes of CO_2 (Huckerby et al., 2012). At a national level, it also appears likely that ocean energy may be able to make a significant contribution. For the UK, the Carbon Trust (Carbon and Trust, 2006) suggests that 15–20 per cent of the country's electricity demand could be met by ocean energy by 2050.

In order to formalize and coordinate this international opportunity, the Ocean Energy Systems Implementing Agreement (OES) was launched in 2001 as a result of a need for a framework for technology development cooperation and international information exchange to coordinate this international opportunity. This need was identified in response to increased research and development activity in the field of ocean wave and tidal current energy. The United Kingdom, Denmark, and Portugal were the founding

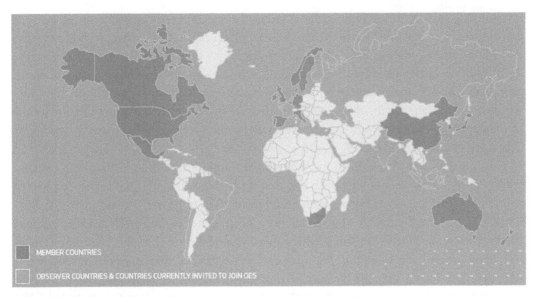

Figure 19.1 OES Membership (as of September 2012)

Source: Huckerby et al., 2012.

signatories to the OES. The purpose of the OES is to bring together countries to advance research, development, and demonstration of conversion technologies, harnessing energy from all forms of ocean renewable resources. OES now consists of twenty member countries, with a number of countries observing the activity of OES and in the process of joining. A global map indicating the location of member and observer states is shown in Figure 19.1 (Huckerby et al., 2012).

Until the late 1990s (with the exception of the wave energy research pioneered in the wake of the 1970s' oil crisis) there was limited innovation, research, and development in the field of ocean energy. However, over the last ten years there has been a renewed political interest—in part due to climate change targets and energy security requirements. This has led to public and private investment, initiation of policy change, and an increase in innovation. In turn this has resulted in the accelerated development of both wave and tidal prototype devices, led initially by small and medium enterprises (SMEs) and universities (Jeffrey et al., 2013). However, recently there has been increasing involvement from large companies such as energy utility companies and multi-national original equipment manufacturers (OEMs), as well as the development of a set of international industry standards for ocean energy within the International Electrotechnical Commission Technical Committee 114 (IEC, 2013).

Many wave and tidal technologies are now at the demonstration and prototype stages, with the sector now on the cusp of small-scale array deployment. However, there is still a requirement for further research, innovation, development, and demonstration in order to overcome technical challenges and cost reduction requirements before it can play a truly significant role in the overall energy mix.

The UK currently holds a leading position in the ocean energy sector, primarily due to its extensive natural resource, existing skills in offshore engineering, research capacity, and a supportive policy and funding framework. That said, there are also an increasing number of countries investing in research and development associated with the ocean energy sector, with formal support mechanisms being put in place at a national level by several international countries. This is evidenced by the growing membership to the Implementing Agreement on Ocean Energy Systems (OES), which has now has members from twenty countries (OES, 2013).

This international growth is in line with the development of international test centres, and the provision of technology push and market pull mechanisms and secure long-term revenue support for emerging wave and tidal technologies. In the UK there are full-scale test centres at the European Marine Energy Centre (EMEC) in Orkney, Scotland, and Wave Hub in south west England, with a drivetrain test facility at the National Renewable Energy Centre (NaREC) in the north east of England. Other such ocean energy test centres are in place and being developed internationally, for example centres such as the Biscay Marine Energy Platform (BiMEP) in Spain, Galway Bay Test Site in Ireland, the Fundy Ocean Research Centre for Energy (FORCE) in Canada, the three Marine Renewable Energy Centres test centres in the United States (Jeffrey et al., 2013), and a number of proposed test facilities in South East Asia. In support of these test centres there are a comprehensive range of model-scale test facilities which are internationally available to support innovation in the sector.

19.1.2 GLOBAL WAVE AND TIDAL RESOURCE

To clarify some of the nomenclature used within descriptions of ocean energy resource, definitions are provided here to identify the difference between each resource assessment value, and the relative importance of each. Resource can be quantified in three ways: 'Theoretical Resource', 'Technical Resource', or 'Practical Resource' represented diagrammatically in Figure 19.2.

The *theoretical* resource is a *high level overview of the theoretical maximum available energy* that is contained within the overall resource at a given location.

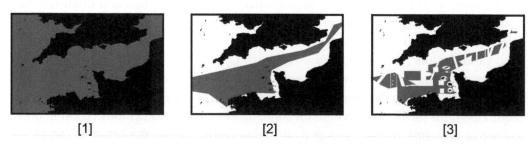

[1]   [2]   [3]

Figure 19.2 Theoretical [1], Technical [2], and Practical Resource [3] Visualization Diagram

The *technical* resource is the *proportion of the theoretical resource that can be technically exploited using existing technology options*, taking account of the limitations of current technology. Constraints such as water depth, estimated spacing requirements, and device capture and conversion efficiency assumptions need to be considered.

The *practical* resource is the *actual value of the technical resource that can be exploited* once military zones; shipping lanes; wind energy developments and fishing (or environmental) constraints have first been accounted for.

19.1.2.1 Tidal Current Resource

There are two different components to the tidal energy resource: energy extracted from the tidal current (the kinetic aspect of the resource), and the tidal range (the different in height between the high and low tide). The response of the resource to both energy extraction techniques is complex and can be non-linear, for example tidal current energy converters act on kinetic forces in the water column, though the overall system response is dominated by a free surface effect, which means that the surface of the seawater is subject to a constant perpendicular normal stress, with zero parallel shear stress.

Tidal current energy is created when a horizontal movement of ocean water takes place in response to a change in tides, altered by coasts, constrictions such as headlands and islands, and bathymetry (a measurement of the depths of the ocean). Tidal current flows are highly predictable, both temporally and in magnitude, with flows influenced only slightly by short-term weather changes. Tidal current energy is manifested in ebb and flood flows within semi-diurnal (approximately 12 hour) or diurnal (approximately 24 hour) tidal cycles specific to each location.

At present, it is only regions with the most extreme tidal currents, such as peak tidal current velocities in the order of 2.5 m/s or greater, that are deemed appropriate for economic exploitation. The global distribution of tidal resource varies geographically and is not evenly dispersed; the most favourable locations for exploitation of tidal energy are located in a relatively limited number of locations, as illustrated in Figure 19.3.

In contrast to a number of other renewable forms of generation, tidal current energy has the advantage that the energy capture is predictable; with harmonic analysis, a prediction of tidal variability can be made indefinitely. Although the real-time power generation will vary corresponding to local changes to the current velocity, the variability is also predictable and is not correlated with other renewable energy sources. This predictable generation enables an easing of integration into the energy network.

Only certain countries have completed tidal energy resource assessments, and fewer still have made an assessment of the technical resource. It has been estimated that the technical tidal current resource (i.e., the energy that can be extracted within reasonable economic and environmental bounds) in Great Britain, Northern Ireland, and the Channel Islands is approximately 29 TWh per year (Carbon and Trust, 2011). A detailed analysis of the practical resource available across a number of identified tidal stream energy locations would allow more informed decisions to be made on the

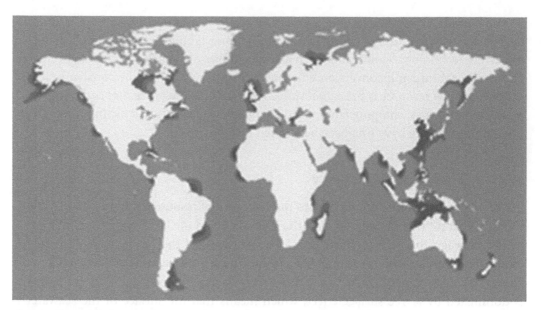

Figure 19.3 Global potentially significant tidal resource
Source: Open Hydro.

suitability for technology deployment, and would in addition allow a comprehensive overview of the resource available globally. In order to address this issue a national resource assessment has been initiated in certain countries where governments and stakeholders have declared an interest in tidal current energy both for energy security and economic development reasons.

19.1.2.2 Wave Resource

Waves are created when wind passes over the surface of the ocean and energy is transferred through air–sea interaction. The energy contained within the waves expresses itself in the form of kinetic motion of the water particles, which follow an orbital motion. This orbital motion decreases with increasing water depth. The wind speed, the fetch (distance of ocean over which the wind blows), and the length of time that the wind blows all contribute to the size and period of the resultant waves, with the wave height primarily being determined by wind speed. Although larger waves are generally more powerful, the overall power within a wave is determined by wavelength, wave period, speed, and water density. Once waves have been generated, they can travel across large distances without significant loss of energy. Unlike tidal currents, the predictability of waves is confined to a relatively short-term forecast, with wave energy generally varying seasonally. However, waves can be anticipated one or two days in advance through direct satellite measurements and meteorological forecasts. These provide a high level of accuracy and allow for adequate network planning.

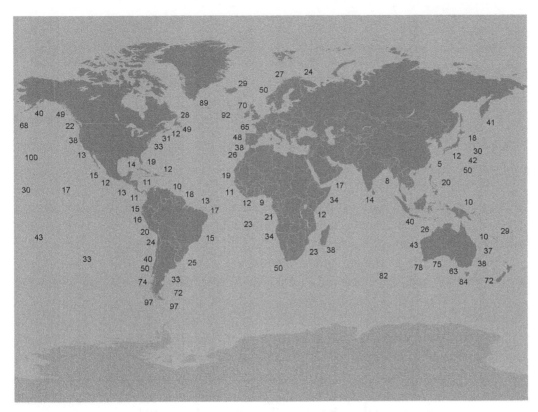

Figure 19.4 Distribution of average annual global wave power level (kW per metre)

Source: Pelamis Wave Power.

The largest ocean wave energy resource exists in deep waters in the offshore environment. With the approach of waves to the shore, the wave length and wave speed decrease, resulting in an increased energy per unit area ('shoaling'), which is measured in kW per metre. However, in comparison with the deep offshore environment, near-shore waves contain a lower overall resource due to the energy loss caused by drag with the seabed.

Figure 19.4 shows the global offshore average annual wave power distribution with increased power levels evident at the west coasts of continents in higher latitudes. This is due to greater fetch and more prominent winds occurring in these regions. The global theoretical wave power potential has been estimated to be 29,500 TWh/year (Mork et al., 2010).

19.1.3 TECHNOLOGY DEVELOPMENT

This section will provide examples of wave and tidal energy technologies that are currently being deployed and developed internationally. The design types vary greatly in concept and power production methods. In the case of some devices, designs have

Technology
Journey

Indicative cost for
activity at this stage

~ € 10k   ~ € 100–500k   ~ € 1–3m   ~ € 10–20m   ~ € 40–100m

Figure 19.5 Likely steps in ocean energy device cost development

Source: Carbon Trust 2011.

evolved over time, with different iterations of the device reaching deployment stages; some technologies are now into their second or third device iteration, incorporating improvements and innovations. Within both the wave and tidal energy sectors, certain device developers are advancing beyond the single device stage and are moving towards the development of arrays.

Although there is a clear growth in interest and development in the ocean energy sector internationally (as evidenced through the membership to the European Energy Research Alliance (EERA) Ocean Energy Joint Programme of an increasing number of member state organizations), it is still, as yet, an emerging sector. The development pathway for these emerging technologies involves modelling, scale testing in tank facilities, and offshore tests before a developer is able to deploy a full-scale prototype design in the sea (Jeffrey et al., 2013). This process is costly both in time and finance, with increased levels of funding required at each step and step-change increases in costs for development associated with each step, as illustrated in Figure 19.5.

When taking into account the challenges associated with the development pathway displayed in Figure 19.5, breakthroughs in devices and components are needed to result in the accelerated development of ocean energy in addition to the establishment of a credible supply chain, thus increasing the market readiness of emerging technologies. Regulatory and institutional reforms including the development of design and test procedures, and protocols which enable an increase in performance and reliability, will help to further increase the rate of the development of the sector. In some cases this process has been further enhanced by the investment and support of major utilities and OEMs, who can provide expertise in manufacturing and procurement, in addition to transferrable knowledge from other sectors.

For any emerging sector to mature and benefit from knowledge transfer, design consensus must be achieved. There currently is limited consensus on the design of wave energy converter technology, with several device categories designed to operate under different wave conditions and site locations. Wave energy converters have been designed to operate in specific bathymetry and conditions, including shallow water, intermediate water, and deep water conditions. The overall design must bear in mind the resource conditions and location in which the converter will operate. Tidal stream energy has achieved greater consensus around the horizontal axis turbine, thus allowing for improved collaboration between technology developers and component suppliers, which could lead to an accelerated trajectory towards an optimized turbine design.

19.1.3.1 Wave Energy Converters

There are a number of different types of technology designed to extract energy from waves. Each classification or type of wave energy converter is designed to extract energy in a specific way using different properties of the motions of the wave. Certain examples of wave energy converter type are shown in Figure 19.6.

Wave energy converters are currently in the demonstration phase of technology development. Although no fully commercial wave devices are yet in production, a number of pre-commercial grid-connected prototype devices have been in intermittent operation for several years. As alluded to above, there are a large number of wave energy device types with over 100 concepts in existence (SI Ocean, 2013). Certain developers have a visible technology development pipeline, which has been aided by the increased availability for grid-connected testing at several test centres, with some devices having experienced several months of at-sea testing. Two such devices that have undergone significant at sea testing are described in Table 19.1.

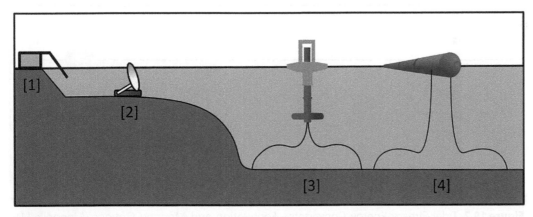

Figure 19.6 Wave energy converter types (Oscillating Water Column [1]; Oscillating Wave Surge Converter [2]; Heaving Buoy [3]; and Attenuator [4])

Table 19.1 Historic wave energy technology examples

Pelamis Wave Power	Aquamarine Power Oyster
• Designed to operate in water depths greater than 50 m	• Designed to operate in water depths between 10 and 15 m
• Attenuator design, five tubular sections linked by universal joints	• Oscillating wave surge converter design, flap moves back and forth in the near shore wave environment
• Hydraulic Power Take Off (PTO) produces grid compliant electricity	• Hydraulic Power Take Off (PTO)
• Rated at 750 kW	• Rated at 800 kW
• The device 'weathervanes' to face the oncoming wave direction	
• Tested at the EMEC test facility, Scotland	• Tested at the EMEC test facility, Scotland

19.1.3.2 Tidal Energy Converters

In a similar manner to that of wave energy devices, tidal energy converters can be classified into a range of 'types'. The types primarily define the technical concept behind the fundamental device operation or the position in the water column in which the device can operate. The different types of tidal energy converters include: horizontal axis turbines, vertical axis turbines, oscillating hydrofoils, and tidal kites amongst a number of other concepts. The tidal energy resource is site specific, and generally lies close to a significant land mass. A greater amount of design convergence is being seen within the tidal energy sector, where most technology developers are now favouring a horizontal axis turbine. however many foundation and mooring options exist, as demonstrated in Figure 19.7. Two tidal energy technologies that have each successfully generated GWh to the electrical grid are described in Table 19.2.

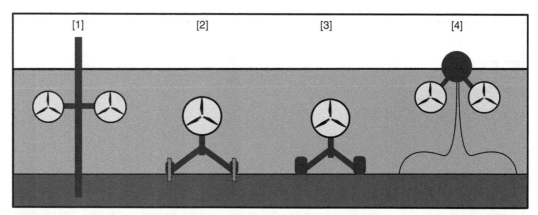

Figure 19.7 Tidal Stream Energy Converters—Foundation and Mooring Options (Monopile [1]; Pinned [2]; Gravity [3]; and Buoyant Moored [4])

Source: SI Ocean, 2012.

Table 19.2 Mature tidal energy technology examples

Marine Current Turbines—SeaGen	Alstom Tidal Generation Limited
• The world's first fully certified tidal stream power station	• Three bladed horizontal axis tidal turbine
• Twin 600 kW horizontal axis turbines mounted on a surface piercing monopole structure	• Buoyant nacelle to allow ease of deployment and recovery using small installation vessels
• Ease of access for operation and maintenance	• Thruster provides full yaw capability, allowing the turbine rotor to face the oncoming flow
• First deployed in Strangford Lough in 2008	• Deployed at the EMEC test facility, Scotland

19.1.3.3 Technology and Research Challenges

The main strategic technology challenges facing the ocean energy sector have been identified (SI Ocean, 2013) and are illustrated in Figure 19.8.

For commercialization, the ocean energy sector needs to make continued and sustained progress in respect of the overall cost and performance of ocean energy technology, from engineering design through to project decommissioning. Some of these challenges may be overcome at the single device level, however new challenges may be identified as the move from single to multiple devices takes place. The integration of arrays into new and existing grid infrastructures will require advances in the predictability of array power outputs. An investigation into device performance at an array level can only augment the understanding of individual device performance. Furthermore, developing greater understanding of the nature of the complex interactions between individual devices within arrays, and more accurately defining the overall impact of multiple energy converter devices within a tidal flow or wave regime, will help to enhance the overall state of the art.

A number of activities and themes have been identified by the European Commission SI Ocean project (MacGillivray et al., 2013) as requiring further research and development in order to allow technology progression towards a more mature ocean energy industry. Within these identified themes, some areas have been identified as more urgent priorities when taking into account the limited funding available to be able to carry out technology development activities. These priorities have been identified through engagement with supply chain, technology, and project developers and are highlighted as areas

Figure 19.8 High level challenges facing the ocean energy sector

Source: SI Ocean, 2013.

which require fundamental underpinning research using the skills, facilities, and cap-abilities of research institutes, as shown in Figure 19.9.

The area of device and system deployment has a number of activities associated with it which are considered to benefit the sector through impacts on cost of energy and reduction of CAPEX and OPEX, or performance and technology improvements, thus facilitating the future growth of the sector in a progression towards array development. Within this area it was deemed that reliability demonstration, design for maintenance, knowledge transfer, and dissemination and novel system concepts were identified as high priority in the research prioritization. Sub systems and sub components within an overall ocean energy system can have a significant impact on overall reliability and performance; however, in order for the overall improvement in optimization of systems within a device, the following areas were identified as high priority and areas that research facilities can help progress knowledge: control systems, intelligent predictive mainten-ance systems, power electronics, Power Take Off (PTO), and foundations and moorings.

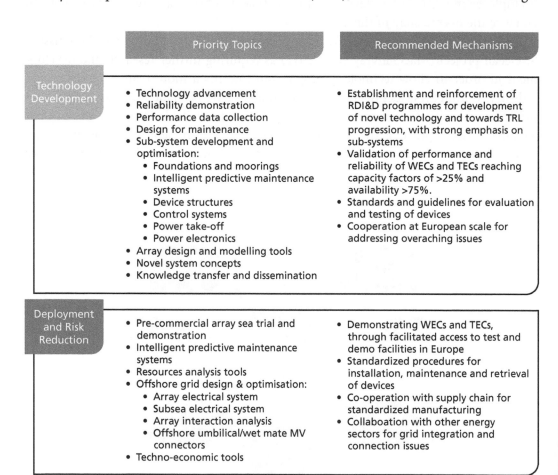

Figure 19.9 Priority topic areas for the areas for the wave and tidal sectors

Source: Magagna et al., 2014.

Tool development and design optimization intends to identify and address some of the tools and optimization processes that will work towards future development, design, and deployment of ocean energy technologies. Areas where further research could work towards solutions are techno-economic tools, resource analysis tools, array design and modelling tools, reliability tools, and site characterization techniques. In order for the ocean energy sector to move from single device deployment to multiple device arrays there needs to be development in the following areas: sub-sea electrical systems, array interaction analysis, and offshore grid design and optimization. These are all areas which have been identified as a high priority for research to work towards overcoming.

19.1.3.4 Cost of Energy

In order for the wave and tidal sector to progress beyond the demonstration phase, it is vital that developers, investors, and policymakers understand the current costs for ocean energy generation and the potential for cost reduction over time. The current cost of wave and tidal energy generation is higher than that of conventional forms of energy, and as discussed throughout this chapter, it is important that wave and tidal can reduce in cost so that the energy produced becomes competitive with other renewable energy sources. For example the Levelized Cost of Energy (LCOE) for ocean energy and off- and on-shore wind are as presented below and in Figure 9.10:

- **Wave:** €340–632/MWh (SI Ocean, 2013)
- **Tidal:** €243– 474/MWh (SI Ocean, 2013)
- **Wind:** €114/MWh (Mott McDonald, 2010)
- **Offshore Wind:** €191– 226/MWh (Mott McDonald, 2010).

An analysis of wave and tidal energy deployment costs was undertaken (SI Ocean 2013) with input provided by industry developers. In order to assess potential cost competitiveness for wave and tidal energy in the future, forecasts were used to provide

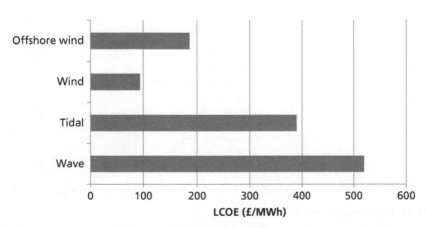

Figure 19.10 Levelized cost of energy (€/MWh) for ocean energy, wind, and offshore wind

a benchmark against which progress can be measured. The baseline cost of energy (CoE) calculations within this section refer to forecasts developed by the SI Ocean (2013) methodology, which takes into account a standard framework for the assessment of wave and tidal energy costs, including all lifetime costs of a marine energy array in relation to a site with a good wave or tidal resource.

Cost data for wave and tidal arrays has been analysed (SI Ocean, 2012, 2013) to show the breakdown of different cost components within the overall Levelized Cost Of Energy (LCOE). Figure 19.9 shows the different sub system costs associated with wave and tidal energy generators; however, it should be noted that the overall costs and proportions per cost-centre component will vary depending on technology type and specific site conditions.

The installation costs of a tidal device form a significant proportion of the overall costs, with the structural and power take-off component costs representing a lower percentage of the overall cost than with wave devices. Additionally, it can be said that that the physical device cost only constitutes a part of the total cost of an array project (as shown in Figure 19.11). Balance of other plant items (e.g. connectors, cabling etc.) in addition to installation, operation, and maintenance costs, add considerable cost to the overall project.

Figure 19.12 presents LCOE cost ranges for wave and tidal energy, reflecting the inevitable uncertainty associated with predicting future costs, and the differences in cost estimates between the source developers questioned and the input resource conditions. A lifetime of twenty years and a discount rate of 12 per cent are assumed for both wave and tidal.

The levelized cost of energy is significantly impacted by the discount rate used within the calculation. For example, a tidal energy installation which produced energy at 32.0 c/kWh with a discount rate of 12 per cent will show a cost decrease to 23.1 c/kWh if a 6 per cent discount rate is used (figures in Euro cents) (SI Ocean, 2013). Perceptions

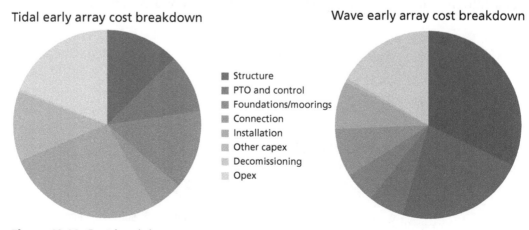

Figure 19.11 Cost breakdown

Source: SI Ocean, 2013.

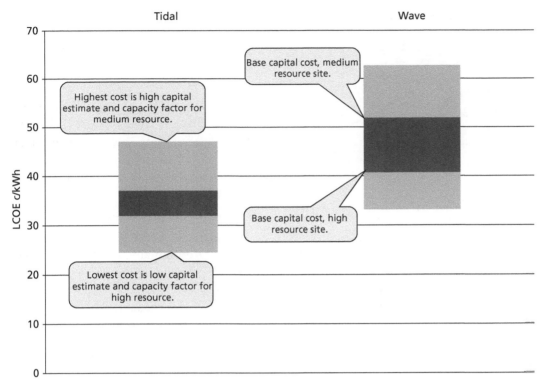

Figure 19.12 Early array costs

Source: SI Ocean, 2013.

of risk have a significant impact on the discount rate used by project developers when assessing a project, and can acutely affect project economics.

19.1.3.4.1 Potential for Cost Reduction. It is undisputed that substantial cost reduction and performance improvement in the wave and tidal sector is necessary in order to ensure that commercialization can take place, so that electricity generated from ocean energy can be produced at a competitive cost to incumbent alternatives. Several mechanisms will help overcome cost barriers. Mechanisms include the concept of learning by experience, through which it is expected that costs will reduce in line with increasing cumulative device manufacture and deployment. Similarly, economies of scale can unlock significant cost savings through series production of modular devices. Other renewable energy technologies, such as wind turbine production and solar photovoltaic cell manufacture, have seen significant cost reductions as cumulative production and deployment has increased. Anticipated cost reduction trends for tidal and wave energy converters are shown in Figure 19.13.

Additional cost reduction can be made through innovation in engineering design. Innovation can take place incrementally, in areas such as enabling technology—

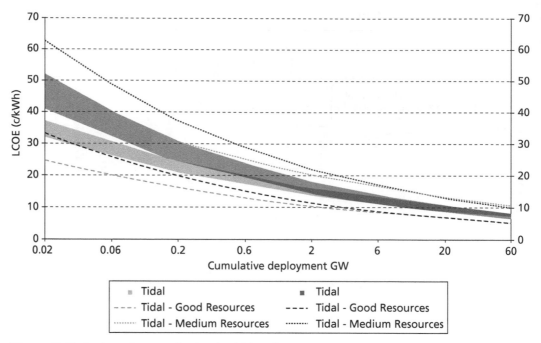

Figure 19.13 Projected cost reduction in tidal and wave energy

Source: SI Ocean, 2013.

for example novel installation techniques, or development and improvement of performance or availability. Innovation can also take place with entirely new concepts opening up increased areas of resource, or creating a step change in device development costs at a system, device, or sub system level.

These methods present cost reduction opportunities; however, technologies do not necessarily follow a linear pathway, and cost reduction often results from a complex interaction of several of the effects outlined above.

19.1.3.4.2 Ocean Energy Relevant Policy. Within the UK, and increasingly at a global level, a number of support mechanisms for marine energy development have been established by governments and intermediary agencies. Due to the implementation of such support measures, there has been an increase in the underpinning research, in addition to the development and deployment of prototype devices, as well as achieving the next step of technology development—array demonstrations.

A range of technology push, market push, and market pull funding mechanisms can be found within the ocean energy sector, with both capital grant support (representing the technology push), and enhanced market payments (representing market push and pull) being implemented at varying rates in several countries worldwide, including some sophisticated mechanisms which are a combination of both.

Examples within the UK of technology push funding include the Technology Strategy Board (the UK's innovation agency) and the Engineering and Physical Sciences Research Council (EPSRC) Grand Challenges, and Wave Energy Scotland funding calls, focusing on individual components, subsystems, and fundamental research. At a market push level, the UK's MEAD (Marine Energy Array Demonstrator) (Carbon Trust, 2011) and MRCF (Marine Renewables Commercialisation Fund) are providing funding to enable the deployment of early array projects. Market pull mechanisms, such as the UK's Renewable Obligation Certificate (ROC), the CfD (Contracts for Difference), or the international Feed-In-Tariffs (FITs) being used in countries such as Canada and France, enable long-term security of revenue, giving confidence to investors and stimulating a market for wave and tidal energy technologies.

While the number of policy support mechanisms available indicates that there is growing support for ocean energy, a number of revisions and iterations have been made to policy mechanisms that had previously been in place. This revision process suggests that policy makers and stakeholders are requiring a period of experimentation in order to reach an optimal support level to facilitate growth within the ocean energy sector (Jeffrey et al., 2013). The level of support provided, and the focus of the support (for example, technology push dominated support mechanisms or market pull dominated support mechanisms), will have significant impact on the development pathway of the sector.

Additionally, a number of national ocean energy deployment targets have been set, reflecting the ambitious goals of governments in achieving significant levels of ocean energy power production, in addition to securing opportunities for domestic manufacture, supply chain formation, and economic benefits associated with potential export markets. There have, however, been examples of political and stakeholder over-optimism, exhibiting unrealistic short-term expectations of the commercial readiness of the emerging ocean energy sector. While it is recognized that early-stage support is necessary in order to allow sector growth, and for investors to gain confidence in the long-term opportunities that exist, the 'correct' balance of support mechanisms will be required in order to achieve an efficient and effective progression from development to deployment.

The process of learning within the policy setting environment requires a delicate balance. With the entry of new potential markets into the global development pipeline, policy intervention will have a significant impact on the attractiveness of a particular location. While there has been opportunity for policy experimentation, there is now a very real requirement for long-term commitment from government and policymakers to ensure that the correct support mechanisms are in place that will stimulate and encourage growth. The significant prize associated with ocean energy remains to be secured, but the future of ocean energy, and the attractiveness of job creation, inward investment, reduced reliance on fossil fuels, and greater energy security that it could bring about, remains buoyant.

19.2 **Hydropower**

19.2.1 INTRODUCTION

Hydropower relies on the water cycle that is driven by the sun. It is a clean and renewable power source that does not pollute the atmosphere like fossil fuel (coal or natural gas) power plants. It is generally available by controlling the flow of water through turbines to produce clean electricity on demand. Impoundment hydropower creates reservoirs that could provide public access, water supply, flood control, and offer a variety of recreational opportunities like fishing, swimming, and boating. For centuries civilizations have taken advantage of the power of water. Once used by the Greeks for grinding wheat into flour, the water wheels of the past have been updated to today's highly efficient turbines that generate electricity by spinning water (UKERC, 2013).

Hydropower accounts for one fifth of the world's electricity supply, and has helped shape and promote economic growth in such countries as Canada, Norway, and the United States. Environmental and social concerns, coupled with financial constraints, resulted in a decade of stagnant investment in the 1990s and critical assessment of the role of hydropower in development. Now, lessons from the past, together with emerging global dynamics, are recasting the role of hydropower and stimulating a renaissance in investment and rehabilitation. The opportunities and challenges are complex, and ultimately dependent on the resources, skills, and will to invest responsibly, with due regard to all aspects of sustainable development (WBG, 2009).

Small hydropower, defined by installed capacity of up to 10 MW, is the backbone of electricity production in many countries in the European Union. Small hydropower is based on a simple process, taking advantage of the kinetic energy and pressure freed by falling water or rivers, canals, streams, and water networks. The rushing water drives a turbine, which converts the water's pressure and motion into mechanical energy, converted into electricity by a generator. The power of the scheme is proportional to the head (the difference between up- and downstream water levels), the discharge (the quantity of water which goes through the turbines in a given unit of time), and the efficiency of the turbine (EU, 2014).

19.2.1.1 **Review of Hydropower Installed Global Capacity and Generation**

Hydropower is an extremely flexible technology for power generation. Hydro reservoirs provide built-in energy storage, and the fast response time of hydropower enables it to be used to optimize electricity production across grids, meeting sudden fluctuations in demands. There is no international consensus on the dividing line between small and large hydro. Today large hydro turbine manufacturers can accommodate up to 30 MW on their small hydro turbine range depending on hydro technology (Andritz, 2014;

Alstom, 2014; Voith, 2014), but for the definition of small hydro power, a capacity of 10 MW total is generally becoming accepted in Europe and is supported by the European Commission and the European Small Hydropower Association (ARE, 2014; BHA, 2014; EREC, 2014; ESHA, 2014; EU, 2014; HEA, 2014; IHA, 2014; NHA, 2014).

Large-scale hydropower projects can be controversial because they affect water availability downstream, inundate valuable ecosystems, and may require the relocations of populations. Despite being a mature technology, in comparison with other renewable energy sources, hydropower has still a significant potential. New plants can be developed and old ones upgraded, especially in terms of increasing efficiency and electricity production as well as environmental performance. In particular, the development of low-head or very low-head small hydro plants holds much promise (WBG, 2009).

Small hydro power plants generate electricity or mechanical power by converting the power available in flowing waters in rivers, canals, and streams with a certain fall (termed the 'head') into electric energy at the lower end of the scheme, where the powerhouse is located. The power of the scheme is proportional to the flow and to the head. Small hydropower schemes are mainly run off-river with no need to create a reservoir. Because of this fact, small hydropower systems can be considered an environmentally friendly energy conversion option, since they do not interfere significantly with river flows and fit in well with the surroundings. The advantages of small hydropower plants are numerous and include grid stability, reduced land requirements, local and regional development, and good opportunities for technologies export (BHA, 2014; EREC, 2014; ESHA, 2014; Gilkes, 2014; IHA, 2014; NHA, 2014; NME, 2014). Approximately 70 per cent of the earth's surface is covered with water, a resource that has always been central to human development (Hekinian, 2014). Hydropower is the most widely used form of renewable energy which is clean and reliable with very good efficiency (EU, 2014). It cannot be depleted over time and can be replenished consistently. The use of hydropower has been characterized by continuous technical development. Hydropower electricity is the product of transforming potential energy stored in water in an elevated reservoir into the kinetic energy of the running water, then mechanical energy in a rotating turbine, and finally electrical energy in an alternator or generator. Hydropower is a mature renewable power generation technology that offers two desirable characteristics in modern electricity systems: a) built-in storage that enables electricity to be provided on demand and b) fast response time that allows reserves to be fed to the grid. Hydropower also has an important role to play in producing renewable electricity. It is low-cost and readily available: power flow is controlled through turbines to produce electricity on demand (Aggidis et al., 2010; LUREG, 2014; Snella et al., 2014). Hydropower can also be applied in combination with other activities, such as flood regulation and wetland management, with no additional water resources or environmental impacts (Subing et al., 2014).

World conventional hydropower potential is considered to be around 2500 GW (7500 TWh in energy terms), of which about 970 GW is already developed (Pazheri et al., 2014). The International Energy Agency (IEA) considers that only 5 per cent of the world small-scale potential is under exploitation. It is thought that small hydropower potential could be increased to 200 GW. Hydropower provides nearly 15.3 per cent of the world's electricity, 3427 TWh of electricity production in 2010, it accounts for about 75 per cent of electricity produced from renewable sources globally and is expected to increase about 3.1 per cent each year for the next 25 years. Hydropower is produced in 150 countries, with the Asia-Pacific region generating 32 per cent of global hydropower in 2010. China is the largest hydroelectricity producer, with 721 TWh of production in 2010, representing around 17 per cent of domestic electricity use. Brazil, Canada, New Zealand, Norway, Paraguay, Austria, Switzerland, and Venezuela have a majority of internal electric energy production from hydroelectric power. Paraguay produces 100 per cent of its electricity from hydroelectric dams, and exports 90 per cent of its production to Brazil and to Argentina. Norway produces 99 per cent of its electricity from hydroelectric sources (IEA, 2014).

19.2.1.2 Technology and Operation

Hydro is a flexible and backup source of electricity adapting quickly to changing energy demands. The early water wheels were replaced from the mid 1800s by a series of new water turbine designs, including reaction turbines like Vortex, Francis, Propeller, and Kaplan and impulse turbines like Pelton, Crossflow, and Turgo (Paish, 2002; Aggidis et al., 2014; BHA, 2014; ESHA, 2014; IHA, 2014; LUREG, 2014; NHA, 2014; Zidonis et al., 2015; Benzon et al., 2015).

From small scale beginnings, hydroelectric power plants continued to increase in size throughout the twentieth century. Hoover Dam's initial 1.3 GW power plant was the world's largest hydroelectric power plant in 1936; it was eclipsed by the 6.8 GW Grand Coulee Dam in 1942. The Guri Dam at 10.2 GW was followed by Itaipu Dam that opened in 1984 in South America as the largest, producing 14 GW, but was surpassed in 2008 by the Three Gorges Dam in China at 22.5 GW (Rai et al., 2014).

World pumped-storage potential is approximately 1 000 GW, or about half the realistic potential. The Ffestiniog Power Station can generate 360 MW of electricity within 60 seconds of the demand arising. Hydroelectricity eliminates the cost of fuel and hydroelectric plants have long economic lives, with some plants still in service after 50–100 years. Operating labour cost is also usually low, as plants are automated and have few personnel on site during normal operation (BHA, 2014; ESHA, 2014; IHA, 2014; NHA, 2014). Where a dam serves multiple purposes, a hydroelectric plant may be added with relatively low construction cost, providing a useful revenue stream to offset the costs of dam operation, like the novel Korean Sihwa 260 MW seawater plant (Aggidis and Benzon, 2013). The sale of electricity from the Three Gorges Dam will cover the construction costs after five to eight years of full generation. Hydroelectric plants are either grid connected or serve specific industrial applications like aluminium electrolytic plants (Grand Coulee in USA, Brokopondo in Suriname, and Manapouri in New Zealand). Hydroelectricity reservoirs and dams often provide water sports facilities, aquaculture, irrigation support for agriculture, flood control, and also become tourist attractions themselves (Rai et al., 2014).

Small-scale hydroelectricity production grew by 28 per cent from 2005 to 2008, raising the total world small-hydro capacity to 85 GW. Over 70 per cent of this was in China (65 GW), followed by Japan (3.5 GW), the United States (3 GW), and India (2 GW). Micro hydro typically produces up to 100 kW of power while Pico hydroelectric power generates under 5 kW (US DoE, 2004; REN21, 2013; IEA, 2014; IHA, 2014; NHA, 2014; Rai et al., 2014; US DoE, 2014).

Generating methods include pumped-storage schemes that currently provide the most commercially important means of large-scale grid energy storage and improve the daily capacity factor of the generation system. The run-of-the-river hydroelectric stations are those with small or no reservoir capacity. Tidal range power plant examples include the French 240 MW La Rance plant. Tidal current hydro schemes use water's kinetic energy, like the Stangford Lock Siemens MCT 1.2 MW plant in Northern Ireland, in a similar way to undammed sources like undershot waterwheels. Underground power stations make use of a large natural height difference between two waterways, such as a waterfall or mountain lake (Aggidis and Benzon, 2013; IEA, 2014; IHA, 2014; NHA, 2014; Rai et al., 2014; Siemens MCT, 2014).

19.2.1.3 The Prospects for Future Hydropower Development

The prospects for future hydropower development include abundant physical and engineering hydropower potential in developing countries. In absolute terms, the total economically feasible potential hydropower capacity in developing countries exceeds 1900 GW, 70 per cent of which (1330 GW) is not yet exploited. This is nearly four times the current installed capacity of 315 GW in Europe and North America, and not quite double the 970 GW installed worldwide (REN21, 2013; IEA, 2014; IHA, 2014). On a regional basis, unexploited potential as a per cent of total potential amounts to: 93 per cent in Africa, 82 per cent in East Asia and the Pacific, 79 per cent in the Middle East and North Africa, 78 per cent in Europe and Central Asia, 75 per cent in South Asia, and 62 per cent in Latin America and the Caribbean. These estimates cover potential new (greenfield) site developments only. Significant additional amounts of energy and capacity are available from rehabilitation of existing energy and water assets, from redesign of infrastructure to meet emerging demands and opportunities (WBG, 2009; REN21, 2013). Notwithstanding the strong development rationale, the enormous technical potential, and the improved understanding of good practices, scaling up hydropower faces important constraints and barriers: emerging trends, driven by more sophisticated energy markets, volatile energy prices, climate change, and increased attention to water management and regional integration, are changing the value proposition of hydropower in development (WBG, 2009; REN21, 2013). The capital required for small hydro plants depends on the effective head, flow rate, geological and geographical features, the equipment (turbines, generators etc.), civil engineering works, and continuity of water flow. Making use of existing weirs, dams, storage reservoirs, and ponds can significantly reduce both environmental impact and costs. Facilities with double purpose that combine power generation for example with flood control, irrigation or drinking water supply have reduced payback periods. The other principal cost element is operation and maintenance (O&M), including repairs and insurance, which can account from 1.5–5 per cent of investment costs. Costs differ considerably depending on the head height of the plant. (Aggidis et al., 2010, UKERC, 2013; ESHA, 2014; IEA, 2014; IHA, 2014; LUREG, 2014). Significant advances have recently been made in hydro-machinery, with important prospects for further improvements. Institutional barriers still exist which hamper development; for example, long lead times to obtain or renew concession rights, concessions locked to a holder that does not actually develop the scheme, and lack of grid connections. In the EU alone administrative procedures take from twelve months (Austria) to twelve years (Portugal). Meeting environmental standards for water management can sometimes limit a plant's capacity, but is also a driver for innovation and improved performance. Hydropower will continue to play a crucial role in the integration of fluctuating renewable energy sources by providing reserve, storage, and balancing capacities for the European electricity grid. Despite being a mature technology, hydropower still has significant untapped potential particularly in the development of new plants (very low-head small hydro plants and pumped storage plants) and also in the upgrading of old ones (increasing efficiency and electricity production and environmental performance). In addition, transformation of conventional into pumped-storage plant has the potential for supporting higher penetration of other renewables, such as wind, and for supporting the grid. Maintaining or upgrading the existing infrastructure is an important focus throughout Europe. R&D activity on hydropower includes pumped storage, low head, maintenance, control methods, and new application-driven hydro related projects like power take-off for ocean energy converters (REN21, 2013; UKERC, 2013; ESHA, 2014).

19.2.1.4 Social and Environmental Constraints on Hydropower Development

Hydropower offers advantages over other energy sources but also faces unique environmental and social challenges. Fish populations can be impacted if fish cannot migrate upstream past impoundment dams to spawning grounds, or if they cannot migrate downstream to the ocean. Upstream fish passage can be aided using fish ladders or elevators, or by trapping and hauling the fish upstream by truck. Downstream fish passage is aided by diverting fish from turbine intakes using screens or racks or even underwater lights and sounds, and by maintaining a minimum spill flow past the turbine (ESHA, 2014). Hydropower can impact water quality, flow, and cause low dissolved oxygen levels in the water, a problem that is harmful to riparian (riverbank) habitats and is addressed using various aeration techniques, which oxygenate the water. Maintaining minimum flows of water downstream of a hydropower installation is also critical for the survival of riparian habitats (ESHA, 2014). New hydropower facilities impact the local environment and may compete with other uses for the land. Those alternative uses may be more highly valued than electricity generation. Humans, flora, and fauna may lose their natural habitat. Local cultures and historical sites may be impinged upon. Some older hydropower facilities may have historic value, so renovations of these facilities must also be sensitive to such preservation concerns and to impacts on plant and animal life (WBG, 2009; IEA, 2012; REN21, 2013; ESHA, 2014).

Identification and management of environmental and social risks is challenged by limited institutional capacity and experience in implementing new standards. This means refining regulatory and policy frameworks at the country/trans-boundary levels, building capacity among developers, as well as electricity companies and government, and enhancing transparency for stakeholders. It also means ongoing research into important environmental issues, such as emissions from reservoirs in shallow tropical sites and continuous improvement in avoiding and mitigating impacts. Infrastructure design based on poor hydrological data can severely compromise performance and decrease the very water management benefits the infrastructure is designed to generate. Climate change accentuates these risks for two reasons: (i) extrapolations of historical data are less reliable as the past becomes an increasingly poor predictor of the future; and (ii) hydrology is ever-changing, placing a premium on designs that maximize flexibility and operations that embrace adaptive management. While the potential for hydropower is known, there is a lack of planning and project prioritization. In particular, engineering studies completed years ago need to be updated with new knowledge (particularly of hydrology) as well as more sophisticated consideration of environmental and social values. As a public good, governments need to undertake strategic assessments and prefeasibility studies in order to develop a pipeline of projects and identify high-value storage sites. Against the demand for hydropower infrastructure is a shortage of financing, exacerbated by the current global financial crisis. This gap is most severe in the poorest countries, where the funds needed well exceed the resources of governments and donors/development banks. Yet increasing resources from the private sector requires a broad range of responses: better policies and institutions; improving payments from energy consumers; clarity in regulations for developing and operating hydro plants; and innovative financial structures that support public–private partnership projects with multiple (public and private) benefits. As future infrastructure will function in an ever-changing hydrology, flexibility and adaptive management skills will be critical (Kumar et al., 2011).

The social and environmental costs of large dams are very important and well documented on the World Commission on Dams final report 'Dams and Development: A framework for Decision Making' (WCD, 2000). The Three Gorges Dam displaced an estimated one million people, many of whom received wholly inadequate compensation (Zoomers, 2011). There is also the possible future effect of lack of rainfall due to climate change on reservoirs and therefore dams' storage capability (Hansena et al., 2014).

Norwegian hydrogeneration has been badly affected by low rainfall in the recent past, and this is true in other parts of the world (Deressa, 2014).

A recent study (Ansar et al., 2014) warns developing countries against large dam projects that typically cost nearly twice as much as first estimated and rarely finish on time which could likely saddle these countries with big debts. Such risks are less of a problem in wealthier countries such as the USA, home of the Hoover dam that is often cited as a success story. There are also many good examples of smaller hydroelectric projects in countries such as Norway and Portugal that made sound economic sense. The response on behalf of the International Hydropower Association stated that: 'Governments in developing countries turn to major hydropower projects for their energy needs because the resource is local, the technology reliable, the scale considerable and the resulting electricity price economical and predictable. That there have been cost overruns with dams has been well publicised long before this research (Ansar et al., 2014) but the fact that governments of developing countries continue to turn to this reliable source of clean energy shows that the advantages can greatly outweigh the challenges' (Taylor, 2014).

To overcome the environmental and social barriers, areas of future focus and research could include:

- The improved understanding of the necessity of hydropower construction and hydropower social and environmental protection.
- Strengthening hydropower social and environmental protection policy research.
- Establishing and improving hydropower social and environmental protection policies and regulations.
- Strengthening the social and environmental management for hydropower construction projects at the planning stage.
- Exploiting and utilizing water resource rationally.
- Water use demand in the upper and lower reaches and on both sides should be taken into full consideration in the construction and operation of dams.
- Improving the biodiversity protection work involved in dam construction.
- Protective measures adopted for rare and endangered plants and animals to be affected according to the natural environmental status and distribution situations.
- Strengthening the social and environmental protection of resettlement that renew ideas, improve policies, enhance standards, create mechanisms, introduce science and technology, and implement management.
- Strengthening the social and environmental protection in the construction preparation and during the construction period.
- The potential social and eco-environmental impacts in the construction preparatory stage should be included in construction management.
- Actively carrying out public participation in dam construction.
- Strengthening scientific research, increasing the understanding of social and environmental impact rules of hydropower and developing hydropower environmental protection technologies.

19.3 **Conclusion**

As has been discussed and examined within this chapter, both ocean energy and hydropower have important roles to play in contributing to the global energy mix. In the case of hydropower, it is already making a significant contribution to the energy needs of countries around the world. Ocean energy, specifically wave and tidal energy, is on the cusp of moving to array scale and commercialization, and although currently a relatively expensive form of energy generation, it is anticipated that it will become more competitive and will form a significant contribution to the world's energy portfolio in the medium to long term.

■ REFERENCES

Aggidis, G.A. and Benzon, D.S., 2013. Operational optimisation of a tidal barrage across the Mersey estuary using 0D modelling, *Journal of Ocean Engineering*, 66, 69–81.

Aggidis, G.A., Luchinskaya, E., Rothschild, R., and Howard, D.C., 2010. The costs of small-scale hydro power production: Impact on the development of existing potential, *Renewable Energy*, 35, 2632–8.

Aggidis, G.A. and Židonis, A., 2014. Hydro turbine prototype testing and generation of performance curves: fully automated approach, *Renewable Energy*, 71, 433–41.

Alstom, 2014. Hydropower, <http://www.alstom.com/power/renewables/hydro/> (accessed on 24/03/2014).

Andritz, 2014. Hydropower, <http://www.andritz.com/hydro.htm> (accessed on 24/03/2014).

Ansar A., Flyvbjerg, B., Budzier, A., and Lunn, D., 2014. Should we build more large dams? The actual costs of hydropower mega project development. *Energy Policy* (article in press). <http://www.academia.edu/6361586/Should_we_build_more_large_dams_The_actual_costs_of_hydropower_megaproject_development> (accessed on 24/03/2014).

ARE, 2014. Alliance for Rural Electrification, <http://www.ruralelec.org/> (accessed on 24/03/2014).

Benzon, D., Židonis, A., Panagiotopoulos, A., Aggidis, G.A., Anagnostopoulos, J. S., and Papantonis, D.E., 2015. Impulse turbine injector design improvement using computational fluid dynamics, *Journal of Fluids Engineering*, 137(4), 041106.

BHA, 2014. British Hydropower Association. <http://www.british-hydro.org/>(accessed on 24/03/2014).

Carbon Trust, 2006. *Future Marine Energy: Results of the Marine Energy Challenge: Cost Competitiveness and Growth of Wave and Tidal Stream Energy*. London: Carbon Trust.

Carbon Trust, 2011. *Accelerating Marine Energy: The Potential for Cost Reduction—Insights from the Carbon Trust Marine Energy Accelerator*. London: Carbon Trust.

Deressa, T., 2014. Climate change and growth in Africa: challenges and the way forward. Top priorities for the continent in 2014, Africa Growth Initiative FORESIGHT. <http://www.subsahara-afrika-ihk.de/wp-content/uploads/2014/01/Foresight-Africa_Full-Report.pdf#page=34> (accessed on 24/03/2014).

EREC, 2014. European Renewable Energy Council, <http://www.erec.org/renewable-energy/hydropower.html> (accessed on 24/03/2014).

ESHA, 2014. European Small Hydropower Association. <http://www.esha.be/> (accessed on 24/03/2014).

EU, 2014. Hydropower. <http://ec.europa.eu/research/energy/eu/index_en.cfm?pg=research-hydropower> (accessed on 24/03/2014).

Gilkes, 2014. Gilbert Gilkes & Gordon Ltd, Hydropower, <http://www.gilkes.com/> (accessed on 24/03/2014).

Hansena, Z., Loweb, S., and Xub, W., 2014. Long-term impacts of major water storage facilities on agriculture and the natural environment: Evidence from Idaho (U.S.). *Ecological Economics*, 100, pp. 106–18.

HEA, 2014. Hydro Equipment Association <http://www.thehea.org/> (accessed on 17/02/2014).

Hekinian, R., 2014. Our Haven, Planet Earth—Sea Floor Exploration, Chapter 2, *Springer Oceanography*.

Huckerby, J., Jeffrey, H., Sedgwick, J., Jay, B., and Finlay, L., 2012. *An International Vision for Ocean Energy—Version ll*. Ocean Energy Systems Implementing Agreement.

IEA, 2012. International Energy Agency Technology Roadmap on Hydropower 2012 <http://www.iea.org/publications/freepublications/publication/TechnologyRoadmapHydropower.pdf> (accessed on 24/03/2014).

IEA, 2014. International Energy Agency, Hydropower <https://www.iea.org/topics/hydropower/> (accessed on 24/03/2014).

IHA, 2014. International Hydropower Association. <http://www.hydropower.org/> (accessed on 24/03/2014).

International Electrotechnical Commission, (IEC) 2013. *Technical Committee 114: Marine Energy—Wave, Tidal and Other Water Current Converters*. [Online] Available at: <http://www.iec.ch/dyn/www/f?p=103:7:0::::FSP_ORG_ID:1316> (accessed on 14/11/ 2013).

Jeffrey, H., Jay, B., and. Winskel, M., 2013. Accelerating the development of marine energy: exploring the prospects, benefits and challenges. *Technological Forecasting and Social Change*, 80, 1306–16.

Kumar, A., Schei T., Ahenkorah A., Caceres Rodriguez R., Devernay J.-M., Freitas M., Hall D., Killingtveit Å., and Liu Z., 2011. Hydropower. In *IPCC Special Report on Renewable Energy Sources and Climate Change Mitigation*, Cambridge University Press, Cambridge, UK.

LUREG, 2014. Lancaster University Renewable Energy Group Hydro Resource Model Tool. <http://www.engineering.lancs.ac.uk/lureg/nwhrm/tool/> (accessed on 24/03/2014).

MacGillivray, A., Jeffrey, H., Hanmer, C., Magagna, D., Raventos, A., Badcock-Broe, A., 2013. *Ocean Energy Technology: Gaps and Barriers*, SI Ocean.

Magagna, D., MacGillivray, A., Jeffrey, H., Hanmer, C., Raventos, A., Badcock-Broe, A., and Tzimas, E., 2014. *Wave and Tidal Energy Strategic Technology Agenda*, SI Ocean.

Mørk, G., Barstow, S.F., Kabuth, A., & Pontes, M.T., 2010. *Assessing the Global Wave Energy Potential*. Proceedings of OMAE2010 29th International Conference on Ocean, Offshore Mechanics and Arctic Engineering June 6–11, 2010, Shanghai, China.

Mott McDonald, 2010. *UK Electricity Generation Costs Update*. London: Mott McDonald.

NHA, 2014. National Hydropower Association, USA. <http://www.hydro.org/> (accessed on 24/03/2014).

NME, 2014. New Mills Engineering, Hydropower. <http://newmillsengineering.com/> (accessed on 24/03/2014).

Ocean Energy Systems, 2013. *About OES.* [Online] Available at: <http://www.ocean-energy-systems.org> (accessed on 6/12/2013).

Ocean Energy Systems, 2012. *Annual Report 2012, Implementing Agreement on Ocean Energy Systems.*[Online] Available at: <http://www.iea.org/media/openbulletin/OES2012.pdf> (accessed on 20/11/2013).

Paish, O., 2002. Small hydro power: technology and current status. *Renewable and Sustainable Energy Reviews*, 6, 537–56.

Pazheri, F.R., Othman, M.F., and Malik N.H., 2014. A review on global renewable electricity scenario. *Renewable and Sustainable Energy Reviews*, 31, 835–45.

Rai, A., Jain, M., and Tomar, N., 2014. Last 50 years of hydro energy–a bibliographic survey. *International Transaction of Electrical and Computer Engineers System* 2, 1, 7–13.

REN21, 2013. Renewables 2013. Global Status Report. <http://www.ren21.net/portals/0/documents/resources/gsr/2013/gsr2013_lowres.pdf> (accessed on 24/03/2014).

SI Ocean, 2013. *Ocean Energy: Cost of Energy and Cost Reduction Opportunities.* [Online] Available at: <http://si-ocean.eu/en/upload/docs/WP3/CoE%20report%203_2%20final.pdf (accessed on 4/12/2013).

SI Ocean, 2012. *Ocean Energy: State of the Art.* [Online] Available at: <http://si-ocean.eu/en/upload/docs/WP3/Technology%20Status%20Report_FV.pdf (accessed on 4/12/2013).

Siemens MCT, 2014. Marine Current Turbines, <http://www.marineturbines.com/>(accessed on 24/03/2014).

Snella, J., Prowsea, D., and Adamsa, K., 2014. The changing role of hydropower: from cheap local energy supply to strategic regional resource. *International Journal of Water Resources Development* Special Issue: Water Infrastructure, 30, Issue 1: 121–34.

Subing, L., Wenchuan, W., and Shiguo, X., 2014. Benefits identification and evaluation of floodwater utilisation in Baicheng region North-Eastern China. *Irrigation and Drainage Irrigation and Drainage*, 63, 1: 38–45.

Taylor, R., 2014. International Hydropower Association. <http://www.ft.com/cms/s/0/effcd742-a82f-11e3-8ce1-00144feab7de.html#axzz2wp6Tlqgj> (accessed on 24/03/2014).

UKERC, 2013. Energy Research Atlas Landscape: Hydropower. <http://ukerc.rl.ac.uk/Landscapes/Hydropower.pdf> (accessed on 24/03/2014).

US DoE Energy Efficiency and Renewable Energy—Hydropower Setting a course for our Energy Future, 2004. <http://www.nrel.gov/docs/fy04osti/34916.pdf> (accessed on 24/03/2014).

US DoE Energy Hydro, 2014. <http://energy.gov/search/site/hydro> (accessed on 24/03/2014).

Voith, 2014. Hydropower, <http://www.voith.com/en/products-services/hydro-power-377.html> (accessed on 24/03/2014).

WBG, 2009. World Bank Group Directions in Hydropower 2009 <http://siteresources.worldbank.org/INTWAT/Resources/Directions_in_Hydropower_FINAL.pdf > (accessed on 24/03/2014).

WCD, 2000. Dams and development a new framework—The Report of the World Commission on Dams. Earthscan Publications 2000. <http://www.internationalrivers.org/files/attached-files/world_commission_on_dams_final_report.pdf> (accessed on 24/03/2014).

Židonis, A. and Aggidis, G.A., 2015. 'State of the art in numerical modelling of pelton turbines', *Renewable & Sustainable Energy Reviews*, 45, 135–44.

Zoomers, A., 2011. Introduction: rushing for land: equitable and sustainable development in Africa, Asia and Latin America development. *Development* 54(1): 12–20.

20 Global wind power developments and prospects

Will McDowall and Andrew ZP Smith

20.1 Introduction

Since the turn of the twenty-first century, wind power has gone from the periphery to the mainstream of energy technologies. Early markets in the USA and northern Europe provided the seeds for a rapid expansion of global wind energy capacity, both within those early markets and expanding well beyond them. As wind energy matures, the challenges facing the industry are changing. Cost reduction remains important, but the focus of innovation is on the offshore sector, and on the challenges posed by integrating wind into power grids at higher levels of penetration.

This chapter explores the global development of wind energy from both a technical and policy perspective. The chapter starts with an overview of historical deployment and future potential, and then follows with discussion of technical developments in wind energy, devoting particular attention to the issues of grid integration and offshore wind development (since onshore wind is a largely mature technology). Subsequent sections examine the major areas of policy relevant to the wind sector and the issues associated with public acceptance.

The chapter concentrates on large-scale wind energy technologies. While the authors note that markets for small-scale wind (turbines below 100 kW capacity) are also growing rapidly, these are not comparable with large wind turbines in terms of contribution to the global energy system as a whole, and are thus outside the scope of the present chapter. Annual deployments of small wind are around 40 MW globally (AWEA, 2010), with projections suggesting a cumulative global deployment potential of less than 4 GW by 2020 (Zhang, 2012). In contrast, annual deployments of large-scale wind are around 35–40 GW, with global cumulative capacity already above 300 GW.

20.2 Wind power historical deployment and future potential

Recent years have seen a rapid acceleration in the rate at which wind power is being deployed globally, and a rapid expansion of wind energy across the world from early markets focused on Europe and the USA. As can be seen in Figure 20.1, the great

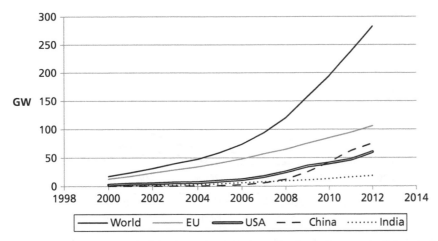

Figure 20.1 Global cumulative deployment of wind power (both onshore and offshore) 2000–12. Data from EWEA and GWEC

majority of wind power deployed by the year 2000 was within the EU, with much of the remainder in the USA. Since then, growth in the USA, India, and China, as well as other smaller markets, has accounted for an increasing share of global deployment and manufacturing, as policies to promote the adoption of wind power have spread worldwide and the technology has matured.

Recent years have also seen the establishment of a maturing offshore wind energy market, particularly focused in the shallow waters of Europe's North Sea. Growth in offshore wind has been rapid, in part because of the large size of most offshore projects, with the record for the largest offshore wind farm being broken recently by Walney phases 1&2 (367 MW combined capacity), then Greater Gabbard (504 MW) and the London Array (630 MW). Yet offshore wind is still a very small portion of the overall wind power installed base in almost all countries except the UK, where it accounts for around 30 per cent of installed capacity. No official data are published on the amount of power generated by offshore wind globally, and the estimates here for leading markets in Europe have been compiled from various sources, including the Danish Energy Agency and consulting firm Lorc.dk, as well as national statistical agencies and transmission system operators (Figure 20.2).

As the technologies have started to mature, and as policy frameworks have been strengthened—thus reducing policy risks for project investors—offshore wind deployments have occurred in ever deeper waters and ever further offshore, both in Europe and in Asia (see Figure 20.3).

20.2.1 FUTURE POTENTIAL

Estimates of the global economic potential for wind energy, onshore and offshore, differ substantially based on different expectations regarding technology costs, support

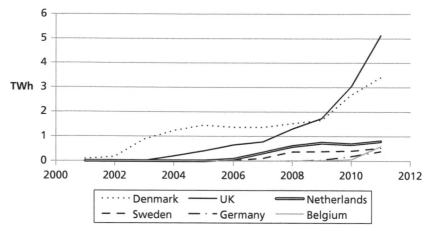

Figure 20.2 Power generated from offshore wind in leading European markets. Data compiled from Lorc.dk, and relevant Danish, UK, Belgian, and Dutch authorities

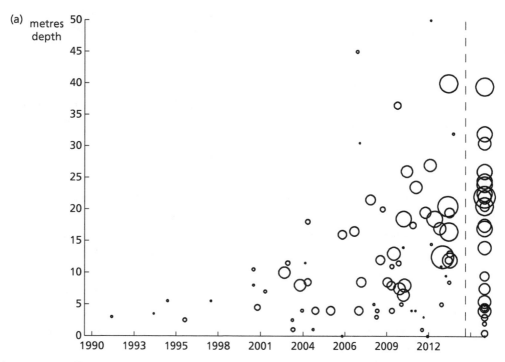

Figure 20.3 Offshore wind energy deployment: (a) depths and (b) distances from shore. Circle areas are proportional to capacity. Circles to the right of the dashed lines represent farms under construction at the time of writing, and scheduled for commissioning 2014–16. Authors' own chart, using data from 4C Offshore, RenewableUK, Danish Energy Agency, Lorc.dk

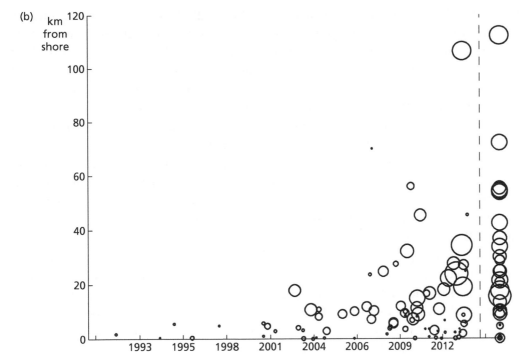

Figure 20.3 Continued

policies, public acceptability of wind energy technologies, and uncertainties concerning the underlying technical potential, as well as uncertainties regarding the evolution of energy demand. Various global scenarios of wind energy deployment over the coming decades have been developed, taking into account physical resource assessments, plausible policies, and expectations around cost trajectories for wind and competing technologies. The IEA's 2°C scenario envisages around 6000 TWh of annual wind generation (both onshore and offshore) by 2050, representing around 15 per cent of global electricity production. Across the range of IEA scenarios, including a scenario in which the world fails to reduce emissions, the contribution of wind to global electricity production is between 5 and 18 per cent in 2050 (IEA, 2012). More ambitious scenarios have also been examined, with Jacobson and Delucchi exploring a scenario in which wind power provides 50 per cent of total global primary power in 2030 (Jacobson and Delucchi, 2011).

Such scenarios show that the potential for wind to continue to expand globally is large. There is a discussion in the literature about the size of the extractable resource, with top-down modelling identifying an onshore resource of 18–68 TW (Miller et al., 2011a). For context, the global rate of primary energy demand is of the order of 18 TW. Marvel et al. (2012) estimate a global resource of 400 TW in the lower atmosphere, and 1800 TW from high-altitude wind power. By modelling the aggregated effects of individual turbines, Jacobson and Archer (2012) estimated an onshore resource of 80 TW, a total of 250 TW globally, and a further 380 TW at altitudes of 10 km. Whereas modelling at the scale of arrays of turbines (Adams and Keith, 2013) suggests that the onshore resource may be smaller, only a quarter of that 80 TW: their modelling predicts a limit to power density per unit land area of about 1 W/m², for wind farms larger than 100 km². Over the next few years, wind farms of that scale will be built, and so this prediction will be tested against empirical data. The climatic effects of large-scale wind extraction are unknown, and contested in the literature. As large-scale in this context refers to tens of terawatts of power, it may be some time before such modelling could be validated, and the implications of such effects thus remain uncertain.

Even falling far short of these technically possible options, the most pessimistic scenario studies envisage wind expanding 2–3 fold as a share of global electricity generation to reach 5 per cent by 2050 (IEA, 2012) or 4.5 per cent by 2040 (EIA, 2013), up from around 1.7 per cent in 2010 (similar figures are found in the baseline cases of the Global Energy Assessment (Riahi et al. 2012)). Since global power demand is rising during this period, these pessimistic scenarios represent at least a quintupling in global deployment of wind between 2010 and 2040 (a rate that would represent a significant slow-down from recent experience, in which wind deployment increased tenfold in the decade to 2010). These scenarios do not include efforts to reduce global emissions beyond those policy measures currently adopted in national legislation.

The importance of wind energy for reaching national, regional, and global decarbonization targets at an affordable cost is well established. Wind is relatively fast to build, is now a well-established technology, and for some countries (notably Britain) wind represents a resource that far exceeds total energy demand, making it highly scalable. It is also suitable for a large number and diversity of market players, as the unit of investment is relatively small—a typical cost for a single 3 MW onshore wind turbine is around £4m. GEA, IEA, and other scenarios that attempt to examine the cost-effective role of wind in reaching global decarbonization objectives typically envisage between 6,000–10,000 TWh per year from wind globally, equating to around 15–25 per cent of global electricity generation.

20.3 Wind power innovation and technological developments

In order to understand the major innovation trajectory of wind energy technologies, it is helpful to sketch out a brief framework for thinking about technological maturity. A simple technology lifecycle model describes how new technologies emerge from a phase of experimentation and learning, characterized by high uncertainty and a plethora of different possible directions for the technology. As experience is gained and greater understanding of the needs of users and the market opportunity is developed, a dominant design emerges (Teece, 1986; Grübler et al., 2012), which is further developed through incremental process innovation.

Onshore wind energy has broadly followed this pattern. Wind technologies passed through an 'era of ferment', with a multitude of competing concepts and designs during the late 1970s and early 1980s. Subsequently, the core elements of a dominant design (three blade, horizontal axis, variable pitch) became established, and innovative efforts have since focused on refinements to this basic design (though the shift towards direct drive has been a more recent substantial change to the design). Since the establishment of the dominant design, innovative efforts have focused on scaling up, on reducing costs and increasing energy yield, and on differentiating turbine products for different market segments (with turbines increasingly designed for specific wind conditions). Patenting activity has intensified substantially since the 1990s, reflecting the take-off of wind energy as a major industry. Patent data reveals a relative shift in the focus of innovative activity from mechanical components towards system optimization for grid integration, software, and control systems, and tailoring turbines to specific conditions (Lee et al., 2009).

Onshore wind can thus be seen as a mature technology, and this chapter does not dwell on the technical details, apart from a discussion of the costs of onshore wind and the trajectories of cost reduction. Offshore wind is at a considerably earlier stage in the technology lifecycle and this chapter therefore devotes special attention to the ongoing innovation and technological challenges within the offshore sector.

20.3.1 COST TRAJECTORIES OF ONSHORE WIND

Until around 2004, the capital costs for onshore wind technologies followed a classic 'experience curve', in which costs fall as a function of cumulative deployment (suggesting the importance of learning by doing) and over time (suggesting cost reductions associated with knowledge accumulation via R&D). Estimates of the learning rate (the percentage reduction in unit cost for each doubling of cumulative installed capacity) varied both over time and in different regions, and different learning rates have been calculated for capital costs and levelized costs, reflecting the fact that manufacturers balance cost reductions against performance improvements. Lindman and Söderholm (2012) provided a recent review of thirty-five studies providing empirical estimates of learning rates in wind power, finding considerable variation in estimates around the mean of 10 per cent. It is worth noting that studies are heavily Euro- and US-centric; few learning-curve estimates have been published that take into account deployment and cost data from China, now the largest wind market.

However, wind power cost trajectories also illustrate the risks of assuming a straightforward relationship of technology costs with either time or cumulative capacity, as discussed in Chapter 2 of this book. Between 2004 and 2010, capital costs (in £/kW) of onshore wind technologies rose in many markets. Since peaking in 2009–10, there is evidence that the costs of wind are again failing. Turbine prices—a key driver of the higher costs—had fallen by around 15–18 per cent by 2012 (IRENA, 2012).

Several factors underpinned this period of cost increase. Rapid growth in global commodity prices played a role, with steel and energy costs estimated to have increased turbine capital costs by around 10 per cent between 2003 and 2008 in the USA, with similar effects in Europe (Bolinger and Wiser, 2011; Lantz et al., 2012). Just as important, however, were supply bottlenecks emerging as a result of surging demand for turbines during this period. Finally, technical improvements were introduced that increased capital costs but also raised energy yields (Lantz et al., 2012).

Recent falls in cost are attributed to falling commodity prices and to increased competition, cutting into turbine manufacturer margins. Various commentators have suggested that there is overcapacity for manufacture of wind turbines, and that this has been a contributing factor in falling turbine prices (BNEF, 2012). Looking to the future, Lantz et al. (2012) argue that capital cost reductions for onshore wind arising from innovation are likely to be modest. Rather, technological changes are expected to result in increasing performance, and lower levelized costs.

20.3.2 OFFSHORE: A NEW FRONTIER FOR WIND INNOVATION

While by 2013 onshore wind had become a mature industry, offshore wind remains a difficult adolescent, showing great potential but also growing pains. In recent years, deployment has accelerated as technologies have improved and policy support strengthened. The first 3 GW of capacity was commissioned in the nineteen years between September 1991 and October 2010; the most recent 3 GW of capacity was commissioned in one year, between September 2012 and September 2013. At the time of writing, just under 7 GW of offshore wind capacity had been installed (Table 20.1), and a further 5 GW was under construction, and expected to be in operation by the time this book is published. Of the installed capacity, 3.7 GW is in UK waters, 1.3 GW in Danish waters, Germany and Belgium each having about 0.5 GW, and the rest distributed in smaller quantities across a dozen other countries.

Table 20.1 Landmarks of cumulative installed offshore wind capacity (Authors' own calculations using data from 4C Offshore, RenewableUK, Danish Energy Agency, Lorc.dk.)

MW_p	Date
1	September 1991
10	May 1995
100	December 2002
1000	October 2007
2000	April 2010
3000	December 2010
4000	September 2012
5000	April 2013
6000	August 2013

As noted previously, the global potential resource is huge, but easy gains are geographically concentrated due to ocean depths. Europe's North Sea provides an exceptional opportunity, with large areas at shallow depths, while North America's Great Lakes and China's Yellow Sea (Bohai) have been argued to provide similar, if smaller, opportunities. Deployments of wind have made use of these relatively easy opportunities, and Figure 20.3 shows how over time deployments have increasingly moved into deeper waters and further offshore. The first offshore turbines were essentially onshore machines planted in the inter-tidal area. A great deal of innovative activity has addressed the challenge of moving into deeper waters. Ultimately, floating turbine designs offer the potential for wind power to be deployed in much deeper waters, and Norway (with a large proportion of its seas at depths beyond 100 metres), has already piloted floating wind turbines. However, floating turbines remain prototypes, with estimates suggesting costs are currently 10–30 per cent higher than for bottom-fixed turbines (Slätte and Ebbesen, 2012).

Not surprisingly, the move to deeper waters and further from the coast, along with higher commodity prices and supply chain constraints discussed above, has led to higher prices (see Figure 20.4). The clustering of recent projects in the range £3–£4 per Watt-peak reflects these moves. Current support regimes are unlikely to support projects above that range: the notable outlier in the top-right corner of each chart is Bard 1, a project whose boldness in terms of depth and distance from shore was matched with its boldness in choosing a novel turbine, and a novel tripod foundation, rather than the monopile, gravity or jacket designs used on earlier projects.

Considerable research and development effort has been dedicated to reducing capital and operating costs of offshore wind. Experimentation has been particularly obvious in foundation structures, with gravity, monopile, tripod, jacket, and 'suction bucket' designs all being tested and deployed, as well as in installation methods and vessels. Effort has also focused on enabling higher levels of reliability and durability in the marine environment, reducing the need for repairs (which are inevitably much more expensive offshore than equivalent repairs would be on land). Given their novelty, the lifetime of offshore turbines remains uncertain, with an impact on the lifecycle costs of power from offshore wind. The first commercial offshore wind farm, Vindeby in Denmark, was commissioned in 1991, and so is only just past twenty years design life now: though it has shown declining production in later years, it has now generated for more than two years longer than its expected twenty-year life, and the additional generation is larger than the reduction in output in years 14–20. The next-oldest commercial farm for which generation data is available, Tunø Knob (also in Denmark), was nineteen years old at the time of writing, and showing no decline in output (using figures from the Danish Energy Agency).

Demonstrator projects—notably at Gunfleet Sands off the UK—have been established to 'piggyback' on larger established farms by using existing connector capacity, maintenance vessels, and crew. These projects enable learning from real-world deployment conditions, while reducing the costs of full-scale experimentation.

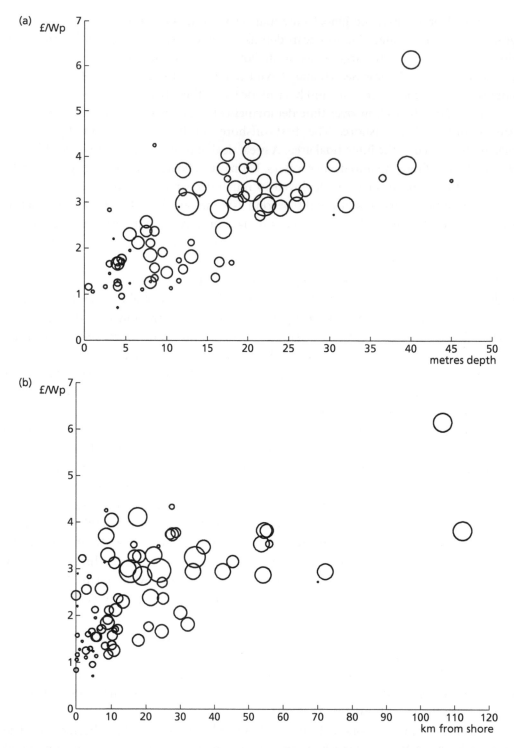

Figure 20.4 Offshore wind unit costs per Watt capacity, by (a) depth and (b) by distance from shore. Circle areas are proportional to capacity. Authors' own chart, using data from 4C Offshore, RenewableUK, Danish Energy Agency, Lorc.dk. Demonstration sites have been excluded

Cost reductions are widely expected as research and experimentation yield results. DONG Energy, one of the major players in European offshore windfarm development, has agreed with the target set by the UK's Offshore Wind Cost Reduction Task Force of a levelized cost of energy of £100 per MWh 'for offshore wind projects seeking financial investment approval in 2020' (Dong Energy, 2012).

Given the tightness across the supply chain caused by the recent expansion of offshore wind, observed learning rates underplay the cost reductions from economies of scale and learning by doing: van der Zwaan et al. (2012) calculate an observed learning rate of 3 per cent, and an underlying learning rate of 5 per cent; they found that cost increases between 2004 and 2009 were largely explained by increases in material prices, consistent with the findings of Heptonstall et al. (2012). However, with few data points to analyse, the underlying learning rate may be some distance in either direction from that estimate.

The growth in capacity, and resulting tightness in the supply chain, has resulted not just in price pressures but also in changes across the wider industry and supply chain. This is well illustrated by the growth of the market for offshore wind installation equipment and services. Several new installation vessels have been commissioned, expanding the fleet that had previously been dominated by vessels from the oil and gas industries. The 4C Offshore database listed, at the time of writing, 70 survey vessels, 412 service vessels, 146 heavy maintenance and construction vessels, 188 construction support vessels, 14 accommodation vessels, and 103 cable installation vessels, in operation or under construction (4C Offshore, 2014). Skills development has been a concern, given the rapid pace of developments in the North Sea. A 2011 report for the UK Commission for Employment and Skills (UKCES, 2011) found specialist skill shortages for Round 2 windfarms (construction ongoing at the time of writing), and further potential skills shortages for Round 3 (construction expected to start in 2014), both in domestic skills (such as planning) as well as skills from the global market (engineers, cable jointers, turbine technicians).

Finally, there have been efforts to ensure that the development of offshore wind does not damage the natural marine environment. Noise pollution from pile-driving is known to harm whale and dolphin populations. Bubble-curtains—a noise-reduction technique borrowed from research for the offshore-oil and bridge-construction industries—have been successfully demonstrated at offshore windfarm installations, including research platforms, where monopiling is one of the most common foundation techniques (Grunau, 2008). Other concerns have been raised about the impact on benthic ecosystems, though recent studies suggest that such fears may be misplaced, at least in terms of short-term impacts. In a study of the Offshore Windfarm Egmond aan Zee (OWEZ) (Lindeboom et al., 2011: 1) found: 'no short-term effects on the benthos in the sandy area between the generators... Overall, the OWEZ wind farm acts as a new type of habitat with a higher biodiversity of benthic organisms, a possibly increased use of the area by the benthos, fish, marine mammals and some bird species and a decreased use by several other bird species.'

20.3.3 INTEGRATING WIND ENERGY INTO THE POWER SYSTEM

As wind energy has become successfully deployed, the challenges of integrating large shares of renewable energy into the grid have started to manifest. Challenges of system integration have become an increasing focus for innovation, debate, and sometimes controversy around the wind industry.

In the near term, transmission constraints have led to increases in 'constraint payments' in many markets, that is payments made by the system operator to curtail wind generation. Constraints have been most acute where good wind resources are found in areas with existing grid bottlenecks: northern Germany, northern China, and West Texas being notable examples. In 2012, 20 TWh of wind were 'constrained off' in China alone, due to inadequate grid connections (Liu, 2012). Part of the problem is that planned and in-progress upgrades to grid infrastructure have not kept pace with the rate of wind deployment, and thus at least some of the most widely reported problems are temporary as grid reinforcement catches up.

In the longer term, there is debate about the extent to which the challenges of integrating variable wind generation into grid operation is likely to act as a barrier to wind deployment, either on technical grounds, or because the costs of grid management become unacceptable. As noted earlier, scenarios of the global contribution of wind under a decarbonization future involve 15–25 per cent of global electricity generation coming from wind energy technologies. Such scenarios imply rather higher levels of wind energy penetration in those regions in which wind energy is most suitable, and the challenges of integrating large shares of wind have become increasingly prominent.

Some of these issues—in particular those related to transmission limits and storage associated with all variable generation—are dealt with elsewhere in this volume (see Chapter 21), and are not addressed further here. Rather, this section addresses three issues that have become increasingly relevant to the technical and policy developments surrounding wind as market shares have grown: impacts on system frequency response, impacts on fossil fuel generation, and impacts on power market prices.

20.3.3.1 Effects of Wind on Grid Stability: Frequency and System Inertia

Power system reliability depends on maintaining the frequency of alternating current within a narrow range around the system target (whether 50 or 60 Hz). In existing systems, sudden loss of a large generator has only limited immediate impact on system frequency, as the inertia of the rotating mass of generators on the system prevents rapid frequency change, providing enough time for generation to ramp up and compensate for the lost generator. However, wind systems have been designed to maximize energy output and efficiency, and are not synchronized with the frequency of the power grid. As a result, the rotating mass of even thousands of wind turbines does not contribute at all to system inertia. This problem is negligible in systems with low shares of wind generation, but becomes more significant as the share of wind increases.

Eto et al. (2010) identify four ways in which challenges arise with high levels of wind generation: reduced system inertia (as introduced above); displacement off the grid of generation sources which incorporate primary frequency response (PFR), that is, balancing services with response times of the order of a small number of seconds; the location of generation sources incorporating PFR may change, creating new transmission bottlenecks; and increased demands on secondary frequency response may require those to be supplemented by PFR sources, making them unavailable for PFR duties.

In small or isolated systems, even fairly low levels of instantaneous generation share by exogenously-variable generators such as wind can cause challenges for grid stability; larger or well-interconnected systems will manifest problems at higher penetrations. Some regulatory systems have already started to address the issue. Delille et al. (2012) report that in the French Islands 'these sources can be required to temporarily disconnect from the grid when their instantaneous penetration level reaches 30%'. Brisebois and Aubut (2011) describe the measures being taken by Hydro-Québec, working together with manufacturers, to incorporate synthetic inertia into the control mechanisms of turbines to provide PFR.

Ruttledge et al. (2012) also describe ways in which the frequency and inertia issue has been mitigated, including adapting grid codes to mandate the provision of some PFR services by grid-connected turbines. In response, 'a number of wind turbine manufacturers have developed emulated inertial response products to respond to significant frequency events' (ibid: 683).

20.3.3.2 Effects of Wind on the Operation of Fossil Fuel Plants

The variability of wind power (along with other variable sources of electricity) tends to result in a greater need for other generation plants to ramp up or down, with more frequent cycling between levels of generation leading to efficiency losses. There has been some discussion in the popular press concerning the extent to which this effect reduces the environmental benefits of wind. Lew et al. (2013) examine a scenario with 33 per cent wind and solar penetration in the Western USA. The costs associated with additional fossil plant cycling amounted to around $0.14–0.67/MWh of wind or solar generation, less than 1 per cent of the levelized cost of wind. The effects on emissions are also small. Reductions in emissions associated with high levels of wind generation far outweigh the additional emissions caused by increased cycling, which are found to be negligible for CO_2 and small for SO_2 and NO_x (with savings of SO_2 around 3 per cent less than would have been expected without accounting for cycling effects).

20.3.3.3 Merit Order Effects on Price

Wind power effectively has zero short-run marginal cost. This means it bids in to power markets as the cheapest provider. This displaces the most expensive producer off the market, and the most expensive producer is the price-setter for the market, so this displacement can reduce power prices. The displaced plant is also, typically, the least efficient plant in a particular class, either the least efficient oil plant, or gas plant, or coal plant (depending on the relative efficiency-adjusted prices of those fuels). This displacement of other forms of generation is known as the 'merit-order' effect, and has been estimated to result in net price decreases in Spain (after accounting for price increases associated with subsidy support; Saenz de Miera et al., 2008). Similar results have been found for Germany (Sensfuß et al., 2008). The merit order effect has implications for the operation of the system and for market design. Where higher levels of wind penetration result in lower market prices for power, incentives for investment in power generation are reduced, resulting in efforts to redesign power markets to ensure sufficient investment in dispatchable plants, such as capacity payments.

20.3.4 LONG-TERM INNOVATION PROSPECTS FOR WIND: THE POTENTIAL FOR TRANSFORMATIVE CHANGE

While the first generation of onshore wind technologies has reached maturity, with offshore wind now at a stage of experimentation and the early phases of growth, a range of more radical wind technologies are being considered, with the potential for transformative change in the longer term. Three trends are noted here, though this is not an exclusive list.

First, despite the dominance of horizontal axis designs, there remains continued experimentation with vertical axis concepts, particularly offshore. An extensive study by Britain's Energy Technology Institute into the viability of large-scale vertical-axis wind turbines is yet to publish a report, and no commercial turbines of this design are being deployed at present. Biomimetic wind-farm design using vertical-axis wind turbines arranged in an array of contra-rotating pairs to capture energy from wake effects in a similar manner to shoaling fish has been proposed (Whittlesey et al., 2010), but remains at the demonstration stage.

Second, a range of high-altitude airborne wind turbines have been developed at the prototype stage. The challenges associated with airborne turbines are obvious, but the allure of high-speed and consistent high-altitude winds have resulted in continued interest in these options (Roberts et al., 2007), with several small firms developing designs or prototypes. However, concerns have been raised about the potential for large-scale deployment of high-altitude wind power to disrupt jet stream winds, and it has therefore been suggested that only a relatively small resource can be sustainably extracted (Miller et al., 2011b).

Finally, while the historic upscaling trend of wind turbines has slowed, there is continued exploration of turbine designs much larger than those in current use. Fichaux et al. (2011) in the Upwind project found solutions, and no design limits, to a wind turbine of 20 MW capacity. Their extrapolated turbine followed the design of almost all extant offshore turbines in having upwind rotors, variable speed with controlled pitch, the rotor diameter would be of the order of 252 metres, the hub height would be 153 metres, and the rotor speed would be 6rpm—compared to their reference 5 MW turbine which has a rotor speed of 12 p.m.

These all remain tantalizing but commercially unrealized prospects that may bring disruptive change: meanwhile incremental evolution of existing designs continues.

20.4 **Wind power policies**

The successful deployment of wind power discussed in the previous section has been driven by supportive public policy. It is worthwhile distinguishing two basic forms of policy support for wind power, though in practice these are often related. First, there are policies supporting the deployment of wind energy technologies; and second, there are policies supporting the development of new wind energy innovations and the establishment of a successful domestic wind energy industry.

20.4.1 DEPLOYMENT POLICIES

Policy support for the deployment of wind energy typically comprises three main elements: targets, subsidy support, and market design.

The establishment of targets—both aspirational and legislated—has been important in providing a signal of policymaker intent, beyond the detail of existing deployment policies. Legislated targets have been particularly powerful in Europe, where the targets for member states enshrined in the Renewable Energy Directive[1] are one of the primary drivers of renewable energy policy. One recent report suggests that more action is necessary if targets are to be met, particularly for accelerating the deployment of wind (Hamelinck et al., 2012), whereas the European Environment Agency (2013) suggests that the EU renewables target for 2020 would be met by current policies, assuming they continue to be implemented unchanged. Beyond Europe, a majority of the world's countries have adopted or announced targets for renewable energy, with most of these countries having adopted supportive policies to meet targets, typically including some form of subsidy (REN21, 2013).

In designing subsidy mechanisms, policymakers have endeavoured to strike the right balance between policies that provide sufficient support to ensure deployment while keeping the costs of subsidy support manageable, and providing incentives for cost reduction through competition and support degression. The mid 2000s were

[1] DIRECTIVE 2009/28/EC OF THE EUROPEAN PARLIAMENT AND OF THE COUNCIL of 23 April 2009 on the promotion of the use of energy from renewable sources and amending and subsequently repealing Directives 2001/77/EC and 2003/30/EC.

characterized by debates about the most effective support measures for renewable energy deployment support, both for renewables in general and for wind in particular. Three basic models competed: a tendering or concession-based model, in which (typically publicly owned) utilities issued competitions to win contracts for providing wind power; tradable certificate markets with requirements that utilities meet portfolio standards or obligations; and feed-in-tariffs. By the end of the decade, it was widely agreed that feed-in-tariffs had tended to be more successful, both in ensuring steady deployment at lower cost, but also in facilitating the development of strong wind power firms (Ragwitz et al., 2007; Bergek and Jacobsson, 2010). Feed-in-tariffs have been the single most popular mechanism for support, though there has been an increase in recent years in the use of competitive bidding mechanisms (REN21, 2013). Finally, many countries (such as Germany and Brazil) facilitate access to finance for wind energy project development, typically through a state-owned investment bank.

Regardless of which precise mechanism of policy support has been adopted, it is widely agreed that the long-term credibility of policy is important in stimulating the development of the market. The political battles over support for wind power in the USA have created a staccato pattern of on-again, off-again support for wind power through the Production Tax Credit, and the resulting pattern of booms and busts has had damaging consequences for the US domestic industry. Similarly, changes to established contracts for onshore wind generation in Spain in 2009–13 created deep uncertainty in the market across Europe, as developers wondered which other countries might follow suit now that a precedent was established (Couture and Bechberger, 2013). These contractual changes in Spain were made despite an earlier study which had shown that the wind payments had actually created net savings, because of merit-order effects (as discussed in Section 12.3.3.3).

In addition to providing subsidy support, policymakers have had to tackle the mismatch between wind power and the regulatory and market structures designed for centralized and dispatchable generation plants. Policy changes to the details of electricity market design, the rules governing grid access, and cost formulae for transmission extension, reinforcements or connections have all been important in enabling the emergence of existing wind markets. As wind power further develops, policy frameworks are increasingly evolving to account for the characteristics of a system with high levels of variable generation that has very low short-run marginal costs.

20.4.2 DEVELOPMENT/INNOVATION POLICIES

20.4.2.1 R&D Support

The seeds of the modern wind industry were planted through the establishment of wind energy R&D programmes in Europe and the USA following the 1970s' oil crises. Public funding for basic research enabled a rapid expansion of designs, ideas, and the networks of scientists and engineers interested in wind energy. Importantly, public funding for R&D has not simply taken the form of research grants to existing research institutions and universities. Public money has also been used to establish test facilities (notably at NREL in the USA and Risø in Denmark, and more recently at NAREC in the UK), and to provide support for networking (as with government funding for the conferences of the American Wind Energy Association in the USA).

Studies of the wind innovation system have provided a detailed picture of the ways in which innovation system dynamics and external conditions enabled the successful development of the technology. The innovation process characterized by an early phase of experimentation and learning eventually led to a dominant design, the three-bladed horizontal-axis turbine. This emerged not as a result of a breakthrough within large scale R&D programmes, though R&D funding was critically important. Rather, the institutional framework that led to success was characterized not by a 'big push' scientific effort, but by fostering entrepreneurial experimentation and learning (through the provision of test facilities), and enabling close links between early users (supported with subsidies) and producers of turbines (Garud and Karnøe, 2003; Jacobsson and Bergek, 2004; McDowall et al., 2013). This framework, found in Denmark, proved successful in facilitating the development of the technology.

During the late 1980s and into the 1990s, the technology was then scaled up, through a gradual process supported by a welcoming market environment that successfully balanced capital cost supports with feed-in-tariffs, particularly in Denmark, Germany, and Spain. These market supports were accompanied by support for testing, monitoring, and networking. Consequently, those three named countries also became centres of manufacturing and deployment.

During this early period, cost reductions emerged, perhaps counter-intuitively, not in those countries that focused on policy supports designed to yield the cheapest possible deployment of wind power, such as the UK and Sweden (Mitchell and Connor, 2004; Bergek and Jacobsson, 2010)—despite the fact that providing incentives for cost reduction was often an explicit part of the rationale for such policies. Rather, it was countries that prioritized strategic development of the industry and that provided support with phased reductions over time that enabled firms to develop and innovate to reduce costs (Jacobsson and Lauber, 2006; McDowall et al., 2013). The insight for policy from this experience is that reliance on 'market pull' (through carbon pricing, for example) or 'technology push' (through R&D funding) are on their own unlikely to be sufficient to bring forward new clean energy technologies. Rather, government has a role in fostering the conditions for an effective innovation system, comprising both market pull and technology push measures, as well as a range of 'systemic instruments', all of which should be appropriate for the stage of technological maturity.

Overall public R&D funding has started to accelerate in recent years, as countries respond to the apparent potential for wind energy, and the increasing competitiveness of the wind sector. Several countries are focusing funding on efforts to develop offshore wind, with the UK's Offshore Renewable Energy Catapult a leading example, together with Germany's alpha ventus test field, and demonstration projects in the East China Sea.

20.4.2.2 Support for Industrial Development

Policies for wind energy deployment and wind energy industrial development have often gone hand-in-hand. This is partly because policymakers have felt that the political legitimacy of wind power will to some extent depend on the perception that domestic jobs and industries are benefiting from the development of wind. As a consequence, recent years have seen many countries establish 'domestic content requirements' embedded within their deployment policies. These typically require that a portion of the physical inputs for a wind energy project must be manufactured domestically if the project is to receive subsidy support. Broadly similar domestic content requirements exist in China, the Ukraine, Brazil, Turkey, India, various US States, two Canadian Provinces, and a range of EU member states. This practice has been controversial, and it is possible that some such policies could be found in breach of international trade law. In 2013, the WTO upheld a ruling that had found the Canadian Province of Ontario in violation of trade rules, because of the domestic content requirements established in Ontario's feed-in-tariff support for renewable energy (Globe and Mail, 2013).

It is perhaps surprising that countries have felt the need to establish domestic content requirement policies for wind technologies, which have high relative transport costs during installation, and for which therefore local manufacturing is often economic without subsidy support. The answer may lie as much in the politics of subsidy support and the implicit or explicit promise that the transition to sustainable energy will not come at the cost of jobs and industrial opportunities, but rather will promote their development. As the industry matures, it seems likely that aspirations to achieve comparative advantage in wind power will decline, since it will become increasingly difficult for new national-champion firms to compete with established and often multinational market leaders.

20.5 Public acceptance, public resistance, and political legitimacy

International opinion polling has consistently suggested solid public support for wind energy (e.g. Council on Foreign Relations, 2012). Even in countries such as Britain where there have been high-profile anti-onshore-windfarm activities, public support is 68 per cent for (support or strongly support), compared to public opposition of 12 per cent (oppose or strongly oppose) (DECC 2014).

However, wind energy is controversial in many countries, with vocal opponents. All radical new technologies—and particularly those that require public support—must establish legitimacy and overcome resistance to change. McDowall et al. (2013) provide an overview of studies that examine how such legitimacy has been established for wind (and note that the literature has focused on Europe and the USA, with less published research from other countries). They highlight the cultivation of domestic manufacturing capacity for the turbine industry; mitigation of objections through cost reductions and technological developments; and the role of policy and market structure in mediating legitimacy: Germany and Denmark's early pattern of distributed ownership (e.g. community-owned turbines) has been credited with fostering legitimacy there, while the UK's policy structures undermined early legitimacy by preventing this model of local ownership from emerging (Mitchell and Connor, 2004; Jacobsson and Lauber, 2006). This raises the importance of policy in mediating the conditions through which wind energy technologies are seen as more or less acceptable, which relate not only to technical and economic characteristics, but also to issues of trust and perceived fairness, both of which can be influenced by the policy structure (both planning procedures and the mechanism of subsidy support).

Though much has been made of the NIMBY (Not In My Back Yard) concept in discussions of public resistance to wind, Devine-Wright has argued that the concept 'has failed to receive empirical support' as an explanation for negative public perceptions of wind (Devine-Wright, 2005). While it is clear that many people dislike wind turbines and fear the effect that local development of wind energy may have, including concerns around potential reductions in house prices (for which there is conflicting evidence from different studies and countries, see, e.g. Gibbons, 2014), there is also good evidence of positive perceptions at a local level, associated with community ownership, engagement, and direct benefits to residents local to the turbines. More nuanced understandings are required that understand and engage with the local and landscape-level impacts that arise from wind, and the concerns of local people.

Anti-wind campaigners have argued that wind farms can damage health, but little evidence has been found to support this claim. Knopper and Ollson (2011: 1) found that 'no peer reviewed articles demonstrate a direct causal link between people living in proximity to modern wind turbines, the noise they emit and resulting physiological health effects'. Chapman et al. (2013: 1) found that 'reported historical and geographical variations in complaints are consistent with psychogenic hypotheses that expressed health problems are "communicated diseases" with nocebo effects likely to play an important role in the aetiology of complaints'.

20.6 Conclusions

At 2 per cent global penetration, wind still seems a small player. But with a 27 per cent annual growth rate, and a pause only just showing up now from the 2008 global financial crisis, it still looks like the opening chapter for wind. The potential is huge, beyond what might be needed to decarbonize the global economy in the foreseeable future. Carefully designed policy support will be important in enabling wind power to meet this potential, enabling the innovation that can reduce costs, facilitate integration into grids, and maintain public support. Particularly important over the coming decade will be innovation and cost reductions in grid integration and in offshore wind.

■ REFERENCES

4C Offshore (2014). Offshore Wind Vessel Database. Retrieved 26 January 2014. <http://www.4coffshore.com/windfarms/vessels.aspx>. (Accessed April 2015).

Adams, Amanda S., and David W. Keith. (2013). Are global wind power resource estimates overstated? *Environmental Research Letters* 8(1) Article number 015021. doi:10.1088/1748-9326/8/1/015021.

AWEA (2010). *AWEA Small Wind Turbine Global Market Study*. Washington, DC: American Wind Energy Association.

Bergek, A. and S. Jacobsson (2010). Are tradable green certificates a cost-efficient policy driving technical change or a rent-generating machine? Lessons from Sweden 2003-2008. *Energy Policy* 38(3): 1255–71.

BNEF (2012). Overcapacity and new players keep wind turbine prices in the doldrums. Press release, 6 March 2012. Available at <http://bnef.com/PressReleases/view/196> (Accessed 5 November 2013).

Bolinger, M. and R. H. Wiser (2011). *Understanding Trends in Wind Turbine Prices Over the Past Decade*. Berkeley: LBNL.

Brisebois, J., and Aubut, N. (2011). Wind farm inertia emulation to fulfill Hydro-Québec's specific need. In *2011 IEEE Power and Energy Society General Meeting* (Vol. 7, pp. 1–7). IEEE. doi:10.1109/PES.2011.6039121.

Chapman, S., St George, A., Waller, K., and Cakic, V. (2013). The pattern of complaints about Australian wind farms does not match the establishment and distribution of turbines: support for the psychogenic, 'communicated disease' hypothesis. *PLOS ONE*, 8(10), e76584. doi:10.1371/journal.pone.0076584.

Council on Foreign Relations, (2012). *Public Opinion on Global Issues*: Chapter 5b: Energy security. New York: Council on Foreign Relations.

Couture, Toby D. and Bechberger, Mischa. 'Pain in Spain: New Retroactive Changes Hinder Renewable Energy', *Renewable Energy World*, 19 April 2013 <http://www.renewableenergyworld.com/rea/news/article/2013/04/pain-in-spain-new-retroactive-changes-hinders-renewable-energy>. (Accessed April 2015).

DECC (2014). Public attitudes tracking survey: wave 9. <https://www.gov.uk/government/publications/public-attitudes-tracking-survey-wave-9> (Accessed 5 May 2014).

Delille, G., Francois, B., and Malarange, G. (2012). Dynamic frequency control support by energy storage to reduce the impact of wind and solar generation on isolated power system's inertia. *IEEE Transactions on Sustainable Energy*, 3(4): 931–9. doi:10.1109/TSTE.2012.2205025.

Devine-Wright, P. (2005). Beyond NIMBYism: towards an integrated framework for understanding public perceptions of wind energy. *Wind Energy*, 8(2): 125–39. doi:10.1002/we.124.

Dong Energy (2012). £100 per megawatt is achievable for offshore wind by 2020. Dong Energy press release, 15 June 2012. Available at <http://www.dongenergy.co.uk/EN/News/UK_news/news/Pages/-100permegawattisachievable.aspx>. (Accessed 29 October 2013).

EIA (2013). *International Energy Outlook*. US Department of Energy, Energy Information Administration.

Eto, J. Undrill, P. Mackin, R. Dashmans, B. Williams, B. Haney, R. Hunt, J. Ellis, H. Illian, C. Martinez, M. O'Malley, K. Coughlin, and K. LaCommare. (2010). Use of frequency response metrics to assess the planning and operating requirements for reliable integration of variable renewable generation. Federal Energy Regulatory Commission, LBNL-4142E, Dec. 2010.

European Environment Agency (2013). Trends and projections in Europe 2013: Tracking progress towards Europe's climate and energy targets until 2020. EEA Report 10/2013. doi:10.2800/93693.

Fichaux, Nicolas, Jos Beurskens, Peter Hjuler Jensen, and Justin Wilkes. (2011). 'Upwind: Design Limits and Solutions for Very Large Wind Turbines.', for the EU Sixth Framework, Contract number: 019945 (SES6).

Garud, R. and P. Karnøe (2003). Bricolage versus breakthrough: distributed and embedded agency in technology entrepreneurship. *Research Policy* 32(2): 277–300.

Gibbons, S. (2014). Gone with the wind: valuing the visual impacts of wind turbines through house prices. Spatial Economics Research Centre Discussion Paper 159, LSE, London.

Grubler, A., F. Aguayo, K. Gallagher, M. Hekkert, K. Jiang, L. Mytelka, L. Neij, G. Nemet, and C. Wilson (2012). Chapter 24 – Policies for the energy technology innovation system (ETIS). In: Johansson, T. B., A. Patwardhan, N. Nakicenovic, and L. Gomez-Echeverri (Eds), Global Energy Assessment - Toward a Sustainable Future, Cambridge University Press, Cambridge, UK and New York, NY, USA and the International Institute for Applied Systems Analysis, Laxenburg, Austria, pp. 1665–744.

Grunau, C. (2008). Drawing the bubble curtain for sound attenuation. *Port Technology International*, Edition 40, p. 18.

The Globe and Mail (2013). Ontario to change green energy law after WTO ruling. 29 May 29 2013.

Hamelinck, C., I. de Lovinfosse, I., M. Koper, C. Beestermoeller, C. Nabe, M. Kimmel, A. van den Bos, I. Yildiz, M. Harteveld, M. Ragwitz, S. Steinhilber, J. Nysten, D. Fouquet, G. Resch, L. Liebmann, A. Ortner, C. Panzer, D. Walden, R. Diaz Chavez, B. Byers, S. Petrova, E. Kunen, and G. Fischer (2012). Renewable energy progress and biofuels sustainability. Report for the European Commission. ENER/C1/463-2011-Lot2. Brussels, Ecofys.

Heptonstall, Philip, Robert Gross, Philip Greenacre, and Tim Cockerill. (2012). The cost of offshore wind: understanding the past and projecting the future. *Energy Policy*, 41 (February): 815–21. doi:10.1016/j.enpol.2011.11.050.

IEA (2012). *Energy Technology Perspectives*. OECD/IEA, Paris.

IRENA (2012). *Wind Power*. Renewable Energy Technologies Cost Analysis Series. Volume 1: Power Sector (Issue 5/5). International Renewable Energy Agency, Bonn.

Jacobson, Mark Z., and Cristina L. Archer (2012). Saturation wind power potential and its implications for wind energy. *Proceedings of the National Academy of Sciences of the United States of America* 109 (39) (September 25): 15679–84. doi:10.1073/pnas.1208993109.

Jacobson, Mark Z. and Mark A. Delucchi (2011). Providing all global energy with wind, water, and solar power, Part I: Technologies, energy resources, quantities and areas of infrastructure, and materials. *Energy Policy* 39: 1154–69.

Jacobsson, S. and A. Bergek (2004). Transforming the energy sector: the evolution of technological systems in renewable energy technology.' *Industrial and Corporate Change* 13(5): 815–49.

Jacobsson, S. and V. Lauber (2006). The politics and policy of energy system transformation–explaining the German diffusion of renewable energy technology. *Energy Policy* 34(3): 256–76.

Knopper, L. D., and C. A. Ollson. (2011). Health effects and wind turbines: a review of the literature. *Environmental Health: A Global Access Science Source*, 10(1): 78. doi:10.1186/1476-069X-10-78.

Lantz, E., R. Wiser, and M. Hand (2012). IEA Wind task 26: the past and future cost of wind energy. National Renewable Energy Laboratory, Technical Report NREL/TP-6A20-53510. Golden, Colorado.

Lee, B., I. Iliev, and F. Preston (2009). *Who Owns our Low Carbon Future? Intellectual Property and Energy Technologies*. London, Chatham House.

Lew, D., G. Brinkman, E. Ibanez, A. Florita, M. Heaney, B.-M. Hodge, M. Hummon, G. Stark, J. King, S.A. Lefton, N. Kumar, D. Agan, G. Jordan, and S. Venkataraman (2013). The Western Wind and Solar Integration Study Phase 2, National Renewable Energy Laboratory, Technical Report NREL/TP-5500-55588. Golden, Colorado.

Lindeboom, H. J., H. J. Kouwenhoven, M. J. N. Bergman, S. Bouma, S. Brasseur, R. Daan, R. C. Fijn, D. De Haan, S. Dirksen, R. Van Hal, R. Hille Ris Lambers, R. Ter Hofstede, K. L. Krijgsveld, M. Leopold, and M. Scheidat. (2011). Short-term ecological effects of an offshore wind farm in the dutch coastal zone; a compilation. *Environmental Research Letters* 6 (3) Article number 035101. doi:10.1088/1748-9326/6/3/035101.

Lindman, Å. and P. Söderholm (2012). Wind power learning rates: A conceptual review and meta-analysis. *Energy Economics* 34(3): 754–61.

Liu, Qi (2012). Keynote speech at the World Wind Energy Conference 2012, 3–5 June, Havana, Cuba.

Marvel, Kate, Ben Kravitz, and Ken Caldeira. (2012). Geophysical limits to global wind power. *Nature Climate Change* 2 (9) (9 September): 1–4. doi:10.1038/nclimate1683.

McDowall, W., P. Ekins, S. Radošević, and L-Y. Zhang (2013). The development of wind power in China, Europe and the USA: how have policies and innovation system activities co-evolved? *Technology Analysis & Strategic Management*, 25(2): 163–85.

Miller, L. M., F. Gans, and A. Kleidon. (2011a). Estimating maximum global land surface wind power extractability and associated climatic consequences. *Earth System Dynamics* 2 (1) (11 February): 1–12. doi:10.5194/esd-2-1-2011.

Miller, L. M., F. Gans, and A. Kleidon (2011b). Jet stream wind power as a renewable energy resource: little power, big impacts. *Earth Systems Dynamics* 2(2): 201–12.

Mitchell, C. and P. Connor (2004). 'Renewable energy policy in the UK 1990-2003.' *Energy Policy* 32(17): 1935–47.

Ragwitz, M., A. Held, G. Resch, T. Faber, R. Haas, C. Huber, R. Coenraads, M. Voogt, G. Reece, P. E. Morthorst, S. G. Jensen, I. Konstantinaviciute, and B. Heyder (2007). Assessment and optimisation of renewable energy support schemes in the European electricity market. Report to the European Commission. Karlsruhe.

REN21 (2013). Global Renewables Status Report. Renewable Energy Policy Network for the 21st Century. Paris.

Roberts, B. W., D. H. Shepard, K. Caldeira, M. E. Cannon, D. G. Eccles, A. J. Grenier, and J. F. Freidin (2007). Harnessing high-altitude wind power. *Energy Conversion, IEEE Transactions on* 22(1): 136–44.

Ruttledge, L., Miller, N. W., O'Sullivan, J., and Flynn, D. (2012). Frequency response of power systems with variable speed wind turbines. *IEEE Transactions on Sustainable Energy*, 3(4), 683–91. doi:10.1109/TSTE.2012.2202928.

Saenz de Miera, G., P. Delriogonzalez, and I. Vizcaino (2008). Analysing the impact of renewable electricity support schemes on power prices: The case of wind electricity in Spain. *Energy Policy*, 36(9), 3345–59. doi:10.1016/j.enpol.2008.04.022.

Sensfuß, F., M. Ragwitz, and M. Genoese (2008). The merit-order effect: A detailed analysis of the price effect of renewable electricity generation on spot market prices in Germany. *Energy Policy*, 36(8), 3086–94. doi:10.1016/j.enpol.2008.03.035.

Slätte, J. and M. Ebbesen (2012). Market potential and technology assessment: floating offshore wind. DNV Kema report for The Crown Estate. Høvik, Norway.

Teece, D. J. (1986). Profiting from technological innovation: Implications for integration, collaboration, licensing and public policy. *Research Policy* 15(6): 285–305.

UKCES (2011). Maximising employment and skills in the offshore wind supply chain. Evidence Report 34.

Van der Zwaan, Bob C. C. C., Rodrigo Rivera-Tinoco, Sander Lensink, and Paul van den Oosterkamp (2012). Cost reductions for offshore wind power: exploring the balance between scaling, learning and R&D. *Renewable Energy* 41 (May): 389–93. doi:10.1016/j.renene.2011.11.014.

Whittlesey, R. W., S. Liska, and J. O. Dabiri (2010). Fish schooling as a basis for vertical axis wind turbine farm design. *Bioinspiration & Biomimetics*, 5(3): 035005. doi:10.1088/1748-3182/5/3/035005.

Zhang, P. (2012). Small wind world report. World Wind Energy Association (WWEA), Bonn, Germany.

21 Network infrastructure and energy storage for low-carbon energy systems

Paul E. Dodds and Birgit Fais

21.1 Introduction

Network infrastructures are the physical assets that move energy between locations, including electricity networks, gas and liquid pipelines, and roads, railways, and shipping. These infrastructures are built and operated in a quite different way to the other parts of the energy system that have been examined in previous chapters. Although existing network infrastructures could mostly continue to operate in a low-carbon energy system, there are opportunities to reduce the cost of delivering energy by adopting revolutionary new network technologies and by repurposing existing assets for low-carbon fuels.

The definition of network infrastructures adopted here is narrower than 'energy infrastructure', used for example by Bolton and Hawkes (2013), which encompasses 'the set of technologies, physical infrastructure, institutions, policies and practices located in and associated with a country which enable energy services to be delivered to customers' (from Skea et al., 2011). In this chapter, only the network parts of energy infrastructures are examined. As the definition suggests, there is a strong social dimension to network infrastructure, as governments are held responsible for supplying energy yet private actors often build and operate the network infrastructure. There are key policy decisions around the roles of governments and private actors in infrastructure development, including the creation of supply chains, regulation of energy markets, and the source of investments. Some of these issues are also explored in Chapter 23.

This chapter examines how network infrastructures have evolved in different countries,[1] with a particular emphasis on electricity and gas networks. Since energy storage is normally integrated into network infrastructure, it is also examined here. Challenges on the horizon are identified, for example the integration of intermittent renewable

---

[1] Countries are categorized in terms of gross national income (GNI) per capita, following the approach of the World Bank, 2013. *How we Classify Countries* [Online]. Available: <http://data.worldbank.org/about/country-classifications> [Accessed 3 Sep 2013]. Three groups are used: high income (greater than $12,616), middle income ($4086–$12,615), and low income (up to $4,085).

electricity generation in high-income countries and expanding access to energy in low-income countries. The technical options and choices to meet these challenges are reviewed. Many of these challenges relate to the energy policy 'trilemma', which is the provision of affordable, secure, environmentally sustainable energy to populations (WEC, 2011). The IEA (2001) define energy security as the availability of a regular supply of energy at an affordable price. Building and operating network infrastructures that are robust to physical and economic disruptions are important steps to achieving energy security (Costantini et al., 2007).

The chapter is structured as follows. Network infrastructures are reviewed in Section 21.2, with a focus on technological options and choices for the future. Section 21.3 examines the challenges faced by high-, middle-, and low-income countries and the suitability of the technological choices to meet these challenges, using case studies of Germany, China, and Kenya. The role of government and of other institutions and stakeholders is considered in Section 21.4. The chapter concludes by briefly examining the economics of network infrastructures in Section 21.5.

21.2 **Options and choices**

There are many different types of network infrastructures that contribute in different ways to the global energy system. This section examines the options and choices for each type of infrastructure, focusing in particular on emerging technologies that could have an important role in the future and the challenges to be overcome to use those technologies.

21.2.1 PRIMARY FUEL TRANSPORT INFRASTRUCTURE

Most primary fuels, including oil (and petroleum products), coal, and biomass, are transported by ship, road, and rail. Many shipping ports have dedicated energy facilities, often including oil storage facilities, and are important hubs for internationally-traded energy commodities. Some roads and railways are constructed specifically to transport energy commodities (for example the railways that link mines with power stations and ports in Australia) but most are designed for public use. For example, while coal comprised 46 per cent of the freight transported on UK railways in 2012/13 (ORR, 2013), trains ran mainly at night as the system is primarily designed to transport passengers.

There are limited options for new technologies to penetrate these sectors in the future. The global shipping industry is likely to change most significantly: LNG (liquefied natural gas) terminals have been constructed in numerous ports over the last two decades (Kumar et al., 2011; Wood, 2012), including storage facilities, and similar facilities would be required if hydrogen were to become a globally-traded energy commodity as part of the move towards low-carbon fuels. While the volume of traded fossil fuels could reduce in the future, with their use restricted to plants with carbon capture and storage (CCS), there is a growing biomass and biofuel trade that uses similar ships and port facilities (Lamers et al., 2014).

21.2.2 PIPELINES

Pipelines have been used for more than two centuries to transport gases and, more recently, oil and petroleum products (Williams, 1981). Although the capital costs of pipelines are substantial, they can efficiently transport large quantities of fuel over long distances, have very long lifetimes, and are generally safer than other network technologies (Furchtgott-Roth and Green, 2013). For these reasons, pipe networks are likely to continue having an important role in low-carbon energy systems in the future.

It is useful to distinguish between transmission and distribution networks. Transmission pipelines are generally larger, operate at much higher pressures (for gases), and are used for long-distance national and international transportation. Distribution networks deliver fuels at a regional and local scale from the transmission networks to consumers. Each type of network faces different challenges in the future.

21.2.2.1 Distribution Networks

In many high-income countries, distribution networks were first constructed to transport locally-produced gas to urban consumers for lighting. They perform the same function today but now mostly distribute centrally-produced natural gas for heating and cooking, which is supplied from national gas transmission networks. The key challenge for these networks is the move to a low-carbon energy system, since the CO_2 resulting from gas combustion cannot realistically be captured so there will probably be a requirement to either decarbonize the gas or to cease using natural gas and hence to decommission the networks. The carbon intensity of the supplied gas could be reduced by injecting a non-carbonaceous gas, such as hydrogen or ammonia, into the gas stream or by converting the network to transport one of these gases instead of natural gas (Dodds and Demoullin, 2013). The distribution networks are substantial infrastructure investments and it is not clear how decommissioning would be funded and how private owners would be compensated were the networks to become stranded assets as a result of the move to a low-carbon economy (Dodds and McDowall, 2013). The loss of economically-viable assets like gas networks in order to fulfil a public good, which in this case is mitigating climate change, is an important issue for governments and the private sector.

21.2.2.2 Transmission Networks

Transmission pipeline networks are used around the world to transport high volumes of both gas and oil. There can be important geopolitical considerations over pipelines when they cross borders, for example the gas pipelines that link Central Asia to Russia and onwards to Western Europe (Afifi et al., 2012). Some pipeline construction projects, for example the Keystone XL oil pipeline linking Alberta, Canada, to refineries in Texas, USA, have been strongly opposed for environmental reasons (Schnoor, 2013), despite the forecast energy security and economic benefits (Shum, 2013).

Transmission pipelines require a very large upfront investment and provide a very large potential flow capacity on commissioning. From the perspective of an investor, these traits lead to a high risk of substantial losses unless there is an immediate large-scale demand for the pipeline. New oil and gas pipelines are generally constructed to satisfy pent-up demand or for energy security reasons, reducing the risk for investors. In contrast, pipelines that are constructed as a necessary prerequisite at the start of a transition to an alternative energy system, transporting, for example, hydrogen or carbon dioxide, will probably have very low initial capacity factors[2] (Dodds and McDowall, 2014). Investors in such infrastructure are faced with almost no foresight about possible future demand levels so there must be a substantial risk attached to such investments. Several strategies have been suggested to deal with this conundrum for hydrogen and a number of these are reviewed by Agnolucci (2007), as discussed briefly below.

Most existing pipelines transport high-carbon fuels and a key challenge is to find a role for them in a low-carbon future, given the long lifetimes of pipelines relative to other infrastructure. It is likely that fossil fuels will continue to be used in the future where emissions can be abated using CCS, so some pipelines will continue in their current role. One opportunity for pipeline operators is to increase the level of integration of pipelines with other parts of the energy system, for example by using networks as storage for intermittent renewable electricity generation during low-demand periods (Teichroeb, 2012) or by supplying fuels to backup generators during high-demand, low-supply periods (Skea et al., 2012).

21.2.3 ELECTRICITY NETWORKS

Electricity networks have a uniform design in almost all countries. Electricity is generated in a small number of centralized fossil fuel, nuclear, or hydroelectric plants, transported to population centres using high-voltage AC transmission networks and

---

[2] The capacity factor of network infrastructure is the ratio of the actual energy throughput to the maximum possible energy throughput.

then delivered to customers using low-voltage distribution networks. Yet both electricity demand and supply are likely to change in a low-carbon economy.

In high-latitude countries, heat is generated mostly through direct combustion of natural gas and oil, but many studies have identified electrification as the most economic low-carbon alternative (e.g. Ekins et al., 2013; Johansson et al., 2012). Similarly, electrification of transport through the deployment of electric vehicles is one strategy to reduce transport emissions (Chapter 10). The increased demands from electrifying heat and transport could result in much greater demand peaks (Robinson et al., 2013), particularly for heat in winter (Wilson et al., 2013), and meeting these peaks is a key challenge for electricity networks in the future.

The portfolio of electricity generation technologies is also likely to be different in a low-carbon economy. While nuclear and hydroelectric generation could continue as now or be expanded, it would be necessary to reduce CO_2 emissions from thermal fossil fuel plants by using CCS or to replace these plants with alternative generation. Low-carbon renewable generation, for example from solar (Chapter 18) and wind (Chapter 20), are the most likely alternatives and are now being deployed in most high-income countries (IEA, 2012). A key characteristic of these technologies is the relatively unpredictable, intermittent nature of their output, and the difficulty of coping with this intermittency increases exponentially as the proportion of renewables in the generation portfolio is increased. In addition, solar power in particular is often decentralized and households can generate and sell electricity, meaning that there can be two-way flows of electricity through distribution networks that were designed for only a one-way flow to households. Germany has a greater proportion of wind and solar generation than most countries and the network issues that have been encountered there are highlighted in a case study in Section 21.3.1.3.

In summary, a low-carbon energy system could have a much higher electricity demand, including much steeper demand peaks, and a much less controllable supply to meet that demand. Both of these factors would complicate and possibly require changes to the current operation of the electricity networks. Grid reinforcement has been the principal strategy in the past to cope with long-term supply and demand variations; in fact, existing electricity networks in high-income countries have largely been developed incrementally over several decades. Yet constructing new overhead power cables is often opposed on environmental grounds and building and operating underground cables instead is much more expensive (Parsons Brinckerhoff, 2012). Since peak electricity flows could greatly increase in the future, and since networks must be designed to cope with peak rather than mean flows, there is much interest in finding alternative, cheaper methods than grid reinforcement. For this reason, most low-carbon network infrastructure research is targeted at electricity networks, particularly transmission networks. Strategies to cope with these changes include grid reinforcement, developing 'super grids' by linking existing transmission networks, introducing energy storage, and eventually developing 'smart grids' that collect information about supply and demand and increase the automation of grid balancing decision-making. Options for super and smart grids are discussed below and energy storage is discussed in Section 21.2.4.

21.2.3.1 Super Grids

One cross-system strategy is to link existing transmission networks to develop 'super grids'. This poses several challenges. First, transmission networks and electricity generation portfolios are normally operated in a country independently of other countries,[3] and few countries are willing to sacrifice their energy policy independence and put a key part of their energy security at risk by creating a regional organization over which they have little control. Second, electricity systems often operate at different phases, frequencies (e.g. 50 Hz or 60 Hz), and voltages, preventing interconnection unless the systems are standardized. The largest synchronous electrical grid in the world is the synchronous grid of Continental Europe, a single phase-locked 50 Hz grid linking twenty-four countries that is underpinned by European Union legislation (ENTSO-E, 2013).

An alternative to full integration is to link networks using high-voltage DC (HVDC) lines. For example, the synchronous grid of Continental Europe is linked using numerous HVDC lines. Since the capacity of individual lines is normally small relative to the flows within each network, numerous lines are required to enable large cross-network flows and to facilitate a competitive single market for electricity across networks. A European super grid has been proposed to connect all European countries as well as neighbouring regions such as North Africa, with the intention of reducing intermittency fluctuations by enabling large power transfers across the continent as required (Energy and Climate Change Committee, 2011). One opportunity would be to integrate new large-scale solar generation in North Africa to the European network, although this would have geopolitical ramifications for European energy security (Lilliestam and Ellenbeck, 2011).

21.2.3.2 Smart Grids

At the most basic level, a smart grid uses information about the network, including consumption patterns, to aid the operation of the network through improved decision-making. A common characteristic of smart grids is the analysis of demand-side data, which began in the 1980s for large industrial customers. Demand-side management (DSM) could be implemented by remotely switching off non-essential industrial electrical devices, such as air conditioners and refrigerators, for short periods. Several high-income countries have programmes to fit smart meters to all customers. Smart meters record consumption data and send it to the grid operators in real time. They can be used

---

[3] Exceptions include cases where large hydroelectric facilities are built and operated with funding from several countries, for example the Itaipu Dam on the border between Brazil and Paraguay, but even here there are independent generators for each country.

to educate consumers about their electricity consumption and can contribute to grid balancing by implementing new electricity pricing structures (for example, by increasing prices during peak periods). There is also the possibility in the future to link smart meters to sockets and other smart grid-enabled devices, enabling DSM for small customers in order to reduce demand peaks. The Open Automated Demand Response (OpenADR) standard for DSM (Holmberg et al., 2012) has already been demonstrated in China, the USA, and the UK.

A fully developed smart grid would use information and communication technology (ICT) to improve the efficiency, reliability, economics, and sustainability of the electricity network. Much of the system balancing would be automated, reducing the importance of human decision-making with the ICT used to transfer, analyse, and act upon data. These systems could eventually control generation and energy storage as well as DSM (Wade et al., 2010).

There are a number of challenges associated with the development of smart grids. There is a high initial capital cost, particularly for deploying smart meters to all customers. Smart grids could be vulnerable to outside interference, for example cyber-terrorism, if they are fully automated. There are privacy issues over the transmission of energy consumption data for individual households, and customers could oppose the loss of control caused by remotely-driven DSM. More fundamentally, the extent to which smart grids can displace physical assets while providing the same system reliability is not clear (Bolton and Hawkes, 2013). Finally, smart grids are implemented within single networks and there are questions, based on early experience in Denmark, over whether they will be incompatible with developing super grids (Blarke and Jenkins, 2013).

21.2.4 ENERGY STORAGE

Energy storage is a critical part of all national energy systems that is normally closely integrated with network infrastructure. There are many types of energy storage that are designed to supply different amounts of energy on different timescales, ranging from pumped hydroelectric storage, which supplies gigawatts of electricity at short notice for short periods to cope with demand peaks, to inter-seasonal gas storage in temperate countries that stores gas for heat generation in winter. Energy storage technologies can be characterized using numerous metrics, the most important of which are the capacity (J), the specific power (W/kg), and power density (W/m^3) for both inputs and outputs, the output energy (J/kg) and energy density (J/m^3), and the nominal cost per unit energy and unit power at the optimum scale (£/J & £/W).

In most current energy systems, the dominant long-term storage energy carriers are fossil fuels—coal, oil, and gas. These precursors are stored in preference to secondary carriers such as electricity and heat because it is much cheaper to store fossil fuels and generate electricity with standby capacity than to store electricity. Other forms of storage (pumped hydro, heat storage) have very low capacities and are used over short time periods.

The principal challenge for energy storage in the future is to contribute to low-carbon energy systems, particularly those with high penetrations of intermittent renewables (Hall, 2008). Electricity generation from renewables will exceed demand for substantial periods in these systems and storing electrical energy can reduce the need for investments in new transmission networks and improve the economics of renewable energy (Strbac et al., 2012). While many electricity storage technologies are designed to feed electricity back into the grid to meet demand peaks, particularly at times of low intermittent supply, it might be more cost-effective to use the stored energy elsewhere in the energy system, for example to meet heat or transport demand. This means that there is uncertainty about the most appropriate method of integrating energy storage technologies into existing energy systems, making it difficult to identify the most effective electricity storage technologies.

21.2.4.1 Existing Energy Storage

Solid fuels, such as coal and biomass, can be easily heaped (although it is necessary to keep biomass dry). Liquid fuels, particularly oil and petroleum products, and biofuels in the future, are stored in large depots, normally close to refineries or in ports. Gaseous fuels use gas holders above ground and, below ground, saline aquifers (for medium-term, higher-power storage) and depleted oil or gas fields (for large-capacity, inter-seasonal storage that requires lower power outputs). Gas transmission pipelines contain a substantial amount of gas, referred to as the linepack, and act as a short-term store. Most gaseous stores contain natural gas but some saline aquifers have been built to store hydrogen and it would also be possible to store hydrogen in used gas fields.

Secondary fuel storage is comparatively rare. Pumped or other hydro is principally used for electricity storage. Heat storage generally takes the form of hot water tanks in buildings or specialized heat storage in district heat schemes.

21.2.4.2 Electricity Storage Technologies

Electricity can be stored in a variety of ways. Electrochemical storage devices, which include conventional batteries, supercapacitors, and flow devices, store electricity directly. If electric vehicles were to become widespread, the vehicle batteries would represent a large potential energy store if they were connected to the grid and under the control of grid operators as part of a smart grid. Other types of storage devices convert the electrical energy into another form, for example flywheels, compressed air, hydrogen, and cryogenic storage. Barton and Infield (2004) discuss the key properties of these technologies. Although most energy storage research is targeted at these technologies, they all currently have high costs per unit energy relative to existing energy storage and none are currently commercially competitive.

The output of these technologies is mostly electricity that is fed back into the grid. An alternative approach is to convert excess electricity into other storage fuels that are not used to return electricity to the grid. For example, car batteries could be charged overnight by a smart grid and the stored energy used as a transport fuel (vehicle-to-grid technologies). Excess electricity could produce heat for storage in boilers, in a demand-side management version of existing storage heaters. Hydrogen could be stored for later electricity production (Lohner et al., 2013; Anderson and Leach, 2004) but could also be used for transport or heat provision (Dodds, 2013). Finding the most appropriate methods of integrating the many different types of energy storage into existing energy systems is a key research question for energy systems researchers.

The economically optimal amount of storage in the energy system will depend on the generation portfolio (with renewables increasing storage requirements much more than CCS) and the rate of technological progress in reducing storage costs (Strbac et al., 2012). Both super grids and smart grids could reduce the need for electricity storage, by balancing excess generation across large regions in the former case and through demand-side management in the latter.

21.2.5 HYDROGEN

Hydrogen fuel cell vehicles have been identified as potentially key low-carbon transport technologies for the future (Chapter 10). Hydrogen would likely be sold from fuelling stations, mimicking the existing petroleum fuel infrastructure. It could be transported to fuelling stations in liquid form, by road tanker, or gaseous form, by pipeline or tube trailer; the most economic delivery system depends on the delivery distance and the quantity of hydrogen (Yang and Ogden, 2007).

Hydrogen delivery infrastructure would be more expensive to construct than the existing delivery infrastructure for petroleum fuels for several reasons (Dodds and McDowall, 2014). First, the energy density of hydrogen is very low compared to hydrocarbon fuels so a greater storage volume is required per unit of delivered energy, and hydrogen storage (particularly gaseous storage) is currently very expensive. Second, hydrogen liquefaction is an expensive process due to the high electricity consumption, and active cooling is required at all times to prevent boil-off. Third, gaseous hydrogen refuelling is slower than hydrocarbon refuelling and the additional time reduces the number of customers that can be serviced by each refuelling station and hence increases the required number or size of refuelling stations.

Hydrogen infrastructure technologies are mature, and numerous examples of hydrogen infrastructure have been built for industrial uses or as test facilities, but efforts are required to reduce the capital costs of hydrogen storage in particular (Zheng et al., 2012).

21.2.5.1 Barriers to Hydrogen Infrastructure

A major barrier to the widespread use of hydrogen as a transport fuel is the lack of an existing fuel delivery infrastructure. A basic network of production facilities, refuelling stations, and delivery mechanisms would need to be created to support the first adopters and this would initially be underutilized during a transition to large-scale hydrogen vehicle deployment. The cost of establishing this infrastructure and bringing it to maturity are substantial (Coalition, 2010). Investors in infrastructure are usually faced with almost no foresight about possible future demand levels and there is a substantial risk attached to such investments. Agnolucci (2007) identifies three broad innovation strategies that have been suggested to deal with this conundrum:

1. Whole system incremental approach: small steps are incrementally taken across different hydrogen markets. This might start with portable power, then buses, government fleet vehicles, commercial and luxury passenger vehicles, and finally ordinary passenger vehicles.
2. Incremental approach: demonstration projects are established to increase the confidence of customers and producers, followed by the introduction of hydrogen vehicles to fleets and then finally to the consumer market.
3. Step-change approach: large-scale investments among all stakeholders, including fuel providers, car manufacturers, government, and consumers, are highly coordinated to bring large numbers of mass-produced, low-cost hydrogen vehicles to the market in a short period of time to reduce infrastructure investment risk. One aim of the 'H2Mobility' consortia that have recently been established in several countries (e.g. UK H2Mobility, 2013) is to achieve such coordination.

21.2.5.2 Opportunities to Facilitate Hydrogen Infrastructure Development

Some new transport infrastructure, for example hydrogen pipeline networks, might only be economically viable if they were to provide energy services to other sectors as well as to the transport sector. Hydrogen is used as an industrial feedstock and could also be used for large-scale electricity storage (Section 21.2.4.2) or for low-carbon heat provision. One option to use hydrogen for heat provision is to convert existing natural gas distribution networks to deliver hydrogen (Section 21.2.2.1). While this could potentially create a large market for hydrogen in a short space of time, it would be necessary to convert or replace other parts of the system, for example boilers, to burn hydrogen (Dodds and Demoullin, 2013).

21.2.6 DEGREE OF ENERGY SUPPLY CENTRALIZATION AND THE IMPACT ON INFRASTRUCTURES

Existing energy supply systems in high-income countries are predominantly centralized, with electricity and petroleum products produced in large plants and delivered to consumers by the infrastructures discussed above. These systems offer economies of scale for fuel processing and electricity generation technologies while keeping the associated pollution away from population centres where electricity is mostly consumed.

Some authors have argued for an alternative decentralized approach in which electricity and heat generation take place within distribution networks, closer to the point of use (Ackermann et al., 2001). Decentralized, small-scale production within a self-reliant local community is sometimes viewed as an important tenet of a green economy, from the perspective of ecological philosophy (Carter, 2001). It is also viewed as a method to educate people to understand energy better and develop a sense of responsibility for the environmental impacts of their energy consumption (Greenpeace, 2005).

21.2.6.1 Advantages and Disadvantages of Decentralized Generation

There are three main technical advantages of decentralization over centralized generation. First, the electricity supply does not depend on the transmission network or on a small number of very large centralized plants, which are susceptible to unexpected faults, adverse weather, and deliberate damage (e.g. terrorism), and this improves energy security. In addition, system costs are lower if the transmission network is not required; however, it might be necessary to use it as backup in the event of a shortfall in local generation (although the required transmission network capacity would be much lower, which would at least partially reduce the cost). Second, decentralized generation utilizing combined heat and power (CHP) has lower energy losses than centralized systems that discharge heat to the environment, potentially reducing overall fuel use (Greenpeace, 2005). Third, transmission energy losses, which for example were around 6 per cent of total electricity generation in Europe in 2010, are not incurred (Index Mundi, 2013).

One of the principal disadvantages of decentralized generation for a low-carbon energy system is the difficulty of integrating some renewable technologies, particularly wind and wave generation, into the energy system. These technologies are normally located far from population centres and the intermittent nature of the electricity supply could be more difficult to deal with in a decentralized system; in fact, integrating existing transmission networks into super grids has been proposed as a solution to this issue (Section 21.2.3.1). Another challenge for decentralized generation is to deal with demand peaks; local energy storage is an option but it is unlikely that smart grids (Section 21.2.3.2) would be viable on such a small scale.

21.2.6.2 Decentralization in Existing Energy Systems

Decentralization in high-income countries is mostly limited to sparsely populated areas where the costs of grid connection are prohibitive. The most notable example of decentralization in highly populated areas is the prevalence of CHP in northern Europe and former Eastern Bloc countries and the recent widespread adoption of solar power generation in Germany. In lower-income countries with low population densities, such as many parts of Africa, where electricity networks are not available, decentralized generation could be the most economic and environmentally-sustainable method of providing electrical energy.

21.3 **Global infrastructure development**

Countries face different network infrastructure challenges depending on their wealth, existing assets, population density, and resources. This section identifies some of the issues faced by high-, medium-, and low-income countries, drawing on some of the discussion above, and presents a case study for each income group.

21.3.1 HIGH-INCOME COUNTRIES

In high-income countries, network infrastructures have traditionally moved primary fuels from mines or ports to processing plants, and secondary fuels from centralized plants to decentralized demand centres. Network infrastructures are often built, owned, and operated quite differently to centralized energy plants, over much longer time-scales: networks are developed incrementally over time to meet changing supplies and demands and are often owned and operated by highly-regulated regional or national monopolies. These traits present unique challenges and opportunities for reducing greenhouse gas emissions because new infrastructures will be required for low-carbon energy technologies, for example to cope with electrification of transport or heat provision, while there is potential for other existing high-carbon infrastructures to become stranded assets.

High-income countries have highly developed electricity transmission and distribution networks linked to almost all of the population. Many of these countries are in temperate climates and often also have extensive natural gas transmission and distribution networks to provide heat in buildings, sometimes via CHP. Energy storage facilities are mainly limited to fossil fuels in most countries.

21.3.1.1 Challenges for the Future

Decarbonizing the electricity supply is a key challenge in many high-income countries. All electricity systems are designed with flexible generation capacity to meet varying demand. With large penetrations of intermittent renewable generation, the future supply portfolio will become less flexible and the network infrastructures will need to be redesigned to cope with this profound change, perhaps through the development of super grids, smart grids, and electricity storage. Simply reinforcing networks to cope with renewable generation peaks is likely to be the most expensive option, so there is much interest in these new network technologies (Strbac et al., 2012).

Another challenge is to find alternative uses for, or to decommission, existing high-carbon infrastructures. For example, extensive gas distribution networks supply natural gas to many homes in temperate countries, but combusting this gas without using CCS is likely to be incompatible with achieving CO_2 emission targets in the future (Dodds and Demoullin, 2013) and we discuss possible alternatives in Section 21.2.2.1. Natural gas heating provides energy diversity compared to relying on only electricity for heat so can improve energy security. It is also the lowest-cost method of providing heat in many countries, so other concerns for policymakers include the potential for increasing fuel poverty if the networks are decommissioned.

Providing infrastructure for low-carbon transport is a significant challenge for high-income countries. A move to electric or hydrogen-powered vehicles would require substantial additional infrastructure. Some of the challenges associated with building such infrastructure are identified in Section 21.2.5. Governments normally heavily regulate and underpin strategies for electricity infrastructure, but their role in the development of transport fuel infrastructure is more unclear, particularly in economies with privatized energy systems. Yet it is difficult for nascent technologies to prosper in the early stages of innovation without either a clear cost advantage, government support, or a niche market to target (Hardman et al., 2013).

21.3.1.2 Case Study: The German 'Energiewende'

Germany has the highest proportion of renewable generation of any large country. Underpinned by a comprehensive feed-in-tariff system, the renewable share in gross electricity generation has risen from 6 per cent in 2000 to almost 23 per cent in 2012. A further increase to at least 35 per cent in 2020 and 80 per cent in 2050 is stipulated in the German Energy Concept from 2011 (BMWi and BMU, 2011). This also entails a growing contribution of intermittent generation that is expected to cover almost 30 per cent of gross electricity consumption by 2020 and more than 50 per cent by 2050 (BMU, 2012).

This 'Energiewende' (energy transition) poses unprecedented challenges for the electricity system. The increasing supply-side volatility is already causing periods of extremely low or even negative wholesale electricity prices when a strong supply of wind or solar generation is combined with low demand. Moreover, while electricity generation was historically located relatively close to the centres of consumption, the expansion of spatially distributed renewable generation requires long-distance transmission, especially from the coastal regions in the north to the south. In response, the German Energy Concept has identified upgrading the network infrastructure as one of the priorities of the energy transition. In 2012, the first national Grid Development Plan concluded that Germany needs to construct 3800 km of new power lines and to optimize a further 4400 km of existing lines in the transmission grid over the next ten years, at a cost of roughly €20bn, in order to realize the Energiewende (Netzentwicklungsplan, 2013a). This includes four high-voltage DC corridors in the north–south direction with a transmission capacity of 12 GW, which might constitute the first step towards a European super grid in the long term. An additional investment of around €22bn is required to connect offshore wind energy to the transmission networks over the next 10 to 20 years (Netzentwicklungsplan, 2013b).

The Energiewende has also affected the electricity distribution networks. High intermittent load fluctuations, particularly caused by the strong expansion of solar PV installations in southern Germany (where some areas already produce more electricity than they consume), have resulted in some distribution grids already operating at their technical limit. Estimates of future investment needs in the distribution grid vary widely, with figures ranging between €10bn and €27bn until 2020. These investments will introduce innovative micro-grid control concepts, for example regulated distribution transformers (BDEW, 2011; DENA, 2012). The regulatory framework and support for the development of smart grids and demand-side management has been strengthened. A central initiative is the E-Energy funding programme, which is currently testing smart grid technologies in six regions across Germany (BMWi and BMU, 2012). Steps are also being taken to integrate renewable sources into the electricity market and to create incentives for generation that is more responsive to demands. The electricity market design needs to provide sufficient incentives for future investments in flexible reserve capacity.

The increasingly intermittent nature of the electricity supply is expected to require an expansion of electricity storage capacity. The potential for pumped storage hydroelectricity has already been almost fully exploited in Germany and researchers are investigating alternative storage systems, particularly compressed air storage, hydrogen storage, power-to-gas as well as a possible contribution from electric vehicles. For example, several power-to-gas demonstration projects are currently being commissioned that use the existing natural gas network to store excess electricity in the form of hydrogen.

Public acceptance is required to enable these infrastructure investments to be realized in the envisaged timeframe. Multilateral and participatory approaches have been used to overcome (mainly) local opposition.

21.3.2 MIDDLE-INCOME COUNTRIES

The proportion of populations connected to electricity networks in middle-income countries varies between 34 per cent (Namibia) and 99 per cent (China) but mostly exceeds 90 per cent (OECD/IEA, 2011). Electricity consumption per capita is consistently much lower in middle-income countries than in high-income countries. Important trends for these countries include rapidly increasing consumption and diversification to cheaper fuels such as natural gas, underpinned by new gas infrastructure. These trends are highlighted in the case study of China that is presented below. The network infrastructures that have been developed in middle-income countries mostly mirror those already present in high-income countries. For middle-income countries, demand growth and decarbonization present twin challenges for infrastructure; further expanding energy services and underpinning economic growth are often higher priorities than decarbonizing the energy system.

21.3.2.1 Challenges for the Future

Many middle-income countries have undergone sustained urbanization over recent decades, creating large cities with high electricity demand for services such as air conditioning, television, and computing (Modi et al., 2005). Supplying electricity to off-grid areas is an aspiration for all countries that have not achieved full population coverage. Security of supply is a major driver in these countries and building new supply capacity, and the associated delivery infrastructure, has been a priority. Improving the electricity supply reliability is also a priority in some countries (Cherni and Kentish, 2007). These trends are likely to continue into the future as urban populations gradually expand their ownership of electrical goods and further increase their demand for electricity.

Decarbonizing supplied energy and reducing the environmental impacts of energy more generally are also drivers in middle-income countries. These countries face comparable challenges to high-income countries: since similar infrastructure has already been deployed, albeit on a smaller scale, there are similar issues for transitioning these to a low-carbon economy, although fewer assets are likely to become stranded as a result of a transition. While some middle-income countries, such as China, have access to the capital and the skilled workers that are required to underpin a transition to a low-carbon economy, other countries have limited resources of either or both of these critical factors and that is a barrier to change in the future (PWC, 2012).

21.3.2.2 Case Study: Energy and Infrastructure Development in China

China is an example of a middle-income country that has undergone rapid industrialization in recent decades. This section examines recent trends using statistics from the China Energy Databook (Fridley et al., 2013).

China has undergone a process of urbanization like many middle-income countries, with the number of rural inhabitants decreasing by 63 million between 2000 and 2005 to 745 million and the urbanization rate reaching 42 per cent in 2005. Yet even before this, in 1998, 99.2 per cent of urban households and 96.9 per cent of rural households had access to electricity, the result of electrification campaigns using a diversity of funding channels, modes and institutional structures (Urban, 2009; Peng and Pan, 2006). Since 1998, electricity generation capacity has increased from 277 GW to 1144 GW (of which 72 per cent is fossil-fuelled), and annual output has increased from 1090 TWh in 1998 to 4459 TWh in 2011. Yet energy consumption is very uneven across the country, with residential energy consumption per capita more than 4 times higher in Beijing than in many poorer provinces, so there is scope for much higher energy service demands in the future. Access in rural areas has been enhanced through targeted decentralized generation, with 51 GW of small hydro schemes commissioned between 2005 and 2011, but the Chinese electricity system mostly resembles electricity systems in high-income countries. The key challenges for the future are to continue servicing the rapidly-growing demand and to improve the reliability of the electricity supply (Cherni and Kentish, 2007). Wind generation capacity increased from 6 GW in 2008 to 62 GW in 2011; while this represents only 5 per cent of the total generation capacity, there will be additional challenges for the networks if the share of wind in the generation portfolio continues to increase.

Gas has become a key energy source for the residential sector over the last 20 years. The population using gas in homes increased from 70 million in 1990 to 176 million in 2000 and to 378 million in 2011, and 55 per cent of the urban population now use gas for heating and/or cooking. China has built more than 22,000 km of natural gas transmission pipelines that supply almost all cities, which have the potential to become stranded assets in the future. In rural areas, there has been some development of decentralized biogas production from animal waste.

High growth trends have also occurred in the transport sector, with total transport energy consumption increasing by 161 per cent between 2000 and 2011. Yet a transition to alternative fuels is more noticeable in China's transport sector than in many high-income countries: electricity has consistently had a 10 per cent market share throughout this period and compressed natural gas (CNG) and liquefied petroleum gas (LPG) have steadily increased their market share to around 6 per cent in 2011. The wider variety of refuelling infrastructures compared to many countries could facilitate the move to alternative low-carbon fuels.

21.3.3 LOW-INCOME COUNTRIES

While road transport fuel is widely available in low-income countries, electricity networks have primarily been restricted to parts of cities and expanding access to energy is the principal concern of governments (Modi et al., 2005). Currently, 1.3 billion people have no access to electricity, 84 per cent of whom live in rural areas (OECD/IEA, 2011; Johansson et al., 2012), and 2.7 billion people rely on traditional biomass for cooking (OECD/IEA, 2006). These numbers are unlikely to change much by 2030 if progress continues along current trends (OECD/IEA, 2010).

21.3.3.1 Challenges for the Future

The high upfront costs of power plants, high grid connection charges, cost-recovery difficulties, unreliable equipment, and a lack of technical capacities for operation and maintenance are all cited in the literature as barriers to expanding access to electricity. Watson et al. (2011) review the evidence on policy interventions and find that few have had demonstrable impact, with interventions targeting political and cultural barriers being particularly weak. They conclude that there can be many complex interactions between different types of barriers, which vary between countries, so a systematic understanding of barriers to access is a prerequisite for identifying effective policies.

Low-income countries with limited existing network infrastructure have an opportunity to follow a different path to high-income countries, for example by developing decentralized electricity generation systems. Decentralized systems are particularly appropriate for rural areas with low population densities in tropical regions, which are found in many parts of sub-Saharan Africa. Decentralized solar power generation can support basic services such as lighting, refrigeration, and information services (Damasen and Uhomoibhi, 2014).

Decarbonization of energy systems is not a priority for low-income countries, particularly as emissions per capita are normally very low. Efforts to develop low-carbon energy systems tend to be promoted through aid programmes and are more likely to be successful when they also offer economic benefits to populations. For example, deploying modern cooking stoves reduces biomass consumption, improves air quality, reduces health impacts, and reduces CO_2 emissions (OECD/IEA, 2006); for these reasons, 7 million people were supplied with advanced biomass cooking stoves in 2009 (OECD/IEA, 2011). Economic benefits can sometimes be realized for larger schemes; for example, Ethiopia has built a 120 MW wind farm and is planning a 1 GW geothermal plant; both facilities take advantage of the local geography and both are economically viable because the generated electricity will be mostly sold to neighbouring countries (Barras, 2013).

21.3.3.2 Case Study: Access to Modern Energy Services in Kenya

In 2009, 84 per cent of the Kenyan population (33 million people) had no access to electricity and a similar proportion of the population relied on the traditional use of biomass for cooking (OECD/IEA, 2011).

Government policies that restrict the generation and distribution of electricity by private companies and cooperatives can prevent the expansion of energy services to off-grid areas. In Kenya, the regulatory and commercial functions of the electricity sector were combined until the Electric Power Act 1997 was passed. Yet the new regulations still restricted electricity distribution activities to the Kenya Power and Lighting Company, which is the national utility (Government of Kenya, 1997), and failed to greatly improve rural electrification (Karekezi and Kimani, 2002). The system was further liberalized in 2006 to establish a rural electrification authority, with the aim of increasing the rural electrification rate to 10 per cent per annum (Kapika and Eberhard, 2013). These changes permitted private generation and distribution and allowed private investors to choose tariffs that would enable a return on investment (Morris et al., 2009). Kirubi et al. (2009) show that these changes have facilitated access to electricity, with many positive consequences for local development. Yet the rural electrification target has not been achieved.

Two options are normally considered for off-grid microgeneration. Solar microgeneration has been used for twenty years (Acker and Kammen, 1996) but there have been difficulties with the reliability of some products (Duke et al., 2002). The second option, micro-hydropower schemes, has lower capital costs than solar generation and is also being actively pursued (Morris et al., 2009).

LPG cooking stoves have been successfully introduced to parts of Kenya, particularly in urban and peri-urban areas, to replace traditional biomass stoves. The Kenyan government has underpinned this transition by developing common standards and LPG delivery infrastructure (OECD/IEA, 2011). Access to credit has also been a barrier and microfinance institutions have had an important role in enabling a rapid transition (Morris et al., 2009).

21.4 The role of government and other stakeholders

Access to energy services and security of supply are critical enablers of economic activity in all countries. Governments are ultimately held accountable for the provision of energy services so have historically had an important role in developing network infrastructures, whether through direct funding or by creating appropriate policy regimes to facilitate private investment. Reliability of energy services is a key concern for governments, leading them to classify network infrastructures as critical national infrastructure in most countries. Energy systems must be designed to minimize faults and to cope with outside events such as adverse weather, accidents or terrorism, and network infrastructures are at particular risk of disruption.

21.4.1 GEOPOLITICAL DRIVERS

There are increasing connections between countries for oil and gas pipelines and more recently for electricity networks. Since these networks are often critical for energy security, they are usually the subject of treaties between countries and have great geopolitical importance, sometimes defining relationships between neighbours. There is a trend towards building ever greater links, even if these are sometimes principally designed to reduce the risk from existing links (e.g. Afifi et al., 2012).

One example is the gas pipelines that link Russian gas fields to European consumers, whose supply has been disrupted several times during winter by disputes between Russia and Ukraine (Costantini et al., 2007). In response, European countries are considering commissioning a new pipeline linked to the Caucasus region that travels through Turkey rather than Russia (Afifi et al., 2012). The geopolitical economy of the gas market is discussed in Chapter 15.

Another example is the potential for a European electrical super grid (Section 21.2.3.1), with existing networks linked using HVDC lines and external links added to North Africa and/or the Middle East. European Union legislation already underpins the synchronous grid of Continental Europe (ENTSO-E, 2013) and the EU would likely be the main political driver towards the formation of a super grid, partly because national energy markets must be compatible with EU single market rules. Yet national governments retain responsibility for energy security and would want to ensure that a super grid would not affect the reliability of their energy supplies. Energy regulators work on a national rather than an international basis and it might be necessary to create a pan-European regulator in the same way that a pan-European consortium of transmission network operators, ENTSO-E, has already been created. Such a move would lead to a loss of sovereignty for individual states, including for some that are outside of the EU but who form part of the super grid. A super grid would not be viable if insufficient countries were to join, particularly if some of those countries were geographically central to the network. Moreover, a disruption in one part of the network could have important energy security ramifications for the whole continent if many countries were reliant on overseas electricity generation.

21.4.2 LOCAL POLITICAL DRIVERS

At the other end of the political scale, local politics can have an important impact on network infrastructures. There are safety implications of having nearby transmission networks. Electricity pylons reduce visual amenity while pipes require access to land. Measures to reduce the environmental impacts of network infrastructures, for example using underground transmission cables, can greatly increase the capital costs (Parsons Brinckerhoff, 2012). Yet impacts are perhaps less onerous than for many energy processing and conversion plants, particularly high-emission fossil fuel plants.

21.4.3 OWNERSHIP AND INVESTMENT IN NETWORK INFRASTRUCTURES

Network infrastructures can be categorized into two types from the perspective of ownership and investment.

The first category contains transmission infrastructures that are built and operated privately, within countries, to meet industrial needs, for example to transport oil between refineries and other industrial facilities. These facilities are regulated by government for health and safety affairs but are otherwise owned and operated privately with little interference, unless they form a critical part of the national energy system.

The second category includes electricity, gas, and any other networks that supply energy to buildings through distribution networks. These infrastructures are normally natural monopolies. Traditionally, they were built, owned, and operated by governmental bodies. For example, in China and Kenya the operation and expansion of the networks is ultimately directed and funded by the government. The wave of privatizations in high-income countries in the 1990s and 2000s led to many networks being privatized. Since the networks operate as monopolies, energy regulators were created in these countries to fix network energy prices.

The challenge with all of these systems is how to encourage investment, innovation, and cost reduction. Issues include the allocation of risk between investors and consumers, and the extent to which existing consumers should pay for an energy system that benefits other consumers in the future (Bolton and Hawkes, 2013). For government-owned networks, costs and risks are socialized as part of public investment programmes. For privately-owned networks, government, regulators, and owners must agree an appropriate funding formula that enables the network to continue operating reliably and sustainably, without incurring unnecessary costs.

For electricity networks, huge investments will be required in the future to integrate low-carbon renewables into existing energy systems. Governments have underwritten such investments in the past and it is not clear how privatized utilities will be able to raise such large funds in the future (The Economist, 2013). A key challenge for governments is to define energy policies that enable these investments to take place in an economically efficient way. In view of the potentially incompatible super grid and smart grid options that are available (Blarke and Jenkins, 2013), it is likely that governments will need to decide how the system should be configured in the future, whether infrastructures are publically or privately owned. This could be achieved, for example, by restructuring energy regulators to take a greater role in system planning and market design. Similarly, the development of new low-carbon infrastructures, for example hydrogen infrastructures, will also require support from governments to give sufficient confidence for supply chains to be developed.

21.5 Economics of network infrastructures

Most network infrastructures require large upfront capital investments but have high capacities and long lifetimes. For transmission networks in particular, this is because it is often necessary to build the entire network before any part of it can be used, so an incremental approach cannot be adopted to reduce project risk. The crucial factor for investors is the capacity factor that is achieved, particularly in the period soon after commissioning. As illustrated for hydrogen in Section 21.2.5.1, a high capacity factor is crucial for an acceptable return on investment and the financial risk associated with the project is greatly increased if there is uncertainty about future demand.

If a high capacity factor is achieved, the infrastructure cost can often represent only a small part of the total cost of the energy that is supplied using the infrastructure. For example, in the UK, transmission and distribution costs represent only 4 per cent and 16 per cent of electricity bills, and only 1 per cent and 15 per cent of gas bills (NERA, 2012). Yet these costs are for mature networks with long lifetimes that were built by public utilities and then privatized in the 1990s. As a result, they represent the short-run marginal costs of operating the infrastructure rather than the long-run marginal costs of constructing the networks, which are higher. The economics of extending existing networks to service off-grid areas will be different because these countries must consider the higher, long-run costs. An important factor here is the large cost discrepancy between transmission and distribution costs. Transmission costs depend on the geography of a country; this makes them particularly low for countries with high population densities like the UK. Distribution costs, while higher, tend to depend on the number of customers such that distribution networks can be constructed incrementally to meet demand, and do not have the same upfront costs as transmission networks.

The cost of energy delivery to the public by gas and electricity networks is almost always heavily regulated by the state. A key facet of energy security is to enable access to energy by avoiding excessive prices and it is particularly important to prevent network operators from making windfall profits in countries where networks are privately owned. For publically-owned utilities, the cost of energy is sometimes subsidized to facilitate energy access. However, as the case study of Kenya in Section 21.3.3.2 illustrates, while such subsidies can help those connected to the grid, they can simultaneously impede access to energy for those that are off-grid by limiting prices to a level that is too low to allow a return on network investments.

21.6 Conclusions

Energy networks are an important yet often overlooked part of the energy system. Most energy network technologies are mature. Even for super grids and smart grids, the subjects of much research in recent years, the technical issues are mostly a question of how to integrate them into existing systems in the most cost-effective way, as the underlying technologies already exist. Both of these technologies are potential options to address the principal technical challenge for electricity networks in high-income countries, which is to incorporate large penetrations of renewables. Energy storage is another option to address this challenge but one that has substantial scope for technological innovation, both to reduce capital costs and to improve the performance of energy storage technologies. Yet cost-effective system integration is still the key issue for energy storage.

A challenge for all governments is to negotiate cost, energy security, and environmental priorities to provide the network infrastructures that are an essential foundation of economic activity. Most network infrastructure issues revolve around geopolitics, ownership, the structure and regulation of networks, and how to fund the required investments to meet energy goals in the future. Governments have a crucial role to achieve energy security by either investing in infrastructure directly through state-owned companies or by setting energy policies and creating effective regulators if networks are privately-owned.

For high-income countries, major investments will be required to underpin the transition to low-carbon economies, particularly where renewable generation is deployed. Other investments might be necessary for new infrastructure, for example for hydrogen, and government intervention will be required to at least underpin these investments. Middle-income countries have three main priorities: transitioning to low-carbon economies; expanding access to electricity across the whole population; and coping with greatly increased energy demands as economic output grows. For low-income countries, access to energy to underpin economic growth is the key policy goal. These countries have the opportunity to take a different approach to building a new network infrastructure, for example by adopting decentralized generation in off-grid areas. Policies to expand access to energy are most likely to be successful when they take a holistic approach to understanding barriers and incorporate a range of measures that work on different scales and target different stakeholders.

■ ACKNOWLEDGEMENTS

This chapter builds on work funded by the RCUK energy programme, including the UK Energy Research Centre and the EPSRC SUPERGEN consortia. The China case study is based on statistics from the China Energy Databook published by the China Energy Group at the Lawrence Berkeley National Laboratory. We are very grateful to Paul Ekins and Jamie Speirs for their insightful comments that helped us to improve the chapter.

■ REFERENCES

ACKER, R. H. & KAMMEN, D. M. 1996. The quiet (energy) revolution: Analysing the dissemination of photovoltaic power systems in Kenya. *Energy Policy*, 24, 81–111.

ACKERMANN, T., ANDERSSON, G., & SÖDER, L. 2001. Distributed generation: a definition. *Electric Power Systems Research*, 57, 195–204.

AFIFI, S. N., HASSAN, M. G., & ZOBAA, A. F. 2012. the impacts of the proposed Nabucco Gas pipeline on EU common energy policy. *Energy Sources, Part B: Economics, Planning, and Policy*, 8, 14–27.

AGNOLUCCI, P. 2007. Hydrogen infrastructure for the transport sector. *International Journal of Hydrogen Energy*, 32, 3526–44.

ANDERSON, D. & LEACH, M. 2004. Harvesting and redistributing renewable energy: on the role of gas and electricity grids to overcome intermittency through the generation and storage of hydrogen. *Energy Policy*, 32, 1603–14.

BARRAS, C. 2013. Ethiopia switches on Africa's largest wind farm. *New Scientist*. London, UK.

BARTON, J. P. & INFIELD, D. G. 2004. Energy storage and its use with intermittent renewable energy. *IEEE Transactions on Energy Conversion*, 19, 441–8.

BDEW 2011. Abschätzung des Ausbaubedarfs in deutschen Verteilungsnetzen aufgrund von Photovoltaik- und Windeinspeisungen bis 2020. Bonn/Aachen, Germany: German Association of Energy and Water Industries.

BLARKE, M. B. & JENKINS, B. M. 2013. Super grid or smart grid: Competing strategies for large-scale integration of intermittent renewables? *Energy Policy*, 58, 381–90.

BMU 2012. Langfristszenarien und Strategien für den Ausbau der erneuerbaren Energien in Deutschland bei Berücksichtigung der Entwicklung in Europa und global. Federal Ministry for the Environment, Nature Conservation and Nuclear Safety. Berlin, Germany: DLR, IWES, und IfnE.

BMWI & BMU 2011. Energy Concept for an Environmentally Sound, Reliable and Affordable Energy Supply. Federal Ministry of Economy and Technology & Federal Ministry for the Environment, Nature Conservation and Nuclear Safety. Berlin, Germany.

BMWI & BMU 2012. Smart Energy made in Germany—Interim results of the E-Energy pilot projects towards the Internet of Energy. Federal Ministry of Economy and Technology & Federal Ministry for the Environment, Nature Conservation and Nuclear Safety. Berlin, Germany.

BOLTON, R. & HAWKES, A. 2013. Infrastructure, Investment and the Low Carbon Transition. *In:* MITCHELL, C., WATSON, J., & WHITING, J. (eds) *New Challenges in Energy Security: The UK in a Multipolar World.* Basingstoke, UK: Palgrave Macmillan.

CARTER, N. 2001. *The Politics of the Environment: Ideas, Activism, Policy.* Cambridge, UK, Cambridge University Press.

CHERNI, J. & KENTISH, J. 2007. Renewable energy policy and electricity market reforms in China. *Energy Policy,* 25, 3616–29.

COALITION 2010. A portfolio of power-trains for Europe: a fact-based analysis. Coordinated by McKinsey & Company.

COSTANTINI, V., GRACCEVA, F., MARKANDYA, A., & VICINI, G. 2007. Security of energy supply: Comparing scenarios from a European perspective. *Energy Policy,* 35, 210–26.

DAMASEN, P. I. & UHOMOIBHI, J. 2014. Solar electricity generation: issues of development and impact on ICT implementation in Africa. *Campus-Wide Information Systems,* 31, 46–62.

DENA 2012. Expansion and Innovation Requirement of the Electricity Distribution Grids in Germany to 2030. Berlin, Germany: German Energy Association.

DODDS, P. E. 2013. A whole systems analysis of the benefits of hydrogen storage to the UK. *World Hydrogen Technologies Convention, 25–28 September 2013.* Shanghai, China.

DODDS, P. E. & DEMOULLIN, S. 2013. Conversion of the UK gas system to transport hydrogen. *International Journal of Hydrogen Energy,* 38, 7189–200.

DODDS, P. E. & MCDOWALL, W. 2013. The future of the UK gas network. *Energy Policy,* 60, 305–16.

DODDS, P. E. & MCDOWALL, W. 2014. Methodologies for representing the road transport sector in energy system models. *International Journal of Hydrogen Energy,* 39, 2345–58.

DUKE, R. D., JACOBSON, A., & KAMMEN, D. M. 2002. Photovoltaic module quality in the Kenyan solar home systems market. *Energy Policy,* 30, 477–99.

EKINS, P., KEPPO, I., SKEA, J., STRACHAN, N., USHER, W., & ANANDARAJAH, G. 2013. *The UK Energy System in 2050: Comparing Low-Carbon, Resilient Scenarios.* London, UK, UK Energy Research Centre.

ENERGY AND CLIMATE CHANGE COMMITTEE 2011. *A European Supergrid.* London, UK, UK Parliament.

ENTSO-E. 2013. *Who is ENTSO-E?* [Online]. *The European Network of Transmission System Operators for Electricity (ENTSO-E), Brussels, Belgium.* Available: <https://www.entsoe.eu/about-entso-e/> [Accessed 3 April 2015].

FRIDLEY, D., ROMANKIEWICZ, J., FINO-CHEN, C., HONG, L., & LU, H. 2013. China Energy Databook Version 8.0. Berkeley, California, USA: China Energy Group, Lawrence Berkeley National Laboratory.

FURCHTGOTT-ROTH, D. & GREEN, K. P. 2013. Intermodal safety in the transport of oil. Fraser Institute, Vancouver, Canada.

GOVERNMENT OF KENYA 1997. The Electric Power Act, 1997. Nairobi, Kenya: Government Printer.

GREENPEACE 2005. *Decentralising Power: An Energy Revolution for the 21st Century.* London, UK.

HALL, P. J. 2008. Energy storage: The route to liberation from the fossil fuel economy? *Energy Policy*, 36, 4363–7.

HARDMAN, S., STEINBERGER-WILCKENS, R., & VAN DER HORST, D. 2013. Disruptive innovations: The case for hydrogen fuel cells and battery electric vehicles. *International Journal of Hydrogen Energy*, 38, 15438–51.

HOLMBERG, D. G., GHATIKAR, G., KOCH, E. L., & BOCH, J. 2012. OpenADR advances. *Ashrae Journal*, 54, B16-B19.

IEA 2001. *Towards a Sustainable Energy Future,* Paris, France, OECD Publishing.

IEA 2012. *Energy Technology Perspectives 2012: Pathways to a Clean Energy System.* Paris, France, International Energy Agency.

INDEX MUNDI. 2013. *European Union—Electric power transmission and distribution losses* [Online]. Index Mundi, North Carolina, USA. Available: <http://www.indexmundi.com/facts/european-union/electric-power-transmission-and-distribution-losses> [Accessed 9 Nov 2013].

JOHANSSON, T. B., NAKICENOVIC, N., PATWARDHAN, A., & GOMEZE-CHEVERRI, L. 2012. Summary for Policymakers. *Global Energy Assessment: Toward a Sustainable Future.* Laxenburg, Austria: International Institute for Applied Systems Analysis.

KAPIKA, J. & EBERHARD, A. 2013. *Power-Sector Reform and Regulation in Africa: Lessons from Kenya, Tanzania, Uganda, Zambia, Namibia and Ghana,* South Africa, HSRC Press.

KAREKEZI, S. & KIMANI, J. 2002. Status of power sector reform in Africa: impact on the poor. *Energy Policy*, 30, 923–45.

KIRUBI, C., JACOBSON, A., KAMMEN, D. M., & MILLS, A. 2009. Community-based electric micro-grids can contribute to rural development: evidence from Kenya. *World Development*, 37, 1208–21.

KUMAR, S., KWON, H.-T., CHOI, K.-H., HYUN CHO, J., LIM, W., & MOON, I. 2011. Current status and future projections of LNG demand and supplies: A global prospective. *Energy Policy*, 39, 4097–104.

LAMERS, P., MARCHAL, D., HEINIMÖ, J., & STEIERER, F. 2014. Global Woody Biomass Trade for Energy. *In:* JUNGINGER, M., SHENG GOH, C., & FAAIJ, A. (eds) *International Bioenergy Trade.* Heidelberg, Germany, Springer.

LILLIESTAM, J. & ELLENBECK, S. 2011. Energy security and renewable electricity trade—Will Desertec make Europe vulnerable to the 'energy weapon'? *Energy Policy*, 39, 3380–91.

LOHNER, T., D'AVENI, A., DEHOUCHE, Z., & JOHNSON, P. 2013. Integration of large-scale hydrogen storages in a low-carbon electricity generation system. *International Journal of Hydrogen Energy*, 38, 14638–53.

MODI, V., MCDADE, S., LALLEMENT, D., & SAGHIR, J. 2005. *Energy Services for the Millennium Development Goals.* New York, USA: Energy Sector Management Assistance Programme, United Nations Development Programme, UN Millennium Project and World Bank.

MORRIS, E., KIRUBI, G., CLEMENS, E., COUTINHO, F., RIJAL, K., SHRESTHA, L. K., & ZONGO, A. 2009. *Bringing Small-Scale Finance to the Poor for Modern Energy Services: What is the Role of Government?* New York, USA, United Nations Development Programme.

NERA 2012. Energy Supply Margins: Update January 2012. London, UK, NERA Economic Consulting.

NETZENTWICKLUNGSPLAN 2013a. Grid Development Plan 2013—Legal Basis, Content, Consultation and Way Forward. Berlin, Germany.

NETZENTWICKLUNGSPLAN 2013b. Offshore Grid Development Plan 2012—Legal Basis, Content, Consultation and Way Forward.

OECD/IEA 2006. Energy for Cooking in Developing Countries. *World Energy Outlook 2006.* Paris, France, International Energy Agency.

OECD/IEA 2010. Energy Poverty: How to make modern energy access universal? Special early excerpt of the World Energy Outlook 2010. Paris, France, IEA-UNDP-UNIDO, International Energy Agency.

OECD/IEA 2011. Energy for all: Financing access for the poor. *World Energy Outlook 2011.* Paris, France, International Energy Agency.

ORR 2013. *Freight Rail Usage: 2013–14 Quarter 1 Statistical Release.* Office of Rail Regulation. London, UK.

PARSONS BRINCKERHOFF 2012. *Electricity Transmission Costing Study.* The Institution of Engineering and Technology, London, UK. Available at: <http://www.theiet.org/factfiles/transmission-report.cfm> [Accessed April 2015].

PENG, W. Y. & PAN, J. H. 2006. Rural electrification in China: History and institution. *China & World Economy,* 14, 71–84.

PWC 2012. *Closing the Talent Gap in the Emerging World.* Delaware, USA, PricewaterhouseCoopers.

ROBINSON, A. P., BLYTHE, P. T., BELL, M. C., HÜBNER, Y., & HILL, G. A. 2013. Analysis of electric vehicle driver recharging demand profiles and subsequent impacts on the carbon content of electric vehicle trips. *Energy Policy,* 61, 337–48.

SCHNOOR, J. L. 2013. Keystone XL: pipeline to nowhere. *Environmental Science & Technology,* 47, 3943.

SHUM, R. Y. 2013. Social construction and physical nihilation of the Keystone XL pipeline: Lessons from international relations theory. *Energy Policy,* 59, 82–5.

SKEA, J., ANANDARAJAH, G., CHAUDRY, M., SHAKOOR, A., STRACHAN, N., WANG, X., & WHITAKER, J. 2011. Energy Futures: The Challenges of Decarbonization and Security of Supply. *In:* SKEA, J., EKINS, P., & WINSKEL, M. (eds) *Energy 2050: Making the Transition to a Secure Low-carbon Energy System.* London, UK, Earthscan.

SKEA, J., CHAUDRY, M. & WANG, X. 2012. The role of gas infrastructure in promoting UK energy security. *Energy Policy,* 43, 202–13.

STRBAC, G., AUNEDI, M., PUDJIANTO, D., DJAPIC, P., TENG, F., STURT, A., JACKRAVUT, D., SANSOM, R., YUFIT, V., & BRANDON, N. 2012. *Strategic Assessment of the Role and Value of Energy Storage Systems in the UK Low Carbon Energy Future.* London, UK, Imperial College London.

TEICHROEB, D. Power to gas: a new electricity storage alternative. World Hydrogen Energy Conference, June 4–7 2012 Toronto, Canada.

THE ECONOMIST 2013. How to lose half a trillion euros. *The Economist.* London, UK.

UK H2MOBILITY 2013. *Phase 1 Results.* UK Government, London, UK.

URBAN, F. 2009. Climate-change mitigation revisited: low-carbon energy transitions for China and India. *Development Policy Review,* 27, 693–715.

WADE, N. S., TAYLOR, P. C., LANG, P. D., & JONES, P. R. 2010. Evaluating the benefits of an electrical energy storage system in a future smart grid. *Energy Policy,* 38, 7180–8.

WATSON, J., BYRNE, R., MORGAN JONES, M., TSANG, F., OPAZO, J., FRY, C., & CASTLE-CLARKE, S. 2011. What are the major barriers to increased use of modern energy services among the World's poorest people and are the interventions to overcome these effective? Brighton, UK, Collaboration for Environmental Evidence.

WEC 2011. Policies for the future: 2011. Assessment of country energy and climate policies. London, UK, World Energy Council.

WILLIAMS, T. I. 1981. *A History of the British Gas Industry.* Oxford, UK, Oxford University Press.

WILSON, I. A. G., RENNIE, A. J. R., DING, Y., EAMES, P. C., HALL, P. J., & KELLY, N. J. 2013. Historical daily gas and electrical energy flows through Great Britain's transmission networks and the decarbonisation of domestic heat. *Energy Policy*, 61, 301–5.

WOOD, D. A. 2012. A review and outlook for the global LNG trade. *Journal of Natural Gas Science and Engineering*, 9, 16–27.

WORLD BANK. 2013. *How We Classify Countries* [Online]. World Bank, Washington D.C., USA. Available: <http://data.worldbank.org/about/country-classifications> [Accessed 3 Sep 2013].

YANG, C. & OGDEN, J. M. 2007. Determining the lowest-cost hydrogen delivery mode. *International Journal of Hydrogen Energy*, 32, 268.

ZHENG, J. Y., LIU, X. X., XU, P., LIU, P. F., ZHAO, Y. Z., & YANG, J. 2012. Development of high pressure gaseous hydrogen storage technologies. *International Journal of Hydrogen Energy*, 37, 1048–57.

22 Metals for the low-carbon energy system

Jamie Speirs and Katy Roelich

The availability of resources critical to the global energy system is a perennial topic with a colourful history. Jevons (1865) raised the question of coal availability in the middle of the nineteenth century. In the middle of the twentieth century Hubbert (1956) raised questions about the future of global oil production. In the twenty-first century a number of authors have begun to question the availability of precious or exotic metals used in the manufacture of some low carbon technologies, often referred to as the critical metals[1] (DOE, 2011; Moss et al., 2011). The list of metals considered in this category is not fixed, but Table 22.1 presents a number of different metals that occur frequently in the critical metals literature.

Demand for a number of lesser known metals has increased significantly in recent decades, driven by the increasing technological sophistication in consumer electronics, the growing demand for these products due to population growth, and economic growth in the developing world (Figure 22.1). The deployment of low-carbon technologies at scale could also contribute significantly to these drivers of critical metal demand. However, it is difficult to make accurate predictions of the demand for these metals in the future, whether supply will be able to meet that demand, and on which factors the availability of these materials is most dependent.

The future supply of and demand for these critical metals is influenced by a number of factors, and these factors are often specific to different metals or technologies. Some studies try to address this granularity by focusing on individual metals or technologies to explore these specific issues in detail. There are also studies which aim to measure how critical the availability of different metals may be to a geographic region, economic sector, or individual company. These studies are often referred to as *criticality assessments*, and involve the combination of a range of metrics into multi-criteria analyses. This chapter examines both the specific factors impacting the future of critical metal supply and demand, and the high level criticality assessments literature.

---

[1] The terms critical materials, strategic materials or strategic metals are also commonly used.

22.1 **Estimating future demand**

Concern over the availability of critical metals is motivated to a significant extent by the anticipated growth in future metal demand. Amongst the drivers of this demand, the anticipated growth in low-carbon technology manufacturing is of particular concern given the very high growth rates implied, and the number of critical metals associated with these technologies. Ten critical metals and the low-carbon technologies they are used in are presented in Table 22.2. Future demand for these technologies is expected to have a significant impact on future demand for critical metals. However, future critical metal demand is also dependent on a number of other factors, all of which are likely to change over time. Future metal demand can be estimated in a number of different ways, from simple use of third-party assumptions to more sophisticated bottom-up, scenario-based estimates. At the centre of these scenario-based estimates is a range of variables that influence the quantity of metal used in manufactured goods, known as *material intensity*, and the change in that material intensity over time. The following discussion first examines the range of approaches used in the literature before examining in more detail the variables driving metal demand over time.

Table 22.1 Five critical metals studies and the critical metals that they include

Metal	NRC	EC	JRC	AEA	Fraunhofer
Indium	✓	✓	✓	✓	✓
Gallium	✓	✓	✓		✓
Rare Earth Elements	✓	✓	✓		✓
Cobalt		✓		✓	✓
Niobium	✓	✓			✓
Platinum Group	✓	✓			✓
Tantalum	✓	✓			✓
Copper	✓			✓	✓
Germanium		✓			✓
Antimony		✓			✓
Lithium	✓			✓	
Titanium	✓				✓
Tin				✓	✓
Phosphorus				✓	
Lead				✓	
Silver					✓
Selenium					✓
Tellurium			✓		
Magnesium		✓			
Tungsten		✓			
Manganese	✓				
Vanadium	✓				

Source: (NRC 2007; Angerer et al. 2009; AEA Technology 2010; EC 2010; Moss et al. 2011).

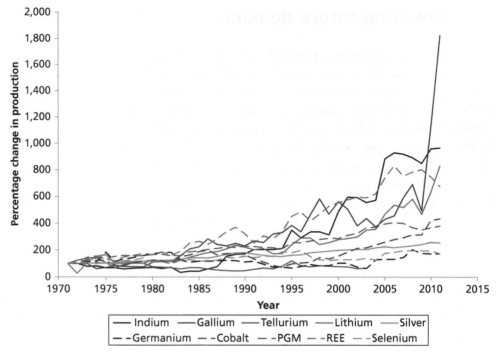

Figure 22.1 Historical production of ten critical metals from 1971 to 2011

Source: USGS.

Notes: The selection of critical metals is based on the metals analysed in Speirs et al. (2013a).

Table 22.2 List of ten critical metals and the low-carbon technologies in which they are used

Metal	Low-carbon technology
Cobalt	Lithium-ion batteries
Gallium	Thin-film photovoltaics (PV)
Germanium	Thin-film PV
Indium	Thin-film PV
Lithium	Lithium-ion batteries
Platinum group metals (PGMs)	Hydrogen fuel cells
Rare earth elements (REEs)	Electric vehicles and wind turbines
Selenium	Thin-film PV
Silver	PV (c-Si), concentrating solar and nuclear
Tellurium	Thin-film PV

Notes: The selection of critical metals is based on the metals analysed in Speirs et al. (2013a).

22.1.1 APPROACHES TO ESTIMATING FUTURE METAL DEMAND

Expert opinion is one of the simplest methods used to provide estimates of future demand (Angerer et al., 2009; Buchert et al., 2009; EC, 2010) (Figure 22.2). This may be conducted in various ways but typically involves asking a number of structured questions to selected individuals with expertise in areas relevant to the future demand for critical metals. However, this process can be compromised by difficulties in posing appropriate questions for the technology-specific aspects affecting demand, or engaging with a sufficient number of experts given the highly specialized nature of these technological issues (Angerer et al., 2009).

Using estimates from third-party sources is an alternative approach. Morley and Eatherley (2008) and Rosenau-Tornow et al. (2009) cite consultancy and market analyst forecasts such as Roskill (Chegwidden and Kingsnorth, 2011). However, with estimates such as these it is difficult to assess the level of sophistication associated with the methodology used and whether this approach is any more or less robust for analysing critical metal availability than other approach.

Finally, studies may wish to conduct bottom-up scenario-based assessments of future metal demand (Angerer et al., 2009; Graedel et al., 2012). A number of important variables can be incorporated into these bottom-up approaches. These are indicated in Figure 22.2 and discussed in more detail below.

22.1.2 KEY VARIABLES INFLUENCING FUTURE METAL DEMAND

The most important variables that factor in the formation of demand include:

- the end-use technologies;
- the projected growth of these technologies;

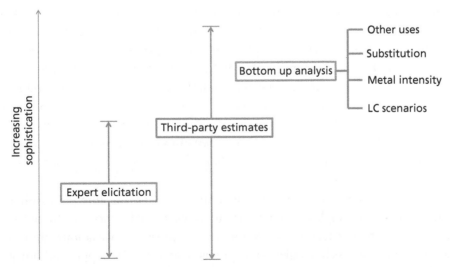

Figure 22.2 Approaches to estimating future metal demand in the literature and their relative sophistication

- the quantity of metal in these technologies (material intensity); and
- the potential substitution of these metals or technologies.

Typically, future demand for critical metals, particularly those used in low-carbon technologies, is expected to be driven by a small number of end-uses. For example, tellurium has several uses in metal alloys and in the chemical industry. However cadmium telluride (CdTe) photovoltaic cells are expected to be the fastest growing and major end-use for tellurium in the coming decades, playing a more significant role in future tellurium demand than all other uses combined (USGS, 2010). Similar cases apply to many of the other critical metals.

Having established these key technological uses for critical metals, plausible scenarios for the growth of these technologies can be examined as a first stage in establishing their impact on future metal demand. For the metals critical to low-carbon technologies, these scenarios are often derived from the literature, of which there are a number of studies covering a wide range of technologies. This literature is broad, with studies applying different methodologies and developing a wide range of scenario outcomes. Choosing a scenario on which to base estimates of future demand is therefore difficult. In the case of metals critical to electric vehicle manufacture, such as lithium or neodymium, a wide range of different vehicle manufacturing scenarios are available. These scenarios are conducted over different time frames, include varying types of low-carbon vehicle, and may or may not be consistent with global GHG emissions targets. These differences coupled with the range of methodological approaches leads to a significant range of estimates of low-carbon vehicle manufacture in any given year (Figure 22.3). This wide range of outcomes is common across many low-carbon technologies.

Once the low-carbon technology demand scenario is established, the *material intensity* or metal demand per unit of these manufactured products is estimated. Material intensity is dependent on several factors and those factors are often technology specific. While individual manufacturers may understand the quantity of metal used in the manufacture of their products, they have little incentive to divulge this information. For complex products involving multiple components produced through global supply chains, the final manufacturer may not even know the material composition of all the components. There is also likely to be variation in the material intensity of products from different manufacturers. Even if the material intensity of all manufacturers' products was known, the variation in the average material intensity over time is still unknown and must be estimated. This variation over time is not always incorporated in critical metals studies and the impact of this omission should be acknowledged when comparing studies (Speirs et al., 2011).

Calculating the material intensity of a gigawatt (GWp) of thin-film photovoltaic (PV) cell provides a good example of the important issues, though aspects of these issues are specific to the particular technology. While there are several variations, thin-film PV consists of three main technologies: amorphous silicon (a-Si); copper indium gallium (di)selenide (CIGS); and cadmium telluride (CdTe). The main metal components of a-Si are reasonably abundant and not generally considered critical to the future of the

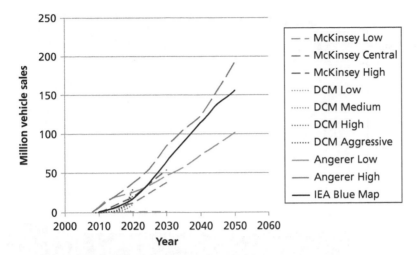

Figure 22.3 Review of different low-carbon vehicle deployment scenarios

Source: Speirs et al. (2013b).

Notes: McKinsey, Dundee Capital Markets (DCM), and Angerer et al. (2009) scenarios include hybrid, plug-in hybrid, and battery electric vehicles. IEA scenarios also include fuel cell vehicles. Some of the data above was extracted from figures using the computer program Engauge Digitizer.

technology. However, CIGS and CdTe are both subjects of criticality concerns due to the speculation about the future availability of indium[2] and tellurium respectively. These two types of thin-film PV are therefore often examined in critical metals studies (Speirs et al., 2011).

Thin-film PV cells consist of an active layer of photoelectrical semiconductor sandwiched between further layers that make up the conductive front and back contacts (Figure 22.4). While both contact layers must be conductive, the front contact layer must also be transparent to allow photons to pass through to the active layer. This transparent conductive layer is often an oxide of zinc or tin.[3] The back contact is often molybdenum, sometimes with an intervening buffer layer (Speirs et al., 2011).

The active layer, consisting of either CIGS or CdTe, constitutes the majority of the thickness of the cell, and is where the critical metal is found. The weight of critical metal per unit area of the cell can therefore be calculated by understanding three things:

1. the thickness of the layer;
2. the density of the active layer material; and
3. the proportion of critical metal in the active layer material.

---

[2] Other metals such as selenium and gallium have been discussed in critical metals studies, but are typically considered of less concern.

[3] Indium is a component of the transparent conductive compound indium tin oxide (ITO). ITO has been used in thin-film PV in the past but is increasingly substituted with other, zinc-based, compounds.

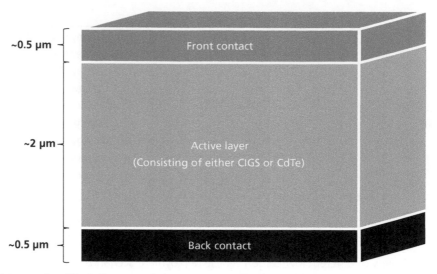

Figure 22.4 A simplified diagram of a generic thin-film layer structure

The *thickness of active layers* in thin-film PV is often stated by manufacturers or analysts (Green, 2010; Candelise et al., 2011). The density of the active layer material is also available in published literature (Green, 2010; Candelise et al., 2011). The proportion of critical metal in the active layer can again be taken from the literature, or calculated where the material has a fixed chemical relationship (stoichiometry). For example, CIGS is referred to as a solid solution of copper indium selenide (CIS) and copper gallium selenide (CGS). The relative proportion of these can vary and is not widely given by manufacturers. CdTe, on the other hand, has a stoichiometric proportionality, and the weight of tellurium in CdTe can be calculated from its chemistry.

The manufacturing processes for depositing these layers of active material vary, and result in some level of wasted material not properly deposited. Some of this waste material can be collected and recycled, but some proportion is lost. The proportion of material retained is sometimes referred to as the *utilization rate*. The utilization rate must be factored in to the calculation of material intensity to fully capture the total weight of metal used to manufacture a unit of thin-film PV. Current utilization rates are in the order of 50 per cent.

Finally, the above factors allow the calculation of critical metal per unit area, but to calculate the material intensity per unit of electricity generating capacity the *efficiency* of electricity conversion must be considered. This is usually expressed as the fraction of solar energy arriving at the panel that is converted to electricity per square metre under standard test conditions[4]. Current thin-film PV efficiencies are in the order of 12–15 per cent.

---

[4] Standard test conditions (STC) are PV test conditions where solar irradiance (1000 W/m^2) solar spectrum (AM 1.5), and temperature (25 °C) are fixed.

These variables can be combined in the following mathematical relationship:

$$M_R = \frac{\rho F \mu}{U S \eta} \tag{22.1}$$

Where M_R is the material requirement in t/GWp, ρ is the density of the active layer material, F is the per cent of material in the layer, μ is the thickness of the layer in microns (μm), U is the utilization rate, S is solar insolation under standard conditions (1,000 W per m^2), and η is the electrical conversion efficiency of the PV cell.

While all these variables can be measured to reasonable degrees of accuracy, when estimating their impact on future demand for critical metal over several decades the rate at which these variables might change over time is harder to estimate. Variables such as efficiency, layer thickness, and utilization rate are all likely to change as manufacturers seek to reduce the cost of PV modules through efficiency improvements, reduction in layer thickness, and improvements in manufacturing processes. The rate at which these variables might change in the future can be informed by examining the historical rate of improvements in these variables and extrapolating into the future, while being bounded by known theoretical limits (Wadia et al., 2009; Green, 2010; Houari et al., 2013). However, much of the literature assumes static values for each of these variables, the impact of which is often not acknowledged. Those who do account for these dynamics over time tend to estimate more optimistic outcomes for the future of PV manufacturing in the face of critical metals availability (Speirs et al., 2011; Houari et al., 2013).

While thin-film PV provides an illustrative example of the issues associated with estimating material intensity, the variables are specific to the technology. When estimating the material intensity of another technology the variables of importance are likely to be different. Lithium intensity in electric vehicles (EVs) for example, is determined by the average capacity in kWh of lithium ion batteries, and is less sensitive to the efficiencies of manufacturing or other performance characteristics. It is also very difficult to calculate the lithium intensity per kWh of battery capacity, and estimates of this variable are typically taken from the published literature. This highlights some of the technology specific issues that are important to consider and are difficult to incorporate into criticality assessment methodologies.

In response to concerns over the availability of a critical metal, or to a rising price of metals due to concerns over scarcity, manufacturers may choose to substitute that metal for another material that provides the same function. The market may also respond to similar pressures by choosing an alternative technology that provides the same utility with reduced or zero use of critical metals. For example, as one of the responses to disruptions in the supply of cobalt in the late 1970s, the magnet industry began developing permanent magnets with non-cobalt chemistries to substitute for the incumbent samarium cobalt magnets of the time. Through this effort neodymium–iron–boron magnets were developed, and this has remained the most powerful magnet alloy to date.

Substitution has the potential to significantly impact future demand for critical metals. However, estimating the impact of metal or technology substitution is difficult. Some criticality assessment studies use expert elicitation to estimate the relative substitutability of technologies using critical metals (Morley and Eatherley, 2008; AEA Technology, 2010; SEPA, 2011). For most of the critical metals used in low-carbon technologies, substitute materials are known (USGS, 2012). However, lithium chemistry batteries and permanent magnet motors are both illustrative exceptions to this. There is no efficient substitute for lithium-based EV batteries given the unparalleled energy density delivered by this chemistry (Tarascon and Armand, 2001; Armand and Tarascon, 2008) and no other chemistry provides magnets of the power achievable with neodymium magnets used in permanent magnet motors (Jones, 2011). Though these metals are hard to substitute for, there are technological alternatives. For example, direct-drive wind turbine designs involve generators with very large permanent magnets. These magnets are typically neodymium–iron–boron alloys and concerns over the supply of rare earth elements from China have precipitated fear that neodymium supply will be limited, impacting wind turbine manufacturing and deployment. However, alternative designs, including geared turbines, do not use neodymium, are well established, and could be deployed in some instances in place of neodymium-based direct-drive turbines (Speirs et al., 2013b).

The low-carbon technologies could account for a significant proportion of expected future demand for critical metals. However, other end-uses not related to low-carbon technologies will remain, and the impact of these other end-uses will not be uniform across all critical metals. In some instances they may still be significant, and discounting them may significantly affect any critical metals analysis. The critical metals tellurium and indium provide two contrasting examples to illustrate this point.

Tellurium has traditionally been used in three different types of end-use: as an alloy additive to improve machinability or fatigue resistance; chemical uses such as to vulcanize rubber, colour glass and ceramics, or as a catalyst; and as a component of electrical applications such as thermal imaging or PV. PV is now the largest end-use of tellurium, and is expected to make up almost all future demand increases (Fthenakis, 2009). Therefore, studies that ignore other end-uses of tellurium in their future demand forecasts are unlikely to be significantly affected by this omission.

Indium, on the other hand, has two significant uses, both of which are likely to drive future demand. In addition to the growing use of indium in CIGS PV modules, indium is also used in the manufacture of flat screen displays where it is a component of the transparent conductive surface layer in the compound indium tin oxide (ITO). Acknowledging this end use and estimating its role in future demand is an important aspect of indium demand forecasting.

22.2 Estimating future supply

A range of factors may influence the future supply of a metal including the quantity of remaining reserves, the cost of extracting those reserves, their geographical location, or the rate of metal recycling. These factors can be separated into two distinct groups: geological or *physical* factors; and economic, social, and geopolitical factors, referred to as *above-ground* factors. While the former may influence the physical production of a resource, the latter may place constraints on the supply chain. Unforeseen constraints to either of these may have detrimental effects on associated markets or economies.

Physical factors of production relate to either the quantity of resource available for production, or the rate at which that resource can be produced. However, estimating either of these is complicated, and both are frequently influenced by above-ground factors such as the market price of a metal, or developments in extraction technology.

The classification of metal resources is defined in the McKelvey Box (Figure 22.5). Metal *resources* include quantities of metal in known ore deposits that are currently technically and economically feasible (*reserves*), and quantities that are potentially producible in the future (including *reserve base*). Quantities of metal in the reserve base and resource categories may cost too much to produce now, but could be produced in the future with a sufficiently high metal price, or if innovation in production technologies drives down future production costs. There are significant uncertainties in estimating reserves, reserve base, and resources, making it difficult to confidently state the quantity of metal available for production in the future. This is compounded by the uncertainty in estimating *undiscovered* resources, some of which will contribute to future production. Undiscovered resource estimates are not readily available for many of the critical metals.

Analysis of critical metals typically includes estimates of global *reserves*, or *reserve base*. Metal reserves are typically estimated by mining companies sampling the ore to determine ore concentration, identifying which quantities of that ore are economically recoverable, and multiplying by the estimated volume of the ore body. The reserve estimates of a region are then aggregated by organizations such as the US Geological Survey (USGS), who present aggregate reserve estimates by country. Given the third-

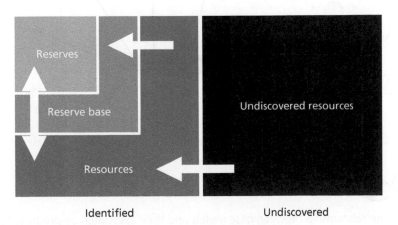

Figure 22.5 McKelvey box presenting metal resource classification

Source: Adapted from EC (2010).

party aggregation of resource estimates, this data is often poor, with omissions, inaccuracies, and a lack of underpinning geological evidence (Speirs et al., 2011, 2013b). However, reserve estimates are likely to be exceeded in many cases as both economics and technology are likely to improve in the future, incentivizing and facilitating production and exploration.

Many metals are not produced on their own, but are by-products of the mining of base metals such as copper, zinc, tin, and aluminium (Figure 22.6). The low concentration of these *by-product* metals makes it uneconomic to produce them individually. However, they become concentrated during the refining of their *host* metals, and can then be recovered economically. For metals exclusively recovered as by-products, reserve estimates are often calculated as a function of the reserve estimates of their associated host metal. For example, the USGS estimate tellurium reserves based on the volume available in copper ore. However, other tellurium deposits do exist, both associated with other metals (such as lead, nickel, and zinc) and as ore potentially mineable for tellurium alone (Houari et al., 2013). This further complicates the interpretation of critical metal reserve estimates when assessing future availability.

While estimates of recoverable resources may provide an indication of potential cumulative production, it is also important to understand the rate at which that resource can be produced year on year. Though data on historical production is commonly available for most metals, estimates of future production rates are less common.

Production rate is considered in several ways. Estimates of current production are often used in indicative metrics such as reserve to production ratios (R/P) (NRC, 2007;

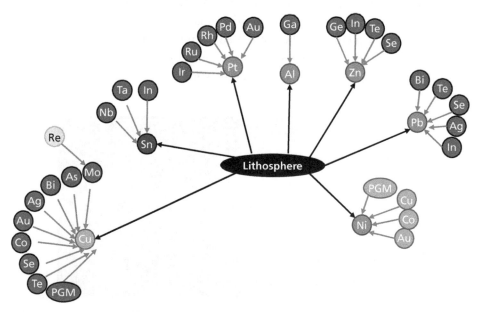

Figure 22.6 The relationship between base metals and their associated by-products found in the same ore deposits, many of which are critical metals

Source: Adapted from Hagelüken and Meskers (2010).

Rosenau-Tornow et al., 2009; AEA Technology, 2010; Achzet et al., 2011; Schüler et al., 2011; SEPA, 2011; Graedel et al., 2012), or future demand to production ratios (Angerer et al., 2009; EC, 2010). The R/P ratio is a common metric used in many extractive industries, including the oil and gas industry, and simply calculates a quantity of reserves for a given company, country or region divided by the annual production from that reserve. From an investor's perspective the R/P ratio can provide a useful indication of the relative endowment of an extractive company's projects, and provide a way to measure companies against each other in order to value them. However, the usefulness of the R/P ratio for indicating the future availability of resources at regional or global scale is questionable, with studies examining regional R/P ratios for oil finding them a poor indicator of future availability (Sorrell et al., 2009). Alternatively current production can be measured against an estimate of demand in a given future year, giving an indication of the increases in production needed to meet future use of critical metals. However, as discussed in the previous section, estimates of future demand are difficult, and subject to significant uncertainty, with no standardized methodology. For example, Angerer et al. (2009) selects fifteen metals or metal groups, and thirty-two associated emerging technologies assumed to drive the demand for these metals in the future. A range of technical and economic factors are then considered for each technology, and global economic conditions accounted for, in order to estimate the increased metal demand between 2006 and 2030. The estimate of future demand is then divided by current production. This kind of supply indicator can convey the relative difference between current production and future demand, but provides no indication of the physical and above-ground challenges associated with increasing production in the future. For some metals, future supply increases may come at a low cost, while for others marginal production may be more economically and technically challenging. The metric is also highly sensitive to the assumptions regarding future demand. The European Commission (EC, 2010) followed a similar methodology to generate indicators of relative future demand. However, given differences in assumptions the resulting factor estimates are significantly different than those derived by Angerer et al. (2009), highlighting the sensitivity of these factors to the uncertainty in future demand estimates (Table 22.3).

Modelling future supply prospects on an annual basis provides a more sophisticated alternative to the simple supply indicators discussed above. However, typically only the material or technology specific studies take this approach and very few critical metals have been examined in this way.

Fthenakis (2009) estimates the annual production rate of indium and tellurium based on third-party projections of the future recovery of their host metals, zinc and copper respectively. Two different host metal cases are defined for each metal based on the literature. From these cases, the quantity of indium or tellurium associated with host metal production is calculated, and a recovery rate assumed, which varies over time. Fthenakis (2009) then estimates the quantity of metal available to thin-film photovoltaic manufacturing by subtracting an estimate of the material demanded for all other uses.

Table 22.3 Estimated future demand in 2030 over current production for ten metals included in two critical metals studies

Metal	Angerer et al.	EC
Gallium	6.09	3.97
Neodymium	3.82	1.66
Indium	3.29	3.29
Platinum	1.56	1.35
Tantalum	1.01	1.02
Silver	0.78	0.83
Cobalt	0.40	0.43
Palladium	0.34	0.29
Titanium	0.29	0.29
Copper	0.24	0.24

Source: Angerer et al. (2009) and EC (2010).

In an alternative approach, Houari et al. (2013) use a system dynamics modelling approach to perform a similar calculation that places more emphasis on the dynamic variability of the key variables over time. This approach also allows for an analysis of the key sensitivities.

The various approaches to estimating future production rates all agree that critical metal production is likely to increase in the future, driven by the price incentive created by increasing demand. This is reasonable given the burgeoning nature of these resource markets, and the relatively underdeveloped nature of the resource.

Since estimates of reserves and incentive to produce those reserves are both a function of metal price and production costs, it is necessary to consider the economic aspects of metal supply. One approach to capture this is the cumulative availability curve. In this approach the geological sources of extractable metal are presented on two axes, with the quantity of the resource along the *x*-axis and the cost at which the resource can be produced on the *y*-axis (Figure 22.7). Different sources are presented in order of cost, with the cheapest sources on the left and the most expensive on the right. This approach provides an idea of the increased cost associated with producing the marginal tonne of metal. Where the cumulative availability curve of a metal is relatively flat, the cost of expanding future production is relatively low, indicating that future metal production might be relatively unconstrained by economic factors. Where the curve appears steep, however, future metal production may be constrained by the high costs of future extraction.

Yaksic and Tilton (2009) present a cumulative availability curve for lithium, highlighting the very large quantities of resource available if consumers are willing to pay the high cost of extracting lithium from sea water. This indicates the physical challenges associated with extracting metal from geological sources with ever decreasing metal concentration, and the ever increasing costs associated with recovering these metals. Being able to compare curves like this for several metals could be a valuable metric for

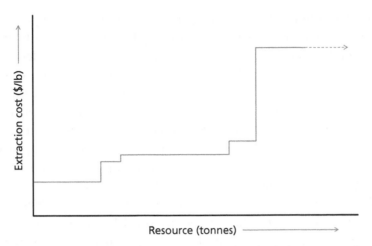

Figure 22.7 Example of a cumulative availability curve with cumulative resources in tonnes on the x-axis and extraction costs on the y-axis

assessing the relative ease with which future production can be increased. However, very few studies employ this approach and cumulative availability curves are not widely published. If published, these curves will be hard to maintain, with costs likely to change over time as technology develops, lowering the cost of extraction for a given geological source. Creating and maintaining cumulative availability curves is also subject to the effects of inflation and the transparency of extraction costs, which operators are not incentivized to divulge. All of these factors are likely to impact on the accuracy of these curves, both immediately and over time.

An increasing concern in recent years is the influence of geopolitics and policy decisions on the availability of critical metals to global markets. In order to analyse the vulnerability of the supply of a particular metal to political influence in any given geographic region, it is first necessary to understand the geographical distribution of the reserves and production of that metal. Where a small number of countries are responsible for a large proportion of a metals production, then policies and events in those countries affecting that metal are likely to have a significant impact on its global availability. Conversely, if the production and reserves of a metal are evenly distributed across a larger number of countries, then geopolitical impacts will have a diminished effect on global metal supply. This assumes no coordination between countries, such as a cartel. No cartel currently exists in critical metals markets but other energy commodities markets such as the global oil market have established cartels, such as the Organisation of the Petroleum Exporting Countries (OPEC).

If supply is concentrated then a number of different events may impact on global metal supply. The first group of impacts relates to industrial policy decisions. A country responsible for a large proportion of global supply of a metal may use that dominant position to its advantage by seeking to control the quantity of metal available to the global market by instituting embargoes on trade with specific countries, or quotas to limit global exports. By doing this countries may simply wish to support or increase the global price of that commodity by creating an imbalance between supply and demand. Alternatively some wider industrial policy goal may be sought such as to encourage the high value end of the supply chain to locate within the country's borders.

Conflict and civil unrest in regions of metal supply concentration can also have a significant impact on global metal availability. In the late 1970s, for example, conflict in the Democratic Republic of Congo (then called Zaire) and neighbouring countries first interrupted cobalt supply routes and then directly affected cobalt mine production (Westing et al., 1986). During the 1970s Zaire was responsible for approximately half of global cobalt production, and supply disruptions precipitated a number of responses, including a significant price increase, strategic stockpiling (Guttman et al., 1983), and concerted efforts towards developing substitute materials (Sichel, 2008). A number of policy options were highlighted in a report by the US Congressional Budget Office (CBO) in response to cobalt supply disruption. These include options to manage mineral availability, such as stockpiling, subsidizing domestic production, and developing novel mineral resources—such as ocean-based resources. Policy options to reduce future cobalt demand include R&D funding for development and supply of substitute materials (CBO, 1982). However, while these responses may be appropriate for the US context, many countries will not have all these options. In particular, many countries will not have domestic resources that can be supported through subsidy.

A number of metrics can be used to quantify the state of the policy environment for metals supply, or the potential for civil unrest to impact on metal exports from specific regions. These include the Fraser Institute's Policy Potential Index (PPI) and related metrics, the UN Human Development Index (HDI) (Graedel et al., 2012), the World Bank's World Governance Indicators (WGI) (Moss et al., 2011), and Fund for Peace's Failed States Index (FSI) (Morley and Eatherley, 2008; EC, 2010; DOE, 2011; Graedel et al., 2012).

Not only are the current geopolitical issues associated with critical metals a concern, but these geopolitics are likely to shift in the future. As concerns arise regarding the exposure of critical metals to supply concentration, the market is likely to seek to diversify supply where possible, by seeking supply relationships for the undeveloped resources of other countries. This has the potential to significantly change the political landscape of resource trade. Rare earth elements (REE), for example, have caused concern to resource markets given the overwhelming concentration of their production in China. As a result new resources are under exploration or development in countries such as Greenland, Australia, Tanzania, Canada, South Africa, and Brazil (Reuters, 2009). This paves the way for a new era of geopolitical negotiation as countries try to gain political influence and ultimately supply contracts with these new REE producers. This has led to concerns that some countries could buy undue political influence and distort longer term REE markets by negotiating the rights to the majority of new resources (Bomsdorf, 2013; Matlack, 2013).

Recycling of metal from products at their end-of-life is a way in which future metal supply may be supplemented, reducing the burden on mine production. However, the

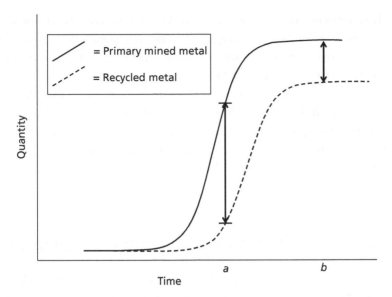

Figure 22.8 Relationship between primary metal production and recycling rate

Note: The solid line represents metal extracted from the lithosphere while the dotted line represents metal recovered from recycling of end-of-life products. At time *a* the relative contribution of recycled metal to total supply is less than at time *b*.

impact of recycling on future metal availability is subject to a number of factors. First, the lifetime of a product delays the availability of its components to the recycling market. In thin-film PV for example, modules may be expected to last for thirty years. Access to recycle the critical metals contained within those modules is therefore delayed by the same period of time. In most cases the recyclable quantity of this metal will be less than 100 per cent, and estimating the future recovery rate[5] is difficult given that it is likely to be a function of technical capability and economic factors thirty years in the future. The relative contribution of recycling during different phases of the primary metal production cycle is also important. Where the rate of production growth is steep, the relative proportion of recyclable material is likely to be smaller than periods where the production rate plateaus. This is due to the recovery rate and the product lifetime delay, and means that periods where demand is growing most quickly coincide with periods where recycling can contribute a smaller relative proportion of production (Figure 22.8). For example, recovery of tellurium from thin-film PV modules in the future might be subject to an 80–90 per cent recovery rate (Fthenakis, 2009) and a module lifetime of twenty-five to thirty years (Houari et al., 2013). As tellurium demand increases, the differential between primary mined metal and recycled metal is greatest (as seen at time *a* in Figure 22.8). As the rate of primary mine production slows, the growth in recycled metal can 'catch up' and the differential decreases (as seen at time *b* in Figure 22.8).

---

[5] The recovery rate is defined as the percentage of metal that can be recycled from end-of-life products.

Despite these limitations, there is an incentive for countries or regions that are net importers of critical metals to encourage recycling as a means to reduce their relative level of imports. This may mean incentivizing the design of products to be easily recycled, and support for recycling capability.

Finally, environmental concerns may result in policy decisions that impact on global metal supply. For example China outlawed artisanal rare earth production, which may have contributed not insignificantly to China's production. The impact of environmental legislation on future metal production is measured in the literature using qualitative scoring, lifecycle analysis, or the UN's Environmental Performance Index (EPI) index (EC, 2010).

22.2.1 DEMAND IN CONTEXT OF FUTURE SUPPLY ESTIMATES

The expected demand increase for critical metals is reflected in estimates in the available literature. Table 22.3, which presents future metal demand as a factor of current production, suggests that by 2030 some metal demand may be up to six times greater than current metal production. However, these results mask the significant uncertainty given the range of variable assumptions found in the literature. To better express this type of uncertainty, ranges of future demand can be estimated based on the full range of plausible assumptions for each of the variables influencing demand (Speirs et al., 2011, 2013b). In addition this range can be presented in the context of historical production and available future supply estimates (Speirs et al. 2011, 2013b). Figure 22.9 presents this type of comparison for lithium. The demand estimate represents the expected future lithium demand from low-carbon vehicles (battery electric vehicles and plug-in hybrid vehicles), while the future supply estimates represent the range of estimates found in the literature. The demand ranges in 2030 and 2050 seem significant in comparison to estimated supply in 2020. However, the top end of these demand ranges assume the highest levels of material intensity found in the available literature, and it is more likely that future demand will be towards the centre of these ranges. Manufacturers are driven by cost pressures to reduce the material intensity of their products where possible, which would reduce the upper end of the ranges presented. It is also likely that future demand will tend towards the lower end of these ranges if perceived scarcity is reflected in a high lithium price signal. In addition there is no evidence to suggest that very high levels of future production might not be achievable.

22.3 Measuring criticality

A number of studies termed 'criticality assessments' attempt to measure metal criticality through a form of multi-criteria analysis. Metal criticality can be defined as the likelihood that future supply of a specific metal will constrain future demand, and the

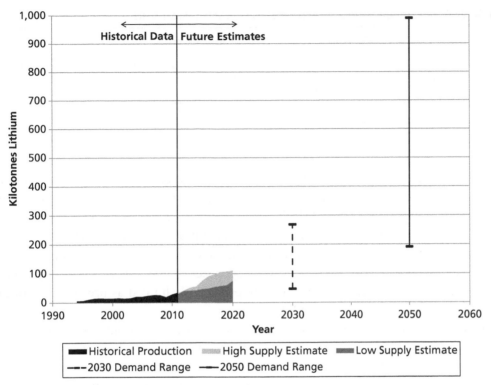

Figure 22.9 Comparison of historical lithium production, forecast supply, and forecast lithium demand from electric vehicles

Source: Speirs et al. (2013b).

vulnerability of consumers to such a supply deficit. The vulnerability to companies, countries, or the global economy is quantified in a number of ways, including estimating the value of the economic sector at risk (Gross Value Added) (EC, 2010), or estimating the capability of the affected sector to respond to supply disruption (Graedel et al., 2012). Recent studies have assessed critical metals in specific geographic regions (Morley and Eatherley, 2008; EC, 2010; DOE, 2011; Graedel et al., 2012), sectors of the economy (DOE, 2011; Moss et al., 2011) or critical metal consuming companies (Duclos et al., 2010). These studies are high level multi-criteria analyses which attempt to: a) quantify the *criticality* of any one of these metals; and b) to rank the critical metals in terms of their relative criticality. However these criticality assessments follow a range of different methodologies and arrive at different conclusions (Erdmann and Graedel, 2011).

Criticality assessments typically gather together a range of metrics or 'factors' representing important determinants of future metal availability. A range of metals or other materials are then assessed and scored against these factors before aggregating scores (with weighting in some cases) to provide a relative measure of criticality. While all assessments follow different methodological approaches, some commonly assessed factors are presented in Table 22.4.

Table 22.4 Factors considered in typical metals criticality assessments

Supply factors	Geopolitical factors	Demand factors	Other factors
Geological availability	Policy and regulation	Future demand projections	Cost reduction
Economic availability	Geopolitical risk	Substitutability	Environmental issues
Recycling	Supply concentration		Media coverage

Source: Adapted from Speirs et al. (2013c).

There are several variations in the application of these criticality factors. For example, geological availability of a metal is measured by some using the ratio of estimated future demand to current production (Morley and Eatherley, 2008; Angerer et al., 2009; EC, 2010; Moss et al., 2011). Other studies use the ratio of known reserves to current production, otherwise known as the R/P ratio (NRC, 2007; Rosenau-Tornow et al., 2009; Schüler et al., 2011). Neither of these approaches fully captures the prospects of future metal supply, and R/P ratios may be particularly limited as predictors of future supply constraint (Speirs et al., 2013c).

Though physical or geological availability is important, criticality assessments have increasingly considered economic and geopolitical drivers of availability (Erdmann and Graedel, 2011). Economic availability is implicit in many of the studies, which include US Geological Survey (USGS) estimates of reserves, an intrinsically economic estimate of available resources. Some studies go further, including a factor for the volatility of metal price, giving an indication of the market sensitivity to availability (AEA Technology, 2010; Duclos et al., 2010; SEPA, 2011). Economics also plays into the potential recyclability of a metal, which is more recyclable as the price of the primary metal increases. Graedel et al. (2012) provide the most sophisticated estimates of future recycling, estimating the useful lifetime of products, and assuming present-day recycling rates, presented in a recent UNEP study (UNEP, 2011).

Geopolitical factors included in criticality assessments are designed to capture the influence of policy, regulation, and political stability on future metal supply. Metals that are produced in only a small number of countries are considered inherently vulnerable to the regulations of those countries, or any socio/political unrest, such as civil war. To measure geopolitical factors, studies first quantify the concentration of sources of metal supply (EC, 2010; Moss et al., 2011; Graedel et al., 2012). This metric can then be combined with internationally collected indices on the regulatory regime, or the relative political stability of the producing countries (EC, 2010).

The future demand for metals can be estimated in a number of ways and is measured in criticality assessment in two factors: demand projections and substitutability. Demand projections can involve assumptions on current demand, use of third-party estimates, or estimates based on expert opinion, and these estimates should account for the factors discussed in Section 22.1. The substitutability of a metal in its common applications can have a significant impact on the development of demand over time. If a metal is highly substitutable then supply constraints are unlikely to arise. Substitutability is difficult to measure as it often arises through innovation in response to cost pressures. Assessments that include a factor for substitutability quantify it through expert opinion.

A number of other factors are included in some studies, including cost-reductions, media coverage, and environmental issues (Speirs et al., 2013c). These are not often included factors and are therefore not discussed further here.

The wide range of different approaches to criticality assessments provides an equally wide range of results (Erdmann and Graedel, 2011). Contributing significantly to this is the differing scope of criticality assessments, which might be assessing the exposure of different countries, regions, or economic sectors to supply risks, all having very different exposures to the supply chains of critical metals. In order to examine the range of variation between these assessments, Speirs et al. (2013c) normalized the criticality scores of eleven different assessments, the results of which can be seen in Figure 22.10. The wide and overlapping ranges of results for each of the metals or metal groups included in Figure 22.10 illustrates the difficulties in drawing general conclusions for a large group of critical metals.

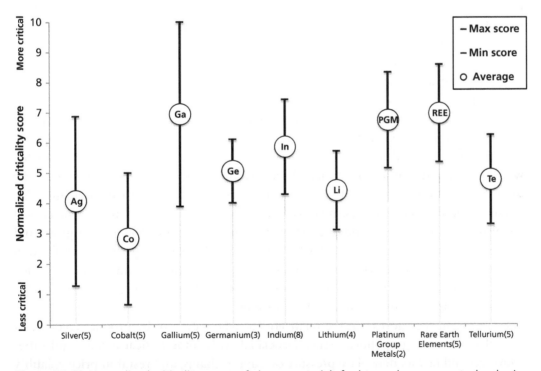

Figure 22.10 Normalized criticality range of eleven materials for low carbon energy technologies found in eleven criticality assessments

Source: Adapted from Speirs et al. (2013c).

Note: Number in parenthesis on the *x*-axis is the number of studies included in the criticality range for each metal.

Many of the factors influencing the criticality of a metal vary over time. However, few assessments include these temporal dynamics. Some have used different assessment frameworks to estimate criticality over different time periods (DOE, 2011; Graedel et al., 2012), and recent assessments have begun to examine the impacts of changing variables over the time period of their assessments (Houari et al., 2013). These and other limitations with a criticality approach are largely a function of the broad nature of their methodologies. In order to assess a large number of metals under a single methodology compromises must be made in the type and detail of assessed criteria. Coupling such assessments with more focused analyses which address a smaller group of metals or technologies in more depth is an appropriate way to address these compromises.

22.4 Conclusions and implications for policy

The concern over critical materials has increased significantly in recent years, reflected in the proliferation of recent research (Speirs et al., 2013c). However, conclusions on the future availability of critical metals, and their impact on low-carbon technology manu-facturing, are difficult for a number of reasons. The available data on resources and future production is poor due to the high cost associated with resource assessment and a lack of incentives to monitor and estimate current and future critical metals supply. The evidence surrounding future metal demand is also unclear, given the significant uncertainties surrounding future uptake of low-carbon technologies, and the 'moving target' associated with the material intensity of these technologies and their development over time.

Several future trends are, however, reasonably certain. Demand for critical metals is likely to grow significantly in the future, driven by low-carbon technology manufactur-ing in addition to the existing economic and technological drivers. Future supply of critical metals is also likely to increase, in response to the rising price incentive associated with increasing demand. The available evidence discussed above does not support the premise that future supply will place significant constraints on growing demand. The relatively recent interest in these metals and the burgeoning nature of low-carbon critical metal demand implies that production of these metals has the potential to grow, and evidence of recent resource estimates for many critical metals suggests that production growth can be supported. There are also a number of mitigating dynamic responses, such as substitution and recycling, which will go some way to softening the impacts of the significant demand growth and resulting price increase.

However, in the short term the relatively inelastic nature of by-product recovery of critical metals, the geopolitical issues associated with some critical metals, and other above-ground factors may place pressure on supply chains, and result in price volatility as markets adapt to the dynamics of these factors.

Criticality assessments provide an interesting way to highlight these issues in an organized multi-criteria analysis. These studies can make some broad conclusions on critical metals' relative exposure to the factors driving future supply and demand, though they may miss some of the nuances of specific technologies or metals. This can be addressed by studies focusing on specific metals, technologies, sectors or other particular issues of relevance.

There are a number of possible policy responses to the concerns over the future availability of critical metals. The issues associated with data are unlikely improve significantly overnight, but for countries that wish to address their exposure to critical metals the maintenance or development of capacity to monitor the supply-side factors is essential. This could be facilitated through support for national geological surveys. Other supply-side measures might include the management of strategic stockpiles of key metals, support for domestic mining where possible, and the support for development of marginal resources such as ocean minerals. Demand-side measures could include support for R&D in substitute materials and support for recycling capacity, including design of products for easy recycling.

Geopolitical intervention may also help to deliver security of supply in the face of short-term geopolitical supply concerns. Strategic partnerships with resource-endowed countries and diplomacy with key metals-exporting nations may mitigate some of these concerns. For larger countries with domestic resources, some of these policy responses will be easily taken. For smaller, less resource-endowed countries policy responses may need to be coordinated through larger political jurisdictions (e.g. the European Union).

All of these policy approaches will necessarily need guidance from research into materials of most concern for a given country or region, the nature of those specific markets, and the policy responses most likely to deliver security of metals supply.

■ REFERENCES

Achzet, B., A. Reller, and V. Zepf (2011). Materials critical to the energy industry: An introduction. Augsburg, Germany, University of Augsburg report for the BP Energy Sustainability Challenge.

AEA Technology (2010). Review of the Future Resource Risks Faced by UK Business and an Assessment of Future Viability. London, UK, Report for the Department for Environment, Food and Rural Affairs (Defra).

Angerer, G., L. Erdmann, F. Marscheider-Weidemann, A. Lullmann, M. Scharp, V. Handke, and M. Marwede. (2009). 'Raw materials for emerging technologies: A report commissioned by the German Federal Ministry of Economics and Technology (English Summary).' Fraunhofer ISI. Retrieved 31/10/2011, from <http://www.isi.fraunhofer.de/isi-en/service/presseinfos/2009/pri09-02.php>.

Armand, M. and J. M. Tarascon (2008). 'Building better batteries.' *Nature* 451(7179): 652–7.

Bomsdorf, C. (2013). Greenland Votes to Get Tough on Investors. *The Wall Street Journal*. New York City, Dow Jones & Company.

Buchert, M., D. Schüler, and D. Bleher (2009). Critical Metals for Sustainable technologies and Their Recycling Potential. Nairobi, Kenya, United Nations Environment Programme (UNEP) and Oko-Institut.

Candelise, C., J. F. Spiers, and R. J. K. Gross (2011). 'Materials availability for thin film (TF) PV technologies development: A real concern?' *Renewable and Sustainable Energy Reviews* 15(9): 4972–81.

CBO (1982). Cobalt: Strategic Options for a Strategic Mineral. Washington DC, Congress of the United States Congressional Budget Office.

Chegwidden, J. and D. J. Kingsnorth. (2011). 'Rare earths - an evaluation of current and future supply.' Roskill and IMCOA, Institute for the Analysis of Global Security.

Retrieved 15/7/2011, from <http://www.slideshare.net/RareEarthsRareMetals/rare-earths-an-evaluation-of-current-and-future-supply>.

DOE (2011). Critical Materials Strategy 2011. Washington, DC, US Department of Energy.

Duclos, S. J., J. P. Otto, and G. K. Konitzer (2010). 'Design in an era of constrained resources.' *Mechanical Engineering* 132(9): 36–40.

EC (2010). Report of the Ad-hoc Working Group on defining critical raw materials. *Critical Raw Materials for the EU*. Brussels, European Commission.

Erdmann, L. K. and T. E. Graedel (2011). 'The criticality of non-fuel minerals: a review of major approaches and analyses.' *Environmental Science & Technology* 45(18): 7620–30.

Fthenakis, V. (2009). 'Sustainability of photovoltaics: The case for thin-film solar cells.' *Renewable and Sustainable Energy Reviews* 13(9): 2746–50.

Graedel, T. E., R. Barr, C. Chandler, T. Chase, J. Choi, L. Christoffersen, E. Friedlander, C. Henly, C. Jun, N. T. Nassar, D. Schechner, S. Warren, M.-y. Yang, and C. Zhu (2012). 'Methodology of metal criticality determination.' *Environmental Science & Technology* 46(2): 1063–70.

Green, M. A. (2010). 'Learning experience for thin-film solar modules: First Solar, Inc. case study.' *Progress in Photovoltaics: Research and Applications* 19(4): 498–500.

Guttman, J. T., J. E. Merrick, V. J. Fenwick, and J. F. Graham (1983). Critical non-fuel minerals in mobilization with case studies on cobalt and titanium. Washington, US National Defense University, Industrial College of the Armed Forces: 140.

Hagelüken, C. and C. E. M. Meskers (2010). *Complex Life Cycles of Precious and Special Metals. Linkages of Sustainability*. Cambridge, MA, MIT Press: 163–97.

Houari, Y., J. Speirs, C. Candelise, and R. Gross (2013). 'A system dynamics model of tellurium availability for CdTe PV.' *Progress in Photovoltaics: Research and Applications* 22(1): 129–46.

Hubbert, M. K. (1956). *Nuclear Energy and the Fossil Fuels*. Houston, Shell Development Co., Exploration and Production Research Division.

Jevons, W. S. (1865). *The Coal Question: An Inquiry Concerning the Progress of the Nation, and the Probable Exhaustion of our Coal-mines. By W. Stanley Jevons*. London; Cambridge, and Cambridge: Macmillan and Co.

Jones, N. (2011). 'Materials science: The pull of stronger magnets.' *Nature* 472: 22–3.

Matlack, C. (2013). Chinese Workers—in Greenland? *Bloomberg Businessweek*. New York City, Bloomberg LP.

Morley, N. and D. Eatherley (2008). Material Security: Ensuring Resource Availability to the UK Economy. Chester, UK, Resource Efficiency KTN, Oakdene Hollins and C-Tech Innovation.

Moss, R. L., E. Tzimas, H. Kara, P. Willis, and J. Kooroshy (2011). Critical Metals in Strategic Energy Technologies. Brussels, Belgium, European Commission Joint Research Centre (EC JRC). EUR 24884 EN—2011.

NRC (2007). Minerals, Critical Minerals, and the US Economy. Washington, DC, National Research Council.

Reuters (2009). Canadian Firms Step Up Search for Rare-Earth Metals. *The New York Times.* New York City, Arthur Ochs Sulzberger, Jr.

Rosenau-Tornow, D., P. Buchholz, A. Riemann, and M. Wagner (2009). 'Assessing the long-term supply risks for mineral raw materials—a combined evaluation of past and future trends.' *Resources Policy* 34(4): 161–75.

Schüler, D., M. Buchert, D. I. R. Liu, D. G. S. Dittrich, and D. I. C. Merz (2011). Study on Rare Earths and Their Recycling. Final Report for The Greens/EFA Group in the European Parliament, Darmstadt, Germany, Öko-Institut.

SEPA (2011). Raw Materials Critical to the Scottish Economy. Edinburgh, Scotland, A report by AEA Technology for the Scottish Environmental Protection Agency (SEPA) and the Scotland and Northern Ireland Forum For Environmental Research (SNIFFER).

Sichel, A. (2008). The Story of Neodymium: Motors, Materials, and the Search for Supply Security, Chorus Motors. Retrieved 12/4/2015, from <http://www.choruscars.com/Chorus_NEO_WhitePaper.pdf>.

Sorrell, S., J. Speirs, R. Bentley, A. Brandt, and R. Miller (2009). *Global Oil Depletion: An Assessment of the Evidence for a Near-term Peak in Global Oil Production.* London, UK Energy Research Centre.

Speirs, J., B. Gross, R. Gross, and Y. Houari (2013a). *Energy Materials Availability Handbook.* London, UK Energy Research Centre.

Speirs, J., R. Gross, C. Candelise, and W. Gross (2011). Materials Availability: Potential Constraints to the Future Low-carbon Economy. Working Paper I: A thin film PV case study. London, UK Energy Research Centre.

Speirs, J., Y. Houari, M. Contestabile, R. Gross, and B. Gross (2013b). Materials Availability: Potential Constraints to the Future Low-carbon Economy. Working Paper II: Batteries, Magnets and Materials. London, UK Energy Research Centre.

Speirs, J., Y. Houari, and R. Gross (2013c). Materials Availability: Comparison of Material Criticality Studies—Methodologies and Results. Working paper III. London, UK Energy Research Centre.

Tarascon, J. M. and M. Armand (2001). 'Issues and challenges facing rechargeable lithium batteries.' *Nature* 414(6861): 359–67.

UNEP (2011). Recycling Rates of Metals: A Status Report. Report of the Working Group on the Global Metal Flows to the International Resource Panel, United Nations Environment Programme.

USGS. (2010). 'Minerals yearbook.' Retrieved 12/4/15, from <http://minerals.usgs.gov/minerals/pubs/commodity/myb/>.

USGS (2012). Mineral Commodity Summaries 2012. Reston, VA, US Geological Survey.

Wadia, C., A. P. Alivisatos, and D. M. Kammen (2009). 'Materials availability expands the opportunity for large-scale photovoltaics deployment.' *Environmental Science & Technology* 43(6): 2072–7.

Westing, A. H. (1986). *Global Resources and International Conflict: Environmental Factors in Strategic Policy and Action.* Stockholm International Peace Research Peace Research Institute (SIPRI) and United Nations Environment Programme (UNEP).

Yaksic, A. and J. E. Tilton (2009). 'Using the cumulative availability curve to assess the threat of mineral depletion: The case of lithium.' *Resources Policy* 34(4): 185–94.

23 Electricity markets and their regulatory systems for a sustainable future

Catherine Mitchell

23.1 Introduction

This chapter makes the argument that the structural design and rules and incentives of electricity networks and markets have a fundamental impact on the way electricity systems work and for their outputs. For example, how secure they are; whether they are reducing their carbon emissions; whether electricity is sold at the lowest price as possible. It also explains that there are a number of new objectives for, and fundamental differences in, the way electricity system operation is thought about. This is in turn driving electricity system change and bringing its own challenges. Because of this, countries need to make sure that the design of their electricity systems and their market and network rules and incentives fit current challenges and future needs.

The relatively recent energy policy goal of decarbonization has led to new low-carbon technology deployment on the energy supply side but also take-up of new technologies to enable improved operation of the energy system and to better incorporate the energy demand side. The decarbonization targets have also altered the relationship between governments and companies, in that governments now have better information and knowledge of the future energy demand and supply mix than energy companies do. Increasingly energy demand will be determined not by consumer behaviour primarily, but by decarbonization and energy efficiency plans and policies.

Within this, it has become clear that early electricity sector decarbonization can play a key strategic role in an economy-wide shift to a zero-carbon energy economy in 2050 because it provides flexibility over transport, industry, and domestic heating/cooling options. Increasing research, development, demonstration, and deployment support into clean energy solutions (supply and demand management technologies) across global supply chains is making the relative prices of different low-carbon solutions highly volatile over the short and medium term. The fall in solar electricity prices due to Chinese investment is a graphic example of this (REN21, 2013). As a result, policy delivery, in energy and wider sectors, and their success or not, has become a major driver of market development and technological uncertainty.

Moreover, while the decarbonization goal is a relatively recent objective of energy policy, it is becoming increasingly necessary for all three energy objectives of decarbonization, security, and affordability to be met to a certain level. It is no longer possible to talk about simple trade-offs. A key aim for policymakers is to manage the energy policy process so as to prevent unacceptable performance of each objective.

At the same time, the decarbonization arc is initiating a shift from a low capital cost, high fuel-based energy system to a high capital and low/zero fuel energy system, which is fundamentally altering the way in which cost reductions and system efficiencies can be made. The energy system is therefore broadly moving from a centralized energy system of a few supply technologies and one-way networks, to one which is made up of multiple different types of supply, demand, and operational technologies that enable the efficient and integrated running of the system. Cost reductions in the former mainly came from economies of scale. Cost reductions in the latter are more to do with the increasingly efficient integration of the operational side. Furthermore, this is also opening up opportunities for major systemic shifts into demand led and decentralized business models and greater international interconnection.

Policies on decarbonization (e.g. smart grid development, storage, interconnection, and so on) have become central for shaping market conditions, for limiting uncertainty, and for encouraging investment. These policy decisions are made outside what is currently seen as the market framework, yet these are central to market outcomes. This is leading to a different way of conceptualizing energy markets and their regulation. Rather than the current two-way approach of a regulator exerting different amounts of regulatory requirements on the electricity system segments (depending on the country and discussed below), the new approach requires a clearer, stronger, more sustained, and more integrated strategic direction from government (and their energy institutions such as system operators, purchasing agencies, etc.) with more open, innovative, and competitive private sector markets on both the supply and demand sides.

There are now detailed experiences of a number of countries of operating electricity systems with high proportions of variable power and attempting to incorporate demand-side response. The assumptions about how electricity systems should be operated are increasingly coming under discussion because of this experience and because of the changing objectives for the electricity system (and its challenges) discussed above. Part of this discussion is what market and network rules and incentives are required to operate these electricity systems. The purpose of this chapter is to explore the details of different electricity system structures, market, and network rules and incentives to encourage appropriate innovation, given the above context.

Section 23.2 provides an overview of the generic segments that make up an electricity system; Section 23.3 provides a brief overview of how electricity markets work; Section 23.4 sets out the challenges these conventional electricity markets are increasingly having to deal with; Section 23.5 explores possible solutions to problems related to (1) electricity market design; (2) capacity versus capability markets; (3) system operator functions; and (4) regulatory reach; and Section 23.6 concludes.

23.2 **Electricity system structure**

Energy systems can be divided into a number of fundamental segments. For example, basic electricity system segments tend to be generators; distribution networks; a transmission network; a system operator; a market and its operators; and retailing to customers of different demand sizes (or supply, as it is known in Europe). In addition, there may be other actors and functions such as a regulator; market information; market monitoring; metering; and data arrangements. The latter functions can all potentially be within the regulatory function, or stand-alone outside it. Gas systems can be divided in the same way as electricity, but with producers and shippers of gas rather than generators. Other energy systems, for example heat, can be divided between providers (similar to electricity generators or gas producers) of heat and then the other segments. This chapter focuses on electricity systems.

Different countries have different combinations of these segments; energy markets and networks can have different rules and incentives; and regulators (which often—but not always—oversee or develop those rules) can also be structured in different ways with different reach and responsibilities, and work to different legal duties. Moreover, this energy system chain can have different types of ownership of different segments (for example, state, private, cooperative or local/municipal) and establish different levels of choice and involvement for customers of different demand types.

The structure, institutions, rules, and incentives in place for markets and their economic regulatory systems therefore have enormous implications for the way in which energy is conceived in everyday life by customers and citizens; for the total amount of, and way, energy is used; and for the change that occurs in the practices of running the electricity system—whether it be to do with the type of technologies used; the amount of innovation stimulated; the momentum or inertia in the system; the type of business models developed; and so on.

Together, therefore, the choices made when constructing or re-building an electricity (or gas/heat) system and establishing its structure, rules, and incentives on use of, or access to, each segment of it have a central importance to the character of the system; its ability to meet environmental, social, and security goals; the cost and distributional impact of doing so; its space for social innovation; and the degree to which customers are able to connect and interact with it.

The conventional description of an electricity system is categorized by the ownership structure and the degree to which market services are valued. There are three broad 'types' of electricity system around the world (taken, and paraphrased, from Cochran et al. 2013).

- Some countries (or regions, such as a US state) operate under a traditional, vertically integrated model of generation, distribution, transmission, and retail. Generally, these assets are owned by one entity (can be private or state owned) and operate under a regulated rate of return.

- The second type is when the main segments of integrated utilities are 'unbundled'—meaning that these segments are separated, and act as an independent economic agent. This can either be complete unbundling—that is different ownership—or 'ring-fenced' unbundling, where the owner is the same but operation is regulated to be separate. Typically, the natural monopolies (i.e. networks) are regulated in some manner, while the other segments are opened up to competition to some degree. Electricity is sold through an energy only market.
- The third type is the same as the second, except that the markets involved are more complex because they value more 'capabilities', which can be turned into streams of revenue beyond only energy. These capabilities could be demand-side response, storage, ancillary services, fast start–stop cycling capability (i.e. turn on and off quickly and safely); regular, dispatchable ramping capability (i.e. an ability to rapidly increase or decrease in rates of energy supply) and so on (Keay Bright,[1] 2013). It is this type which enables a valuing of the demand side and flexibility, which is important for systems with high proportions of variable power.

23.3 A brief overview of how electricity markets work

This chapter provides a very brief overview of how a conventional electricity market works (see Sioshansi 2008a and 2013 for a good introduction). This is rather dry but necessary. The next section explains the challenges that electricity markets are facing. The following section then proposes some solutions. In order to understand these challenges it is important to understand the basics of electricity market operation.

There is a very confusing vocabulary about electricity markets—often because different countries have different terms for the same thing. The wholesale market for electricity is an overarching term that is used to describe all the various markets where wholesale electricity can be bought and sold, whether in the short term through to the far future. Moreover, electricity can be bought and sold in many situations and these are variously called markets, platforms, exchanges, and mechanisms. It is therefore very important when talking about electricity markets that discussion is kept very specific and avoids excessive generalization.

Because electricity supply has to balance demand at all times, generators have to sell electricity in 'real time', when their physical generation is matched to the technical operating requirements of the network. Other names for 'real-time' markets tend to be 'balancing', or 'clearing', but even these terms can differ between countries. Most countries have at least two markets, this 'real-time' market and a day-ahead market. However, countries with actively traded electricity tend to have a real-time market;

---

[1] See Keay Bright for different technology ramping and cycling times.

multiple forward markets (i.e. markets which sell electricity at any point in the future i.e. for a day ahead, a week, month, year, two years ahead etc); and intraday markets.

These forward markets can either be physically based—meaning that their bids and sales end up being related to actual electricity—or they can be financially based—which means that they are essentially financial instruments to hedge the price of the actual electricity that generators and suppliers have bought and sold. These financial instruments do not finally trade out to actual electricity but to money lost or gained. These forward markets can be traded in directly from a company's trading arm (which will have paid to be a member of the market) or they can be accessed via a platform (which the company will also have paid to access). A company like Bloomberg has a platform through which electricity trades can be made in many different markets around the world, whether physical or financial. Moreover, electricity markets are not necessarily country or state based. So, for example, Nordpool includes Denmark, Sweden, Finland, Norway, Lithuania, and Estonia and parts of Germany. The Central Western European Pool (CWEP) includes Austria, Belgium, France, the Netherlands, and Switzerland (Europa, 2013).

Markets can also be 'mandatory'—that is where all participants who want to sell electricity in a country are required to sell through them—although there are many different versions of how this is set up; or they can be 'voluntary'. Voluntary markets tend to refer to forward markets where participants can choose to participate in them. Similarly, a 'gross' market indicates that all physical generation of an electricity system has to be bought and sold through them at a certain time, whereas a 'net' market indicates that not all generation has to go through them, as with the UK's Balancing Mechanism (BM) (and the reason why it is called a mechanism not a market) where only the balancing portion of the electricity is traded. The latter is a bilateral, net, mandatory market.

In this case of a bilateral market in Great Britain, the buying and selling takes place between two parties. The generator or buyer has to tell the BM market operator (Elexon) what their contracted agreement is and what they want physically to buy and sell to get themselves in balance with that agreement, and they have to tell the National Grid (the system operator) what their actual output is going to be at gate closure (just before real time) relative to that contracted agreement.

An alternative to the bilateral market is a 'pool' real-time market where generation volume is offered to, and demand volume is bid from, a central operating body at certain prices. These markets tend to be 'gross', 'mandatory', and physical.

Electricity has to be balanced in a moment to moment way, and frequency response capability is generally divided into three response times[2] (National Grid, 2014; Borggrefe and Neuhoff, 2011). The balancing of the system is done via a combination of a market

---

[2] For example, NG achieves this via primary response (within 10 seconds and sustained for a further 20 seconds); secondary response (within 30 seconds and can be sustained for a further 30 minutes); and high-frequency response (the reduction of power within 10 seconds and can be sustained indefinitely).

operator (i.e. the balancer in a balancing market or mechanism (BM)) and a system operator. These two functions are usually separate but they can be combined together. This balancing occurs in 'real time' in all types of electricity markets, whatever their details. All BMs require generators to notify the BM market operator of their output for a balancing period (and this is made legal through codes and licenses of that country, and is therefore mandatory).

Buying and selling electricity is a central part of the electricity business. Generators can sell into the future and buyers can buy ahead. Gradually time moves on to catch up with these forward trades, and the BM market operator balances all of the contracts for selling and buying electricity in one particular period of time—whether they were originally traded for a day ahead or two years ahead. Electricity participants tell the market operator and/or the system operator (depending on country) what their 'actual' position is at gate closure relative to their contracted position. Electricity system rules differ, but in the UK, Gate Closure is 30 minutes ahead of 'real time'. And 'real time' is when scheduling occurs (in other words, when the decision of what generation to bring on line, or take off—otherwise known as constraining off—is implemented). These intervals can also differ between countries (see Table 23.1).

Table 23.1 Summary of key features in various markets (taken from Riesz et al., 2013; Sioshansi, 2008; Borggrefe and Neuhoff, 2011). BETTA = British Electricity Trading and Transmission Arrangments; Nordpool = Northern European Electricity Market; PJM = Pennsylvania, New Jersey and Massachusetts Electricity Market; NEM = North Eastern Market of Australia; SWIS = South West Interconnected System of Australia.

	BETTA, GB	Ireland	Nordpool	PJM	Brazil	NEM	SWIS
Market Speed							
Dispatch Interval	30 min	30 min	60 min	5 min	Centralised dispatch by SO	5 min	30 min
Scheduling to Dispatch	I hour	20 hours	I hour	5 min	–	5 min	Day ahead
Multiple Markets/ platforms	Multiple (bilateral + balancing)	Single	Multiple	Multiple	Multiple	Single	Multiple
Gross (mandatory)/ net (voluntary) exchange	Net	Gross	Gross	Net	Gross	Gross	Net
Capacity mechanism versus capability mechanism	Will have demand side in capacity mechanism via Energy Act 2013	Capacity mechanism	Will have demand side via capability mechanisms	Will have demand side via capability mechanisms	no	Demand side through capacity market	Will have demand side through capacity market

The point to understand in electricity system operation is that the sale of physical electricity happens in a market at a snapshot, real-time moment after gate closure. Choice of these set times to 'clear' the market are necessary to reconcile trades, to establish 'system' costs of operating and managing the electricity system, and to establish electricity prices. The reality is that the electricity system is being managed all the time in a second-by-second manner by the system operator so that the voltage/ frequency and so on is at the right level. This electricity system management is continuous but, in effect, the real-time market, which occurs every half hour, hour and so on, is superimposed on that technical requirement. The linking of the contracted positions from all the trades occurs when the market bids and offers in the 'real-time' market are reconciled by an algorithm, and it is from this algorithm that the 'price' of electricity is established for this period of time. The algorithm decides which generation from which generator at which price should be taken first, and so on until generation matches the required demand of that time period (discussed in more detail below).

Buyers and sellers of electricity need to make sure that whatever contracted position they have made (in whatever futures market and/or with whatever bilateral contracts) 'unwinds' in such a way that they are in balance (i.e. buying and selling what they are contracted to do) and that they have not lost money in that process (for example, if they sold at £6/MWh for a certain amount in the future, it would not be good if all they could buy to balance that amount when that time period finally arrives was at £8/MWh). Buyers and sellers use various forward or intraday electricity markets to balance their contracted positions. It is therefore important that markets are 'liquid,' meaning that there is enough electricity for sale for different time periods and in different 'clip sizes' to enable the buyers and sellers to match their needs.

The needs of different generators will differ. Generators of variable[3] power require liquid markets far more than generators of firm[4] power because they are more uncertain about what their actual output will be at any one time. Smaller generators need access to smaller 'clip sizes' and markets that don't have large up-front transactions costs for users. Limited liquidity works to the advantage of incumbents because it adds risk for potential new entrants. However, for fossil generators in a market with increasing amounts of low marginal cost carbon and/or zero marginal cost variable renewable power, they have to ensure that they do not find themselves too expensive for a particular time period. This is known as being 'out of the money', and discussed further below. In brief though, in times of very windy or sunny resource, zero marginal cost wind or solar electricity will be dispatched before fossil generation. This may mean that less (or even zero) fossil generation will be required depending on demand levels.

---

[3] Meaning that the output from the power plants changes depending on weather conditions and cannot be counted on to be dispatched when the system operator wants it.

[4] Power is known as firm if it can be dispatched when the system operator wants it.

23.4 **Challenges to conventional electricity system structures and operation**

As introduced in the sections above, there are a number of challenges to conventional electricity system operation and regulatory structures (please see multiple chapters in Sioshansi 2008a, Sioshansi, 2008b, and Sioshansi, 2013; Cochran et al., 2013; Borgreffe and Neuhoff, 2011; Joskow, 2008a and 2008b; Riesz et al., 2013; Bauknecht et al., 2013; Binz et al., 2012; Milligan and Kirby, 2010).

These overarching challenges are:

- the need to reduce their carbon output while at the same time ensuring security and affordability;
- the increasingly strategic role electricity plays in the economy-wide shift to a zero carbon energy in 2050;
- the need to refocus electricity system structure and market and network design and operation to stimulate and enable the necessary innovation and change needed to transform from the current fossil-dominated electricity system to a more low-carbon one;
- increasing technical and market uncertainty;
- the increasing importance of policies and strategic decision-making (such as smart grid development, storage, interconnection) on technological and market conditions;
- increasing amounts of variable power renewable electricity (together with low marginal cost nuclear power) undermining the traditional electricity market marginal cost pricing model and increasing concerns about sufficient investment in flexible system capabilities to secure electricity system operation.

This section focuses on the last three of these bullet points.

The current low-carbon electricity supply policies of different countries are broadly based on nuclear power or renewable energy or both (REN21, 2013; WNISR, 2013). This chapter does not focus on the relative benefits of one technology or the other. In practice, most countries have 'chosen' a technology pathway either because of historical legacy (i.e. France and nuclear power); because of geographical resources (i.e. Norway and hydro); or because of social preferences (i.e. non-nuclear Denmark, Germany, and New Zealand). Countries which have had large amounts of both nuclear and renewable energy for a long time (i.e. Switzerland or Sweden) tend to have renewable electricity supplied from hydropower or biomass electricity, which are both very flexible generating technologies and which can 'load-follow' the less flexible nuclear power plants.[5]

---

[5] A load-following capability is when a power plant is flexible enough to match the required electricity demand on the system.

There are various countries that still have fossil-dominated electricity systems and which are, to a lesser or greater degree, supporting both nuclear power and renewable energy as a means to reduce their carbon emissions (e.g. Britain, China, India). At the moment, for most electricity systems with both renewable energy and nuclear power, there are no major operational issues from mixing the technologies. Most of these countries are reasonably pragmatic about future technology pathways: in a sense, they are waiting to see how the technologies and their costs develop before they choose which one they want to support in particular. Britain is unusual in its clarity of support for nuclear power (UK Parliament, 2013).

However, if both nuclear power and renewable electricity develop in proportion then, at some point, a choice will have to be made about which technology has priority when operating the electricity system (Verbruggen, 2008). This is because both nuclear power and variable power renewable electricity plants are relatively inflexible. Electricity system operation, which is trying to manage the voltage and frequency on a second-by-second basis, requires a certain proportion of the electricity power plant stock to be flexible so that it can change supply to match the changing demand. As the relative proportions of nuclear and renewable electricity rise, and if there is limited demand flexibility in the system, operational questions of which inflexible technology to 'constrain'[6] off first will increasingly occur—and indeed already occurs (Ofgem, 2009; Utility Week, 2013). This will increasingly affect the economics of the different supply technologies and electricity systems. This is because the electricity system can *either* pay double for electricity—as is the case in the UK where both the generating and the constrained off electricity is paid for within the electricity market balancing mechanism—*or* not pay for the constrained off power plant, which is therefore negatively economically affected by system operation procedures. Given that the conventional operational norm in a fossil-based electricity system is to constrain off the smaller power plant, it is renewable energy that routinely loses out economically.

This is where a de facto choice occurs when operating an electricity system. If *either* an active *or* a passive decision is taken to continue with the current operational norm (i.e. not change the operational code) then the electricity system will favour nuclear over renewable electricity. If renewable electricity is preferred because of cost (or other reasons) then the operational rules of the system will have to change to reflect this. At the moment, in most countries this is not a decision that is being forced. It is one reason, however, why countries such as Germany or Italy, which have said they are not supporting nuclear power, find it easier to make both short- and long-term electricity system infrastructure planning decisions.

Much has been written about the theoretical concerns of adding renewable electricity to the energy system. Most of that literature has been overly negative and/or conservative

---

[6] 'Constraining off' occurs when more generation is being generated at a particular place and time for the transmission capacity available, or when there is a lack of demand for the available generation and the SO has to choose which generation to take and which to constrain off.

about the consequences, meaning that the costs to the system of greater levels of variable renewable power has been overestimated (as shown by e.g. Milligan, 2010; SSREN, 2011: ch 8; Borggrefe and Neuhoff, 2011; IEA, 2011). However, evidence from electricity systems where there is a high proportion of variable renewable electricity has shown that there are three serious market or operational challenges that need to be addressed. These challenges are that a move from a low-capital, high fuel-cost technology system to one which is based on high-capital, low fuel-cost technologies is undermining of the traditional electricity market marginal cost pricing model; is leading to concerns of possible 'missing money' and lack of investment; and the increasing need to harness a flexible demand side within the electricity system to match the relative inflexibility of nuclear and/or renewable electricity.

Hitherto, most electricity systems incorporate markets that are based on marginal cost pricing. A market operator via an algorithm takes the cheapest offer for a volume of generation first,[7] and then the next lowest offer is taken and so on until there is enough generation to meet demand for any particular time period (for example, every half hour throughout a day).[8] The price of the last kWh of electricity supplied becomes the price paid for all generation bought in that half hour—marginal cost pricing. In this system, demand changes over the day in different half-hour slots and so the marginal cost of electricity changes over the day. Periods of greater demand lead to higher marginal prices than periods of low demand. The economics of electricity power plants has been based on the average of all these half hourly prices covering the total cost of the power plant over the year, even if some power plants generate most of the time (base load) or for only some of the time (peak load). This basic building block of conventional electricity economics is undermined by the combination of zero marginal cost renewable electricity and low marginal cost nuclear power. This electricity effectively increases supply and shifts the supply curve to the right, bringing down both average and peak prices (Bauknecht et al., 2013; Cochran et al., 2013). An increasing share of nuclear power and renewable electricity in the absence of increased demand means that conventional generators will be displaced more often by low or zero marginal cost sources and sell electricity at lower prices when they are selected—and they therefore will run at lower and less predictable capacity factors and therefore earn less revenue.

This is a general result of increased low-carbon electricity capacity without a matching increase in demand which, on the whole, is the intention of low-carbon energy policies. There are only a few countries where there is sufficient low or zero marginal cost plant deployed to really make a difference to electricity prices—and this is primarily in Denmark and Germany. Nevertheless, other electricity system stakeholders, including fossil generators, are watching the impacts of this with interest (Business Week, 2014). In one sense, the challenge of marginal cost pricing is a successful result of bringing on new low-carbon capacity, thereby increasing supply. In the absence of increased demand one

---

[7] This is necessarily rather simplistic and generalized.
[8] This can vary, see Table 23.1.

would expect prices to drop. This is part and parcel of moving to a low-carbon energy system with high carbon sources being displaced, as one would expect, by low-carbon energy sources.

It is also to be expected that the period of transition from a high to a low-carbon energy system has winners and losers. However, appropriate electricity market design can both make this transition easier for those power plants finding themselves 'out of the money' and cheaper for customers who have to pay for it (Bauknecht et al., 2013). Closing a fossil fuel plant may be the economic answer if its average costs are greater than average revenues. However, it is not necessarily in the interests of the electricity system for fossil power plants to close. An electricity system may not use generation from a power plant often, but it may be very important to have access to that power plant from a system point of view for a few hours a year. Efforts are therefore being made to keep capacity, which otherwise might be uneconomic from a power plant point of view, on the system until such a time that it 'naturally' retires; until new forms of cheaper, cleaner, more flexible capacity come on line; or as a way to manage the electricity system securely.

Thus, a growing challenge, and a side product of the move to lower carbon generation, is that of 'missing money' for individual power plants and what to do about it (Joskow, 2008). As said above, the price paid for electricity is, typically, no higher than the marginal cost of operation of the highest cost production, particularly if there is an electricity price cap (e.g. the Australian market NEM has a rule which set a maximum market price cap of AUD\$13,100/MWh for the 2013–14 financial year). This means that the highest cost producers which may be needed in periods of high electricity demand may not receive enough revenue to recover their fixed costs. Over time, this missing money may lead to generation being taken offline or reduce the incentive to invest in new electricity capacity.

In tandem to this, as more countries have policies and targets for low carbon or renewable electricity, the investment risk in fossil generation is increasing. An investor in a fossil power plant has to be confident that they can sell their power plant output for enough years at a sufficient price to make their investment worthwhile. As more and more low-carbon or variable renewable power comes on the system, so there is less and less market demand to be met by fossil generation and the average system marginal cost may fall. Just as individual fossil generators are questioning their economics in electricity systems with high proportions of variable power, so the availability of flexible capacity and suitable ancillary services becomes even more important in order to complement their uncertain output. This suits gas power plants because of their flexible capabilities, but less so coal or nuclear because of their ramping abilities (Keay Bright, 2013). There are therefore concerns that insufficient investment will come forward to provide the 'right' type of capacity and capabilities for a secure electricity system.

Finally, a further challenge to conventional electricity systems is that related to balancing. Conventional electricity systems balance *uncertain* demand (i.e. customers around the system turning their lights, appliances, and load requirements on and off: for example electric trains stopping and starting, which has certain regular characteristics from day to day) with *certain* supply to match it (i.e. supply from firm, dispatchable coal, gas, and nuclear power plants). As greater proportions of variable power are introduced into the electricity system, the system operator and/or balancer has to balance the uncertain demand with more uncertain supply. This is a fundamental change in the operation of the electricity system, and has major knock-on effects for the economics of energy. One way to respond to this is to make the demand side more flexible by a greater use of storage and demand-side response mechanisms in markets, including load shifting via smart grids. While demand may be hard to predict, variable power is harder so trying to make demand more flexible is helpful, and also (once appropriate market rules are established) has many operational and economic benefits (Riesz et al., 2013). There are still very few countries which operate markets to match supply and demand by changing demand, with the Pennsylvania, New Jersey, and Massachusetts (PJM) market possibly the best-known example. This new approach is sometimes called a demand-focused electricity system.

Cochran et al. (2013) concur with the above but add in the challenge of minimizing complexity of design in order to attract extensive market participation, and ensuring market depth in forward markets to increase flexibility in, and reduced risk for, investment.

23.5 Key solutions for electricity markets, industry structure, and regulatory reach

The sections above have set out the key issues facing electricity markets. It is remarkable how little literature, academic or otherwise, there is describing a 'best practice' low-carbon electricity system. Sioshansi has two useful edited volumes: Competitive Electricity Markets ((CEM): Sioshansi, 2008) and the Evolution of Global Electricity Markets ((EGEM): Sioshansi, 2013) which explore these areas. The Regulatory Assistance Project (RAP) has global[9] best practice reviews for market and regulatory issues of moving to a low carbon future (<http://www.raponline.org>). For example, there is a best-practice way to incorporate demand-side response into markets (Hurley et al., 2013); reviews and critiques of capacity mechanisms described in more detail below (e.g. Keay Bright, 2013). On RAP's website and elsewhere there are useful reviews of best-practice wholesale market design (Milligan and Kirby, 2010; Binz et al., 2012; Cochran et al., 2013); a review of balancing and intraday market design (Borggrefe and Neuhoff, 2011); and a review of designing distribution tariffs (Linvill et al., 2013). These critically explore a power system in one country or one aspect of a power system across countries.

Discussions about the design and operational solutions to these electricity system challenges revolve around a combination of new operational and management rules

---

[9] Taken from their four regions of work: the USA, Europe, India, and China.

within electricity markets; capacity and/or capability payments to overcome issues of 'missing money' and concerns about under-investment; new roles for system operators; and issues about regulatory reach.

This section explores these potential solutions.

23.5.1 COMPLEMENTARY MARKET RULES AND INCENTIVES FOR LOW-CARBON AND VARIABLE POWER

When taken together, Cochran et al. (2013), Bauknecht et al. (2013), Riesz et al. (2013), Hogan (2012), Keay Bright (2013), Borggrefe and Neuhoff (2011), and IEA (2011) provide a basic framework for progressive electricity market rules and incentives.

Bauknecht et al. (2013) focus on the efficient integration of renewable electricity into electricity systems with other sources of power—flexible and inflexible. They argue that (1) market design should support the integration of renewable electricity into the system at least cost—in other words, that thinking is system orientated rather than concerned about single plants; (2) that the abilities of renewables are recognized, and that they are not expected to do what they are no good at. An example of this would be exposing variable renewable electricity to a balancing market which would be likely to lead to individual storage solutions to increase flexibility, rather than the much more efficient system-wide solutions; (3) that system-wide capability solutions of storage, demand-side, smart grids, flexibility in general is what markets should be incentivizing in a system-wide manner; (4) that variable power renewables should not be exposed to market risk before markets are adapted to their characteristics; (5) and that short-term and balancing markets should be adapted to ensure inclusion of renewable electricity by implementing helpful features, for example, a short-gate closure, intraday markets, and so on.

Riesz et al. (2013) broadly agree with Bauknecht et al. (2013) and add that (1) markets should be designed to access the full degree of technical flexibility in an electricity system (as described in detail in IEA, 2011); (2) that there should be 'fast' markets (i.e. gate closure is shortly before scheduling rather than an hour or longer and that the space between scheduling and dispatch is also short), as is the case in PJM and the Australian North Eastern Market (NEM, see Table 23.1); and (3) that large, liquid markets should be developed, including through interconnection and inclusion of the demand side.

Again, Hogan (2012) agrees with Riesz et al. (2013) and Bauknecht et al. (2013) but also argues that electricity systems should have the following complementary characteristics to capability markets (described in the next section). There should be: (1) the ability to aggregate services across large balancing areas; (2) increasing interconnectivity with neighbouring markets; (3) transmission investments to mitigate internal congestion and constraints (which otherwise usually lead to renewable shut down); and (4) improved, centralized weather forecasting and better operationalization of forecasts, as they do in Spain (Federico, 2010).

23.5.2 LARGE, LIQUID, INTERCONNECTED MARKETS

This importance of large, liquid markets to balance individual contracted positions is an aspect of electricity markets which is growing in importance as proportions of variable power increase. For example, a generator of variable power will have a contracted position to sell physical generation at a certain time. Despite best efforts, variable generators can expect to want to top up (i.e. to buy some generation to reach their contracted position) or spill (i.e. sell some generation, again to reach their contracted position) at some point before the time they have contracted for. For example, a wind generator may have contracted in the day ahead market to sell 1 MWh of wind electricity at a certain time but then find they have 2 MWh at real time. In this situation, they will want to sell that extra 1 MWh at the best possible price. They will try and sell this in the intraday markets. It is very important that these are liquid enough for their needs. Intraday markets can be at fixed times (say six times a day as in Spain) and therefore catch more liquidity at the market times (Weber, 2010). Other intraday markets occur continuously (e.g. Germany) but then lose liquidity, which can be difficult for small players and so on (Borggrefe and Neuhoff et al., 2011).

If a generator has not managed to sell the extra 1 MWh in the intraday markets, they have to tell the SO/BM market operator that they are 1 MWh over their contracted position at gate closure. It is at this point that the design of the physical, real-time electricity market becomes so important. In a bilateral type market, which has penal buying and selling prices for the out-of-balance physical electricity, there is a great deal of risk involved for all generators, but particularly variable power generators. The SO/BM reconciliation occurs and the generator will sell their generation at the system buy price. In times of tight supply, that can lead to a high system buy price, but if there is oversupply, as there often is for variable power at a windy or sunny time, then the price they can sell at may be very low. Thus, buying and selling in the balancing market is a major risk for variable power generators—they will always prefer to do this in less risky markets.

A pool mechanism generally has lower risk for variable power, depending on the rules. If a generator is 1 MWh over, they can sell into the real-time 'pool' and they will receive the marginal system price. Again, at times of high supply and low demand this can be a low price, but it is much easier to track the expected trend in prices over time and it is much easier to establish a contract for difference price against the pool price for any output.

In conclusion, the design of wholesale markets can be made much easier for variable power, without diminishing their efficiency or adding difficulties for fossil or nuclear generation. This is because the necessary design rules to complement variable power also increase flexibility, liquidity, transparency, and so on which also improves competition (Sioshansi, 2013; Cochran et al., 2013; and Borggrefe and Neuhoff, 2011). These rules reduce risk and therefore have an impact on the ability of new entrants, small players or variable power to invest, develop, and then exist, more or less easily, in a market.

23.5.3 CAPABILITY MECHANISMS

A second solution to the challenges in this changing electricity world is to ensure that sufficient capacity and necessary services are available going to the future to ensure security of supply. The concerns of undermined marginal cost pricing and 'missing money' have led to calls by some for capacity payments, whereby expensive capacity which would not otherwise be competitive in electricity markets is paid some sort of fixed cost to make it worth the while of the owner to keep it going and available rather than to shut it down.

RAP has said:[10]

... the real challenge facing the wholesale market today [is] ensuring that the market properly reflects the value of those dispatchable capacity resources providing the system with the flexibility it needs to fully utilise low marginal cost resources while at the same time ensuring the value of those resources unable or unwilling to adapt their operating profile as needed shifts downward in line with their actual value to the power system. What is needed, in other words, is not investment in capacity resources per se, but rather a realignment of investment based on those resource attributes of most value to the power system.

Energy only markets (i.e. electricity markets without a capacity or capability mechanism) rely on wholesale electricity prices to reward capacity through infra-marginal rent (i.e. the difference between peak prices and the short run marginal costs of each power plant). Because of the concerns discussed above (the undermining of marginal cost pricing as a result of more zero marginal cost renewable energy and potential 'missing money'), a great deal of discussion has been undertaken about what (and if) a mechanism is needed to overcome these problems. The Regulatory Assistance Project has produced a very useful set of papers in this area (Gottstein, 2011; Gottstein and Skillings, 2012; Hogan, 2012; Baker and Gottstein, 2013; Keay Bright, 2013), and the following discussion derives from synthesizing those papers.

However, it is not always the case that an electricity system needs a capacity mechanism—because there already might be sufficient flexible capacity available. The first question a decision-maker should ask themselves is: do we need a capacity market? Keay Bright (2013: 24) puts forward a checklist of questions to discover whether there is a need for a capacity mechanism and, if so, whether its design is robust. If the answer to these questions is YES, it suggests the design is robust, but if it is NO, it suggests there are significant concerns about the need for it. Keay Bright posed and answered these questions in relation to the capacity mechanism suggested in the Electricity Market Reform of the UK. She decided that the answer was NO to most of the questions, and as such there is no need for a capacity mechanism in Britain. These questions are:

---

[10] RAP, 2013, RAP response to the EC Consultation on Generation Adequacy, Capacity Mechanisms and the Internal Energy market in Electricity.

Does the mechanism:

a. Seek to deliver the range of capabilities that the system will actually need to meet *net demand* with an increasing proportion of renewables?
b. Maximize the potential for existing resources (generation, demand, and storage) to deliver the necessary capabilities before resorting to incentivizing more expensive new resources?
c. Seek to secure comparable services from all potential resources, in particular, the demand side? Are the eligibility criteria for the demand side and for distributed or aggregated resources fair and reasonable, on an equal footing to centralized conventional generation?
d. Ensure that resources that cannot provide the necessary range of capabilities (e.g. inflexible generation) are put at an appropriate competitive disadvantage to those resources that do provide the capabilities?
e. Charge the costs of reliability services in a way that avoids creating earnings risks that are difficult to manage for renewable generators? To the extent that these risks are increased, does the proposal address how the potential adverse impact on the deployment of renewables can be addressed in other ways?
f. Deliver reliability in a manner that promotes future cost reductions and innovation in the provision of flexible capabilities and avoids foreclosing the market to future providers?
g. Create a potentially scalable design, including the future integration of neighbouring balancing areas and the sharing of capability resources? Consider potential effects on market coupling and available mitigating measures?
h. Recognize the carbon content of resources procured to provide the range of capabilities?

Having decided that there is a need for capacity mechanism, then the design of the mechanism has to be chosen. There have so far been different means of setting capacity mechanisms: via centralized procurement with a system operator; through a quantity or price mechanism (i.e. the system operator determines the quantity and asks for price bids or the SO sets the price of capacity, and asks for quantity bids); the capacity mechanism can be market wide or targeted. Irrespective of the type of mechanism, the design of traditional capacity mechanisms so far has been based on the purchase of a fixed quantity of capacity sufficient to meet highest total system demand (peak demand) plus a reserve margin. Such a capacity mechanism generally rewards all resources (i.e. firm MWs) the same way. These mechanisms are typically based on conservative, high projections of the amount of firm capacity required to meet future demand—meaning that the capacity mechanism supports an unnecessarily high amount of capacity paid for by the consumer. Countries which do have large amounts of variable power (German, Denmark, and Spain) are discovering through operational learning that lower amounts of firm capacity, whilst maintaining security levels, are required in practice (Keay-Bright, 2013).

As described above, systems with a high proportion of variable renewable power will need, firm, dispatchable capacity with flexible capabilities to be able to meet the net demand at least cost. These capabilities include the demand-side response, storage, ancillary services discussed above but also flexible, fast start–stop cycling capability (i.e. turn on and off quickly and safely); regular, dispatchable ramping capability (i.e. an ability to rapidly increase or decrease in rates of energy supply); and ramping capability reserved now to be used in the future (known as ramping service). Together, these capabilities provide the ability to 'flex' the supply side to meet net energy demand. These are central requirements, which should be the characteristics that are paid for in a capacity market (Keay Bright,[11] 2013). In this sense, countries should be thinking of '*capabilities*' markets (which can include a payment for straightforward capacity) rather than a '*capacity*' market—since capacity is only one dimension of the capability needs of a secure electricity system (Gottstein and Skillings, 2012; Keay Bright, 2013).

Following on from this argument, while capacity mechanisms have long been argued as a way to cover fixed costs, they can have downsides (Keay Bright, 2013). Traditional approaches to capacity mechanisms may extend the life of high-carbon, inflexible assets—which is a threat both to decarbonization targets and also to the reliability of the system with a high proportion of variable renewables.

Increasing shares of variable renewable energy mean that the net energy demand (i.e. total demand minus variable power) is becoming more challenging to anticipate and serve over different time-scales. A high degree of flexibility in the portfolio of dispatchable resources is called for within generation, demand response, and storage (RAP publications, including Hurley et al., 2013; Cochran et al., 2013; Riesz et al., 2013). Insufficient flexible capability within the available generation mix might lead system operators and regulators to resort to balancing challenges by curtailing renewables or it can lead system operators to deal with resource adequacy problems by bringing on significant amounts of little used capacity. This is both extremely costly and the opposite of most energy policy goals, which tend to be to increase renewable generation, keep costs down, and run an efficient system.

Electricity markets are obviously extremely complex and adding in a capacity and/or capabilities market adds to that complexity. However, provided that (1) markets and network use and costs are transparent; (2) markets and networks are set up so that all technical flexibility capabilities (discussed below) can be accessed; and (3) market actors (whether regulators or market operators) are set up and incentivized to process, and use, up-to-date information—in other words, keep up with the rapid technological and operational change at work in our energy systems—then capabilities markets are simply part of an efficient electricity system (Sioshansi, 2013; Cochran et al., 2013 and Borggrefe and Neuhoff, 2011; IEA, 2011; all RAP capacity papers).

Hogan (2012) also puts forward an alternative to a capacity mechanism—the enhanced forward services market (EFSM). This follows a similar approach to the tranched capacity mechanism and essentially tries to build up a capability service market, that is for ramping, cycling, and full participation of the demand side, thereby driving investment in improving flexibility capabilities of what is already on the system.

---

[11] See Keay Bright for different technology ramping and cycling times.

In conclusion, once it has been decided that a capabilities market is needed, a well-designed capability mechanism should be supporting the type of flexible capacity or service that is wanted in a low-carbon electricity system which has increasing amounts of low or zero marginal cost electricity and which wishes to tap the flexibility availability in the system, including from the demand side. In this way, a capability market can be expected to become an important integral part of a flexible electricity system.

23.5.4 SYSTEM OPERATOR RATHER THAN FOR-PROFIT SYSTEM OPERATOR

The system operator can have different functions, the main ones being managing gate closure and system operation, capabilities markets, and forward markets. They can also manage the market operation in conjunction with system operation. System operators can work to all sorts of structures undertaking different combinations of these functions. They can be not-for profit (i.e. state owned or owned by electricity system stakeholders) or private. However, while not always the case (i.e. in the National Grid, England and Wales), operator function is generally separate from the transmission business functions (Borggrefe and Neuhoff et al., 2011).

Given the increasing complexity of markets, made possible by new ICT technologies, and the myriad choices available for short through longer term development of infrastructure (meaning type of generation and its scale; networks, interconnectors, storage, smart grids, prosumers, etc.), the SO role is likely to become more integral to efficient system operation. Moreover, as market size, interconnection, and liquidity increases—and as capability markets and services increase—so it is possible to imagine a combining of both the market operator and the system operator role. For example, in Denmark, a not-for-profit SO responsible for security and the technical transition to a low-carbon and secure electricity system has been instituted, while at the same time, the regulator has become smaller (DERA, 2004). Thus, the division between the system operator role and the regulator role is altering as the move from low-capital, high-cost fuel to high-capital, low-cost fuel alters the focus of cost reduction from individual power plants to system operation and efficiencies.

23.5.5 ALTERING THE REGULATORY REACH

In general, whether in the USA, Europe, or Australasia, economic energy regulators are responsible for the rules concerning the sale of energy and the regulation of the means of getting it to customers (i.e. networks). One of the obvious factors in energy systems is how inter-related their structure, rules, and incentives are to their outcomes (whether it be prices for customers, carbon emissions, the level of innovation, or a change in practices). Markets will encourage a low-carbon and secure energy system if the structure, rules, and incentives are set up to reflect and value low-carbon characteristics and benefits and if different segments are complementary. As has been explored above, this complexity is only becoming greater because of ICT enabling more trades; accessing more flexibility within the electricity system; enabling more variable power; linking bigger and bigger markets; and so on. This has implications for the economic regulator, which has to be set up in such a way that they are able to keep up with the change enabled by the ICT revolution (Perez, 2013).

Moreover, because the structure, rules, and incentives of an energy system are a social construct and have major distributional impacts (Sayer, 2013; Walker, 2013), there should be a clear line of responsibility for the decisions taken when constructing, re-building or operating an energy system.

The role of regulators, and their reach, is central to the design, construction, and operation of electricity systems. Regulators can be very big and have a very broad reach, as in the UK, or they can be small and have a very narrow, economic reach, as they are in Denmark. As argued above, if the challenges of the electricity system imply a bigger role for the SO, this may imply a lessening role for the regulator.

23.6 **Conclusion**

This chapter has illuminated the developing challenges for key electricity market design and electricity system structure for a low-carbon electricity system. These challenges for electricity systems are particularly obvious in countries with high proportions of low-carbon and/or variable power. However, market and operational designs that enable efficient incorporation of low-carbon and variable power also enable efficient operation of an electricity system without variable power because of their inclusion of the demand side. It also provides an up-to-date review of literature of this area for those who wish to understand more of the details.

■ **REFERENCES**

P Baker and M Gottstein, 2013, Capacity Markets and European Market Coupling—can they co-exist? Discussion Paper, RAP, March 2013; <http://www.raponline.org/document/download/id/6386>. Accessed April 2015.

D Bauknecht, G Brunekreeft, and R Meyer, 2013, From Niche to Mainstream: the Evolution of Renewable Energy in the German Electricity Market, Chapter 7 in F Sioshansi (ed.), 2013, *Evolution of Global Electricity Markets: New Paradigms, New Challenges, New Approaches*, Academic Press.

R Binz, R Sedano, D Furey, and D Mullen, 2012, Practicing Risk-Aware Electricity Regulation: what every state regulator needs to know, A Ceres/RAP report; <http://www.ceres.org/resources/reports/practicing-risk-aware-electricity-regulation>. Accessed April 2015.

F Borggrefe and K Neuhoff, 2011, Balancing and Intraday Market Design: options for wind integration, German Institute for Economic Research, DIW Berlin No 1162.

Business Week, 2014, <http://www.businessweek.com/news/2014-03-04/rwe-reports-first-loss-in-more-than-60-years-on-power-generation>. Accessed April 2015.

J Cochran, M Miller, M Milligan, E Ela, D Arent, and A Bloom, 2013, Market Evolution: Wholesale Electricity Market Design for 21st Century Power Systems, NREL/TP-6A20-57477, October.

DERA, 2004, Act no. 1384 of December 20, 2004 on Energinet Denmark: Unauthorised Translation <http://www.ens.dk/en/node/2210>. Accessed April 2105.

Europa, 2013, Quarterly Update of European Electricity Markets, <http://ec.europa.eu/energy/observatory/electricity/doc/20130611_q1_quarterly_report_on_european_electricity_markets.pdf>.

G Federico, 2010, The Spanish gas and electricity sector: regulation, markets and environmental policies, IESE Business School, University of Navarra, <http://www.iese.edu/es/files/IESE_num%205_Energy_web%20final%20version_tcm5-58079.pdf>. Accessed April 2015.

M Gottstein, 2011, Capacity Markets for a Decarbonised Power Sector: Challenges and Lessons. A case study of forward capacity markets, April <http://www.raponline.org/search/site/?q=gottstein%202011>. Accessed April 2015.

M Gottstein and S Skillings, 2012, Beyond Capacity Markets—Delivering Capability Resources to Europe's Decarbonised Power System, <http://www.raponline.org/document/download/id/4854>. Accessed April 2015.

M Hogan, 2012, What Lies Beyond Capacity Markets? Delivering Least Cost Reliability Under the New Resource Paradigm, regulatory Assistance Project, <http://www.raponline.org/document/download/id/6041>. Accessed April 2015.

D Hurley, P Peterson, and M Whited, 2013 Demand Response as a Power System Resource, 2013, www.raponline.org/document/download/id/6597.

IEA, 2011, Harnessing Variable Renewables- A Guide to the Balancing Challenge, <http://www.iea.org/publications/freepublications/publication/Harnessing_Variable_Renewables2011.pdf>. Accessed April 2015.

P Joskow, 2008a, Lessons learned from electricity market liberalisation. *Energy Journal*, 29, 2, <http://economics.mit.edu/files/2093>. Accessed April 2015.

P Joskow, 2008b, Capacity payments in imperfect electricity markets: need and design, *Utilities Policy* 16 (2008) pages 159–70.

S Keay Bright, 2013, Capacity Mechanisms for Power System Reliability—why the traditional approach will fail to keep the lights on at least cost and can work at cross-purposes with carbon reduction goals, <http://www.raponline.org/search/site/?q=keay-bright%202013>. Accessed April 2015.

C Linvill, J Shenot, and J Lazar, 2013, Designing Distributed Generation Tariffs Well—fair compensation in a time of transition, A RAP Report; <http://idahocleanenergy.org/wp-content/uploads/2013/01/2013-12-rap-carl-dg-tariffs-6898.pdf>. Accessed April 2015.

M Milligan and B Kirby, 2010, Market Characteristics for Efficient Integration of Variable Generation in the Western Interconnection, A NREL Report; <http://www.nrel.gov/docs/fy10osti/48192.pdf>. Accessed April 2015.

M Milligan, 2010, Wind Power and the Power System—wind power myths debunked, Webinar Presentation, NREL, <http://www.uwig.org/awea_webinar_with_milligan-power_and_energy-01-2010.pdf>. Accessed April 2015.

National Grid, 2014, Mandatory Frequency Response, <http://www2.nationalgrid.com/uk/services/balancing-services/frequency-response/mandatory-frequency-response/>. Accessed April 2015.

Ofgem, 2009, Letter to Alison Kay, National Grid: Managing constraints on the GB Transmission System; <https://www.ofgem.gov.uk/ofgem-publications/52935/20090217managing-constraints.pdf>. Accessed April 2015.

C Perez, 2013, Unleashing a golden age after the financial collapse: Drawing lessons from history, *Environmental Innovation and Societal Transitions* 6: 9–23.

REN21, 2013, Renewables 2013- Global Status Report, <http://www.ren21.net/portals/0/documents/resources/gsr/2013/gsr2013_lowres.pdf>. Accessed April 2015.

J Riesz, J Gilmore, and M Hindsberger, 2013, Market Design for the Integration of Variable Generation, Chapter 25 in F. Sioshansi (ed.), 2013, *Evolution of Global Electricity Markets: New Paradigms, New Challenges, New Approaches*, Academic Press.

A Sayer, 2013, Power, Sustainability and Well Being—An Outsider's View, in E. Shove and N. Spurling (eds) 2013, *Sustainable Practices—Social Theory and Climate Change*, Routledge.

F Sioshansi (ed.), 2008a, *Competitive Electricity Markets: Design, Implementation and Performance*, Elsevier (referred to as CEM).

F Sioshansi, 2008b, Competitive electricity markets: questions remain about design, implementation, performance, *The Electricity Journal*, 21: 74–87.

F Sioshansi (ed.), 2013, *Evolution of Global Electricity Markets: new Paradigms, New Challenges, New Approaches*, Academic Press (referred to as EGEM).

UK Parliament, 2013, Energy Act 2013, <http://services.parliament.uk/bills/2013-14/energy.html>. Accessed April 2015.

Utility Week, 2013, Anaylsis: The media lays into 'rip-off' wind <http://www.utilityweek.co.uk/news/Analysis-The-media-lays-into-rip-off-wind/911522#.VR0syvk7vLQ>. Accessed April 2015.

Verbruggen, 2008, Nuclear power and renewable energy: A common future? *Energy Policy* 36, 11, 4046–7, <http://www.sciencedirect.com/science/article/pii/S0301421508003030>. Accessed April 2015.

G Walker, 2013, Inequality, Sustainability and Capability—Locating Justice in Social Practice, in E. Shove and N. Spurling (eds) 2013, *Sustainable Practices—Social Theory and Climate Change*, Routledge.

C Weber, 2010, Adequate intraday market design to enable the integration of wind energy into the European power systems, *Energy Policy*, 38, 7: 3153–63.

Part III
Global Energy Futures

24 Global scenarios of greenhouse gas emissions reduction

Christophe McGlade, Olivier Dessens,
Gabrial Anandarajah, and Paul Ekins

24.1 Introduction

In March 1994, the UNFCCC entered into force and recognized the necessity of preventing dangerous anthropogenic interaction with the climate system via increases in greenhouse gases (article 2). Defining the 'dangerous' level is a value judgement; however, a combination of insights from climate science and precautionary policy has since concluded that the warming should be limited to below 2 °C compared with pre-industrial times. The 2 °C target was subsequently explicitly included in the Copenhagen Accord of 2009 (UNFCCC, 2009) and is still included in the text adopted by the UNFCCC (2012) in Durban. To limit the long-term temperature rise to this level, deep cuts in global greenhouse gas emissions are needed.

The mitigation commitments (or 'Copenhagen pledges') made to date are inconsistent with achieving a 50:50 chance of limiting the surface temperature rise to this 2 °C objective (UNEP, 2010). Although many authors argue that with further ambitious global policies the target could still be reachable (Hare et al., 2010; den Elzen et al., 2011; Rogelj et al., 2011), others suggest it is too late, due to the momentum of energy demand growth in the current fossil-fuel based energy system (Joshi et al., 2011; New et al., 2011).

Part of the reason for this dichotomy of views is that the complexity of the climate system, and the extent of uncertainties embedded within it, gives rise to a wide variety of possible emission trajectories that may be consistent with a 2 °C temperature rise. The additional uncertainties and complexities with modelling the global energy system lead to an even wider range of views on whether the necessary cuts in emissions are possible and, if so, how they may be achieved.

In this chapter we present a new study examining the feasibility of limiting the average global temperature rise to 2 °C, which has been undertaken using the latest version of the TIAM-UCL global integrated assessment model.

Following a description of TIAM-UCL in Section 24.2, we analyse four specific areas. First, given the ongoing delay in implementing a global agreement on greenhouse gas

(GHG) emissions reduction, we examine the window of opportunity for continuing to increase emissions while still limiting the average surface temperature rise (Section 24.3). Second, we construct a number of emissions mitigation scenarios to explore the specific resource and technology needs implied by each (Section 24.4). We focus in particular on the dynamics of coal-to-gas switching (Section 24.4.2.1) and on the decarbonization of the electricity sector (Section 24.4.2.2). In Section 24.4.3, we then discuss the differences in mitigation and costs between high-, middle-, and low-income regions (as defined by the World Bank (2013)). Finally, in Section 24.5 we examine how the rates of technology deployment in the mitigation scenarios compare with deployment rates observed historically.

24.2 **TIAM-UCL**

The TIMES Integrated Assessment Model (ETSAP-TIAM) is a linear programming partial equilibrium model developed and maintained by the IEA's Energy Technology Systems Analysis Programme (ETSAP) (Loulou and Labriet, 2007).

ETSAP-TIAM is a technology-rich, bottom-up, whole-system model that computes the energy system that either minimizes total energy system cost or maximizes social welfare under a number of imposed constraints. It models all primary energy sources (oil, gas, coal, nuclear, biomass, and renewables) from resource production through to their conversion, infrastructure requirements, and finally to sectoral end-use.

The new sixteen-region TIAM-UCL model breaks out the UK from the previous Western Europe region in the fifteen-region ETSAP-TIAM model. Resources and costs of primary energy extraction are specified separately within each of these sixteen regions, the names and abbreviations of which are presented in Table 24.1.

Within this work we will examine in particular the differences between mitigation in high-, middle-, and low-income groups of countries. These groups follow the classification given by the World Bank (2013).[1] Since TIAM-UCL has only sixteen regions, the designation of regions within these groups is necessarily simplified: not all countries within one region will belong to a single group. Groups are nevertheless chosen to represent the majority of countries with that region. The group to which each region has been assigned is also given in Table 24.1.

There are a total of forty-three energy service demands (such as billion vehicle kilometres or production of iron and steel) specified in TIAM-UCL, split between transport (with fourteen individual energy service demands), residential (ten), industrial (ten), commercial (eight), and agricultural (one) sectors.

---

[1] We combine the World Bank's low and lower-middle categories into our 'low-income' group, and keep the upper-middle and high categories as our 'middle-income' and 'high-income' groups respectively.

Table 24.1 TIAM-UCL regions, abbreviations, and economic groups to which each have been assigned

Region	Abbreviation	Assigned group
Africa	AFR	Low
Australia	AUS	High
Canada	CAN	High
Central and South America	CSA	Middle
China	CHI	Middle
Eastern Europe	EEU	High
Former Soviet Union	FSU	Middle
India	IND	Low
Japan	JAP	High
Middle East	MEA	Middle
Mexico	MEX	Middle
Other Developing Asia	ODA	Low
South Korea	SKO	High
United Kingdom	UK	High
United States	USA	High
Western Europe	WEU	High

Regional energy service demands are initially calibrated to 2005 levels. Future demand is then projected based on an assumed relationship of each energy service demand with exogenously specified rates of regional GDP growth, regional population growth, household sizes, or, for industry sub-sectors, sectoral contributions to GDP. For example, personal transport use (in billion vehicle kilometres) is assumed to grow as a function of the growth in GDP/capita.

For all the scenarios run in this work, a 'reference' or 'base case' is first formed that incorporates no GHG emissions abatement requirements. This base case uses the standard version of the model that relies upon minimizing the discounted system cost: this is used to generate base 'supply prices' for each energy service demand in the model. These supply prices are the marginal costs, including commodity and technology costs, of meeting the demands. For the GHG abatement scenarios, constraints are introduced and the model then re-run using the elastic-demand version. This version of the model maximizes social welfare (the sum of consumer and producer surplus) and allows the energy service demands to respond to changes in the supply prices that result from these new constraints. All scenarios in this work are run with perfect foresight.

The climate module in TIAM-UCL contains equations that model the concentrations of three different greenhouse gases (GHGs) with strong global warming potential: CO_2, CH_4, and N_2O. The module tracks the accumulation of anthropogenic emissions in the atmosphere and calculates the change in radiative forcing resulting from the modification of atmospheric concentrations of these GHGs. Finally, on a global scale, equations calculate the realized temperature change resulting from the change in radiative forcing. In this work, the climate module is used to constrain the model to certain bounds on temperature rise (along with other constraints as discussed in Section 24.3). TIAM-UCL is calibrated to the MAGICC climate model (Meinshausen et al., 2011). MAGICC provides the 60 per cent probability of long-term stabilization at average surface temperature rises. The results reported by, or constraints imposed using, the TIAM-UCL climate module will thus also correspond to a 60 per cent probability of being achieved.

The base year of TIAM-UCL is 2005, the model is run in full to 2100, and thereafter the climate module is run to 2200. Most of the results in this chapter are presented only to 2050 (and are reported in five-year increments), but running the model for longer periods than reported in results helps to mitigate many of the near and end-term effects associated with finite modelling horizons.

Two other key assumptions in TIAM-UCL that are relevant for this analysis are the availability of carbon capture and storage (CCS), and the carbon intensity of biomass resources.

In all scenarios, CCS can first be applied to electricity and industrial technologies from 2025. Assumptions are optimistic on the rate at which it can be deployed: in 2025 in each region CCS can be applied to a maximum of 15 per cent of total electricity generation, while in the industrial sector it can capture between 10–20 per cent of process emissions and emissions from generating process heat (depending on the specific technology and sector). After 2025 all CCS technologies can grow at a maximum rate of between 10–15 per cent per year. Some maximum levels of CCS penetration are, however, applied in certain sectors. For example, a maximum of 54 per cent of emissions can be captured from process heat technologies in the iron and steel industries in each region (it is assumed that CCS, which has a 90 per cent capture rate, can be applied to 60 per cent of total emissions).

The maximum availability of bio-energy (solid biomass, bio-crops, municipal waste, and industrial waste) grows steadily over the model horizon from 56 EJ/year in 2005 to 136 EJ/year in 2050. When solid biomass or bio-crops are used to generate heat or electricity, it is assumed that 3.5 kg CO_2 are emitted for every GJ used (around 5 per cent of their embodied CO_2).

A more detailed explanation of input assumptions, approaches, and data sources can be found in McGlade and Ekins (2014), Anandarajah and McGlade (2012), and Anandarajah et al. (2011).

24.3 **Windows of opportunity for mitigating temperature rise**

If there are no explicit constraints placed on GHG emissions in TIAM-UCL, the average global surface temperature increases by 4.2 °C in 2100 compared with pre-industrial times. Annual GHG emissions reach 86 GtCO$_2$-eq in 2100 (up from around 45 Gt in

2010), and cumulative greenhouse gases emissions between 2010–2100 total 6,650 $GtCO_2$-eq.

As noted in the introduction, with the continuing delay in implementing a global agreement on GHG emission reduction, TIAM-UCL can be used to examine the window of opportunity for continuing to increase emissions along this 'high' emissions pathway while still limiting the average surface temperature rise.

However, before doing so, it is important to recognize that, as discussed above, the results from TIAM-UCL are based on finding the cost-minimal (or surplus maximizing) energy system that satisfies the energy-service demands subject to the imposed constraints. The levels of temperature rise to which we constrain the model in this work are usually binding, and so the model often produces solutions with strong and rapid reductions in global GHG emissions.

While these reductions are technically feasible, and economically desirable, rates of emissions reduction are not only dependent on economic or technological criteria. In reality political decisions and social acceptance will heavily influence the emissions pathways and rates of emission reduction that can be achieved. It is therefore important to examine what rates of emissions reduction could be practically achieved on a global level.

While there is little empirical evidence on this, some of the available literature base has suggested certain maximum possible global reduction rates that could be achieved. den Elzen et al. (2011) for example suggest maximum annual rates of 3.5 per cent, while Davis et al. (2010) provide a figure of 4.3 per cent/year. While these take into account assumptions about technological development, economic costs, and socio-political factors, such estimates are not particularly well founded. Therefore in this work as well as imposing constraints on temperature rise, we also study the implications of imposing different maximum rates of emissions reduction.

Figure 24.1 thus presents a variety of scenarios that have been run using TIAM-UCL with different constraints placed on GHG emissions levels and rates of emissions reduction. We first restrict the average global temperature rise to the levels shown (2 °C, 2.5 °C, or 3 °C), and secondly permit a maximum annual percentage reduction in global emissions (1, 2, 3, 4 per cent/year) or impose no constraint (i.e. allow the model to decarbonize the energy system at the rate it calculates to be optimal, however large, to reach the temperature targets). Finally, we fix the model to the 'reference' scenario (the scenario that requires no emissions reduction) for different periods of time before permitting the model to run freely, subject to the other two GHG constraints.

Since no emissions reduction is required up to a certain period, but the model must subsequently limit the temperature rise to different targets, global emissions will therefore be at a maximum level (i.e. peak) in that period. The combination of these constraints therefore allows us to investigate the relationship between the year in which global emissions can peak (the x-axis of Figure 24.1), the temperature rise (y-axis), and the slowest possible rate of annual post-peak global emissions decline (the circles of different shades).

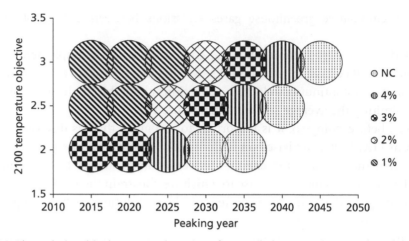

Figure 24.1 The relationship between the rate of annual decrease in emissions for a specific peaking year (*x*-axis), level of temperature rise (*y*-axis), and the maximum rate of post-peak emissions reduction

Note: 'NC' stands for no constraint (i.e. emissions can decline at any rate required to meet the temperature constraint).

Figure 24.1 shows, for example, that a 3 °C temperature rise can be achieved with emissions peaking delayed to 2025, with emissions needing to fall no faster than 1 per cent p.a. thereafter. If peaking is delayed to 2030, 2035, and 2040, then emissions need to be able to fall by at least 2 per cent, 3 per cent, and 4 per cent p.a. respectively. It is apparent from Figure 24.1 that allowing a greater temperature rise or a more rapid post-peak reduction in annual emissions affords more time before global emissions must peak.

It can also be seen that, regardless of the peaking year, the 2 °C target is now only achievable if annual CO_2 emissions are permitted to fall by at least 3 per cent per year. In this case the peak in global emissions must occur between 2015 and 2020. With a 4 per cent permitted annual reduction in global emissions, this peak is pushed back to 2025, and only if there is no restriction on the rate of reduction in annual emissions can the peak be extended to 2035. In this case emissions must fall by 6 to 10 per cent every year after the peak year. The emissions and energy system lock-in mean that it is not possible for emissions to peak after 2035 and still restrict the temperature rise to 2 °C.

If the maximum post-peak reduction in emissions is reduced to 1 per cent or 2 per cent, it is not possible to keep the temperature rise to below 2 °C. With a maximum permitted 1 per cent fall in post-peak emissions, a 3 °C limit to temperature rise will only be achieved if emissions peak before 2025 (as already noted), or before 2020 if there is to be only a 2.5 °C rise. Figure 24.1 also shows that a permitted 2 per cent p.a. emissions reduction tends to delay the latest peaking year by five years compared with a 1 per cent fall.

The Copenhagen Accord (UNFCCC, 2009) stipulates only that the long-term[2] warming must remain below 2 °C above the preindustrial mean, without regard to exceeding

---

[2] We have interpreted 'long-term' here to mean 2100.

Table 24.2 Changes in the latest year in which global emissions can peak for different annual post-peak emissions reductions if overshooting of 2 °C is permitted or not

Post-peak emissions reduction	1%	2%	3%	4%	5%	NC
Overshoot permitted	NA	NA	2020	2025	2030	2035
No overshoot	NA	NA	2015	2020	2020	2025

this level in the short to medium term. Therefore, although the above results require the temperature rise not to exceed the level shown in 2100, they allow an 'overshoot' of this level within the model timeframe. For example, as shown in Figure 24.2, the temperature rise reaches 2.25 °C in 2060 in the most delayed action 2 °C scenario (with emissions peaking in 2035).

To investigate the implications of not permitting any overshoot, we can re-run the 2 °C scenarios from Figure 24.1 but forbid the temperature rise to exceed 2 °C in any time period. The differences between these sets of scenarios are presented in Table 24.2. With no overshoot, it is impossible to keep the temperature rise below 2 °C if emissions peak after 2025, even when there is no constraint placed on maximum rates of post-peak emission reduction, whereas with a permitted overshoot, this situation is delayed by ten years. Allowing an overshoot therefore permits greater flexibility with regards to achieving the long-term goal of the Copenhagen Accord. The climate risks of overshooting the target, and therefore whether it should be allowed, should be a key consideration in discussions on global emissions mitigation pathways.

Regardless of whether an overshoot is permitted or not, these results suggest that, while it may still be possible to limit the average global temperature rise to the 2 °C target, the window of opportunity for realizing this is closing rapidly. Further, as will be discussed in more detail below, the required transition of the energy system that is implied in *all* of the 2 °C scenarios would be extremely challenging to realize.

24.3.1 SCENARIOS IMPLEMENTED IN THIS WORK

In the following sections we concentrate our study on only a subset of the above scenarios. These are used to describe and highlight the key changes in the energy system under different levels of CO_2 emissions mitigation. The first case is simply the reference scenario introduced above that has no emissions reduction requirements. As noted above, this reaches a 4.2 °C temperature rise in 2100, with the temperature continuing to increase thereafter. The two other scenarios are constrained to achieve around a 60 per cent chance of a 2 °C and 3 °C rise in the average surface temperature. These temperature rise constraints must not be exceeded in any time period out to 2200, that is they do not allow a temperature overshoot in any year.

In contrast to the scenarios included within Figure 24.1, it is assumed in these 2 °C and 3 °C scenarios that prior to 2020 only the individual countries' emissions reductions that were pledged as part of the Copenhagen Accord are realized. After 2020, the model is free to mitigate emissions at any decline rate, but must ensure that emissions are reduced sufficiently to satisfy the temperature constraint.

These mitigation scenarios are not designed to be forecasts. Rather, they have been constructed to simulate possible emissions pathways that are commensurate with the pledges made to date, and either with the current nominal political goals (in the 2 °C scenario),[3] or in the circumstance that emissions reduction pledges are not sufficiently ambitious (in the 3 °C scenario).

The key assumptions in TIAM-UCL concerning the availability of CCS and bio-energy were discussed above, but it is evident that the potential availability and effectiveness of these technologies is currently very uncertain. In the runs described above, bio-energy and CCS can be used in conjunction, for example in the power sector or to produce hydrogen or biofuels. This results in net-negative emissions. Such negative-emission technologies compound the uncertainties surrounding CCS and bio-energy. We therefore also run a sensitivity scenario that uses the same emissions constraints as in the 2 °C scenario but in which these negative emission technologies are not allowed. The four scenarios therefore implemented are:

(i) reference case (REF)—no GHG emissions reductions required in any period;
(ii) 3 °C scenario (3DS)—up to 2020 emissions reductions follow the Copenhagen Accord pledges and subsequently there must be around a 60 per cent chance of keeping the average temperature rise below 3 °C in all time periods to 2200;

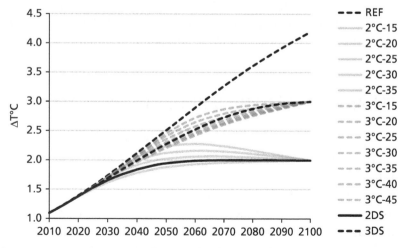

Figure 24.2 Temperature changes in the scenarios from Figure 24.1 and in the three temperature scenarios implemented

---

[3] These two overarching goals are (i) for a global agreement to come in force from 2020 that (ii) will limit the surface temperature rise to 2 °C.

(iii) 2 °C scenario (2DS)—up to 2020 emissions reductions follow the Copenhagen Accord pledges and subsequently there must be around a 60 per cent chance of average temperature rise below 2 °C in all time periods to 2200; and

(iv) 2 °C scenario (2DS-nobioCCS)—identical emission reductions to 2DS, but with no negative emissions technologies (bio-energy with CCS) allowed.

Figure 24.2 presents the long-term changes in temperature over the model timeframe from these scenarios, and places these into context with the scenarios from Figure 24.1 above (the temperature pathways in 2DS and 2DS-nobioCCS are identical).

24.4 **Scenario results**

24.4.1 COMPARISON WITH OTHER INTEGRATED ASSESSMENT MODELS

To provide context for our three scenarios, it is useful first to examine the outlooks provided by other energy systems models. Since fossil fuels still comprise 85 per cent of total primary energy supply, we focus here on the cumulative production levels of coal, gas, and oil projected by a wide variety of models (van Vuuren et al., 2011; Thomson et al., 2011; Masui et al., 2011; Riahi et al., 2011; IEA, 2012; Shell, 2013; Morita, 1999; IIASA 2009, 2012; World Energy Council, 2013) under different scenarios over the timeframe 2010–50.

Since these scenarios have very different future GHG emissions trajectories, we classify each into one of three groups according to their approximate projected 2100 temperature rise. For a similar exercise the Intergovernmental Panel on Climate Change (IPCC) identifies six 'categories of scenarios' (Metz et al., 2007) that cover the full range of potential GHG emissions futures. These six categories are principally classified according to their radiative forcing levels, but the IPCC also uses MAGICC to provide approximate ranges of temperature rise in 2100 within each. To aid comparability, we have therefore chosen our three groups to each contain approximately two of the IPCC's categories. The scenarios are thus grouped into those that provide a 60 per cent chance of limiting the 2100 temperature rise to: (a) equal to or below 2 °C, (b) between 2 °C and 3 °C, and (c) above 3 °C. There are twenty-six individual scenarios in each of the first two of these groups, and thirty-four in the third. The three scenarios run as part of this work also broadly fit within each of these groupings.

Production of coal, gas, and oil from each of these groups is displayed in Figure 24.3. Black lines indicate the median value, boxes the interquartile range, and dashed lines span the maximum and minimum values.

In the grouping with the lowest temperature rise, the median cumulative production of coal between 2010 and 2050 is 3,600 EJ, with all scenarios agreeing that no more than 5,500 EJ are produced. For context, coal reserves total just over 20,000 EJ (Federal

Institute for Geosciences and Natural Resources (BGR) 2012), and so within this bracket of temperature rise mitigation all sources agree that at least 70 per cent of current reserves are not required to be produced prior to 2050.

Production of oil and gas in this grouping are approximately similar at a median of just over 6,000 EJ, around 70 per cent greater than the utilization of coal. Again for context, reserves of gas and oil are 7,100 EJ and 7,400 EJ respectively (McGlade, 2013; McGlade and Ekins, 2014).

Cumulative production of coal increases significantly in the two higher temperature brackets, reaching a median of 8,000 EJ in the scenarios that lead to a greater than 3 °C temperature rise by the end of the century. Therefore a median of just over 60 per cent of current coal reserves remain unused even in this highest temperature grouping. The increase in gas and oil production is somewhat more muted: median cumulative production of gas in the two higher temperature bands is similar at around 7,700 EJ, and production of oil 8,000 EJ and 8,800 EJ in each. These scenarios thus suggest that cumulative production of oil will likely exceed gas and coal under all temperature rises.

The black triangles in Figure 24.3 are the cumulative production values from the 2DS, 3DS, and REF scenarios provided by TIAM-UCL. In the 2 °C bracket, for example, cumulative coal production is 3,200 EJ, suggesting that 85 per cent of the existing coal reserves are not produced prior to 2050. The TIAM-UCL scenarios can all be seen to lie within the range of the estimates provided by other scenarios, and are generally within the inter-quartile range.

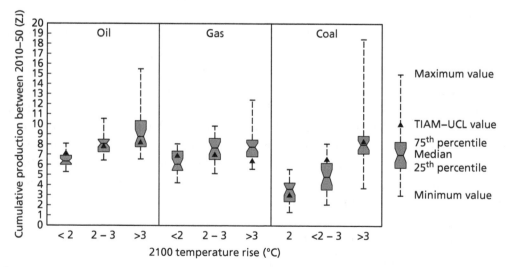

Figure 24.3 Cumulative production of fossil fuels in the three emissions mitigation scenarios generated in this work and comparison with scenarios generated by a variety of other modelling groups

Notes: There are twenty-six individual scenarios in the first two of these groups, and thirty-four in the third. The three scenarios run in this work are indicated by black triangles.

Sources: (van Vuuren et al., 2011; Thomson et al., 2011; Masui et al., 2011; Riahi et al., 2011; IEA, 2012; Shell, 2013; Morita, 1999; IIASA 2009, 2012; World Energy Council 2013; this work).

24.4.2 RESOURCE AND TECHNOLOGY NEEDS

This section now seeks to investigate in more detail some of the energy system changes that occur between the four scenarios (including the 2 °C scenario that does not allow bio-energy to be used with CCS). There are a wide variety of results that could be explored, however to narrow the focus we concentrate on two specific factors, namely the level of coal-to-gas switching, and electricity system decarbonization.

24.4.2.1 Coal to Gas Switching

Figure 24.4 presents primary energy production in each of the four scenarios between 2010 and 2050. In the two 2 °C scenarios, primary energy production shifts significantly: from comprising 85 per cent of primary energy in 2010, fossil fuels' share drops to below 60 per cent by 2050. The significant reduction in coal production in 2025 in the two low-carbon scenarios (2DS and 2DS-nobioCCS), and the difference in coal production between these and the higher-carbon scenarios (3DS and REF), are the most obvious changes in Figure 24.4.

In contrast, gas consumption up to 2035 in 2DS is greater than in either REF or 3DS (Figure 24.5). Indeed, consumption is around 25 per cent greater than REF throughout

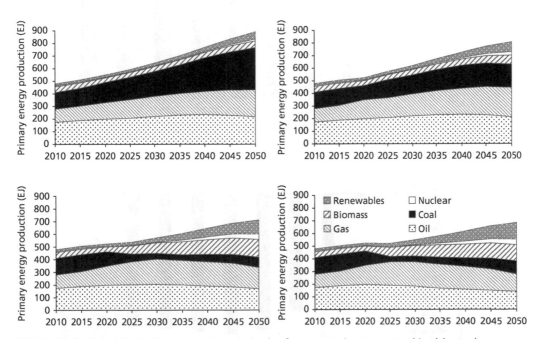

Figure 24.4 The global primary energy mix in the four scenarios generated in this work

Note: Clockwise from top left: REF, 3DS, 2DS-nobioCCS, 2DS.

Primary energy equivalent of renewable technologies and nuclear is based on the electricity generated from these sources that is final energy is equal to primary energy.

the 2020s. This difference reduces after 2030, as consumption continues to climb in the reference case while starting to decline slightly in the 2 °C scenarios. Consumption is consequently lower by 2040. The difference in gas consumption between 3DS and REF is less significant in early periods but overall gas consumption in 3DS is close to 0.5 Tcm/ year greater up to 2050.

The greater levels of gas consumption in the low-carbon scenarios suggests that on a global level, gas may be able to play an important role as a bridging fuel on a medium timescale (2015 to 2035). There are a number of important caveats to this interpretation, however.

First, the difference between 2DS and 2DS-nobioCCS highlights the importance of CCS to the role of gas. Consumption falls much more rapidly in 2DS-nobioCCS than in 2DS, a pattern that is observed in all regions.

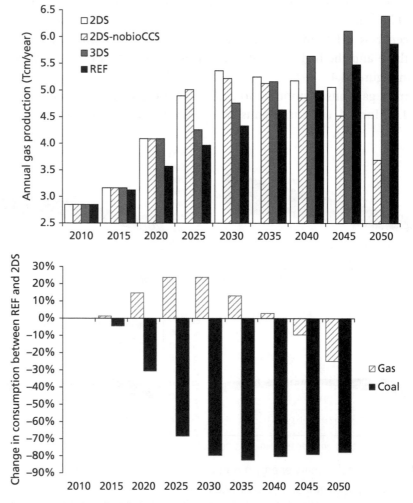

Figure 24.5 Gas production in the four scenarios (top) and changes in gas and coal production between 2DS and REF over time (bottom)

Second, as was shown in Figure 24.4, global consumption of coal must be significantly reduced in a 2 °C scenario in all time periods. This can be observed more clearly in Figure 24.5, which places the percentage change in coal consumption alongside the percentage change in gas consumption between REF and 2DS. It is evident that coal production must be severely curtailed in all periods. It would therefore perhaps be more appropriate to state that the increase in gas consumption between 2015 and 2040 helps to fill some of the gap left by the drop in coal production, rather than that gas is 'displacing' coal over this period. After 2040 gas consumption is also lower in 2DS, as electricity production by renewable sources continues to expand rapidly, alongside a slight upturn in coal consumption (used with CCS).

Third, there are significant regional differences in the role that gas can play in decarbonizing the energy system. In the high-income regions, gas consumption remains broadly constant up to 2050 in the reference case, but is much higher (around 35 per cent between 2020 and 2040) in the 3 °C and 2 °C scenarios. The change in consumption between the scenarios in middle- and low-income regions is much more muted: gas increases by only 10 per cent for a period of ten years between the 2 °C and reference cases. It is also important to note that even these regional relationships do not necessarily translate to all countries within the groups.

24.4.2.2 Electricity Sector Decarbonization

Figure 24.6 presents the share of total emissions from each sector over the model time-frame in 2DS and 3DS. This shows emissions from the electricity sector dropping extremely rapidly in 2DS in the 2020s, so that they are almost zero by 2030. Electricity-sector emissions also fall by the largest degree compared with all other sectors in 3DS, so that even though the sector is not GHG-neutral by 2050, its contribution to global emissions has more than halved from 2010 levels. It is evident that it is most cost-effective to decarbonize the energy system by first rapidly decarbonizing the electricity sector; this applies within all the regional income groupings and indeed in all of the regions in Table 24.1 above. Comparing Figure 24.6 with Figure 24.1 indicates that, for 3DS, the peaking of emissions in 2045 implies that emissions fall very rapidly (i.e. without constraint) thereafter.

Figure 24.7 presents global electricity generation in the four scenarios and the GHG intensity of electricity in each. As noted above, in 2DS electricity is on average GHG-negative globally after 2030. GHG-negative electricity is the most cost-effective manner by which to decarbonize many end-use sectors and consequently electricity production is noticeably higher. When bio-energy cannot be used with CCS (2DS-nobioCCS), it is not possible to achieve GHG-negative electricity; however electricity production is around 15 per cent higher than in 2DS after 2035. This additional electricity is predominantly used in the industrial and residential sectors, in both cases to aid the decarbonization of heat (high temperature process heat in the industrial sector and via heat pumps in the residential sector).

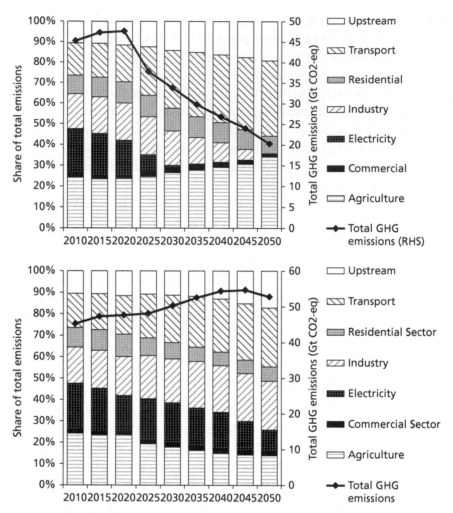

Figure 24.6 Contribution of each sector to global emissions in 2DS (top) and 3DS (bottom)

Total electricity production levels are approximately similar in REF and 3DS. Nevertheless, the GHG intensity of electricity is considerably lower in 3DS, particularly in later periods. The decarbonization of electricity, along with the coal-to-gas switching discussed above, is generally sufficient to meet the emissions reductions required to achieve a 3 °C temperature rise without the need for much additional electrification of end-use sectors. There is a further discussion of the deployment rates of electricity generation technologies in Section 24.5.

24.4.3 **Where The Mitigation Occurs**

This section explores regional differences in per capita emissions and costs of mitigation in the different temperature scenarios. It is worth emphasizing that, as with all results

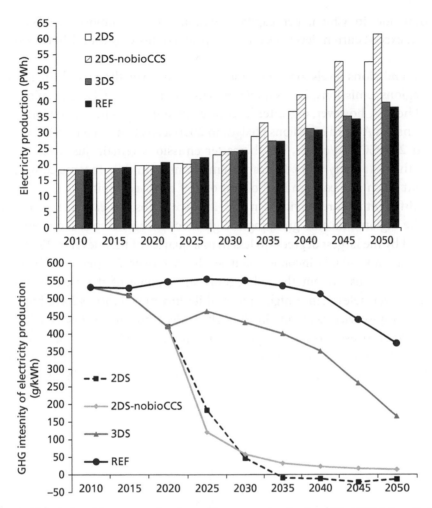

Figure 24.7 Global electricity generation in the four scenarios (top) and its GHG intensity (bottom)

presented, these are based upon the global surplus-maximizing solution and not upon exogenously apportioning emissions reduction rates or levels to different regions.

24.4.3.1 PER CAPITA EMISSIONS

Figure 24.8 presents changes in regional per capita emissions over time in the four scenarios. With the large increase in coal consumption shown previously in Figure 24.4, it can be seen that in REF average per capita emissions in high-income regions rise to almost 20 tCO_2-eq/capita—close to the current level in the United States and Australia. Per capita emissions in low-income nations stay much more constant over the model timeframe, remaining under 4 tCO_2-eq/capita in all periods, while emissions in middle-income economies rise to the largest extent. Driven primarily by coal production in China, per capita emissions rise to almost 12 tCO_2-eq/capita, and indeed exceed current levels seen in European countries (around 10 tCO_2-eq/capita) by 2035.

In 3DS, emissions levels are much more constant over the model timeframe. Low-income regions' emissions, for example, remain at around 3 tCO_2-eq/capita in all time periods. There is, however, a greater level of change in high-income regions. Satisfying the Copenhagen Accord emissions pledges in 2020 leads to a reduction in average levels of around 2.5 tCO_2-eq/capita, but thereafter emissions remain just over 12 tCO_2-eq/capita for the next thirty years. Per capita emissions in middle-income countries do rise slightly, but reach a maximum level of just over 9 tCO_2-eq/capita.

A very different picture is evident in both 2 °C scenarios. Per capita emissions fall in all three of the regional income groupings. High-income regions' per capita emissions fall steadily and by the largest degree—to less than a quarter of their level in 2010—while the drop in low and middle-income countries is just over 50 per cent between 2010 and 2050. It is thus evident that all regions share some of the burden of emissions reduction. Nevertheless, since high- and middle-income countries have higher relative levels of energy-service demands in difficult-to-decarbonize sectors (such as aviation and industrial processes respectively), they maintain a higher level of emissions. This means that while there is much less disparity in per capita emissions than at present, the 2050 value in low-income nations is less than half the values in middle- and high-income regions.

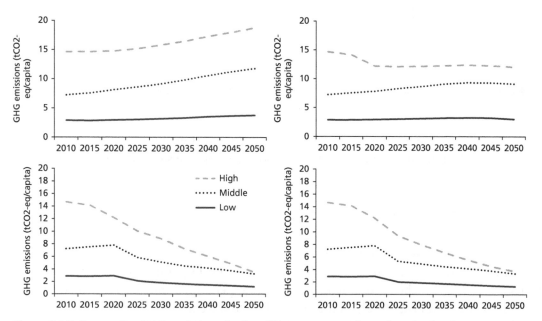

Figure 24.8 Per capita GHG emissions in the different economic regions in the four scenarios implemented in this work

Note: Clockwise from top left: REF, 3DS, 2DS-nobioCCS, 2DS.

24.4.3.2 ECONOMIC IMPLICATIONS

We can also examine some of the economic implications of these global decarbonization pathways. We focus on the CO_2 prices generated endogenously in each of the mitigation scenarios (Figure 24.9), and the changes in total annual energy system costs between these and REF (Figure 24.10).

As mentioned in Section 24.3.1, the energy service demands in the mitigation scenarios can respond to changes in the supply price of meeting these demands that result from the imposed emissions reduction requirements. The total annual energy system costs presented here therefore include both the capital and operating costs of the energy system itself and the costs of this demand reduction (Kesicki and Anandarajah, 2011).

The top of Figure 24.9 shows the annual CO_2 prices generated under the emissions mitigation scenarios. We assume that full trading of emissions credits is possible from 2025 onwards and so there is an identical CO_2 price in all regions. Prior to this there is a much wider regional distribution, and so for illustrative purposes in Figure 24.9 we show the CO_2 price generated for China.

In 2DS, the CO_2 price reaches over a \$100/tonne[4] in 2025 and increases at approximately 4 per cent annually between 2025 and 2050. When bio-CCS is not allowed, the resulting CO_2 price more than doubles, reaching over \$600/tonne by 2050. In contrast, the CO_2 price in 3DS is considerably lower: following a peak in 2020, caused by the need to meet the imposed Copenhagen Accord emissions reductions, the CO_2 price remains below \$30/tonne in all time periods.

To give an idea of how variable these CO_2 prices are to assumptions on the maximum rates of emissions reduction and the year in which emissions peak, the bottom of Figure 24.9 presents box and whisker diagrams of the 2050 CO_2 prices generated in all of the 2 °C, 2.5 °C, and 3 °C scenarios that were included in Figure 24.1. Higher CO_2 prices in 2050 are generated the longer the delay in implementing global emissions reduction (i.e. the later the date on which global emissions peak) and the greater the required level of emissions reduction. All of the CO_2 prices in the 3 °C scenarios remain just below \$100/tonne, a level below the lowest price in any of the 2 °C scenarios. There is nevertheless a very wide spread in the possible 2050 CO_2 prices that are generated in the 2 °C scenarios, ranging from \$120/tonne to \$650/tonne.

Figure 24.10 presents the percentage change in total annual energy system costs between 2DS and REF. On a global level, the overall energy system is smaller in 2DS than in REF: this is because the supply prices of meeting the energy service demands in 2DS are greater, and so, responding elastically, the energy service demands are reduced. The total annual energy system costs, which as noted above include the costs of this demand reduction, are nevertheless greater.

---

[4] All prices and costs here are presented in real terms in 2005 US\$.

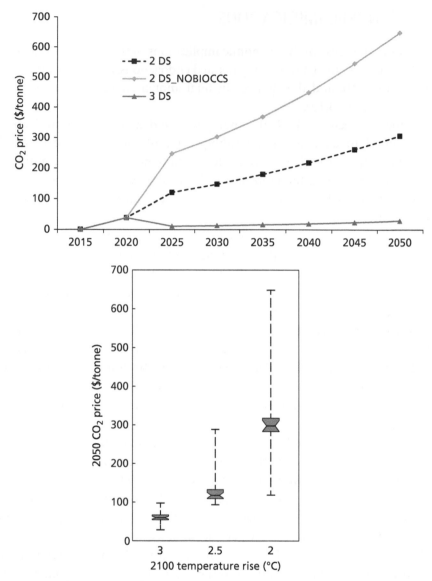

Figure 24.9 CO_2 prices generated in 3DS, 2DS, and 2DS-nobioCCS (top), and ranges of the 2050 CO_2 prices in the 2 °C, 2.5 °C, and 3 °C scenarios presented in Figure 24.1 (bottom)

There is some variation in total energy system costs at a regional level, however. In 2020, overall costs are higher in high-income regions since these are required to meet their Copenhagen Accord emissions reductions.

Fossil fuel consumption in higher-income regions consequently falls (as seen previously), and there is resultant downwards pressure on fossil fuel prices. Cheaper resources are therefore available to the middle- and low-income countries and so there is almost no additional cost to these regions in 2020. While marginal, the change in cost is still positive, however. The assumption of perfect foresight means that some of the

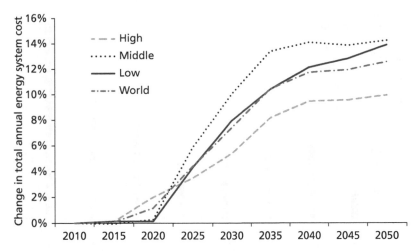

Figure 24.10 Percentage increase in total annual energy system cost in 2DS compared with REF in the different economic regions

middle- and low-income regions do implement some emissions reduction (albeit at a much lower level of ambition than in high-income regions).

This switches after 2020, with a greater proportional cost in middle- and low-income regions. For middle-income regions, the rapid increases in energy-service demand in REF are predominantly met through an increase in coal consumption. Given that coal consumption needs to be severely restricted in 2DS until CCS is fully available, these regions require a greater level of investment to meet the emissions reductions required. In low-income regions it is assumed that the unit capital costs of many of the low-carbon technologies (such as CCS) are higher than in the high-income regions. These are consequently more expensive to deploy and overall costs are higher.

In summary, despite the fact that high-income regions are responsible for a majority of the cumulative emissions emitted since industrialization, mitigation costs are proportionately higher for middle and low-income regions. Nevertheless, given the absolute increase in coal consumption in middle and low-income regions in REF, these regions are responsible for more than two thirds of GHG emissions in 2050. It is therefore again evident that any attempt to keep the surface temperature below $2\,°C$ requires a truly global effort.

24.5 **Are technology development rates reasonable?**

In this section we seek to interrogate further some of the specific rates of technology deployment that are implied by TIAM-UCL. As shown previously in Figure 24.7, the increase in electricity production, particularly in the low-carbon scenarios after 2030, is extremely rapid: generation in 2DS grows at a compound average of 4.2 per cent/year between 2030 and 2050, and at 4.8 per cent/year between these years in 2DS-nobioCCS. Since the GHG intensity of this electricity is very low (and negative in 2DS), we narrow our focus here on the required

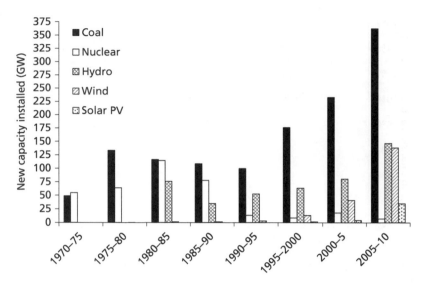

Figure 24.11 Historical rates of installation of new electricity capacity globally for a selection of technologies

Sources: (BP, 2013; EIA, 2011; IEA, 2013).

Note: No data is available for rates of hydro electricity installed prior to 1980. Coal capacity prior to 2005 has been derived using IEA data and the assumption that coal power plant utilization is 65%.

rates of installation of new capacity of the low-carbon electricity technologies, specifically nuclear, solar, CCS (by any fuel source), and wind under the different scenarios.

It is useful to compare these rates with those that have been observed historically. These are presented in Figure 24.11, which includes the new capacities that have been installed globally for a selection of electricity technologies in five-year periods since 1970. Installation rates of new (unabated) coal power have been by far the largest, and indeed have been increasing since 1990. The largest single increase in any five-year period was just over 360 GW between 2005 and 2010. Since 2010, the deployment of new solar-PV and wind capacity has also increased noticeably, with over 90 GW solar and 190 GW wind capacity installed globally between 2007 and 2012.

The global rates of new capacity installation in 2DS are presented in Figure 24.12. It is immediately apparent that these rates are entirely unprecedented. The five-year installation rate of new solar-PV[5] capacity exceeds 1,000 GW from 2035 onwards, and reaches over 1,500 GW between 2045 and 2050. This is over four times the fastest installation rate ever observed historically for a single electricity technology.

Putting this into context, if we assume a land-use efficiency of 150 W/m,[2] 1,500 GW new capacity is equivalent to covering 10,000 km² in high efficiency solar panels. This is approximately equivalent to covering half of Wales with solar panels in five years. It is also important to note that the total global installed capacity in 2050 reaches almost 5,000 GW; this is equivalent to 34,000 km,² or approximately the land area of the Netherlands.

---

[5] There is very little deployment of concentrated solar power (CSP) in these results.

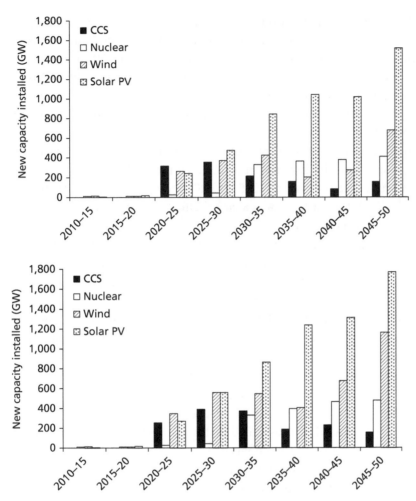

Figure 24.12 Rates of installation of new electricity capacity globally for a selection of technologies in 2DS (top) and 2DS-nobioCCS (bottom)

It should, however, be noted that the total installed capacities of PV are not distributed evenly between the income regions. In 2050, the majority is found in middle (47 per cent) and low-income (38 per cent) countries, many of which have large areas that could be well suited to large solar PV installations.

The installation of CCS technologies shown in Figure 24.12 have biomass, coal, and gas as feedstock fuels. In early periods biomass and gas are preferred, but there is little investment in new gas-CCS plants after 2035, and a greater proportion of coal-CCS capacity is installed. The five-year CCS installation rates from all fuels are broadly in line with the highest rates observed for coal historically, a rate of installation for an as-yet unproven technology that will be very challenging to achieve in practice.

In 2DS-nobioCCS, there is an increased level of gas-CCS installed, but more noticeably much greater installation levels of new wind capacity. There is an average of 500 GW new wind capacity installed in each five-year period between 2025 and 2045, over 65 per cent more than the 300 GW average seen in 2DS over the same period. As noted above, the largest increase in new wind capacity in any five-year period was 190 GW between 2007 and 2012. The rate suggested by TIAM-UCL in 2DS for new wind installation is about 50 per cent greater than that observed recently, but it would need to more than double if bio-energy with CCS is not available.

While these installation rates required in the 2 °C scenarios are therefore extremely rapid compared with historical rates, we can make two additional observations to put these in context.

First, the maximum historical rate of new capacity installed shown in Figure 24.11 largely stemmed from the increase in a single country. China alone installed 295 GW new coal power capacity between 2005 and 2010, over 80 per cent of the 360 GW rise seen globally. Therefore, if there was a global effort to install these levels of low-carbon technologies, as indeed it is assumed there will be in these scenarios, then the required rates do not appear so much greater than historical precedent. In 2DS in China, for example, the highest rate of installation of new capacity of a single low-GHG technology in a five-year period is 295 GW (solar-PV between 2045 and 2050), that is equal to its rate of installation of coal power between 2005 and 2010.

Second, part of the need for the necessary rates of installation of new low-carbon sources of electricity in later periods stems from the absence of high rates of investment in early periods (prior to 2020).

Actual installation rates have, however, been faster between 2005 (the base-year of the model) and 2012 than those given by TIAM-UCL. For example, total installed wind capacity in 2012 is already over 280 GW, whereas in 2DS there is only 110 GW in 2015.

As previously discussed, 2DS assumed that no GHG mitigation was required between 2005 and 2010 and only the Copenhagen Accord emissions reductions were required prior to 2020. Under these levels of emissions mitigation, TIAM-UCL suggests it is more cost-effective to invest in efficiency measures and use less coal (and more gas) than invest in renewable technologies. If the current levels of investment that are being witnessed globally are maintained or continue to increase, by installing greater levels of renewable technologies in earlier periods, the required rates of installation in later periods would be somewhat lower than those suggested in these scenarios.

This does not, however, imply that the investments into low-GHG electricity technologies that have been made to date are sufficient to provide a global emissions pathway commensurate with a 2 °C temperature rise. If TIAM-UCL were required to provide a 2 °C temperature rise but was allowed to mitigate emissions cost-optimally from 2005 rather than from 2020 (as assumed in 2DS), investment into new renewable technologies would exceed those observed in reality.

24.6 **Conclusions**

Given the continuing delay in implementing a global agreement on GHG emission reduction, we initially examined the window of opportunity to continue increasing emissions while still limiting the average surface temperature rise. We found that while it may still be possible to limit the average global temperature rise to the nominal 2 °C target agreed on by policy makers, there is rapidly shrinking opportunity chance to do so.

We next generated four scenarios to examine resource use and the energy system at different levels of emissions mitigation, and first compared these to scenarios generated by a wide range of other modelling groups. In our 2 °C scenario, 85 per cent of existing coal reserves remain unused prior to 2050, similar to the median level suggested by all other models. More generally, it was found that all scenarios produced by other modelling groups that result in a temperature rise of less than or equal to 2 °C in 2100 use less than 30 per cent of existing coal reserves by 2050.

Looking at the level of coal-to-gas switching in our scenarios, we found that on a global level, gas could play an important role during the transition to a low-carbon energy system. A number of important caveats were discussed, however. Existing levels of coal production must fall by a much greater proportion than gas can rise, and also just because gas is useful for a period of time at a global level, it does not follow that increased gas consumption aids decarbonization efforts in all regions and countries—for some countries, unconstrained gas use could substitute for renewables, rather than coal, resulting in an emissions increase.

We next examined the electricity sector and found that it is most cost effective to decarbonize this sector first in all regions globally. This decarbonization effort occurs rapidly so that electricity globally has zero net emissions of GHGs by 2030 in our 2 °C scenarios.

Per capita emissions were found to fall in all regions globally in the 2 °C mitigation scenarios: in high-income regions these drop to less than a quarter of their 2010 level by 2050, while in middle- and low-income regions they fall by 50 per cent. Nevertheless, between a 2 °C scenario and a scenario with no emissions reduction, it was found that middle- and low-income regions have a higher percentage increase in total annual energy system costs (including the costs of demand reduction) than high-income regions. For such an outcome to be politically acceptable to the countries concerned, some proportion of these costs will probably need to be financed at concessionary rates by high-income countries.

We finally examined the implied deployment rates of the low-carbon electricity technologies in our 2 °C scenarios. Both the overall level and rate of installation of new capacity that is required globally was found to be entirely unprecedented. We noted, however, that extremely rapid rates of new electricity capacity installation have been observed historically within individual countries. With a global effort to mitigate emissions we consider that it is not unreasonable to suggest that such a substantial and rapid deployment of low-carbon electricity technologies is feasible, but achieving it would depend on appropriate policy incentives being put in place sooner rather than later.

It is also worth bearing in mind that the global effort required to attempt to limit the temperature rise to 2 °C is in itself entirely unprecedented. While the continuing delay in implementing a global deal on emissions mitigation, and the consequently narrowing window of opportunity to prevent a temperature rise of more than 2 °C, means that such investment on the required scale is perhaps increasingly unlikely, we believe that it remains possible, and indeed the socially prudent and responsible response to the challenge, and threat, of anthropogenic climate change.

■ **REFERENCES**

Anandarajah, G., Pye, S., Usher, W., Kesicki, F., & McGlade, C., 2011. TIAM-UCL Global Model Documentation. London, UK: UCL Energy Institute. Available at: <http://www.ucl.ac.uk/energy-models/models/tiam-ucl/tiam-ucl-manual>. Accessed April 2015.

Anandarajah, G. & McGlade, C.E., 2012. *Modelling Carbon Price Impacts of Global Energy Scenarios*, London, UK: UCL Energy Institute. Available at: <http://www.theccc.org.uk/wp-content/uploads/2012/04/UCL_2012_Modelling_carbon_price_impacts_of_global_energy_scenarios.pdf>. Accessed April 2015.

BP, 2013. *BP Statistical Review of World Energy*, London, UK: BP.

Davis, S.J., Caldeira, K., & Matthews, H.D., 2010. Future CO_2 emissions and climate change from existing energy infrastructure. *Science (New York, N.Y.)*, 329, pp. 1330–3.

EIA, 2011. *International Energy Outlook 2013*, Washington, DC, USA: U.S. Energy Information Administration. Available at: <http://www.eia.gov/forecasts/ieo/pdf/0484(2011).pdf>.

Den Elzen, M.G.J., Hof, A.F., & Roelfsema, M., 2011. The emissions gap between the Copenhagen pledges and the 2°C climate goal: Options for closing and risks that could widen the gap. *Global Environmental Change*, 21, pp. 733–43.

Federal Institute for Geosciences and Natural Resources (BGR), 2012. *Energy Study 2012. Reserves, Resources and Availability of Energy Resources*, Hannover, Germany: BGR. Available at: <http://www.bgr.bund.de/DE/Gemeinsames/Produkte/Downloads/DERA_Rohstoffinformationen/rohstoffinformationen-15e.pdf?__blob5publicationFile&v53>. Accessed April 2015.

Hare, W., Lowe, J. A., Rogelj, J., Sawain, E., & van Vuuren, D., 2010. Twenty-first century temperature projections associated with the pledges. In *The Emissions Gap Report: Are the Copenhagen Accord Pledges Sufficient to Limit Global Warming to 2oC or 1.5oC? A Preliminary Assessment*. Nairobi, Kenya: United Nations Environment Programme, pp. 46–52.

IEA, 2012. *Energy Technology Perspectives 2012: Pathways to a Clean Energy System*, Paris, France: International Energy Agency.

IEA, 2013. *Energy Balances of OECD and Non-OECD Countries*, Paris, France: International Energy Agency. Available at: <http://dx.doi.org/10.5257/iea/ebo/2013>. Accessed April 2015.

IIASA, 2009. GGI Scenario database Ver 2.0. Available at: <http://www.iiasa.ac.at/Research/GGI/DB/. Accessed April 2015.

IIASA, 2012. Scenarios database. *Global Energy Assessment*. Available at: <http://www.iiasa.ac.at/web-apps/ene/geadb/>. Accessed April 2015.

Joshi, M., Hawkins, E., Sutton, R., Lowe, J., & Frame, D., 2011. Projections of when temperature change will exceed 2°C above pre-industrial levels. *Nature Climate Change*, 1(8), pp. 407–12.

Kesicki, F. & Anandarajah, G., 2011. The role of energy-service demand reduction in global climate change mitigation: Combining energy modelling and decomposition analysis. *Energy Policy*, 39, pp. 7224–33.

Loulou, R. & Labriet, M., 2007. ETSAP-TIAM: the TIMES integrated assessment model Part I: Model structure. *Computational Management Science*, 5(1–2), pp. 7–40.

Masui, T., Matsumoto, K., Hijioka, Y., Kinoshita, T., Nozawa, T., Ishiwatari, S., Kato, E., Shukla, P., Yamagata, Y., & Kainuma, M., 2011. An emission pathway for stabilization at 6 Wm-2 radiative forcing. *Climatic Change*, 109(1), pp. 59–76.

McGlade, C. & Ekins, P., 2014. Un-burnable oil: An examination of oil resource utilisation in a decarbonised energy system. *Energy Policy*, 64(0), pp. 102–12.

McGlade, C. E., 2013. *Uncertainties in the outlook for oil and gas*. PhD thesis, University College . London. Available at: <http://discovery.ucl.ac.uk/1418473/2/131106%20Christophe% 20McGlade_PhD%20Thesis.pdf>. Accessed April 2015.

Meinshausen, M., Raper, S. C. B., & Wigley, T. M. L., 2011. Emulating atmosphere-ocean and carbon cycle models with a simpler model, MAGICC6—Part 1: Model description and calibration. *Atmospheric Chemistry and Physics*, 11(4), pp. 1417–56.

Metz, B., Davidson, O. R., Bosch, P. R., Dave, R., & Meyer, L. A., 2007. *Contribution of Working Group III to the Fourth Assessment Report of the Intergovernmental Panel on Climate Change, 2007*, Cambridge, UK: Cambridge University Press.

Morita, T., 1999. *Emission Scenario Database prepared for IPCC Special Report on Emission Scenarios convened by Dr Nebosja Nakicenovic*, Ibaraki, Japan: National Institute for Environmental Studies.

New, M. Liverman, D., Schroeder, H., & Anderson, K., 2011. Four degrees and beyond: the potential for a global temperature increase of four degrees and its implications. *Philosophical Transactions. Series A, Mathematical, Physical, and Engineering Sciences*, 369, pp. 6–19.

Riahi, K., Rao, S., Krey, V., Cho, C., Chirkov, V., Fischer, G., Kindermann, G., Nakicenovic, N., & Rafaj, P., 2011. RCP 8.5—a scenario of comparatively high greenhouse gas emissions. *Climatic Change*, 109(1), pp. 33–57.

Rogelj, J., Hare, W., Lowe, J. A., van Vuuren, D., Riahi, K., Matthews, B., Hanaoka, T., Jiang, K., & Meinshausen, M., 2011. Emission pathways consistent with a 2°C global temperature limit. *Nature Climate Change*, 1(8), pp. 413–18.

Shell, 2013. *New Lens Scenarios*, London, UK: Royal Dutch Shell. Available at: http://www.shell. com/global/future-energy/scenarios.html.

Thomson, A., Calvin, K., Smith, S., Kyle, G., Volke, A., Patel, P., Delgado-Arias, S., Bond-Lamberty, B., Wise, M., Clarke, L., & Edmonds, J., 2011. RCP 4.5: a pathway for stabilization of radiative forcing by 2100. *Climatic Change*, 109(1), pp. 77–94.

UNEP, 2010. *The Emissions Gap Report—Are the Copenhagen Accord Pledges Sufficient to Limit Global Warming to 2°C or 1.5°C? A Preliminary Assessment*, Available at: <http://www.unep. org/publications/ebooks/emissionsgapreport/>. Accessed April 2015.

UNFCCC, 2009. *Report of the Conference of the Parties on its fifteenth session, held in Copenhagen from 7 to 19 December 2009. Part Two: Action taken by the Conference of the Parties at its fifteenth session*, Copenhagen, Denmark: UNFCC.

UNFCCC, 2012. *Report of the Conference of the Parties on its seventeenth session, held in Durban from 28 November to 11 December 2011. Addendum. Part Two: Action taken by the Conference of the Parties at its seventeenth session*. Durban, South Africa: UNFCC.

Van Vuuren, D., Stehfest, E., den Elzen, M., Kram, T., van Vliet, J., Deetman, S., Isaac, M., Klein Goldewijk, K., Hof, A., Mendoza Beltran, A., Oostenrijk, R., van Ruijven, B., 2011. RCP

2.6: exploring the possibility to keep global mean temperature increase below 2°C. *Climatic Change*, 109(1), pp. 95–116.

World Bank, 2013. *How We Classify Countries*. Available at: <http://data.worldbank.org/about/country-classifications>. Accessed April 2015.

World Energy Council, 2013. *World Energy Scenarios: Composing Energy Futures to 2050*, London, UK: World Energy Council.

25 Energy and ecosystem service impacts

Eleni Papathanasopoulou, Robert Holland, Trudie Dockerty, Kate Scott, Tina Blaber-Wegg, Nicola Beaumont, Gail Taylor, Gilla Sünnenberg, Andrew Lovett, Pete Smith, and Melanie Austen

25.1 Introduction

In Chapter 6 we considered the contribution ecosystem services make to human well-being (Millennium Ecosystem Assessment, 2005; Mace et al., 2012) and explored how a range of emerging techniques are allowing us to examine the implications of energy policy on the delivery of these services (e.g. Wiedmann et al., 2007; United Nations Environment Programme, 2009; Lenzen et al., 2012; Yu et al., 2013). It was argued that to design effective and sustainable energy policy we must consider the full implications of different strategies on ecosystem services and subsequently for society both within our national borders and overseas (van der Horst and Vermeylen, 2011). This consideration should be based on both economic measures (Hamilton 2006; TEEB 2010) and measures that assess the broader environmental and societal costs associated with energy production (Gasparatos et al., 2011; van der Horst and Vermeylen 2011).

Understanding the implications of different strategies is complex as they will be occurring against a background of environmental and demographic change. First, evidence presented in the Fifth Assessment Report of the Intergovernmental Panel on Climate Change describes significant changes in global patterns of temperature and precipitation over the coming century (IPCC, 2014). Such change will be associated with disturbance to ecosystems (Fischlin et al., 2007) that will impact biodiversity and ecosystem function. Secondly, human activity over the last century has increased the rate of species extinction to between 1,000 and 10,000 times the background level (Vié et al., 2009) and led to the degradation of ecosystems globally, such that the Millennium Ecosystem Assessment (MEA) estimates that 60 per cent of ecosystem services are degraded or being used unsustainably (Millennium Ecosystem Assessment, 2005). Finally, the global population is expected to undergo substantial demographic changes over the coming century (Metz et al., 2007), with a growing affluent middle class. It is likely that there will be interactions between these processes, accelerating the rate of species and habitat loss and influencing the provision of ecosystem services at a time when human society would benefit from these services as a way of adapting to environmental change (Fischlin et al., 2007).

The need to understand the wider implications of UK energy policy and its impacts on ecosystem services has therefore never been greater. This chapter shows how impacts on ecosystem services can be used as a metric to assess different energy technologies and future scenarios. A global energy assessment framework (GEAF) is described and applied to four energy technologies: nuclear, gas, wind, and biomass. The results show how the different energy systems impact ecosystem services, while the challenges of applying such an assessment framework are highlighted in the discussion on further improvements. These are particularly associated with the issues of scale, intensity, and duration of impact. To begin the chapter, a depiction of the global footprint of the UK electricity demand is presented to illustrate the global reach of just one country's energy demand.

25.2 The global reach of UK electricity production

The demand for electricity in the UK generates activity beyond its national borders. Using global trade data from the GTAP8 (Global Trade Analysis Project 8) database it can be calculated that in 2007 (the most recent year for which data are available) UK final demand for electricity generated direct and indirect economic activity valued at $36.5 billion. Figure 25.1 illustrates that around 86 per cent ($31.5 billion) of this total occurred within the UK. Outside of the UK, significant contributions came from other European countries (particularly the Netherlands and Norway), Asia, the Russian Federation, and North America.

The relationship between economic activity and environmental pressures is complex (Grossman and Krueger, 1995) with impacts dependent on the context under which production is occurring. Trade data allows us to identify those sectors that contribute most to economic activity. This is central to understanding implications for the environment and ecosystem services as some sectors (e.g. mining, agricultural production) will be more strongly associated with negative impacts than others. Further examination of the GTAP8 database shows that across the fifty-seven economic sectors defined, demand for electricity in the UK drives significant economic activity associated with those with a clear link to energy production (e.g. gas, coal, and oil sectors), and those that support this activity (e.g. construction, transport, manufacture of machinery). Other than electricity itself ($22.1 billion) by far the largest value occurs in the 'other business services' sector ($3.1 billion) which is an amalgamation of activities such as banking, insurance, recycling, and government.

By decomposing activities associated with energy production geographically, we can consider the environmental and social context in which activities occur. Similarly, the implications for ecosystem services will differ between sectors as noted above. This is illustrated in Figure 25.2 which, for UK electricity demand, depicts the linkages between geographic regions and sectorial activities. Understanding the implications of such

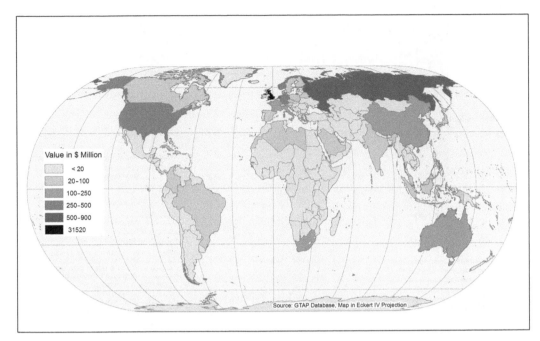

Figure 25.1 Value of economic activity derived from UK electricity demand for 2007 demonstrating the international reach of demand in the UK

patterns is central to evaluating the environmental consequences of different policy options to meet future energy demands. In this chapter we therefore assess in broad terms what type of ecosystem service impacts can be expected globally, dependent on the type of energy demanded in the UK.

In the next section we propose a novel assessment framework to identify the local and global impacts on ecosystem services associated with the lifecycle stages of different energy technologies.

25.3 Global energy assessment framework

The global energy assessment framework (GEAF) presented in this chapter illustrates how impacts on global ecosystem services by different technologies can be assessed by following a number of steps. Firstly, separate matrices were constructed to record the impacts on the terrestrial and marine environments for each technology. The technologies were described using a lifecycle stage framework from the IPCC (2011) Special Report on Renewables which groups lifecycle stages into upstream, fuel cycle, operation, and downstream categories. Subsequently, specific processes in each of these categories for each of the technologies were then specified. For example, in the upstream stage activities such as resource extraction, material and component manufacturing, and construction were distinguished. The fuel cycle stage listed activities related to fuel

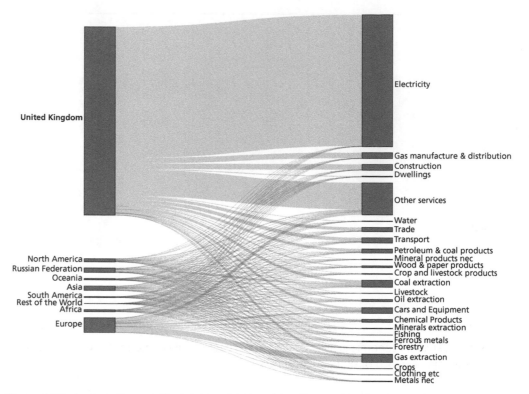

Figure 25.2 Sankey diagram illustrating global activity ($) associated with UK electricity demand. The dark grey bar on the left divides total economic activity driven by UK electricity demand based on the proportion of activity in each region. The grey bar on the right divides activity based on the proportion within each broad economic activity. The light grey bars link the region to the sectors indicating the flow of activity between the two

production and extraction, processing and delivery. Operations included the generation of electricity, maintenance, and operation of infrastructure and equipment, while the downstream stage consisted of dismantling, decommissioning, disposal, and recycling activities (see Table 25.1). Taking the example of nuclear the upstream stage would include activities such as the mining of iron ore and aggregates (resource extraction) which are used in the construction of UK energy plants; the fuel cycle stage includes extraction of the energy feedstock (uranium), its processing and transportation; while the decommissioning stage includes dismantling and decommissioning of the nuclear energy plant.

To capture the impacts on ecosystem services, each of the lifecycle steps were then assessed in terms of impacts on four ecosystem services: three ecosystem service categories were taken from the Common International Classification of Ecosystem Services (CICES) framework (Haines-Young and Potschin, 2012) and one from the MEA (Millennium Ecosystem Assessment 2005). CICES classifies ecosystem services as: provisioning services which include nutrition, materials, and energy provided by ecosystems; regulating services which include mediation of wastes, mediation of flows, and

Table 25.1 Lifecycle stages and impacts of nuclear power on global marine ecosystem services. Black identifies a negative impact; diagonal stripe a neutral or no impact; and grey identifies an inconclusive impact. The results for each of the energy systems are aggregated up to Level 1 of the generalized lifecycle stage and the main heading for the Ecosystem Service classification

Lifecycle Stages		Ecosystem Services						
Generalized lifecycle stage	Detail of lifecycle stage	Intermediate	CICES classification					
		Supporting	Provisioning		Regulating		Cultural	
Level 1	Level 2		Nutrition	Materials	Regulation of bio-physical environment	Flow regulation	Symbolic	Physical
Upstream	Resource extraction	black	black	stripe	black	black	stripe	black
	Material manufacturing	black	black	black	black	black	stripe	grey
	Component manufacturing	black	black	black	black	black	black	black
Fuel Cycle	Resource extraction/production	black	black	black	stripe	stripe	grey	grey
	Processing/conversion	black	black	black	black	black	grey	grey
	Delivery to site	black	black	black	black	black	grey	black
Downstream	Dismantling	grey	grey	grey	grey	grey	grey	grey
	Decommissioning	grey	grey	grey	grey	grey	grey	grey
	Disposal and recycling	grey	grey	grey	grey	grey	grey	grey

maintenance of physical, chemical, and biological conditions; and cultural services which include physical, intellectual, and spiritual interactions. In addition to these three major headings, our assessment also included the intermediate or 'supporting' services (Millennium Ecosystem Assessment 2005). As they are an underlying element of the other three categories of service (provisioning, regulating, and cultural) to include them in any analysis creates a risk of double counting. However, in this case they were included alongside the other categories as many of the impacts from energy generation could only be related to supporting services. Additionally, the impacts are not being summed or valued across ecosystem service categories so the issue of double counting is not a serious concern.

Table 25.1 shows only a section of a completed assessment matrix for nuclear power. Each matrix in reality was made up of twenty-four columns detailing the different ecosystem services as defined by CICES and MEA, and between twenty-eight and forty-three rows depending on the number of detailed activities included in the energy technology lifecycle. During the process of completing the impact assessments, each process and impact is firstly identified as either local or global in nature depending on where the activity occurs. A local impact would signify that the activity occurs in the UK, whereas a global impact signifies that it occurs abroad. On review of the location of these impacts in the terrestrial and marine environment and across technologies, it was noted that the operation stage was predominately described as local in its impact. As this chapter is interested in the global impacts, the operation stage was omitted from further discussion here.

An impact category was assigned to each process across each of the detailed ecosystem services based on a combination of expert judgement and literature reviews. This approach was employed to gain a comparable broad brush overview of the different global impacts of each of the energy systems. Within each matrix the different types of impacts were scored and coded according to whether the lifecycle process was considered to have a negative impact (black), a positive impact (crosshatch), a neutral or no impact (diagonal stripe), or an inconclusive impact (grey) on global ecosystem services. The results shown in Table 25.1 for the offshore impacts of nuclear power, for instance, notes that the upstream activities associated with nuclear power have predominately negative impacts on supporting services based on the opinion that water transport pollution, invasive species introduced through ballast waters, and dredging of ports will be detrimental to the proper functioning of the environment.

It proved necessary to find a means of summarizing the large amounts of information in each matrix, and for this purpose a graphical technique was employed. Continuing with the example of nuclear power, the results in Table 25.1 were translated into a result graph such as that shown for Nuclear–Marine (Figure 25.3), by using the percentage occurrence of a global impact in a group of cells associated with a particular lifecycle stage and ecosystem service. For example, in the upstream lifecycle stage of the Nuclear–Marine graph, the different types of global impacts associated with each of the four ecosystem services were identified and calculated as a percentage of the total cells within each ecosystem service group (see Table 25.1). For example, in the case of supporting services where there is one column, if all upstream activities were expected to have a negative impact, then this would be shown by a column in the results graph within the "supporting service–upstream stage" coloured 100 per cent black. The results shown in Figure 25.3 are taken from the complete matrix and therefore do not completely coincide with percentage calculations based on the smaller Table 25.1 sample. The same approach is continued for the remaining services and cells. For instance, if the total number of cells in the "provisioning service–upstream stage" totals six, of which five cells are coloured

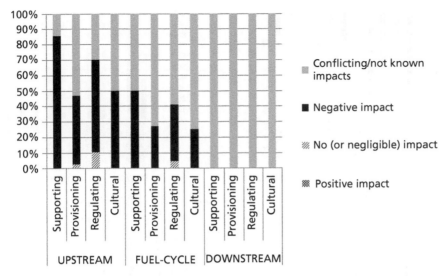

Figure 25.3 Nuclear impacts on global marine ecosystem services

black (see Table 25.1 and Figure 25.3), this would indicate that there was a predominate negative impact (83 per cent). If one cell showed no impact (diagonal stripe) this would then account for the remaining 17 per cent. These calculations were carried out for all the technologies being considered covering both terrestrial and marine impacts.

25.4 Impacts of energy technologies on global ecosystem services

Four contrasting energy systems: nuclear, gas, wind, and biomass to CHP were assessed using the GEAF. These energy systems were chosen because they feature strongly in many scenarios of UK energy futures (e.g. Ekins et al., 2013), represent a range of carbon intensities, and encompass both the marine and terrestrial environments. Figure 25.4 presents a graphical summary of the results.

Upstream impacts across all energy systems are associated with activities which occur overseas related to both mining and transportation of materials (iron, steel, aluminum, concrete) imported to the UK to construct power stations or wind farm infrastructure. Such impacts are predominantly negative across the whole range of ecosystem services (supporting, provisioning, regulating, and cultural). Terrestrial impacts include construction impacts due to land take (impacts on biodiversity etc.) and conflicting cultural impacts relating to location of infrastructure/planning issues and public acceptability. Marine impacts include damage done to marine habitats in international waters due to the dredging that occurs in foreign ports to accommodate larger ships as well as the invasive species introduced into these local environments by ships'

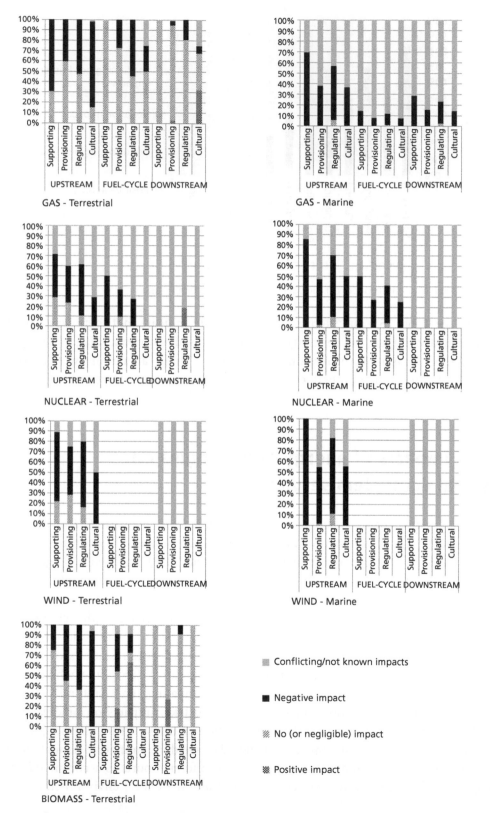

Figure 25.4[1] Global ecosystem service impacts associated with nuclear, gas, wind, and biomass

ballast waters. These invasive species then compete and often replace existing native marine species. Provisioning services can be negatively affected due to the release of antifouling and waste discharges from shipping in ports of exporting countries. Additionally, dependent on the environmental laws and governance in countries in which metals (such as iron ore, aluminium, and graphite) are mined, tailings deposited offshore could damage natural marine habitats through smothering of benthic communities, the release of metals subsequently taken up by marine organisms as toxins, and increased levels of turbidity which decreases the ability of some marine organisms to photosynthesize, causing mortalities and affecting the ability of ecosystems to provide supporting services.

Fuel cycle stage impacts relate to the extraction/production and processing of feedstock. Wind is a freely available resource and as such has no fuel cycle stage impacts. In this project consideration of biomass was limited to *Miscanthus* and Short Rotation Coppice production in the UK. Global impacts shown in the fuel cycle stage of the graph in Figure 25.4 relate to overseas production and importation of root stock. Hence the benefits (positive impacts) indicated during the growth cycle (e.g. reduced need for fertilizers and pesticides compared to conventional crops) accrue overseas. With respect to natural gas, global onshore impacts relate to water use and pollution issues during processing and transmission, and global nuclear impacts relate to the mining of uranium.

The fuel cycle stage impacts associated with the marine environment vary across offshore gas and nuclear. For offshore gas the negative impacts outside UK waters occur predominately during the construction of the offshore gas rigs and pipelines in international waters (i.e. in order to extract natural gas which is then imported into the UK to generate electricity). Damage caused to the benthic (seabed) communities as well as the displacement of species in the immediate areas surrounding these non-UK gas rigs, reduces biodiversity and natural habitats that affect the delivery of ecosystem services. Change in these natural environments can also increase stress levels of marine organisms making them more susceptible to pests and diseases resulting in increased mortalities. The negative impacts associated with increased pollution levels in exporting countries' waters are due to transportation, and collision with and disturbance to cetaceans (i.e. whales and dolphins) which have high cultural value. The fuel cycle for offshore wind is excluded as it does not require mining or fuel processing.

It is noticeable for the marine environment, that there are a number of conflicting or unknown impacts which occur in the *fuel cycle and downstream stages* across all energy technologies. In the context of offshore gas, these are associated with the transportation and dismantling of non-UK gas rigs (supplying fuel to the UK) which are known to have immediate negative effects, however there is the possibility that habitats return to some pre-construction level after a period of time–but this is uncertain. With regards to nuclear

---

[1] Local impacts are not included in these results and the operation stages have been deleted as the impacts were local/national in nature.

energy, decommissioning and spent-fuel storage often occurs on land which means it is unlikely that there are any global marine impacts associated with the downstream stage of nuclear. The unknown impacts noted for the offshore-wind downstream stage is linked to the uncertainty of how these wind farms will be decommissioned and dismantled and to which countries the deconstructed wind turbines will be exported.

25.5 **Next steps**

The results that have been discussed so far provide a general overview of the global impacts of four energy technologies. There are, however, improvements that can be made. Firstly, the types of energy technologies should be expanded to include coal, geothermal, hydro, oil, solar, tidal as well as carbon capture and storage processes that will be significant in future energy scenarios. The same assessment approach could be used to attribute different global ecosystem impacts across the lifecycle of each of these energy systems providing comparative results that can then be used to analyse different future scenarios.

Secondly, in the current discussion each of the impacts along the lifecycle stages of the energy systems have been weighted equally, which may not be appropriate. Here the scale of the effect on an ecosystem service is a major factor that should be considered. For example, nuclear power may have adverse consequences for the ecosystem services where uranium is mined, but the spatial footprint of areas where this resource is mined will likely be very limited. This contrasts with other sources of energy, such as biomass which may be more benign (range from damaging to beneficial) but potentially affect very large areas leading to greater direct and cumulative impacts. Similarly, the smaller spatial footprint of some technologies presents options for avoiding or minimizing conflict with societal demands for ecosystem services, which could be lost for those technologies that require larger areas for deployment. One way of dealing with this issue is to create an exposure/impact coefficient which will weigh the different types of impacts associated with the extent of a particular lifecycle process. Two elements are therefore required to create the coefficient (or weighting factor): the relative intensity of an impact of a process in the lifecycle stage of the energy system and the degree of exposure of the natural environment.

Linked to this second point, the analysis in its present form shows that impacts are occurring globally but is unable to identify exactly where these impacts are occurring geographically. Such analysis could employ techniques such as those which describe flows of global trade already discussed, or could draw on more detailed local analysis for production stages identified as being critical for the provision of services. Identifying the geographical location provides insight into the social and environmental context in which the ecosystem service impacts are occurring, enabling the intensity and exposure to be refined. This final point represents a significant challenge for the ecosystem services community as a whole, not only in the context of energy production, but more broadly in understanding reliance on ecosystem services for different individuals and communities, at different spatial scales. As our ability to address this question increases, the broad scale impact framework developed here can be integrated with geographic data to help inform energy policy with a more detailed focus on the needs of society.

25.6 Conclusions

This chapter has drawn on the findings of research conducted for the UKERC Global and Local Impacts on Ecosystem Services project,[2] which presents a global assessment of the implications of UK energy technologies (natural gas, biomass, nuclear, and wind) on ecosystem services. Mapping global economic activity attributable to UK electricity demand highlighted the fact that its footprint extends beyond national borders; however, it is uncertain how this translates into impacts on ecosystem services given the difficulty in linking economic activity to specific environmental impacts in all but a few cases. A global energy assessment framework was therefore created to record these impacts by creating matrices that captured the lifecycle stage and ecosystem service impact across the technologies. These matrices were completed through expert knowledge and literature reviews to provide comparable results.

Ultimately, although further work is needed to understand the implications of energy strategies more explicitly, the assessment presented here supports the case for energy policy to consider the global implications of different strategies on ecosystem services and societies that rely on them. In seeking to design energy policy that addresses climate regulation, and so preserves the integrity of natural systems, care must be taken to ensure that, globally, the costs associated with each strategy do not exceed the benefits. Incorporating an assessment of impacts across a full range of ecosystem services both at the national and global scale permits the identification of strategies that present the best pathways for future energy production, so that mechanisms to mitigate negative environmental and social impacts may be designed.

■ **REFERENCES**

Ekins P, Keppo I, Skea J, Strachan N, Usher W, and Anandarajah G. 2013. The UK Energy System in 2050: Comparing Low-Carbon, Resilient Scenarios. UK Energy Research Centre.

Fischlin, Andreas, G. F. Midgley, J. T. Price, Rik Leemans, Brij Gopal, Carol Turley, M. D. A. Rounsevell, O. P. Dube, Juan Tarazona, and A. A. Velichko. 2007. 'Ecosystems, Their Properties, Goods, and Services.' In: Parry, M. L.; Canziani, O. F.; Palutikof, J. P.; van der

---

[2] More information on the project can be accessed from: <http://www.ukerc.ac.uk/support/tiki-index.php?page=RF3LImpactsonEcosystem+Services>.

Linden, P. J.; Hanson, C. E., eds. Climate change 2007: impacts, adaptation and vulnerability. Contribution of Working Group II to the Fourth Assessment Report of the Intergovernmental Panel on Climate Change. Cambridge, UK: Cambridge University Press: 211–72.

Gasparatos, Alexandros, Per Stromberg, and Kazuhiko Takeuchi. 2011. 'Biofuels, Ecosystem Services and Human Wellbeing: Putting Biofuels in the Ecosystem Services Narrative.' *Agriculture, Ecosystems & Environment* 142 (3–4) (August): 111–28. doi:10.1016/j.agee.2011.04.020.

Grossman, Gene M., and Alan B. Krueger. 1995. 'Economic Growth and the Environment.' *The Quarterly Journal of Economics* 110 (2): 353–77.

Haines-Young, R., and M. Potschin. 2012. 'CICES Version 4: Response to Consultation.' Centre for Environmental Management, University of Nottingham.

Hamilton, Kirk. 2006. *Where Is the Wealth of Nations?: Measuring Capital for the 21st Century.* World Bank Publications. Washington, DC. USA.

IPCC, 2011: IPCC Special Report on Renewable Energy Sources and Climate Change Mitigation. Prepared by Working Group III of the Intergovernmental Panel on Climate Change (O. Edenhofer, R. Pichs-Madruga, Y. Sokona, K. Seyboth, P. Matschoss, S. Kadner, T. Zwickel, P. Eickemeier, G. Hansen, S. Schlömer, C. von Stechow (eds)). Cambridge University Press, Cambridge, United Kingdom and New York, NY, USA.

IPCC, 2014: Climate Change 2014: Synthesis Report. Contribution of Working Groups I, II and III to the Fifth Assessment Report of the Intergovernmental Panel on Climate Change (Core Writing Team, R. K. Pachauri and L. A. Meyer (eds)). IPCC, Geneva, Switzerland.

Lenzen, M., D. Moran, K. Kanemoto, B. Foran, L. Lobefaro, and A. Geschke. 2012. 'International Trade Drives Biodiversity Threats in Developing Nations.' *Nature* 486 (7401) (June 6): 109–12. doi:10.1038/nature11145.

Mace, Georgina M., Ken Norris, and Alastair H. Fitter. 2012. 'Biodiversity and Ecosystem Services: A Multilayered Relationship.' *Trends in Ecology & Evolution* 27 (1) (January): 19–26. doi:10.1016/j.tree.2011.08.006.

Metz, B., O. R. Davidson, P. R. Bosch, R. Dave, and L. A. Meyer (eds). 2007. Contribution of Working Group III to the Fourth Assessment Report of the Intergovernmental Panel on Climate Change. Cambridge University Press, Cambridge, United Kingdom and New York, NY, USA.

Millennium Ecosystem Assessment. 2005. 'Ecosystems and Human Well-Being: Synthesis.'. Island Press, Washington, DC, USA.

TEEB. 2010. The Economics of Ecosystems and Biodiversity: Mainstreaming the Economics of Nature: A synthesis of the approach, conclusions and recommendations of TEEB. Accessed 06/04/15. Available from <http://www.teebweb.org/publication/mainstreaming-the-economics-of-nature-a-synthesis-of-the-approach-conclusions-and-recommendations-of-teeb/>.

United Nations Environment Programme. 2009. 'Guidelines for Social Life Cycle Assessment of Products.'. UNEP. Accessed 06/04/15. Available from <http://www.unep.org/pdf/DTIE_PDFS/DTIx1164xPA-guidelines_sLCA.pdf>.

Van der Horst, Dan, and Saskia Vermeylen. 2011. 'Spatial Scale and Social Impacts of Biofuel Production.' *Modelling Environmental, Economic and Social Aspects in the Assessment of Biofuels* 35 (6) (June): 2435–43. doi:10.1016/j.biombioe.2010.11.029.

Vié, Jean-Christophe, Craig Hilton-Taylor, and Simon N. Stuart. 2009. *Wildlife in a Changing World: An Analysis of the 2008 IUCN Red List of Threatened Species^{TM}.* IUCN, Gland, Switzerland.

Wiedmann, Thomas, Manfred Lenzen, Karen Turner, and John Barrett. 2007. 'Examining the Global Environmental Impact of Regional Consumption Activities—Part 2: Review of Input–output Models for the Assessment of Environmental Impacts Embodied in Trade.' *Ecological Economics* 61 (1) (February 15): 15–26. doi:10.1016/j.ecolecon.2006.12.003.

Yu, Yang, Kuishuang Feng, and Klaus Hubacek. 2013. 'Tele-Connecting Local Consumption to Global Land Use.' *Global Environmental Change* 23 (5) (October): 1178–86. doi:10.1016/j.gloenvcha.2013.04.006.

26 Policies and conclusions

Paul Ekins

26.1 Introduction

Chapter 1 of this book introduced the energy trilemma: energy security, decarbonization of energy, and energy affordability. This book has shown this trilemma to be a 'wicked problem' (Rittel and Webber, 1973) facing policymakers and impacting on businesses and citizens.[1] Each of the individual components of the trilemma is complex, and each component has basic and multiple connections to the others, so that the trilemma has no final solution. Efforts to improve the situation in respect of one component may make the others better or worse, in multi-dimensional ways that are hard to predict.

However, policymakers, and hence the businesses and citizens who both influence them and are affected by the policies they implement, have little choice but to engage with these issues. While having 'no policy' may be a theoretical policy option, it would certainly lead to an undesirable and perhaps a chaotic social outcome. This book has set out many of the social, technological, economic, and policy options related to global energy, and the context within which these need to be addressed. This chapter will summarize some of the essential parameters of these options, in the hope of indicating which choices may lead to better social outcomes for more people now and in the future.

This concluding chapter both draws together the main themes and conclusions from the earlier chapters in this book, and touches on other important themes which have not been dealt with in detail in this volume, with references as to where further information on these may be found if desired. It starts with what are here characterized as 'global imperatives': particular aspects of the trilemma which, if they are not energetically addressed, will be likely to ensure poor energy outcomes for very many people. These imperatives have been the subject of detailed discussion earlier in the book. This chapter seeks to pull these discussions together and move towards some conclusions.

The chapter then moves on to consideration of the different energy futures that have been explored, especially in respect of their greenhouse gas (GHG) emissions, and more particularly the carbon dioxide emissions from the energy system. This is in fact the major differentiating, and most contested, factor in different energy futures. If

---

[1] 'Wicked' problems are characterized by incomplete or contradictory knowledge, different opinions, which may be based on different value systems, held by large numbers of people, substantial economic implications, and complexity, both internally and in their relationship with other issues. Such problems are not amenable to definitive solution, although some resolutions of them may be judged better than others.

climate science was not pointing unequivocally at future climate change with possibly devastating impacts on human societies, few would be questioning an energy outlook to the middle of this century and beyond that was dominated by fossil fuels. That is the natural trajectory of past energy developments, which have huge momentum from past investments in knowledge, technology, and infrastructure. Such an outlook would raise issues of energy security, resource scarcity, and affordability for the future, as it has in the past, but policymakers would be comfortable about addressing those issues in that context, as they have done before.

The imperative to decarbonize the energy system has changed all that and greatly enlarged the range of energy futures that are being considered, and the range of energy technologies that are now being developed and deployed through the implementation of public policies and new institutional arrangements that are the outcome of vigorous political and social debate. Consideration of these issues has comprised the core of this book, and forms the bulk of this concluding chapter.

26.2 **Global imperatives**

26.2.1 THE GOVERNANCE OF ENERGY SYSTEMS: MARKETS AND REGULATION

It is not surprising that energy systems around the world exhibit a great variety of institutional, economic, and social arrangements. All of them entail the operation of markets to some extent. All of them entail some kind of regulation, at one or more levels of government. But how these institutional arrangements are defined, formed, and interact is subject to huge variation. It is neither possible nor desirable to specify ideal market-regulation relationships, because such relationships are inescapably context-dependent. But, in any context, some arrangements unquestionably work better than others, and it is the policymakers' responsibility to identify how these arrangements might be improved in the situation in which they are working and then seek to bring such improvements about.

Chapter 23 explored in some detail how a relatively mature electricity system that has evolved in the fossil fuel era could be regulated to shift towards a system that incorporates more renewable energy sources and lower carbon emissions. This is an issue that is relevant to all industrialized countries and many emerging economies that already have substantial power grids. It is also relevant, but in a different way, to countries that are still putting in place their first electricity systems, with less-developed market and regulatory capacities and arrangements. Such countries may not wish to put in place a universal power grid, just as telephone land lines now seem unnecessary in many places, as decentralized technologies become more affordable and available, and improve their performance, as discussed in Chapter 8 and, more briefly, below. An important aspect of facilitating the transition to a low-carbon electricity system will be managing—through the right combination of policy, markets, and regulations—the changing role of fossil fuels in the system, which, though their share of power generation will be decreasing, will have a crucial part to play for many years to come.

Electricity is, of course, not the only energy vector. Others are liquid fuels, natural gas (methane), and hydrogen, and these require their own, different systems of regulation. Electricity grids, needing to link primary energy suppliers, generators, and end-users through complex transmission and distribution systems, are normally subject to more elaborate regulatory arrangements than the simpler distribution systems of, say, liquid transport fuels. But energy is such a fundamental input into economies and human life more generally that governments are generally not content to leave its provision entirely to market operations. And one of the most important reasons for this is the desire of all countries and communities for energy security.

26.2.2 ENERGY SECURITY

Energy security relates to the desire of governments, businesses, and citizens to have access to energy services when, where, and in the quantity that they need and want. It also means that such energy services can be accessed at an affordable price. It is one of the horns of the energy trilemma. It has a number of dimensions, including: the ability, economic and geopolitical, to access primary and refined fuels as necessary; the availability of the technologies and infrastructure to generate, transmit, and distribute the chosen energy carriers; the capacity of these technologies and infrastructures to withstand or recover from accidents or natural extremes; the possession of the end-use appliances that convert these into desired energy services; and the institutional and regulatory arrangements that ensure that all these components of the energy system are maintained, renewed, and operated safely, as required by the end-users. Cherp and Jewell (2011) have summarized these factors as sovereignty (the ability to control or respond to external agents and factors), robustness (of technologies and infrastructures to a range of circumstances), and resilience to unforeseen events (technological, climatic or market-related).

The processes of globalization described in Chapter 3 of this book, both of energy systems but also more broadly, have complex implications for energy security. On the one hand, countries without their own indigenous energy resources are obviously dependent on imports, and the extension and liberalization of energy markets can increase their energy security; on the other, the increasing use of energy encouraged by these open markets may introduce a new vulnerability to rising and volatile prices, and a new dependence on their continued and orderly functioning (Wicks, 2009). Energy-exporting countries are often no less dependent on these markets, for their economies often come to rely on the revenues from energy exports. As 2014 events in Ukraine have shown, the importing and exporting of energy is both a source of vulnerability and a potential economic weapon and can be deployed by both sides in a dispute.

However, there is not a straightforward relationship between the energy security of a given country and its dependency on imported energy (Mitchell et al., 2013). Some countries are relatively energy secure even though they are highly import dependent because they can access resources from a variety of sources, via a diversity of supply routes (e.g. for the case of gas supplies to the UK). Conversely, the energy security of some countries has been affected by disruptions to their domestic energy supplies or infrastructure. Examples include the Fukushima incident in Japan, the inability of French nuclear stations to run at full capacity in hot weather in 2003, and the fire at the UK's largest gas storage facility in 2006.

Energy security is deeply connected to and influenced by the other two horns of the energy trilemma. Climate change and the challenge of decarbonization encourage countries to move to low-carbon energy sources. For fossil fuel exporters, this may be economically deeply threatening. For importers, this may give an opportunity to diversify away from fossil fuels to indigenous renewable energy sources, or to nuclear power, both of which have their own, different, implications and challenges for energy security. Energy security may also be increased by countries making more efficient use of energy, thereby becoming less vulnerable to future fossil fuel price rises or shocks.

Energy security is also intimately connected to affordability. Energy systems may be engineered to very high standards of robustness, but at a cost. Their resilience to climate extremes can be increased, but at a cost. Low-carbon energy systems are, for the present at least, more expensive than their fossil-fuelled counterparts. Yet there is still a large population in the world which has no access to modern energy services, which is perhaps the worst energy security situation of all.

26.2.3 ENERGY ACCESS: INCREASING ACCESS TO MODERN ENERGY SERVICES

'Energy is eternal delight' wrote the eighteenth-century English poet William Blake. Certainly having to live in the modern age without access to modern energy services is a very great constraint on delight. Equally certainly, the 1.4 billion people without access to electricity and the 2.7 billion people who rely on biomass for cooking and heating (GEA, 2012: Exec Sum, p.36) will give priority to gaining modern energy services rather than worrying about greenhouse gas (GHG) emissions. If those who already have such access do not help those without it to gain this access without laying down a lot of long-lived infrastructure to burn high-carbon fossil fuels, then attempts to reduce global GHG emissions in the short to medium term are likely to prove fruitless.

For many of these people, especially those in urban areas, the route to this access will be through a central grid, and such grids could also spread quite widely into rural areas in developing countries, as they have in industrialized countries.

But, as Chapter 8 made clear, technologies in this area are changing extremely fast, and it is quite possible that decentralized energy, with or without a central grid, becomes the energy source of choice, just as mobile phones have replaced landlines as the aspirational means of communication.

But decentralized energy is certainly no panacea for the issue of widening energy access. There is no universal model for its provision, which is crucially dependent on local institutions, markets, skills, and social arrangements. Bangladesh is the poster-child of solar PV diffusion, with over 1.6 million solar home systems (SHSs) distributed by 2012 through its government-owned financial institution Infrastructure Development Company Limited (IDCOL), which has a number of partner organizations, the best-known and most effective of which is Grameen Shakti (Sharif and Mithila, 2013). IDCOL has a target of over 4 million SHSs by the end of 2015. The SHSs are cheaper on average than the lighting alternative of kerosene, but poor households (which use little kerosene too) still require an SHS subsidy—this has fallen from US$90 to US$25 over the course of the programme, and is expected to be withdrawn altogether if costs go on falling as anticipated. In addition, Mondal and Islam (2011) have calculated that Bangladesh has a grid-connected PV potential, with a benefit–cost ratio of greater than 1, of around 1.7 TWh (compared with a 2011 electricity consumption of 0.26 TWh).

However, even as countries like Bangladesh achieve mass diffusion of both off-grid and, potentially in the future, on-grid solar PV, other countries are struggling to find a diffusion model that works for them. In many places coal is still the default generating option, especially as populations urbanize, and in these places substituting natural gas for coal would lead to enormous health benefits as well as reducing carbon emissions substantially. Substituting gas for coal in power generation should be a major priority in all countries on the grounds of public health as well as climate change mitigation.

Electricity is important but, as GEA 2012 stresses, it is not all-important, and 'energy policies addressed at fuel switching and improving heating and cooking systems, especially in rural areas, have received very little policy attention in energy sector reform' (GEA, 2012: 185). This needs to change. LPG (liquid petroleum gas), for example, can provide a much cleaner and more efficient alternative to biomass for heating and cooking, providing a greatly enhanced energy service with, again, large health benefits as well.

Finally, it is worth remembering that the energy access requirement is often articulated very modestly in terms of the quantity of energy that needs to be supplied. Bazilian and Pielke (2013) have compared for electricity the International Energy Agency (IEA) definition of energy access per capita with the average actual amounts of electricity consumed calculated by the World Bank and the amounts for 2035 projected by the US Energy Information Administration (with the 2035 projections for the USA, Germany, and Bulgaria assumed to be the same as their 2010 values). From Figure 26.1 it can be seen that the 'energy access' definition is a small proportion even of current Bulgarian consumption. It may confidently be assumed that those with such low levels of electricity consumption will aspire to the much higher levels of developed countries, emphasizing still further the challenge of providing the energy without greatly increasing carbon emissions, the topic of the next section.

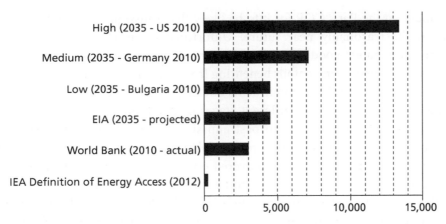

Figure 26.1 Actual and projected global per capita electricity consumption (kWh/year)
Source: Bazilian and Pielke 2013, p. 76.

26.2.4 REDUCING GREENHOUSE GAS EMISSIONS

26.2.4.1 The Global Negotiations

The Conference of the Parties (COP) to the UN Framework Convention on Climate Change (UNFCCC) will meet in Paris in December 2015. The outcome will determine whether the rhetorical commitment of the international community of nations to keeping average global warming to below 2 °C has any last vestige of credibility.

Chapter 4 in this book sets out clearly both the context and the major issues that will need to be addressed at the Paris COP. It is worth noting at this stage that maintaining climate stability is an example of what Barrett (2007: 74) calls an 'aggregate effort' public good. Its delivery depends on sufficient aggregate effort from those involved. In the case of climate change that would seem to mean substantial emission reduction commitments from countries that are responsible for at least 80 per cent of emissions.

In terms of carbon dioxide emissions from fossil fuel use and cement production, the six largest emitting countries/regions (with their share in 2012 between brackets) were: China (29 per cent), the United States (15 per cent), the European Union (EU27) (11 per cent), India (6 per cent), the Russian Federation (5 per cent), and Japan (4 per cent) (69 per cent in total, Olivier et al., 2013: 10). The members of the G20 (Argentina, Australia, Brazil, Canada, China, France, Germany, India, Indonesia, Italy, Japan, Republic of Korea, Mexico, Russia, Saudi Arabia, South Africa, Turkey, the United Kingdom, the United States, and the European Union) are collectively responsible for about 78 per cent of global emissions. Adding in India, Taiwan, Thailand, and Iran, the percentage rises to around 87 per cent (Olivier et al., 2013: 16–17). This gives an idea of the kind of minimum coalition of countries that would need to make credible emissions reduction commitments at the Paris COP for it to be likely that emissions globally would start to decline in the reasonably near future. Unfortunately, as is always the case with aggregate effort public goods, countries have powerful incentives to free-ride on the efforts of others (free-riding entails benefiting from the public good provision of others without making comparable commitments oneself). In addition, countries have widely diverging interests, are at very different stages of development, and have made very different historical contributions to past emissions, which have accumulated in the atmosphere—all considerations that make global agreement more difficult. For any agreement at the Paris COP to be delivered on, especially if the reduction commitments go much beyond what is likely to be achieved with business-as-usual (which they will need to do to have any chance of meeting the 2 °C target), the agreement will need to contain a mixture of benefits and penalties respectively for those countries which meet or miss their commitments (or which make inadequate commitments). Penalties, in particular, were conspicuously absent from the Kyoto Protocol. There is not as yet much sign that the above group of countries is willing to embrace them.

Benefits, on the other hand, were part of the Kyoto Protocol, in the form of the various trading mechanisms that were provided for, including the Clean Development Mechanism (CDM). Mechanisms of this sort are certainly likely to be required in any new agreement, to provide for the transfer of both technology and finance for both emission reduction and adaptation to climate change, if emerging economies and less-developed countries are to be prepared to commit to significant emission reduction.

As Chapter 4 makes clear, while hope may spring eternal, there are few grounds for optimism from the Paris talks. The grounds for hope are that both the climate science and the evidence of climate change on the ground, as well as the reducing cost of renewables, will persuade countries to make the aggregate effort that will be required if this issue is to be effectively addressed.

26.2.4.2 Urbanization

The world is currently undergoing a 'second urbanization wave', which, by 2050, is likely to add around 3 billion people to the world's cities (UNEP, 2013). GEA (2012: 1310) estimates that 60–80 per cent of global energy use is by city dwellers. Given the long-lived nature of urban infrastructure, energy use in the future will be strongly influenced by whether or not the new or extended cities of the next three to four decades facilitate efficient energy consumption in their delivery of energy services through their buildings and transport systems, and other infrastructures (including water, waste, and information and communication technologies). The dynamics of city development, which include both rise and decline, are complex, and result in very different trajectories of energy use, as shown in Figure 26.2.

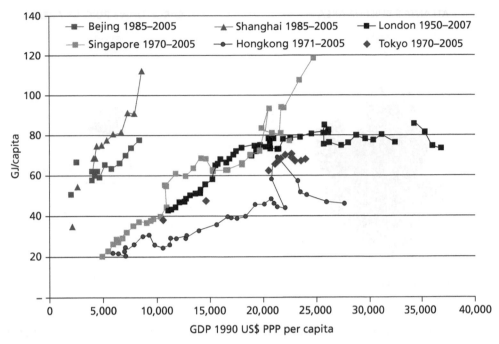

Figure 26.2 Longitudinal trends in final energy (GJ) versus income (at PPP, in Int 1990 $) per capita for six megacities

Source: GEA (Global Energy Assessment), *Global Energy Assessment: Toward a Sustainable Future* (2012), Cambridge University Press.

GEA (2012: 1311) stresses that '*Systemic* characteristics of urban energy use are generally more important determinants of the efficiency of urban energy use than those of individual consumers or of technological artefacts', so that 'the share of high occupancy public and/or non-motorized transport modes in urban mobility is a more important determinant of urban transport energy use than the efficiency of the urban vehicle fleet', and the compactness of the urban form with less need for mobility can result in less total energy use than dispersed dwellings, even if they are individually very energy-efficient. The importance of urban design in energy use is illustrated by the very different footprints of Atlanta and Barcelona, cities of comparable population but very different urban area, as shown in Figure 26.3. The CO_2 emissions in Atlanta related to urban transport are 11 times those in Barcelona (Lefèvre, 2009: 9)

Other imperatives for urban areas in this century are to become essentially pollution-free (GEA, 2012: 1312) by strengthening pollution controls in cities that already have universal access to modern energy services, and providing such access through clean fuels, as outlined in the first section of this chapter, to those cities that do not yet have it; and to build new buildings or refurbish old ones such that they use energy much more efficiently. This point is discussed in more detail in the relevant section below.

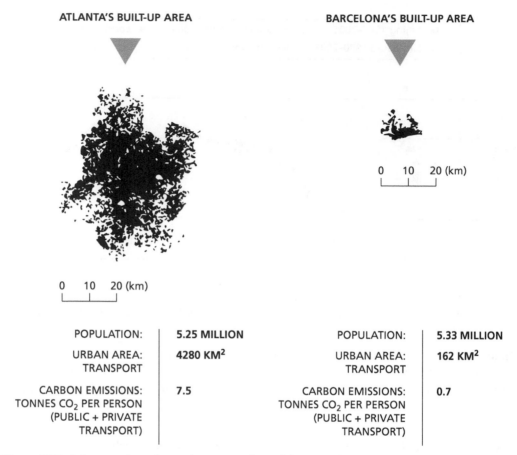

ATLANTA'S BUILT-UP AREA BARCELONA'S BUILT-UP AREA

POPULATION:	**5.25 MILLION**		POPULATION:	**5.33 MILLION**
URBAN AREA: TRANSPORT	**4280 KM²**		URBAN AREA: TRANSPORT	**162 KM²**
CARBON EMISSIONS: TONNES CO₂ PER PERSON (PUBLIC + PRIVATE TRANSPORT)	7.5		CARBON EMISSIONS: TONNES CO₂ PER PERSON (PUBLIC + PRIVATE TRANSPORT)	0.7

Figure 26.3 Atlanta or Barcelona, the range of possible urban futures
Source: Lefèvre 2009, p. 10.

26.2.5 SUSTAINING GLOBAL ECOSYSTEM SERVICES: MITIGATING LOCAL AND GLOBAL IMPACTS

While the emission of greenhouse gases leading to climate change is the principal global environmental concern related to energy systems, and local air pollution can cause damage to human health as well as the environment, other environmental concerns include a wide range of impacts on ecosystems, as Chapters 6 and 25 showed. The desirability of reducing the impact on human health from local air pollution, especially in fast-growing emerging economies, provides a powerful additional rationale for moving towards low-carbon energy sources in those countries. However, the major insight from Chapter 6 was that the impacts of the energy system on ecosystem goods and services can arise not only from production and consumption to meet local needs, but also from extraction or production activities associated with exports, the consumption of which takes place elsewhere. The same point in relation to greenhouse gas emissions is forcefully made in Chapter 5. It is essential that importing countries are both aware of these impacts, and, in cooperation with the exporting countries where the impacts are taking place, take action to mitigate them.

26.3 **Global energy scenarios in 2050: high GHG emissions**

26.3.1 PROJECTING CONTINUING RELIANCE ON FOSSIL FUELS

As noted in Chapter 1, there are a number of global energy scenarios that envisage substantial continued reliance on fossil fuels for the majority of energy use well into the future. These scenarios come nowhere near the stated policy ambition of reducing GHG emissions such that average global warming remains below 2 °C. Rather, as Chapter 1 makes clear, the likely average global warming emerging from such scenarios is 4 or 6 °C.

Chapters 12, 13, and 14 allow various conclusions to be drawn about the feasibility and desirability of these kinds of scenarios, quite apart from their climate impacts. First, it is quite clear that there are sufficient oil and gas reserves to sustain the levels of production that are envisaged in these scenarios. Second, there seems no reason to imagine that technological progress will not make oil and gas resources more accessible in the future, as it has in the past. However, it is also clear that unconventional and harder-to-access resources will be increasingly required to satisfy the levels of demand projected in the scenarios. This means that fossil fuel prices are unlikely to fall to historically low prices, and may well increase in price over time, depending on energy demands. It also means that geopolitical or other factors that prevent timely investment in these less accessible resources, or which prevent the extraction and export of more accessible resources, such as the current disruptions in the Middle East, may cause serious price volatility, as has occurred in the past.

Under these circumstances the fossil fuel industry finds itself in something of a double bind. If demand for fossil fuels continues to grow, then prices will remain high. Although this will give a return to exploration and production investment in more remote resources, it will also give greater incentives to develop low-carbon substitutes for fossil fuels. On the other hand, lower demand for fossil fuels could cause prices to ease, reducing the competitive pressure from low-carbon sources, but this would also make the investment in less accessible resources less profitable and, at some point, would render it altogether uneconomic. This would be challenging, especially for the private oil majors, who are increasingly turning to these less accessible resources to replenish their reserves.

More generally, high-GHG scenarios do not usually provide guidance as to the kinds and stringency of policies that would be required to adapt to the level of climate change that these emissions seem likely to generate, or what the impact of this would be on economic growth and development. For example, in the IEA's 6DS scenario, economic growth globally remains such as to increase global primary energy demand by 69 per cent over the period 2011–50, while CO_2 emissions increase by 62 per cent to 54.6 $GtCO_2$ over the same period, leading to average global warming of 6 °C, which gives the scenario its name (IEA, 2014). ExxonMobil's scenarios to 2040, with significant energy efficiency gains, project global energy demand to increase by only 35 per cent over 2000–40, driven by population growth and an 80 per cent increase in GDP per capita over that period. Energy-related CO_2 emissions increase from 30.6 $GtCO_2$ in 2010 to a peak of 36.8 $GtCO_2$, a 20 per cent increase, when they peak and decline slightly to 36.3 $GtCO_2$ by 2040, because of the implementation of advanced technologies and lower-carbon fuels (ExxonMobil 2014: 7,11 and underlying database). This is closer to the IEA's 4DS (4 °C) scenario, in which 2030 emissions are 39.4 $GtCO_2$, rising to 41 $GtCO_2$ by 2045, when they level off and decline very slightly by 2050. Perhaps because of the very considerable uncertainties involved, none of these scenarios project that the climate damages from these enhanced levels of global warming would have any impact on the economies that produce them, or require policies to adapt to these impacts. Such an assumption is likely to be very optimistic, even by 2040 or 2050, and certainly thereafter. However, such adaptation policies, considered in detail in IPCC WGII (2014) are outside the scope of this book.

26.3.2 POLICIES FOR A HIGH-CARBON FUTURE

Both the IEA 4DS and the ExxonMobil 2014 scenarios do contain some carbon abatement policies, the former a realization of the carbon reduction commitments made at the Cancun Conference of the Parties to the UNFCCC in 2010, the latter a combination of unspecified energy efficiency policies, which reduce energy demand, and an implied cost of carbon in OECD countries that reaches around US$80 per tonne CO_2 'with developing nations gradually following, led by China' (ExxonMobil, 2014: 32).

A feature of current energy consumption is the continuing high level of subsidization of fossil fuels, which have increased from US$311 billion in 2009 to US$544 billion in 2012, more than five times the level of subsidy paid to renewables (IEA, 2013a). IEA (2013b) estimates that at the time of writing 15 per cent of global CO_2 emissions were receiving an incentive of $110 per tonne in the form of fossil fuel subsidies, while only 8 per cent of emissions were subject to a carbon price. While the justification for fossil fuel subsidies is often that they give energy access to low-income households, in fact IEA (2013a: 93–98) reports that only 7 per cent of fuel subsidies in poor countries go to the bottom 20 per cent of households, while 43 per cent go to the richest 20 per cent. The proportion for low-income households is even lower with petrol subsidies, because people from richer households are more likely to own and drive cars. G20 countries have a commitment from 2009 to phase out 'inefficient' subsidies to fossil fuels in the 'medium term', but since then such subsidies have grown substantially as noted above, and with no definition as yet of the 'medium term', the 2009 commitment seems somewhat hollow.

Another area where support for fossil fuels continues to outstrip that in low-carbon energy sources is research and development (R&D), where private R&D spending related to fossil fuels is more than public R&D spending on all energy sources (see Chapter 1, Table 1.4). This is, perhaps, not wholly surprising, given that private companies are market-driven and fossil fuels still comprise such a large proportion of energy markets, while low-carbon energy sources still generally require public subsidy. But the fact remains that this continuing direction of large-scale subsidy and R&D funding into fossil fuels can only reinforce current emission trends that seem likely to bring about widespread and potentially very disruptive climate change.

26.4 **Global energy scenarios in 2050: very low GHG emissions**

26.4.1 PROJECTING A LOW-CARBON FUTURE

Chapter 1 noted that the IPCC's 5th Assessment Report from Working Group III reviewed around 900 scenarios, published in the scientific literature, that projected varying degrees of carbon emissions reduction. Table 26.1 shows the macroeconomic cost, in terms of a reduction in consumption from the baseline, and a reduction in the rate of economic growth, that derived from some of those scenarios projecting different levels of carbon abatement.

Table 26.1 Global mitigation costs in cost-effective scenarios to meet different GHG concentrations in 2100

	Consumption losses of cost-effective scenarios			
	[% reduction in consumption relative to baseline]			[percentage point reduction in annualized consumption growth rate]
2100 Concentration (ppm CO_2eq)	2030	2050	2100	2010–2100
450 (430–480)	1.7 (1.0–3.7) [N: 14]	3.4 (2.1–6.7)	4.8 [2.9–11.4]	0.06 [0.04–0.14]
500 (480–530)	1.7 (0.6–2.1) [N: 32]	2.7 (1.5–4.7)	4.7 (2.4–10.6)	0.06 [0.03–0.13]
550 (530–580)	0.6 (0.2–1.3) [N: 46]	1.7 (1.2–3.3)	3.8 (1.2–7.3)	0.04 [0.01–0.09]
580–650	0.3 (0–0.9) [N: 16]	1.3 (0.5–7.0)	2.3 (1.2–4.4)	0.03 [0.01–0.05]

Source: Based on IPCC, 2014: Summary for Policymakers. In: Climate Change 2014: Mitigation of Climate Change. Contribution of Working Group III to the Fifth Assessment Report of the Intergovernmental Panel on Climate Change (Edenhofer, O., R. Pichs-Madruga, Y. Sokona, E. Farahani, S. Kadner, K. Seyboth, A. Adler, I. Baum, S. Brunner, P. Eickemeier, B. Kriemann, J. Savolainen, S. Schlömer, C. von Stechow, T. Zwickel, and J.C. Minx (eds.)). Cambridge University Press, Cambridge, United Kingdom and New York, NY, USA.

Table 26.1 shows that the median global mitigation cost, compared with the model baseline, of keeping atmospheric GHG concentrations to 450 ppm by 2100, is 4.8 per cent of consumption (with a range of 2.9–11.4 per cent), which represents a reduction in the average annual consumption growth rate of 0.06 per cent (with a range of 0.04–0.14 per cent). It may be noted in comparison that these reductions in consumption are from baselines in which consumption is assumed to grow in the different scenarios between four-fold and over ten-fold compared to today's levels. Even with mitigation, therefore, on these assumptions, societies in 2100 would be at least 390 per cent richer than they are now.

The mitigation costs would be roughly halved by relaxing the carbon constraint to an atmospheric concentration of 580–650 ppm. It should be stressed that these costs do not take into account the climate damages that would occur from the emissions levels in the baseline or in the mitigation scenarios (these damages would be less in the 450 ppm than in the 580–650 ppm scenarios and highest in the baselines, which have less or no GHG abatement). Arguments for reducing emissions arise from perceptions (for example, in Stern 2007) that mitigation costs are likely to be considerably outweighed by climate damages, which are extremely difficult to incorporate fully in integrated assessment models. The estimates in Table 26.1 also do not take into account any co-benefits, such as reducing local air pollution, that may result from reducing GHG emissions (or co-disbenefits, where these arise).

Because the model (TIAM-UCL) used in the scenarios described in Chapter 24 of this book does not incorporate a macroeconomic model, it is not possible to compare directly the mitigation costs reported in the IPCC survey (see Table 26.1) with those from TIAM-UCL. However, it was seen that the carbon prices in the TIAM-UCL 2 °C scenarios ranged from around US$120–650 per tonne CO_2 (where bio-CCS is included in the model, but the highest value results from scenarios in which mitigation action is delayed until 2035, which makes it much more expensive to reduce emissions to the extent required thereafter; the next highest, also a delayed action scenario, is US$450). This compares with the 2050 carbon price range in the IPCC 430–480 ppm scenarios of around US$70–900 per tonne, median around US$230, and the 25–75 percentiles US$130–300 per tonne (IPCC WGIII, 2014b: Figure 6.21, p. 47), suggesting that the model runs in Chapter 24 are broadly comparable with the majority of those scenarios surveyed by the IPCC, which reach roughly the same level of abatement.

The next section considers the investment which has been estimated to be required to bring about a low-carbon energy system by mid century and how such investment may be incentivized to flow in this direction.

26.4.2 FINANCIAL REQUIREMENTS

Large-scale investment in energy systems in industrialized countries is required simply to maintain them as infrastructures age and need to be replaced. Emerging and developing economies require very large energy system investments for their industrialization and to give their populations access to modern energy services. GEA (2012: xiii) estimates current investments in the energy sector to be around 2 per cent of global GDP, or US$1.4 trillion per year (this comprises investment in both the supply side and the energy components of the demand side of the energy system—see GEA 2012, pp. 408–410 for a discussion). The number estimated by the IPCC WGIII (2014c: 14) is similar, around US$1.2 trillion. For the future, IEA (2012: 137) estimates that such 'business as usual' investments will need to total around US$105 trillion by 2050, or US $2.6 trillion per year (but note that these estimates do not take account of the extra costs likely to be required to adapt the energy system to climate change on a business-as-usual path, or the costs of climate damage to the energy system, which could be significant).

GEA (2012: 1619) estimates the additional investment required to give all populations access to modern energy services at US$36–41 billion per year until 2030, although IPCC WGIII (2014c: 40) arrives at an estimate of over double that, US$72–95 billion. It may be that the difference is due to the latter figure comprising only low-carbon (more expensive) energy investments, but the sources do not make this clear.

Estimates of the net additional annual investment cost to move from the current global emissions trajectory to one consistent with the 2 °C global warming limit include US$360 billion over 2010–29 (calculated from IPCC WGIII 2014c, pp. 3–4, although the upper end of the range is US$1.0 trillion)[2] and EUR 317 billion over 2011-15 rising to EUR 811 billion by 2026-30 (McKinsey, 2009, p.43), with the IEA (2012: 137) estimating a total extra US$36 trillion needing to be invested by 2050—roughly US$1 trillion per year. IEA (2012: 138) notes: 'Average annual investments in the 2DS, from 2010 to 2020, are US$2.4 trillion, 25% higher than in the 6DS. From 2020 to 2030, annual investment requirements under the 2DS rise to US$3 trillion. This 36% increase over the 6DS is due to higher investments in renewable power, retrofits of residential and commercial buildings and CCS in the power and industry sectors.' Figure 26.4 illustrates the sectoral breakdown of this extra US$36 trillion over the period to 2050.

With global GDP in 2012 at around US$70 trillion, and assuming average annual global economic growth of around 2 per cent, global GDP reaches $100 trillion by 2030, so that the IEA (2012) estimate of additional investment costs are of the order of 1 per cent global GDP. If it is the case that there are diverse risks of severely adverse effects of climate change, especially for low-income communities with their limited ability to cope, as climate science suggests (see IPCC WGII: 12–13), then the 'insurance premium' represented by this additional investment is very modest in relation to the potential costs that are being avoided. Indeed, millions of young, healthy people across the world pay a much greater premium for health insurance to hedge the risk of paying for expensive treatment in the event of an accident or serious illness.

---

[2] It is noteworthy that the lower bound of the IPCC WGIII (2014c) estimate suggests that a low-carbon energy system could actually require less investment than its high-carbon alternative, partly, of course, because investment in energy efficiency reduces the supply capacity that is required.

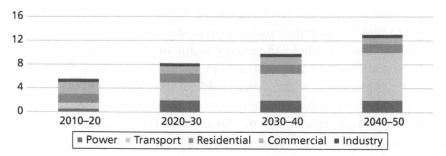

Figure 26.4 Additional investment needs in the IEA's 2DS, compared to its 6DS, scenario

Source: © OECD/IEA (2012) Energy Technology Perspectives 2012 - how to secure a clean energy future, IEA Publishing. Licence http://www.iea.org/t&c/.

Incentivizing an extra US$1 trillion per year investment in clean energy sources is a substantial policy ambition, but a much greater task is the *redirection* of the US$100 trillion that is projected to be required for the global energy system even in the absence of the decarbonization imperative. Currently, much of this US$100 trillion seems destined to flow into discovering and developing fossil fuel resources to meet the world's expanding energy needs, as indeed is made explicit in the IEA (2012)'s 6DS scenario. To avoid such a climate outcome, the challenge is to bring about a redirection of *existing* financial flows, away from carbon-intensive to low-carbon technologies, and to enable energy to be used much more efficiently than at present in both societies dependent on fossil fuel and those relying very largely on low-carbon energy sources.

Although these sums seem, and are, very large, the problem is not one of basic availability of capital. Figure 26.5 shows that the kind of institutional investors shown manage funds that total US$76 trillion, any of which is available in principle for clean energy investment.

At present clean energy is only receiving a fraction of the required investment if the world's energy system is to be decarbonized over a timescale that comes anywhere near policymakers' 2 °C limit aspiration. Figure 2.5 in Chapter 2 showed that global new investment in clean energy grew rapidly through the early 2000s, and reached US$317 billion in 2011, but then declined to US$254 billion in 2013, only around a quarter of the *extra* US$1 trillion investment in clean energy that is required, and this does not begin to touch the US$100 trillion that needs to be switched from fossil fuels.

The issue is that private investors will only invest in clean energy if the risk–reward profiles of the investments meet their investment criteria. Where low-carbon investments are more expensive than the high-carbon alternatives, the low-carbon investments will only be made if public policy makes these more attractive financially than the high-carbon alternatives. Furthermore, some investors are not prepared to take on the risks associated with new low-carbon technologies since they are looking for 'safer' low-risk investments (Blyth et al., 2014). Many governments now have substantial experience with instruments that seek to reward low-carbon investors appropriately, including feed-

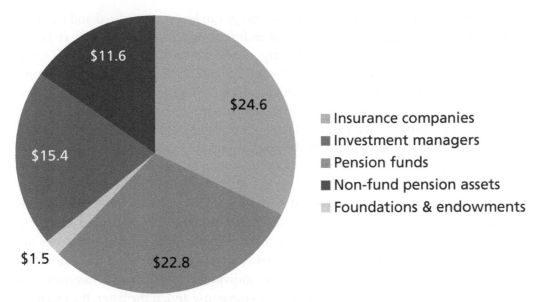

Figure 26.5 Total assets by type of institutional investor
Source: Fulton and Capalino 2014, p. 3.

in-tariffs, portfolio standards or obligations, capital grants, enhanced capital allowances, or other forms of subsidy. If the costs of low-carbon efficiency and energy technologies decline with increasing deployment, as has been the experience with many of them, and if fossil fuel prices stay high or increase, then the level of subsidy required will decrease over time. This is the assumption that lies behind global energy scenario projections, some of which were discussed above, that incur relatively low macroeconomic costs compared with high-carbon alternatives. It will require credible and consistent public policy that gives adequate risk-adjusted returns for private investors to ramp up their investments and achieve a step-change in deployment of clean energy technologies for there to be a realistic chance for these scenarios to be realized.

26.4.3 INNOVATION FOR LOW-CARBON ENERGY SYSTEMS

26.4.3.1 Energy demand: aspirations, behaviours, and technologies

Change in the energy sector since the industrial revolution has been rapid and dramatic, with a huge range of energy demand technologies and associated energy consumption practices being invented, developed, and adopted as new, more convenient, and versatile energy sources became widely available—from wood, through coal, oil, natural gas, nuclear, renewables, and electricity. Going forward, as Chapter 2 made clear, further innovation has a key role in making low-carbon scenarios affordable, and making very low-carbon scenarios feasible. This section summarizes and concludes on innovation in energy demand. The next section does the same for energy supply.

Chapter 7 showed that the demand side of energy can be conceptualized and analysed in a number of different ways: a physical-technical and economic model (PTEM), beloved of energy system modellers, an energy services approach, theories about social practices, and socio-technical transitions theory. All give different insights—some complementary, some conflicting—into how energy demand has evolved in the past and how it might develop in the future. The different approaches to considering influences on energy demand reveal a wide range of factors that play a part in that influence, which introduce considerable uncertainty in the level of energy demand that will exist in the future. In emerging and low-income countries it may be confidently predicted that energy demand will increase, perhaps very greatly, as incomes in these countries rise and new populations acquire effective demand for modern energy services. However, even in these circumstances, the extent of energy demand increase could vary very greatly, depending on the efficiency of the technologies installed, whether policies are adopted to counteract any rebound effects, whether new infrastructure facilitates low-energy lifestyles (see Figure 26.3), and whether individual attitudes and behaviours, and social practices, favour extensive or conserving energy use and, if the latter, the extent to which public policies will be enacted to support these, as described in Chapter 9 in relation to buildings.

Similar arguments apply to road transport, shipping, and aviation, the areas explored in Chapters 10 and 11. Considerable innovation and change may be confidently expected in all these areas, as technologies, social institutions, and attitudes respond to developments in science, economies, and culture. It is currently uncertain as to whether such developments will increase or reduce energy use, and by how much. There are important implications for consumers if, as is projected in many scenarios (see, for example, Ekins et al., 2013), the need for winter heating in cold countries is reduced through increasing the energy efficiency of buildings, the remaining demand for heating is significantly provided by electricity (such as by heat pumps), the need for heating and cooling in all countries is reduced through building design, and both electric and hydrogen fuel cell vehicles become important in road transport. Chapters 9, 10, and 11 have raised and discussed a number of these issues, which are briefly touched on further below.

26.4.3.2 Energy Supply: Technological Research, Development, and Deployment

Chapter 2, and many of the chapters which have explored individual low-carbon energy supply technologies (Chapters 15–21), have stressed the importance of innovation in respect of these technologies, particularly innovation that reduces their costs, if low-carbon energy supply is to become competitive with that from fossil fuels. The extent of cost-reducing innovation is often described through learning or experience curves (see for example, Chapter 18 in this book, Figure 18.5, p. 365), and associated 'learning rates', the percentage reduction in unit cost for each doubling of installed cumulative capacity.

Table 26.2 gives learning rates for different electric power generation technologies from a literature review of different studies. Learning rates may be calculated for one aggregate factor, or split out into two factors, LBD and LBR, as shown in the table. Nuclear and coal have relatively low learning rates. Of the renewables, the narrowest range of estimates is for the best established renewable technology, hydropower. High rates of learning have been estimated for natural gas, on-shore wind, solar PV, and bio-power. Perceptions of the affordability of low-carbon energy systems will be greatly influenced by whether the main low-carbon sources—nuclear, wind, solar, hydro, ocean, and bioenergy—can achieve and sustain high rates of learning into the future.

However, as the range of learning rates in Table 26.2 makes clear, learning is not a given, although it is doubtless very desirable and important, even if large investments or strong policies for deployment are put in place. Chapter 2 also made clear that innovation happens in complex innovation systems that are context specific, and that require appropriate policy frameworks as discussed further below.

26.4.4 TRANSITION TO LOW-CARBON ENERGY SYSTEMS

Energy as it is finally used by consumers can take a number of forms. Modern energy services—power for a range of services, heat, mobility—are now overwhelmingly delivered by one or a combination of four energy vectors or carriers: electricity, natural gas, liquid fuels, and, on a small scale now but perhaps more in the future, hydrogen, all of which can be used for both heat and power generation. All these energy carriers have to be produced from primary energy sources, and all of them can also be produced by carbon-intensive or low-carbon means, except for natural gas, which is itself a primary source of energy that is subject to the least processing to make it suitable for final demand.

If the world's energy system is going to achieve large reductions in carbon emissions, then the delivery of energy services, and the energy carriers associated with this, will have to become less and less carbon-intensive. They will need to be decarbonized.

It would be very surprising, given the very different characteristics of low-carbon technologies from the fossil fuel technologies they are replacing, if a low-carbon energy system was much the same as the predominant energy systems of today. The systemic changes that are to be expected will affect all of the main systems that deliver energy services (heat, power, and mobility), but in very different ways. A low-carbon heat system will not be able to rely on the building-level combustion of fossil fuels, from which the carbon emissions cannot be captured and stored, and will need to switch to heating systems relying on low-carbon electricity, hydrogen or biomass. A low-carbon transport system will not be able to use petrol (gasoline) or diesel derived from oil, but will need to use biofuels or, again, low-carbon electricity or hydrogen. Both heat and transport systems will need to transform energy more efficiently than today into the desired energy services, through more efficient buildings and vehicles. This efficiency

Table 26.2 Range of reported one-factor and two-factor learning rates for electric power generation technologies

Technology and energy source	No. of studies with one factor[a]	No. of studies with two factors	One-factor models[b] Range of learning rates	One-factor models[b] Mean LR	Two-factor models[c] Range of rates for LBD	Two-factor models[c] Mean LBD rate	Two-factor models[c] Range of rates for LBR	Two-factor models[c] Mean LBR rate	Years covered across all studies
Coal									
PC	4	0	5.6–12%	8.3%	–	–	–	–	1902–2006
PC+CCS[d]	2	0	1.1–9.9%[d]	–	–	–	–	–	Projections
IGCC[d]	2	0	2.5–16%[d]	–	–	–	–	–	Projections
IGCC+CCS[d]	2	0	2.5–20%[d]	–	–	–	–	–	Projections
Natural gas									
NGCC	5	1	–11 to 34%	14%	0.7–2.2%	1.4%	2.4–17.7%	10%	1980–1998
Gas turbine	11	0	10–22%	15%	–	–	–	–	1958–1990
NGCC+CCS[d]	1	0	2–7%[d]	–	–	–	–	–	Projections
Nuclear	4	0	Negative to 6%	–	–	–	–	–	1972–1996
Wind									
Onshore	12	6	–11 to 32%	12%	3.1–13.1%	9.6%	10–26.8%	16.5%	1979–2010
Offshore	2	1	5–19%	12%	1%	1%	4.9%	4.9%	1985–2001
Solar PV	13	3	10–47%	23%	14–32%	18%	10–14.3%	12%	1959–2011
Biomass									
Power generation[e]	2	0	0–24%	11%	–	–	–	–	1976–2005
Biomass production	3	0	20–45%	32%	–	–	–	–	1971–2006
Geothermal[f]	0	0	–	–	–	–	–	–	–
Hydroelectric	1	1	1.4%	1.4%	0.5–11.4%	6%	2.6–20.6%	11.6%	1980–2001

[a] Some studies report multiple values based on different datasets, regions, or assumptions.
[b] LR=learning rate. Values in italics reflect model estimates, not empirical data.
[c] LBD=learning by doing; LBR=learning by researching.
[d] No historical data for this technology. Values are projected learning rates based on different assumptions.
[e] Includes combined heat and power (CHP) systems and biodigesters.
[f] Several studies reviewed presented data on cost reductions but did not report learning rates.

Source: Rubin et al. 2015.

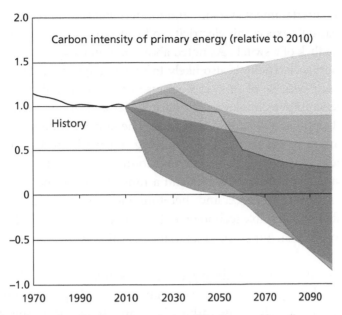

Figure 26.6 Carbon intensity of primary energy in mitigation and baseline scenarios, normalized to 1 in 2010

Source: Clarke L., K. Jiang, K. Akimoto, M. Babiker, G. Blanford, K. Fisher-Vanden, J.-C. Hourcade, V. Krey, E. Kriegler, A. Löschel, D. McCollum, S. Paltsev, S. Rose, P.R. Shukla, M. Tavoni, B. van der Zwaan, and D. van Vuuren, 2014: Assessing Transformation Pathways. In: Climate Change 2014: Mitigation of Climate Change. Contribution of Working Group III to the Fifth Assessment Report of the Intergovernmental Panel on Climate Change (Edenhofer, O., R. Pichs-Madruga, Y. Sokona, E. Farahani, S. Kadner, K. Seyboth, A. Adler, I. Baum, S. Brunner, P. Eickemeier, B. Kriemann, J. Savolainen, S. Schlömer, C. von Stechow, T. Zwickel, and J.C. Minx (eds)). Cambridge University Press, Cambridge, United Kingdom and New York, NY, USA. Figure 6.16

would make these services more affordable even when the fuels providing them were more expensive. Lifestyle and behaviour changes that require less heating and air conditioning, and favour walking, cycling, and public transport, supported by appropriate infrastructure, over private vehicles, could reduce energy expenditures further.

Figure 26.6 illustrates the increasing decarbonization of primary energy that will be required in order to stay within certain ranges of atmospheric concentration of GHGs. It can be seen that a number of the 430–530 ppm scenarios require negative net carbon emissions after 2050 (and after 2070 for some of the 530–650 ppm scenarios). To achieve this, energy technologies must be installed that result in carbon emissions being sucked out of the atmosphere. The most commonly cited of such technologies is power generation using biomass that captures and stores the resulting carbon emissions, which is discussed briefly below.

26.4.4.1 Electricity

Electricity is the most versatile of energy carriers. For many applications it is non-substitutable. Household appliances, modern information and communication technologies and entertainment systems, and modern lighting all require electricity, and consumers increasingly expect this electricity to be instantly available when and where it is required, at the flick of a switch. As noted above, and in Chapters 9, 10, 11, and 24, in low-carbon scenarios electricity is also likely to play a major contribution in delivering heat and transport. Because of this, and because of the availability of a range of low-carbon power generation technologies, the early decarbonization of electricity is usually a feature of cost-optimal low-carbon scenarios. The current state of and prospects for different low-carbon power generation technologies have been described in Chapters 15–20. How these are combined in different countries will naturally depend on the resource endowments of those countries and a range of economic, social, and cultural considerations. A further resource, and therefore energy security, consideration that is raised by a number of renewable technologies in particular, as described in Chapter 22, is the availability of the sometimes rare materials on which these technologies sometimes depend.

It is in the electricity system that the biggest changes from today may be expected as and when societies decide to go low carbon. Even with efficiency improvements, the envisaged role of electricity in providing low-carbon heat and transport means that electricity consumption is likely to increase in industrial countries that already have a high level of energy use, as well as in less-developed countries that still have a high level of unmet demand for electricity. As described in Chapter 21, the integration of, especially, large quantities of renewable energy sources into electricity grids will fundamentally change their nature and operation, raising significant challenges for their continuing smooth operation during a low-carbon transition. The challenges arise from at least three characteristics of renewable power that are at variance with systems that have been based on fossil fuels or nuclear power. First, renewables can generate power locally, at a small scale, requiring a fundamental redesign of centralized generation, transmission, and distribution systems if such power is to be accommodated. Second, renewables can be variable and intermittent, requiring greater back-up capacity, storage, interconnection and demand response than fossil fuel systems. This is challenging both economically and in terms of balancing the system. Third, the generation of renewable electricity tends to have high upfront capital costs but low running costs, with a marginal cost of fuel essentially zero. Market structures based on marginal pricing related to the cost of fossil fuels are completely unsuited to such marginal cost characteristics.

However, with challenges often come new opportunities and advantages. Zero-marginal cost, non-polluting electricity, based on indigenous, inexhaustible resources, will surely be regarded as a great blessing by future generations once the technologies have declined in cost and been well integrated into electricity systems. Decentralized electricity systems give opportunities for local awareness, control, and ownership of energy consumption and production that many people will regard as empowering and more secure. Electricity systems connected across continents to take advantage of the differing availability of renewable energy sources at different times can increase overall energy security. New information and communication technologies that facilitate networks that are more interactive between consumers and producers offer enormous potential for new energy services. And the value of new means of energy storage in low-carbon electricity systems may stimulate technological breakthroughs involving heat, hydrogen, batteries, or means not yet thought of, that again could offer quite new opportunities for market developments. Like many of the technological transformations of the twentieth century, from motor cars to computers, the low-carbon electricity system of the twenty-first is likely to offer benefits and advantages many of which have not yet even been imagined.

26.4.4.2 Transport Fuels

While the dominant transport fuels are currently liquid fuels derived from fossil fuels, the extensive continued use of these fuels is incompatible with a low-carbon future. Low-carbon liquid fuels may be made from biomass (biofuels: bio-diesel and bio-ethanol). The other main fuels considered for low-carbon transport are electricity and hydrogen. The challenges and opportunities relating to bioenergy are the subject of Chapter 17, while low-carbon road transport, shipping and aviation options are discussed in detail in Chapters 10 and 11, with hydrogen discussed further briefly below.

26.4.4.3 Heating and Cooling

High-temperature heat is required for a wide range of industrial processes and is currently overwhelmingly provided by the combustion of fossil fuels. In principle the carbon emissions arising from this combustion could be captured and stored, although this very rarely happens at present. In the Chapter 24 2DS scenario industrial CCS stored 4.8 $GtCO_2$ in 2050, so the availability of this technology is of considerable importance to reduce emissions from industry. Chapter 15 showed that there is still some way to go before the contribution of CCS to decarbonization in this or other sectors is clear. On the efficiency of heat use in industry, there are substantial opportunities for improvement in some sectors in some countries, but this is strongly dependent on the sector and country, as are the policies that would be required to realize them. Some of the opportunities and related policies are reviewed in GEA 2012, Chapter 8.

In many warm developing countries, cooking provides the principal demand for low-temperature heat. As noted in Chapter 8 and in Section 26.2.3 in this chapter, substituting modern energy sources for traditional biomass in cooking can lead to considerable improvements in energy efficiency and indoor air quality, and therefore human health. In higher-income countries and households far larger absolute demands for low-temperature heat are for warmth in buildings in colder climates or seasons, and in warmer climates or seasons for cooling, or air conditioning. Energy efficiency technologies for buildings now exist to reduce both requirements substantially, by 50–90 per cent through 'holistic retrofits', according to GEA (2012: 653), and by 75 per cent or more through application of the most advanced energy building methods for new buildings, in most jurisdictions (GEA, 2012: 675). On the basis of a major new modelling exercise in this area, GEA (2012: 715) concludes:

> Assuming that today's state-of-the-art becomes standard practice in new construction and retrofit, world space heating and cooling energy use can decline by 46 per cent in 2050 from 2005 levels, in spite of the approximately 126 per cent growth in global floor space, elimination of fuel poverty, and significant increases in thermal comfort levels. The implementation of such a scenario requires an approximately US$14.2 trillion undiscounted investment (US$18.6 trillion without technology learning), but results in US$58 trillion savings in undiscounted energy expenditures. However, while this scenario is achievable at net profit, it does require significant policy effort.

As is clear from the discussion in the Energy and Buildings chapter in GEA 2012 (Chapter 10), as well as from Chapters 7 and 9 in this book, the existence of cost-effective opportunities for increased energy efficiency in buildings does not necessarily mean that they will be taken up by building owners and occupants, nor is it always easy to stimulate their take up through public policy. In addition, there are many important issues to be considered if energy efficiency technologies are to be installed in such a way that they do not create other problems, such as lack of ventilation leading to poor indoor air quality, moulds, and damp, or overheating in summer, the latter of which is especially important in the context of the urban heat island effect and climate change. Policies for energy efficiency in buildings need to be framed in a context that takes simultaneous account of the three 'domains' of decision-making discussed further below, involving some mix of pricing, regulation, and support for innovation and new technology. Given the projected scale of urbanization over the next few decades, it is imperative if GHG emissions are to be controlled that both new buildings, as they are constructed, and retrofits, as existing buildings are upgraded and refashioned, are subject to energy efficiency policies that are implemented effectively.

26.4.4.4 Hydrogen

Elemental hydrogen is very reactive, so that hydrogen exists in nature overwhelmingly as compounds, most commonly in hydrocarbons (fossil fuels and biomass) and water. Hydrogen as an energy source needs to be produced from these compounds. If produced from fossil fuels (the dominant current mode of hydrogen production is through the reforming of methane, natural gas), hydrogen is only a low-carbon fuel if the resulting carbon dioxide is captured and stored. Low-carbon hydrogen may also be produced from biomass and through the electrolysis of water. The environmental attractiveness of hydrogen is that, at the point of end-use, its only emission is water or water vapour. This makes it a very attractive option for urban areas where local air pollution is a problem, and for a low-carbon society where the hydrogen has been produced in a low-carbon way.

One of the major projected uses of hydrogen is as a transport fuel in fuel cell vehicles, and this is discussed in Chapter 10. Hydrogen can also be used to generate heat and electricity, separately or as combined heat and power. These uses are discussed in Dodds and Hawkes (2014). Chapter 21 in this book considers many of the infrastructure issues related to hydrogen. Some the wider technological, economic, and social issues related to hydrogen are explored in detail in Ekins (2010), which concludes that whether hydrogen becomes a major energy carrier in the future will depend on the achievement of cost reductions through scientific and technological progress; sound engineering that produces reliable and safe products; performance improvements that make the products attractive to consumers; the strength of policy drivers to reduce energy-related emissions; and public engagement processes that engender confidence in the technologies.

26.4.5 POLICIES TO DEVELOP AND DEPLOY LOW-CARBON TECHNOLOGIES

Many of the earlier chapters in this book have discussed policies related to their topic in some detail, and there is no need, nor intention, to repeat or even summarize these policies here. Rather this section seeks to provide a framework for thinking about energy policy, and particularly about the energy policies that are required in order to decarbonize the energy system.

There are a number of classifications of energy and environmental policies. One typology groups instruments under four generic headings (see Jordan et al., 2003): market/incentive-based (also called economic) instruments, two of the most important types of which are environmental taxes and charges, and emissions trading; classic regulation instruments, which seek to define legal standards in relation to technologies, and environmental performance, pressures or outcomes; voluntary (also called negotiated) agreements between governments and producing organizations, which amount to agreed self-regulation; and information or education-based instruments, such as eco-labels, which promote awareness of the relevant issue or of more eco-efficient products.

In practice a policy instrument may be a hybrid in that it may have the characteristics of more than one type. It has also been increasingly common in more recent times to seek to deploy these instruments in so-called 'policy packages', which combine them in order to enhance their overall effectiveness across the three (economic, social, and environmental) dimensions of sustainable development.

An omission from the above list is innovation policies, which are included in the 2008 OECD policy framework (OECD, 2008), which comprises: direct regulatory instruments; taxes and charges; tradable permit systems; public financial support, including payments for ecosystem services; policies to promote innovation and technological development; information measures; and voluntary schemes.

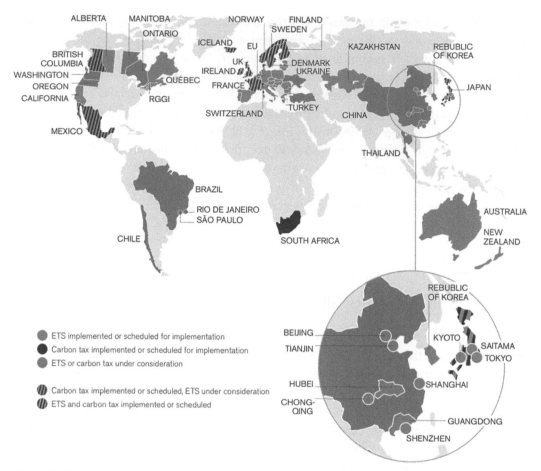

Figure 26.7 Summary map of existing, emerging, and potential regional, national, and sub-national carbon pricing instruments (ETS and Tax)
Source: World Bank, 2014.

Perhaps the most common policy prescription to address climate change is carbon pricing, whether through carbon taxes, emissions trading or some combination of the two. Figure 26.7 shows that, contrary to many perceptions, this is a prescription that has actually been implemented in a number of countries.

Figure 26.7 shows that, globally, 39 national and 23 sub-national jurisdictions have implemented or are scheduled to implement carbon-pricing instruments, including emissions trading systems and taxes. Among these, the emissions trading schemes are valued at about $30 billion. China has the world's second largest carbon-trading system, covering the equivalent of 1115 million tons of carbon dioxide emissions (World Bank, 2014).

Goulder and Schein (2013) have conducted an assessment of the relative advantages and disadvantages of carbon taxes and emission trading systems. On a number of grounds carbon taxes seem to be preferred, one of the most important of which is that additional climate change mitigation policies do not reduce emissions in a cap-and-trade system (unless the cap is adjusted downwards, which then undermines the principal feature of an emissions trading system, which is that it gives assurance over the quantity of emissions), whereas under a carbon tax additional policies do reduce emissions further. This is an important consideration if in fact policy mixes are desirable or necessary in seeking emission reductions, as will be argued below. However, Figure 26.7 suggests that there are political advantages to emission trading systems, such as the ability to allocate emissions permits for free, which have led to them being introduced more frequently than carbon taxes, despite the theoretical advantages of the latter.

Figure 26.7 illustrates another point about carbon pricing and, indeed, climate policy more generally, which is that it does not necessarily have to be implemented at the national level. The Canadian province of British Columbia has a successful carbon tax, and Québec has a cap-and-trade system, whereas there is nothing of the kind at the Canadian Federal level; RGGI (the Regional Greenhouse Gas Initiative) is a cooperative effort among the states of Connecticut, Delaware, Maine, Maryland, Massachusetts, New Hampshire, New York, Rhode Island, and Vermont to cap power sector emissions, and there is an emissions trading system in California, but there is no federal carbon trading system in the USA; similarly, China has emissions trading initiatives in five provinces and two cities, but no national scheme. Beyond carbon pricing, the C40 Cities Climate Leadership Group is involved in a range of actions to reduce greenhouse gas emissions, emphasizing the imperative of urban action on climate change, both to reduce emissions and making cities resilient against climate change, especially in the current context of rapid urbanization.

The Stern Review (Stern, 2007: 349) considered that a policy framework for carbon reduction should have three elements: *carbon pricing* (for example, through carbon taxes or emission trading, as discussed above); *technology policy* (this includes innovation policy, to promote the development and dissemination of both low-carbon energy sources and high-efficiency end-use appliances/buildings); and the *removal of barriers to behaviour change* (to promote the take-up of new technologies and high-efficiency end-use options, and low-energy/low-carbon behaviours).

Chapter 7 showed that changing behaviour in respect of efficiency and other energy technologies is likely to be more complex than simply a matter of removing barriers, but Stern's three-part classification maps closely onto the three policy 'pillars' of Grubb (2014), which in turn correspond to three different 'domains' of risk, economic theory and processes, and opportunity (Grubb, 2014: 47).

Figure 26.8 reproduces Grubb's illustration of his conception. The headings of the three domains—satisfice, optimize, transform—correspond respectively to behavioural, neoclassical, and evolutionary economics. The policy approaches, or 'pillars' most relevant to these domains are respectively standards and engagement (which include regulation, information, and voluntary agreements in the classifications above), markets and pricing (which include economic instruments in the classification above), and strategic investment. The outcomes from these policy approaches are what Grubb calls 'smarter choices', which essentially mean cost-effective increases in efficiency, 'cleaner products and processes' resulting from price-induced changes in behaviour and technology, and 'innovation and infrastructure', which cause the economy to shift to a new production possibility frontier, resulting in this context in much lower carbon emissions. As can be seen from Figure 26.8, both standards and engagement and strategic investment have a medium relevance in the delivery of cleaner products and processes, and markets and prices have some effect on smarter choices and innovation and infrastructure. Because the three policy approaches are aimed at tackling different problems, and have different objectives, it seems likely that, in line with the contention of the Stern Review (Stern, 2007) cited above, all three types of policy are required simultaneously if the required scale and speed of decarbonization are to be achieved.

In the context of the different approaches to and sectors of energy demand discussed in Chapters 9–11, and the technologies of low-carbon energy supply discussed in Chapters 15–20, the different policy pillars may be seen to encourage greater energy efficiency in the use of energy in buildings and transport (Pillar 1), encourage the development and purchase of more efficient machines and appliances (Pillars 1, 2, and 3), accelerate the development and deployment of low-carbon supply technologies (Pillars 2 and 3), and create entirely new technologies (Pillar 3). The combined result of these changes will be new patterns of energy consumption and lifestyles; new business models taking advantage of these combined with the new technologies; and new institutional and regulatory structures that integrate the technologies and behaviours into a new functioning system.

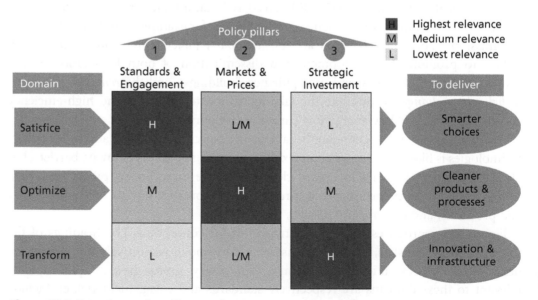

Figure 26.8 Domains, policy pillars, and outcomes
Source: Grubb 2014, Figure 2.5, p. 69.

26.5 **Conclusions**

Capitalism thrives on flux and challenge, especially when they have a strong techno-logical component. As a wicked problem, there is no 'right' answer or easy resolution to the energy trilemma. It cannot be put into a mathematical formulation to be solved. But wicked problems can have good answers, and this book has shown what some of the necessary ingredients of these answers are. On energy security, people first and foremost need physical access to modern energy services. On energy affordability, they need that access to be sufficiently inexpensive for them to keep healthy and participate in the society and economy of which they are a part. And on decarbonization, those that already have high levels of energy access and affordability need to demonstrate that these conditions can be maintained with fast-declining carbon emissions, while those still building their first integrated energy systems need to take best advantage of available and emerging low-carbon technologies.

Many of the technologies required to make progress in all these areas exist and are being deployed, some times more successfully than others. Mistakes are inevitable, for such a social transformation, at the global level, has never before been attempted. Even at national level, this purposeful, policy-driven energy transition is very different from previous energy transitions that were largely driven by markets. What is crucial is that mistakes are learned from, rather than leading to the abandonment of attempts to address all aspects of the trilemma. The achievement of the prize of a low-carbon, secure energy system, accessible and affordable to all, is without doubt within the collective human capability. It will require international cooperation leading to political will, and political will leading to sustained public policy at different levels, leading to changed institutions and behaviours in all countries, that make the best use of new energy technologies, that in turn lead to new opportunities for employment and wealth creation. There are many indications, documented in these pages and elsewhere, including improvements in technological performance, reductions in cost, increases in investment, growth in awareness of climate change and understanding of public policy complexities, that suggest that a tipping point towards a good resolution of the energy trilemma may be in prospect. If this book can help bring that prospect closer, then it will have served its purpose.

■ REFERENCES

Azevedo, I., Jaramillo, P., Rubin, E., and Yeh, S. 2014 (Centre for Climate and Energy Decision Making, Carnegie Mellon University) 'Technology Learning Curves and the Future Cost of Electric Power Generation Technology', presentation to the 37th IAEE International Conference, June, New York, <http://www.usaee.org/usaee2014/submissions/Presentations/IAEE%20Learning%20Azevedo%20Jaramillo%20Rubin%20Yeh%202014.pdf>. Accessed April 2015.

Azevedo, I., Jaramillo, P., Rubin, E., and Yeh, S. 2013 *PRISM 2.0: Modeling Technology Learning for Electricity Supply Technologies* EPRI, Palo Alto, CA. 3002000871, <http://www.epri.com/abstracts/Pages/ProductAbstract.aspx?ProductId=000000003002000871>. Accessed April 2015.

Barrett, S. 2007 *Why Cooperate? The Incentive to Supply Global Public Goods*, Oxford University Press, Oxford.

Bazilian, M. and Pielke, R. 2013 'Making Energy Access Meaningful', *Issues in Science and Technology*, Summer, pp. 74–9, <http://sciencepolicy.colorado.edu/admin/publication_files/2013.22.pdf>. Accessed April 2015.

Blyth, W., McCarthy, R., and Gross, R. 2014 'Financing the Power Sector: Is the Money Available?', Working Paper from the UKERC Uncertainties project, April, UKERC, London, <http://www.ukerc.ac.uk/support/Energy+Strategy+Under+Uncertainty+External>. Accessed April 2015.

Cherp, A. and Jewell, J. 2011 'The Three Perspectives on Energy Security: Intellectual History, Disciplinary Roots and the Potential for Integration', *Current Opinion in Environmental Sustainability*, Vol. 3 No.4, pp. 202–12.

Dodds, P. and Hawkes, A. Eds 2014 'The Role of Hydrogen and Fuel Cells in Providing Affordable, Secure Low-Carbon Heat', a White Paper from H2FC SUPERGEN, London, <http://www.h2fcsupergen.com/whitepaper/lowcarbonheat/>. Accessed April 2015.

Ekins, P. Ed. 2010 *Hydrogen Energy: Economic and Social Challenges*, Earthscan, London.

Ekins, P., Keppo, I., Skea, J., Strachan, N., Usher, W., and Anandarajah, G. 2013 'The UK Energy System in 2050: Comparing Low-Carbon Resilient Scenarios', UKERC Research Report UKERC RR/ESY/2013/001, February, UKERC, London.

ExxonMobil 2014 *The Outlook for Energy: a View to 2040*, Exxonmobil, <http://www.exxonmobil.com/Germany-German/PA/Files/energy_outlook_2014.pdf>. Accessed April 2015.

Fulton, M. and Capalino, R. 2014 *Investing in the Clean Trillion: Closing the Clean Energy Investment Gap*, January, CERES, <http://www.ceres.org/resources/reports/investing-in-the-clean-trillion-closing-the-clean-energy-investment-gap/view>. Accessed April 2015.

GEA (Global Energy Assessment) 2012 *Global Energy Assessment—Toward a Sustainable Future*, Cambridge University Press, Cambridge, UK and New York, NY, and the International Institute for Applied Systems Analysis, Laxenburg, Austria.

Goulder, L. and Schein, A. 2013 'Carbon Taxes versus Cap and Trade: a Critical Review', *Climate Change Economics*, Vol.4 No.3, p. 28, DOE: 10.1142/S2010007813500103.

Grubb, M. 2014 *Planetary Economics: Energy, Climate Change and the Three Domains of Sustainable Development*, Routledge, London/New York.

IEA (International Energy Agency) 2012 *Energy Technology Perspectives 2014*, IEA, Paris.

IEA (International Energy Agency) 2013a *World Energy Outlook 2013*, IEA, Paris.

IEA (International Energy Agency) 2013b *Redrawing the Energy-Climate Map*, World Energy Outlook Special Report, June, IEA, Paris.

IEA (International Energy Agency) 2014 *Energy Technology Perspectives 2014*, IEA, Paris.

IPCC WGII 2014 'Summary for Policymakers', in: *Climate Change 2014: Impacts, Adaptation, and Vulnerability. Part A: Global and Sectoral Aspects. Contribution of Working Group II to the Fifth Assessment Report of the Intergovernmental Panel on Climate Change* (Field, C.B., V.R. Barros, D.J. Dokken, K.J. Mach, M.D. Mastrandrea, T.E. Bilir, M. Chatterjee, K.L. Ebi, Y.O. Estrada, R.C. Genova, B. Girma, E.S. Kissel, A.N. Levy, S. MacCracken, P.R. Mastrandrea, and L.L. White (eds)). Cambridge University Press, Cambridge, United Kingdom and New

York, NY, USA, This and the full reports are at <http://ipcc-wg2.gov/AR5/report/>. Accessed April 2015.

IPCC WGIII, 2014a 'Summary for Policymakers', in: *Climate Change 2014, Mitigation of Climate Change*. Contribution of Working Group III to the Fifth Assessment Report of the Intergovernmental Panel on Climate Change (Edenhofer, O., R. Pichs-Madruga, Y. Sokona, E. Farahani, S. Kadner, K. Seyboth, A. Adler, I. Baum, S. Brunner, P. Eickemeier, B. Kriemann, J. Savolainen, S. Schlomer, C. von Stechow, T. Zwickel, and J.C. Minx (eds)). Cambridge University Press, Cambridge/New York, <http://report.mitigation2014.org/spm/ipcc_wg3_ar5_summary-for-policymakers_approved.pdf>. Accessed April 2015.

IPCC WG III (Intergovernmental Panel on Climate Change Working Group III) 2014b 'Assessing Transformation Pathways', Ch. 6 in IPCC, 2014 *Climate Change 2014: Mitigation of Climate Change*. Contribution of Working Group III to the Fifth Assessment Report of the Intergovernmental Panel on Climate Change (Edenhofer, O., R. Pichs-Madruga, Y. Sokona, E. Farahani, S. Kadner, K. Seyboth, A. Adler, I. Baum, S. Brunner, P. Eickemeier, B. Kriemann, J. Savolainen, S. Schlömer, C. von Stechow, T. Zwickel, and J.C. Minx (eds)). Cambridge University Press, Cambridge/New York, <http://mitigation2014.org/report/final-draft/>. Accessed April 2015.

IPCC WG III (Intergovernmental Panel on Climate Change Working Group III) 2014c 'Cross-Cutting Investment and Finance Issues', Ch.16 in IPCC, 2014 *Climate Change 2014: Mitigation of Climate Change*. Contribution of Working Group III to the Fifth Assessment Report of the Intergovernmental Panel on Climate Change (Edenhofer, O., R. Pichs-Madruga, Y. Sokona, E. Farahani, S. Kadner, K. Seyboth, A. Adler, I. Baum, S. Brunner, P. Eickemeier, B. Kriemann, J. Savolainen, S. Schlömer, C. von Stechow, T. Zwickel, and J.C. Minx (eds)). Cambridge University Press, Cambridge/New York, <http://mitigation2014.org/report/final-draft/>. Accessed April 2015.

Jordan, A., Wurzel, R., and Zito, A. Eds 2003 *'New' Instruments of Environmental Governance?: National Experiences and Prospects*, Frank Cass, London.

Lefèvre, B. 2009 'Urban Transport Energy Consumption: Determinants and Strategies for its Reduction', S.A.P.I.E.N.S [Online], Vol. 2 No. 3 (2.3, Cities and Climate Change), <http://sapiens.revues.org/914>. Accessed April 2015.

McKinsey 2009 *Pathways to a Low Carbon Economy: Version 2 of the Global Greenhouse Gas Abatement Cost Curve*, <http://www.mckinsey.com/client_service/sustainability/latest_thinking/greenhouse_gas_abatement_cost_curves>. Accessed April 2015.

Mitchell, C., Watson, J., and Whiting, J. Eds. 2013 *New Challenges in Energy Security—The UK in a Multipolar World*, London, Palgrave Macmillan.

Mondal, M.A.H. and Sadrul Islam, A.K.M. 2011 'Potential and viability of grid-connected solar PV system in Bangladesh', *Renewable Energy*, Vol. 36, pp. 1869–74.

OECD (Organisation for Economic Cooperation and Development) 2008 'An OECD Framework for Effective and Efficient Environmental Policies', OECD, Paris.

Olivier, J.G.J., Greet, J-M., Muntean, M., and Peters. J. 2013 *Trends in Global CO2 Emissions: 2013 Report*, PBL/JRC, Netherlands, <http://edgar.jrc.ec.europa.eu/news_docs/pbl-2013-trends-in-global-co2-emissions-2013-report-1148.pdf>. Accessed April 2015.

Rittel, H. and Webber, M. 1973 'Dilemmas in a general theory of planning', *Policy Sciences*, Vol. 4, pp. 155–69.

Rubin, E.S., Azevedo, I., Jaramillo, P. and Yeh, S. 2015 'A review of learning rates for electricity supply technologies' *Energy Policy*, Vol. 86, pp. 198–218.

Sharif, I. and Mithila, M. 2013 'Rural Electrification using PV: the Success Story of Bangladesh', presented at PV Asia Pacific Conference 2012, *Energy Procedia* 33, pp. 343–54.

Stern, N. 2007 *The Economics of Climate Change: the Stern Review*, Cambridge University Press, Cambridge.

UNEP (United Nations Environment Programme) 2013 *City-Level Decoupling: Urban Resource Flows and the Governance of Infrastructure Transitions*, a Report of the Working Group on Cites of the International Resource Panel (Swilling, M., Robinson, B., Marvin, S. and Hodson, M.), UNEP, Nairobi.

Wicks, M. 2009 *Energy Security: A National Challenge in a Changing World*, August, DECC, <http://130.88.20.21/uknuclear/pdfs/Energy_Security_Wicks_Review_August_2009.pdf>. Accessed April 2015.

World Bank 2014 *State and Trends of Carbon Pricing*, World Bank Group, Washington DC, <http://www.worldbank.org/en/news/feature/2014/05/28/state-trends-report-tracks-global-growth-carbon-pricing>. Accessed April 2015.

■ AUTHOR INDEX

Aasness, J. 202
Achzet, B. 463
Acker, R. H. 442
Adams, A. 408
Adams, J. G. U. 201
Adjei, A. 172
Afifi, S. N. 429, 443
Aggidis, G. A. 396–8
Agnolucci, P. 429, 435
Agostini, A. 332
Aguilera, R. 248, 249
Alcock, C. 223
Allsopp, C. 263
Allwood, J. M. 126–7, 131, 136, 140
Anandarajah, G. 502, 515
Andaloro, A. P. F. 172
Anderson, B. 176
Anderson, D. 434
Anderson, K. 224
Andrew, R. M. 95
Angerer, G. 361, 453, 455, 457, 463, 470
Ansar, A. 400
Archer, C. L. 407
Armand, M. 460
Arnold, L. 306
Arnstein, S. R. 178–9
Arthur, B. 37
Arvesen, A. 100
Atanasi, E. 276
Aubut, N. 415
Aufhammer, W. 176
Awerbuch, S. 151
Axon, C. J. 178
Ayres, R. 53, 61, 68
Azpurua, R. 276
Azzopardi, B. 361

Babikian, R. 216
Bacon, R. W. 153
Baer, P. 141
Bahn, O. 197
Baiocchi, G. 61
Baker, P. 490
Balvanera, P. 114
Banister, D. 193, 202, 204
Barbose, G. 368
Barnes, D. 151, 153
Barras, C. 442
Barrett, S. 543
Barrett, J. 94, 102, 106
Barrett, M. 202

Barros, V. R. 1
Barth, B. 371
Bartiaux, F. 137–8, 141, 173
Bartis, J. 270, 272
Barton, J. P. 433
Bates, J. 156
Bauknecht, D. 483, 486, 488
Bayar, T. 335
Bazilian, M. 542
Bechberger, M. 418
Bell, C. J. 171
Benzon, D. 396–7
Berechman, Y. 202
Bergek, A. 38, 418, 419
Berghof, R. 215
Berndes, G. 340, 343, 346
Besant-Jones, J. 153
Bettencourt, L. M. A. 364
Bhakar, J. 193
Bickerstaff, K. 323
Bidder, B. 299
Bilir, T. E. 1
Binz, R. 483, 487
Blaikie, N. 164
Blake, W. 541
Blarke, M. B. 432, 445
Blumstein, C. 135, 165
Blyth, W. 552
Boccard, N. 320, 324
Bodansky, D. 85
Boehland, J. 177
Bolinger, M. 410
Bolton, R. 426, 432, 445
Bomsdorf, C. 466
Bond, T. 149
Borgeson, M. 171
Borgreffe, F. 483, 485, 487, 488, 489, 492, 493
Bourdieu, P. 137
Bowles, A. 295
Bowman, C. 274
Bows, A. 224–5
Bradford, P. A. 311
Bradshaw, M. 55, 269
Brandt, A. 248, 276
Breyer, C. 364
Bridge, G. 58, 60, 64, 66
Brisebois, J. 415
Broderick, J. 314
Brophy Haney, A. 168
Brown, M. 168, 170–1
Bryngelsson, M. 46

Bryson, J. 56
Brännlund, R. 203
Buchert, M. 455
Buhaug, Ø. 210, 214, 223
Bulkeley, H. 67, 117
Busch, J. F. 174
Byakola, T. 156
Byrne, R. 149

Caldeira, K. 92, 115
Callon, M. 178
Candelise, C. 361, 365, 367, 368, 370, 458
Cantor, R. 324
Carlsson, B. 138
Carlé, J. E. 361
Castán Broto, V. 67
Castello, S. 368
Cervero, R. 192
Chapman, L. 193
Chapman, S. 421
Charpin, J.-M. 325
Chateauraynaud, F. 323
Chatterjee, M. 1
Chatterton, T. 178, 179
Chaurey, A. 362
Chazan, G. 299
Chegwidden, J. 455
Cherni, J. 440–1
Cherp, A. 540
Chhina, H. 279–80
Chilingar, G. 270
Chindo, M. 269
Chisholm, M. 58, 59
Chitnis, M. 142
Chopra, K. L. 360
Clark, K. 273
Clerides, S. 200
Cleveland, C. 66
Clough, L. 149
Cochran, J. 478, 483, 487, 488, 489, 492
Cochran, T. B. 307, 317
Conde, M. 321
Connor, P. 419, 421
Cook, I. 60
Coombe, D. 192, 193
Cooper, D. 279
Cooper, G. 175
Cooremans, C. 181
Corner, A. 323
Costantini, V. 427, 443
Costanza, R. 112, 113, 114
Couture, T. D. 418
Cowley, S. 318
Creutzig, F. 332
Crooks, E. 235
Crossa, J. S. 310
Cullen, J. M. 126–7, 131, 136, 140

Daily, G. C. 113
Daly, H. E. 196, 199
Damasen, P. I. 442
Dammer, A. 270, 272
Darby, S. 180
David, P. 55
Davis, L. 56, 174
Davis, S. J. 92, 115
De Clercq, G. 299
De Gouvello, C. 152, 154
De Lillo, A. 368
Dean, M. 59
Delille, G. 415
Delucchi, M. A. 407
Demaison, G. J. 273
Demoullin, S. 428, 435, 438
den Elzen, M. G. J. 499
Denholm, P. 360
Depledge, J. 74–5, 78–9
Deressa, T. 400
Devine-Wright, H. 178
Devine-Wright, P. 421
Dewald, U. 38
Dicken, P. 52, 56–7, 59, 62, 64
Dietz, T. 133, 170
Dimitropoulos, J. 142
Dodds, P. 428–9, 434, 435, 438, 561
Dokken, D. J. 1
Dooley, J. J. 42
Doornbosch, R. 42
Dornburg, V. 335
Dray, L. M. 222, 225
Druckman, A. 141
Drupady, I. M. 153
Dröge, S. 77
Duclos, S. J. 470
Duke, R. D. 156, 443
Duncan, D. 270–1
Durix, L. 152
Dutton, J. 269, 283
Dyni, J. 271

Eatherley, D. 455, 460, 466, 469, 470
Eberhard, A. 442
Ebi, K. L. 1
Edens, B. 95
Edmonds, J. A. 315
Eide, A. 246, 332
Eide, M. 223
Ekins, P. 430, 502, 508
El-Katiri, L. 262
Elam, M. 322–3
Eliasson, J. 195
Ellegård, A. 148, 152
Ellenbeck, S. 431
Elliott, D. 307, 310, 315, 318
Emmert, S. 170

Enflo, K. 55
Enkvist, P.-A. 135, 166
Erdmann, L. K. 469, 470, 471
Estrada, Y. O. 1
Evans, A. D. 225
Ewing, B. R. 115
Eyre, N. 142

Faaij, A. P. C. 332, 334, 335
Falcan, J. 325
Farrell, A. 248, 276
Fast, S. 117
Fawcett, T. 142
Federico, G. 488
Feenstra, R. C. 214
Ferioli, F. 45
Fichaux, N. 416
Field, C.B. 1
Figueres, C. 82
Fischer, C. 107
Fischer, G. 344
Fischlin, A. 525
Flannery, B. 85
Fleiter, T. 167
Flynn, D. F. 114
Flynn, H. 366
Foley, G. 151
Fontaras, G. 201
Fouquet, R. 36, 37, 53, 128, 129, 130, 141
Fox, A. K. 107
Foxon, T. 37
Francu, J. 270
Frank, L. D. 202
Frantzeskaki, N. 39
Freeman, C. 35, 36, 39
Freidberg, S. 60
Fridley, D. 174, 440
Froggatt, A. 314, 315, 316, 323
Fthenakis, V. 463, 467
Fulton, L. 190, 193
Furchtgott-Roth, D. 428

Gales, B. 55
Gallagher, K. S. 36, 37
Garman, J. 297
Garud, R. 419
Gasparatos, A. 113, 525
Geels, F. W. 8–9, 13, 38, 199, 218
Genova, R. C. 1
Gibbons, S. 421
Giddens, A.137
Gifford, R. 88
Gillingham, K. 142, 167
Girma, B. 1
Girod, B. 190
Givens, N. 283
Goldberg, S. 307, 313, 320

Goldemberg, J. 310, 320
Gomez-Baggethun, E. 113
Goodrich, A. 369
Gottstein, M. 490, 492
Gowing, M. 306
Graedel, T. E. 455, 463, 466, 469, 470, 471, 472
Graham, W. R. 221
Gram-Hanssen, K. 137, 172
Green, J. D. 211
Green, K. P. 428
Green, M. A. 458, 459
Grimes, R. W. 317, 318, 320–2, 326
Gross, R. 45–6, 200
Grossman, G. M. 526
Grubb, M. 87, 167, 563–4
Grubler, A. 34, 52, 125–6, 129, 131, 136, 138, 139, 324, 408
Grunau, C. 413
Gustavsson, M. 148, 152, 283
Guthrie, H. 282
Guttman, J. T. 466
Guy, S. 169

Ha, H.-K. 215
Haas, R. 133, 142, 324
Hagelüken, C. 462
Haines-Young, R. 528
Hall, C. 53, 54
Hall, D. O. 331
Hall, P. J. 433
Hallerman, T. 235
Hamelinck, C. N. 333
Hamilton, J. 178
Hamilton, K. 525
Hanaoka, T. 166, 168
Hansen, J. 83
Hansena, Z. 399
Hansson, A. 46
Hardman, S. 438
Hare, W. 499
Hargreaves, T. 137
Harris, J. 142
Harris, P. G. 88
Harvey, D. 52
Harvey, L. D. D. 174
Hawkes, A. 426, 432, 445, 561
Hay, R. K. M. 343
Healy, S. 174
Hecht, G. 306
Hekinian, R. 395
Hekkert, M. P. 38
Held, D. 52
Helm, D. 62
Hemmelskamp, J. 38
Heptonstall, P. 413
Herring, H. 306
Hertwich, E. G. 92, 101

Hewitt, C. N. 116
Hewlett, J. 324
Hickman, R. 193, 204
Hiesl, A. 324
Hogan, M. 488, 490
Holmberg, D. G. 432
Honoré, A. 262
Hoogwijk, M. 343
Houari, Y. 459, 462, 464, 467, 472
Howard, D. 67
Hrdomadko, J. 297–8
Hubbert, M. K. 452
Huber, M. 61
Huc, A.-Y. 276
Huckerby, J. 378–9
Hudson, L. A. 164
Hultman, N. 307, 308, 325
Hummels, D. 59–60
Hurley, D. 487, 492
Hutton, M. Z. 270
Høyer, K. G. 198

Illich, I. 141
Infield, D. G. 433
Ismer, R. 107
Iyer, M. 174

Jaccard, M. 239
Jackson, T. 141
Jacobson, A. 407
Jacobson, A. D. 153, 156
Jacobsson, S. 38, 418, 419, 421
Jacoby, H. 190, 200
Jaffe, A. B. 36, 166
Jager-Waldau, A. 364
Jakob, M. 102–3
Janda, K. B. 135, 172, 174, 178, 179, 180, 181
Jeffrey, H. 379–80, 384
Jenkins, B. M. 432, 445
Jevons, W. S. 68, 452
Jewell, J. 321, 540
Jiusto, S. 66
Johansson, T. B. 131, 343, 344, 430, 441
Jones, C. 234, 235
Jones, N. 460
Jordan, A. 82, 561
Joshi, M. 499
Joskow, P. 62, 483
Jotzo, F. 87
Junginger, M. 45–6
Juniper, T. 112
Jørgensen, M. 361

Kahn Ribeiro, S. 189
Kahya, D. 297
Kainuma, M. 166, 168
Kallis, G. 321
Kama, K. 269, 271–2

Kammen, D. M. 42, 153, 156, 442
Kandpal, T. C. 362
Kapika, J. 442
Karekezi, S. 149
Karnøe, P. 419
Karvonen 171
Kasarda, J. D. 211
Keay Bright, S. 479, 486, 487, 488, 490–2
Keith, D. W. 408
Kemp, R. 138, 139
Kempener, R. 40
Kempton, W. 176
Kentish, J. 440–1
Kenworthy, J. 192
Kern, F. 38
Kesicki, F. 515
Kessides, J. N. 307, 308, 317, 319, 324
Khodyakova, Y. 299
Khrennikova, D. 295
Kilpatrick, J. 343
Kim, S. H. 315
King, S. 175
Kingsnorth, D. J. 455
Kirby, B. 483, 487
Kirkegaard, J. 63
Kirubi, C. 153, 442
Kissel, E. S. 1
Kithyoma, W. 149
Klaassen, G. 45
Klare, M. 256
Knopper, L. D. 421
Kockelman, K. 192
Kohler, J. 223
Komatsu, S. 153
Koomey, J. 307, 308, 325
Kordjamshidi, M. 175
Kozloff, K. 153
Krammer, P. 224
Krause, M. 153
Krebs, F. C. 361
Krewitt, W. 354
Krijgsman, R. 292
Krueger, A. B. 526
Krupa, J. 235
Kruzner, K. 176–7
Krzyzanowski, M. 202
Kumar, A. 399
Kumar, M. 149
Kumar, S. 427
Kuuskraa, V. 232, 282
Kuznetsov, V. 319
Köhler, J. 38

Lachapelle, U. 202
Laherrère, J. 270
Lamers, P. 428
Lammi, H. 307
Langhelle, O. 239

Lantz, E. R. 410
Larsen, E. R. 202
Lattanzio, R. 276
Lauber, V. 419, 421
Le Billon, P. 58, 64
Le Quéré, C. 92
Leach, M. 434
Lechner, N. 176
Leduc, G. 204
Lee, B. 43–4, 409
Lee, D. S. 210, 224
Lefèvre, B. 192, 193, 545–6
Leggett, J. 19, 256
Lehtonen, M. 323
Leidreiter, A. 335
Leiringer, R. 134, 163, 164, 181
Lemaire, X. 153–4
Lenoir, A. 135
Lenzen, M. 95, 96–7, 115, 116, 133, 525
Leontief, W. 94
Levine, M. 136
Levy, A. N. 1
Lew, D. 415
Lewis, B. 201
Li, Y. 116
Lilliestam, J. 431
Lin, J. 174
Lindeboom, H. J. 413
Lindman, Å. 409
Linvill, C. 487
Liu, H. 61
Lohner, T. 434
Loorbach, D. 39
Louçã, F. 36
Ludwig, J. 331
Lundvall, B.-A. 36
Luo, L. 38
Lutzenhiser, L. 133, 135, 168, 176, 177, 179
Lévêque, F. 311, 324, 325
Lyman, E. S. 319
Lynas, M. 83
Lynd, L. R. 346
Lysen, E. 340

Ma, L. 131
Maas, J. 114
Macalister, T. 322
McCormack, D. P. 172
MacCracken, S. 1
McDowall, W. 419, 420, 428–9, 434
Mace, G. M. 112, 525
MacGillivray, A. 387
McGlade, C. 249, 260
McGlade, C. E. 270–1, 502, 508
McGowan, F. 9
Mach, K. J. 1
Machado, G. 61, 116
MacKerron, G. 314, 320

McKinley, R. 254
McManus, B. 274
McNeill, J. 68
Madden, P. 276
Maigne, Y. 154
Makan, A. 298
Mander, S. 215
Mann, P. A. 137
Marigo, N. 367, 370
Markard, J. 139
Marschinski, R. 102–3
Martinez-Alier, J. 61
Marvel, K. 407
Marx, K. 141
Mastrandrea, M. D. 1
Mastrandrea, P. R. 1
Masui, T. 507
Matlack, C. 466
Matson, P. A. 113
Matthews, R. 347
Mauthner, F. 357
Maycock, P. D. 365
Meadowcroft, J. 239
Mecklin, J. 311
Mehos, M. 360
Mei, M. 274
Meier, A. K. 135, 165
Meinshausen, M. 502
Melillo, J. M. 197
Meskers, C. E. M. 462
Metz, B. 507, 525
Metz, D. 190, 193, 203
Meyer, I. 199
Meyer, R. 276
Mez, L. 307
Mfuh, W. 151, 155
Michael, K. 234
Michal, C. 176
Millard-Ball, A. 190
Mille, L. M. R. 407, 416
Miller, T. 180
Milligan, M. 483, 487
Millington, D. 273–4, 276, 280
Minx, J. 61
Mitchell, C. 10, 419, 421, 541
Mitchell, D. 246
Mitchell, J. 257
Mithila, M. 542
Modi, V. 440–1
Moezzi, M. 134, 135, 142, 172, 174, 177, 178
Mokhtarian, P. 194
Moll, H. C. 106
Mondal, M. A. H. 542
Moner-Girona, M. 151
Monteith, J. L. 343
Morawski, J. 276
Morita, T. 507
Morley, N. 455, 460, 466, 469, 470

Morris, E. 153, 442, 443
Moss, R. L. 453, 466, 469, 470
Mourik, R. 180
Mrk, G. 383
Muirhead, J. 358
Mulhall, R. 56

N'Guessan, M. 150
Naik, S. N. 112
Negro, S. O. 38
Neij, L. 45–6
Nelson, K. 138
Nelson, R. R. 138
Nemet, G. 42
Nemet, G. F. 365
Nesheiwata, J. 310
Neuhoff, K. 107, 483, 485, 487, 488, 489, 492, 493
New, M. 499
Newell, P. 77
Newman, P. 192
Nonhebel, S. 106
Nordström, S. 153
Nuttall, W. J. 317, 318, 320–2, 326

Ó Gallachóir, B. P. 196, 199
Obama, B. 236, 281
Ogden, J. M. 434
Ogunsola, O. 270
Oliver, C. 235
Olivier, J. G. J. 73, 543
Ollson, C. A. 421
Ong, S. 354
Orcutt, M. 279
Otieno, D. 156
Owens, S. 194, 204
Ozanne, J. L. 164

Palit, D. 361
Palmer, K. 167
Pan, J. H. 440
Parag, Y. 142, 179, 181
Park, Y. 215
Pascal, L. 276
Paulsson, E. 77
Pavitt, K. 36
Pawson, R. 180
Pazheri, F. R. 395
Pearson, P. 36, 37, 53
Peffer, T. 174
Peng, W. Y. 440
Penner, J. E. 210, 224
Perez, C. 494
Peters, G. P. 61, 92, 94, 95, 96, 101, 115
Pickett, K. 118
Pielke, R. 542
Piore, A. 316
Plotkin, S. 196

Potschin, M. 528
Poudenx, P. 195
Power, A. G. 114
Prentice, J. 235
Pretty, J. 114
Princen, T. 142, 178
Prins, G. 75

Rabl, A. 45
Ragwitz, M. 418
Rai, A. 396–7
Rai, V. 46
Ramana, M. V. 310, 324
Rao, N. D. 141
Raudsepp-Hearne, C. 114
Rayner, S. 75
Reckwitz, A. 137
Rehman, I. H. 150
Rehmatulla, M. 223
Reiche, K. 362
Remme, U. 248
Renard, C. 317
Rethinaraj, G. 310, 312, 315, 317, 324
Riahi, K. 136, 408, 507
Riesz, J. 483, 487, 488, 492
Riley, A. 299
Rip, A. 138
Rittel, H. 538
Roberts, B. W. 416
Roberts, J. T. 141
Robinson, A. P. 430
Rogan, F. 200
Rogelj, J. 499
Rogers, H. 295, 301
Rogner, H.-H. 19
Roop, J. M. 116
Rosa, E. A. 133
Rosenau-Tornow, D. 455, 463, 470
Rosenfeld, A. 134
Rosenow, J. 170
Rosner, R. 307, 313, 320
Rothwell, R. 35
Rotmann, S. 180
Rubin, E. S. 45, 46
Rubin, J. 66
Rudin, A. 142, 177
Rudkin, E. 368
Ruhl, C. 54–5, 64
Ruiz-Perez, M. 113
Russell, P. 270–1
Ruttledge, K. 415
Ryckmans, Y. 337

Sadrul Islam, A. K. M. 542
Saenz de Miera, G. 416
Sagar, A. D. 45
Samaras, Z. 201

Samiullah, S. 174
Sanchez, T. 153
Santos, G. 193
Saunders, M. 128, 141, 193
Sausen, R. 224
Sayer, A. 494
Schaeffer, M. 83
Schatzki, T. 137
Scheutzlich, T. 153
Schipper, L. 190
Schneider, M. 308, 310, 311, 312–13, 315, 324, 326
Schnoor, J. L. 429
Schäfer, A. 59, 189, 190, 200, 212, 215, 217, 225
Scholl, L. 190
Schot, J. 139
Schrattenholzer, L. 344
Schreurs, M. A. 307, 310, 312, 326
Schultem, B. 153
Schultz, S. 299
Schwanen, T. 172
Schweber, L. 134, 163, 164, 181
Schwägerl, C. 320
Schüler, D. 463, 470
Scott, A. 36, 151, 314
Scrase, J. I. 46
Searchinger, T. 332
Sebastia-Barriel, M. 59
Senger, F. 223
Sensfuß, F. 416
Seth, P. 151
Seyfang, G. 178
Sharif, I. 542
Sharp, J. 239
Shove, E. 134, 137, 168–9, 173–4, 175–7, 179
Shum, R. Y. 429
Sichel, A. 466
Siegrist, M. 323
Sioshansi, F. 479, 483, 487, 489, 492
Skea, J. 43, 47, 426, 429
Skillings, S. 490, 492
Slade, R. 342
Smeets, E. 344
Smil, V. 53, 55, 57–9, 67–8
Smith, A. 139
Smith, J. B. 81, 83
Smith, T. W. P. 218, 225
Smokers, R. 199
Socolow, R. 133
Soete, L. 39
Solli, C. 96
Sonvilla, P. M. 370
Sorrell, S. 19, 53, 128, 142, 203, 249
Sovacool, B. K. 130, 153, 177, 179
Speight, J. 272
Speirs, J. 454, 456, 459, 460, 462, 468–9, 470, 471, 472
Spirov, P. 279

Springmann, M. 107
Srinivasan, T. S. 310, 312, 315, 317, 324
Stankiewicz, P. 320
Stankiewicz, R. 138
Stavins, R. N. 166
Steinberger, J. K. 141
Stephenson, J. 180
Sterman, J. 200
Stern, J. 10, 37, 179, 263, 295, 301
Stern, N. 563, 564
Stevens, P. 283, 292
Steyn, G. 36
Stoft, S. E. 135, 165
Stopford, M. 217
Strange, S. 52, 53
Strbac, C. 433, 437
Strengers, Y. 176–7
Struben, J. 200
Strzelecki, M. 297
Subing, L. 395
Sukhdev, P. 112
Sundqvist, G. 322–3
Sunikka-Blank, M. 135
Sunstein, C. R. 178
Susskind, L. E. 85
Sutherland, R. J. 204
Swanson, V. 270–1
Szabó, S. 148, 156
Szarka, J. 308
Söderholm, P. 409

Tarascon, J. M. 460
Taylor, R. 400
Teece, D. J. 408
Teichroeb, D. 429
Templeton, M. 149
Thaler, R. 178
Thirumurthy, N. 362
Thomas, C. E. 198, 204
Thomas, S. 308, 313, 317, 319, 320, 324, 325, 326
Thomson, A. 507
Thrän, D. 340
Tickell, O. 318
Tilton, J. E. 464
Topouzi, M. 180
Torny, D. 323
Trembath, A. 282
Trezona, R. 37
Truffer, B. 38
Truninger, M. 137
Tukker, A. 95
Turkenburg, W. C. 333–4
Twena, M. 307

Uhomoibhi, J. 442
Unruh, G. C. 37, 139
Upton, S. 42

Urban, F. 440
Ürge-Vorsatz, D. 136, 142, 168, 180

van Alphen, K. 236
van Campen, B. 152
van Dam, J. 348
van den Berg, W. 44
Van den Brink, R. M. 199
van der Horst, D. 114–15, 525
van der Slot, A. 44
van der Zwaan, B. 45, 413
Van Noorden, R. 201
van Sark, W. 365
van Vliet, B. 139
van Vuuren, D. 507
Van Wee, B. 199
Vera Morales, M. 221
Verbong, G. 38, 138
Verhoest, C. 337
Vermeylen, S. 114–15, 525
Vine, E. 170
Visschers, M. 323
Vié, J.-C. 114
von Stechow, C. 239
Vuchic, V. R. 192

Wade, N. 432
Wadia, C. 361, 459
Wahab, G. 339
Walker, A. J. 343
Walker, G. 310, 326, 494
Walker, G. P. 117
Wamukonya, N. 149
Warde, A. 137
Warr, B. B. 53, 61, 68
Watson, J. 46, 230, 234, 238, 314, 325, 441
Webber, M. 538
Weber, C. 489
Weber, M. 38
Wee, B. 193
Weiss, J. 170
Weiss, M. 197
Weiss, W. 357

Werner, C. 368
Wessely, S. 199
Westing, A. H. 466
White, L. L. 1
Whittlesey, R. W. 416
Wicke, B. 346
Wicks, M. 540
Wiedmann, T. 93, 94, 96, 100, 115, 525
Wilhite, H. 175–6
Wilk, R. 141, 180
Wilkinson, P. 167
Wilkinson, R. 118
Williams, T. I. 428
Wilson, A. 171–2, 177, 178, 179
Wilson, C. 139
Wilson, I. A. G. 430
Winskel, M. 46, 307
Winter, S. G. 138
Wiser, R. 368
Wolfram, C. 130–1, 176
Wood, D. 427
Woodcock, J. 202
Woods, J. 346
Wrigley, E. 58, 68
Wyckoff, A. W. 116

Yaksic, A. 464
Yamamoto, H. 344
Yamin, F. 74, 78
Yang, C. 434
Yen, T. F. 270
Yeo, S. 87
Yeung, H. 52, 65
Yu, Y. 115, 525

Zachariadis, T. 199, 200, 201
Zerriffi, H. 149
Zhang, T. 133
Zhao, H. 283
Zheng, J. Y. 434
Zomer, A. N. 153
Zoomers, A. 399
Zweibel, K. 360

■ GENERAL INDEX

affordability *see* energy
agriculture 16, 152, 299, 397
  and bioenergy 331–2, 341, 346
  energy for 54–5, 61
  environmental degradation 113
air–conditioning 176
Airbus Global Market 215
aircraft:
  contrails 224
  energy intensity 216–17
  technology 221, 222
  see also aviation
airlines, operational changes 221–2
airport capacity 218–19
airspace congestion 219
Alaska 257
Albania 274
Algeria 232–3
anaerobic digestion 334, 337
Angola 274
appraisal optimism 46
Argentina 116, 261, 285, 296, 339
Australia 427, 466
  emission inventories 95
  fossil fuels 252, 262–3, 285
  LNG 295, 296
  regulation 493
  solar energy 363
Austria 396
aviation 209–25, 554
  biofuel 220
  demand for 5, 210–12
  energy demand 209–10
  freight 211
  hydrogen 220
  missing from emission targets 78–9
  passenger 211–12
  see also aircraft
Azerbaijan 274

Bangladesh 153
  Infrastructure Development Company Limited
      (IDCOL) 542
behaviour:
  barriers to change 563
  change efforts 178–80
  four dimensions of 179–80
Belgium 172–3, 296, 310
bioenergy 5, 331–50, 502
  future demand for 29

and sustainability 345–9
  technologies 332
biofuel:
  advantages of 197
  aviation 220
  bioethanol 118–22, 339
  disadvantages of 67, 346
  global use 337–9
  supply chain 118–19
biogas 337, 338
biomass 10, 15
  availability modelling 339–45
  certification schemes 347–8
  definition 331
  feedstocks 333–4
  global use 331
  for heating 334–5
  increasing production 342
  land and resource conflicts 332
  marine and terrestrial impacts 531–3
  potential 343–4
  for power generation 335–7
  small appliances 334–5
Bloomberg 480
Boeing Current Market Outlook 215
Border Tax Adjustment (BTA) 107
BP 64, 233, 262
  model 27–30
  Statistical Review of World Energy 250, 251,
      252, 257
Brazil 261, 270, 296, 466
  bioenergy 335, 339, 345
  bioethanol 118–22
  hydropower 396, 431
  impacts imported 116
  R&D 30
  renewable energy 40
  wind power 420
buildings:
  cooling 176–7
  energy efficiency 163–81, 560
  energy use in 4–5
  increasing size 177
  passive designs 176–7
  residential sector 16–17
  solar energy 357
Bulgaria 172–3, 542

Canada 14, 466, 563
  capital cost financing 170

Canada (*cont.*)
  CCS 231–2, 234–7, 239–40
  emission inventories 95
  gas 262, 263
  hydropower 394, 396
  national targets 85
  ocean energy 380, 393
  oil 256, 274
  wind power 420
capability:
  frequency response 480
  markets 492–3
  mechanisms 490–3
capacity payments 490
capital cost 133, 249
    bioenergy 335
    CCS 230, 237, 239–40, 517
    conventional energies 150
    financing 165, 169, 170–2
    from low to high 477, 558
  networks 428, 432, 434, 443–4, 446
    nuclear 319
  oil sands 380
    transport 197, 199–200
    wind 409–10, 419
carbon:
  debt 347
  intensity 67, 217, 557
  leakage 101, 107
  pricing 3, 67, 239, 562–3
  taxes 562–3
  see also CO_2; taxes
carbon capture and storage (CCS) 5, 229–41
  capital grants 235, 239
  in global scenarios 502, 506, 510, 559
  global status 231–2
  incentives 239–40
  Large-scale Integrated Projects (or LSIPs) 231,
      233, 241
  new capacity installation 519–20
  technology demonstration 234–40
  under the CDM 80
cars:
  occupancy 194, 195
  on–road efficiency 199
  ownership 191, 192
  taxes 200
  see also transport; vehicles
cement production 543–4
Central Western European Pool (CWEP) 480
Channel Islands 381
Chevron 64, 271–3, 286, 301
China 44, 563
  airports 219
  bioenergy 337
  car ownership 191–2
  CCS 232, 520
  climate change regime 79

coal 15, 252, 253
  electricity networks 432
  emissions 76, 104–5, 514, 543
  energy demand 12–14
  energy services 131
  gas 22, 261, 263, 285
  hydropower 397
  LNG 296
  network infrastructures 439–41, 444
  nuclear power 312–13, 325
  oil 257, 270
  political leadership 87
  R&D 30, 232
  rare earths 460, 466, 468
  regulation 174
  renewable energy 40–1
  solar energy 357, 358, 363, 367, 369, 370
  transport 215
  wind power 405, 411, 414, 420
Clean Development Mechanism (CDM) 74, 77–8, 80,
      87, 108, 544
  certified emission credits (CERs) 77–8, 80
climate change 4, 10, 25–7
  national targets 84–6
  windows of opportunity 502–7, 520–1
climate change regime 73–88
  political leadership 86
  regulation 87
  and technology 79–80
  temperature target 81–7
CO_2 1–2, 26–7
  concentration 83
  growth of emissions 30
  intensity 190, 191
  marginal abatement cost curves (MACCs) 135
  prices 515–16
  storage reservoirs 241
  top emitters 75–6
  transported 229
  see also carbon; emissions; indirect emissions
coal 297–300
  availability 19–20
  categories of 252
  China 15
  cumulative production levels 507–8
  emission performance standard (EPS) 235
  future demand for 29
  learning rates 555
  non–power uses 251
  prices 2–3, 24, 297
  production and consumption 253–5
  reserves 246, 251–2
  storage 432
  trade 23
  see also transitions
coal–bed–methane (CBM) 269
coal–fired power plants 46
cobalt 466

cognitive rules 139
Colombia 274
combustion 333, 334–5
Common International Classification of Ecosystem
    Services (CICES) 528, 530
compressed natural gas (CNG) 196–7
concentrating solar power (CSP) 354, 356, 358–60
Congo Brazzaville 274
conservation supply curves (CSCs) 165–7
consumption:
    emissions 79
    and energy 53, 60–1
    growth 17
    limits to 142
    policy 106–7
    reduction 106
cooking 2, 137, 149, 152, 428
    biomass 2, 15–16, 151, 331, 441–2, 541, 559
    modern stoves 442, 443, 542, 559
cooling 559–60
critical metals 452–73
    availability 464–5, 470
    by-products 462
    criticality 452, 468–73
    end-use technologies 455, 456
    future demand 453–60, 463, 470–1
    future supply 460–8, 472
    and geopolitics 465–6, 470
    material intensity 453, 456, 459
    policy decisions 465–6, 473
    production costs 464
    production rate 462–4
    recycling 461, 466–8, 470, 472–3
    reserves 461–2
    resources 461
    substitution 456, 459–60, 470
    utilization rate 458
cultural services 529
cycling 195, 202
Czech Republic 172–3, 254, 364

dams 399
Danish Energy Agency 405
decarbonization see electricity decarbonization; energy
decarbonization
Democratic Republic of Congo 274, 466
Denmark 172–3
    bioenergy 335
    electricity system 491, 493
    emission inventories 95
    ocean energy 378
    regulation 494
    wind power 411, 419
DONG Energy 413

ecological debt 61
economic development
    and energy availability 55
    and energy demand 11–14
    and energy use 68
    and environmental pressures 526
    and transport 202–3
economic integration see globalization
economies of scale 59, 477
ecosystem services 6, 112–22
    cultural 112
    and energy production 114–15
    and energy technologies 531–3
    global impact drivers 116
    impacts exported 115–17
    impacts on 525–35
    marine 529–33
    provisioning 112
    regulating 112
    social impacts 117–22
    supporting 112
    sustaining 546–7
    value of 113–14
Edison 299
electric vehicles (EVs) 459–60
electricity 17–18
    changing demand 487
    decarbonization 437, 511–12, 521, 557–8
    demand-side management (DSM) 431–2, 439
    from solar energy 356, 358, 370–1
    global reach 526–7
    HVDC (High Voltage Direct Current) 358
    levelized costs of electricity (LCOE) 359, 365, 369,
        389, 390
    prices 482, 485, 490
    production costs 230
    renewable 67, 483, 488
    storage 396–8, 429, 432–4, 437, 439
electricity markets 476–94
    nature of 480, 489
    overview 479–82
    pool mechanism 489
    real time and forward 479–80, 481, 482
    rules and incentives 488
    state intervention 63
    wholesale 479
electricity networks 429–32
    China case study 440–1
    decommissioning 437
    Germany 438–9
    grid 541–2
    high-voltage DC (HVDC) lines 431
    reinforcement 430
    smart grids 431–2, 434
    super grids 431, 443–4
electricity supply:
    access to 441–3
    security 314, 440
electricity system:
    balancing 480–2, 486–7
    challenges to 483–7

electricity system: (*cont.*)
  decarbonization 511–12, 521
  decentralized 150, 477, 558
  fossil–dominated 484
  management 370–1
  operator 493
  ownership 64, 478–9
  policies for 483
  regulation 539–40
  renewables in 558
  storage 371
  structure 478–9
electrification:
  and development 149
  and the grid 148
  public–private partnerships 153–4
  rates of 131
  strategies 152–3
  subsidies for 149
emissions 1
  cement production 543–4
  consumption–based 79, 115
  drivers of 133–4
  embodied 79
  fossil fuels 543–4
  future scenarios 98–9
  global 26–7
  global negotiations 543–4
  lifecycle 93, 99–101
  methodologies 78–9
  peak 503–5
  production–based 79, 115
  trading systems 562–3
  see also carbon; CO_2; indirect emissions
emissions inventories:
  consumption–based 93–101
  definitions 94
  production–based 93, 96
  territorial–based 93–4, 95–6
emissions mitigation 499–521
  achievable 503
  economic implications 515–17
  global costs 549–50
  investment 550–3
  regional per capita 513–14, 521
  targets 81–6
  technology deployment 517–20, 521
energy:
  access to 148–56
  affordability 9, 477, 538, 541
  availability 19–20, 55, 68
  and consumption 53, 60–1
  decarbonization 476, 477, 483, 538, 541,
    555, 557
  decent living approach 141
  environmental impacts 10
  as factor of production 52, 53–8
  global context 9–31

  global scenarios 547–9
  levelized cost 413
  market integration 62–5
  measuring 11
  needs approach 140
  per capita use 12–14
  prices 2–3, 24–5, 56, 476
  relative decline in costs 53
  renewable 40, 87, 299, 430, 483–5
  security 2, 10, 18–25, 477, 538, 540–1
  storage 6, 396–7, 559
  sufficiency 142
  trade in 20–4
  and transportation 53, 57–80
  well–being approach 140–1
  see also bioenergy; renewable energy
energy demand 4, 125–42
  Africa 12–13
  aviation 209–10
  and development 130–1
  drivers of 129, 133–4, 476
  and economic development 11–14
  models 553–4
  primary 12, 14
  shipping 210
  trends 16–18, 125–34
energy efficiency 67, 476, 520, 559
  aviation and shipping 221
  buildings 163–81, 560
  gap 166, 167–9, 189
  positivism in 165–9
  project choice 178
  regulation 173–5
  techno–economic potential 134–5
  technology investment 44
  vehicles 195–201
Energy Efficiency Design Index (EEDI) 224
energy futures models 27–31
energy intensity 190
  aircraft 216–17
  convergence 64, 69
  decline in 55
  of exports 105
  of production 67
  shipping 217–18
energy production:
  decentralized 436–7, 440, 442
  and ecosystem services 114–15
energy services 135–6
  access to 2, 4, 60–1, 541–2
  decentralized 150, 154, 477, 542, 558
  demand reduction 136
  elasticities 128–30
  theory 209
energy supply 15–16
  conservation curves 135
  global trends 125–34
  innovation needed 554–5

energy systems:
  costs 515–16
  decarbonization policies 105–8
  global integration model 6
  and globalization 52–69
  and innovation 34–48, 553–5
  markets 539–40
  regulation 539–40
energy transformation sector 16
energy trilemma 427, 565
  affordability 9, 477, 538, 541
  decarbonization 476, 477, 483, 538, 541, 555, 557
  security 2, 10, 18–25, 477, 538, 540–1
energy use:
  in buildings 4–5
  and economic expansion 68
  variation within nations 131–2
enhanced forward services market (EFSM) 492
enhanced coal bed methane recovery (ECBM) 237
enhanced oil recovery (EOR) 231, 235–6, 237, 240
ENI 286
Environmental Protection Agency (EPA) 237
Environmentally Extended Multi-Region
      Input–Output (EE–MRIO)
Analysis 94, 97
E.ON 297, 298
Eritrea 14
Estonia 270–1
European Commission, NER300 competition 233
European Emissions Trading System (ETS) 67,
      222, 223
European Energy Program for Recovery 233
European Energy Research Alliance (EERA) Ocean
      Energy Joint
Programme 384
European Environment Agency 417
European Photovoltaic Industry Association
      (EPIA) 369
European Union (EU) 11
  aviation and shipping 224
  climate change regime 79
  critical metals 463
  electricity networks 431
  electricity sector 62–3
  emissions 201, 514, 543
  Energy Performance of Buildings Directive
      (EPBD) 172–3, 357
  Energy Performance Certificates 138
  Fuel Quality Directive (FQD) 235, 273
  gas markets 297–300
  hydropower 394
  impacts exported 116
  Industrial Emissions Directive (IED) 254, 300
  Large Combustion Plants Directive (LCPD) 254,
      297–300
  NAMEAs 93
  national targets 86
  nuclear power 310–11

  oil 259
  regulation 493
  renewable energy 40, 299
  Renewable Energy Directive 347, 417
  renewable fuel obligations 201
  roadmap for emissions reduction 311, 314
  SI Ocean project 387
  super grid 443
  transport White Paper 192
  wind power 404–5, 420
Exxon 64
ExxonMobil 27–30, 271, 286, 548

feed-in-tariffs 47, 241, 362, 393, 418–20, 438
Finland 172, 325, 326, 335
fish populations 398
  see also marine
food:
  as energy foodstock 5, 334, 338–40, 343, 347–8
  energy requirements 132
  price spikes 346
  supply 332, 341–2, 344
  taste in 132, 137
foreign direct investment (FDI) 62, 63–4
forestry 344, 347
formal rules 139
fossil fuels 1–2, 244–65, 415
  emissions 543–4
  future resources 5
  importance of 15
  prices 516
  R&D 549
  reliance on 547–8
  subsidies 548
  unconventional 5, 268–87
  see also coal; gas; oil
France 296, 300
  bioenergy 339
  emission inventories 95
  nuclear power 307, 323, 325, 326, 541
  ocean energy 393
  transport 215
Fraser Institute, Policy Potential Index (PPI) 466
Friends of the Earth 347
fuel cycle 527–8, 533

gas:
  availability 19–20
  backup for renewable energy 300
  China 441
  compressed (CNG) 196–7
  cumulative production levels 507–8
  future demand for 29
  globalizing market 5, 291–303
  hydrates 269
  international market 263–4
  marine and terrestrial impacts 531–3
  network decommissioning 437–8

gas: (*cont.*)
    prices 2, 24–5, 298
    reserves 244–5, 547
    storage 432–3
    supply cost curves 247–51
    system 478, 540
    tight 269
    traded 22–3
    transmission networks 428
    wholesale prices 300–2
    see also liquefied natural gas (LNG); natural gas;
        shale gas
gasification 333, 334
Gazprom 295, 298
Georgia 274
geothermal power 16
German Environment Agency 354
German Federal Institute for Geo-Sciences and
    Mineral Resources
(BGR) 250, 252, 256
Germany 139, 172, 274
    bioenergy 335, 338, 339
    CCS 234
    coal 254
    emission inventories 95
    Energiewende 438–9
    gas to coal 297–8
    network infrastructures 430
    nuclear power 307, 310, 315, 320, 484
    renewable energy 299
    solar energy 356, 362
    wind power 414, 419, 420
Ghana 151
Global Alliance for Clean Cookstoves 151
Global Alliance for Productive Biogas 151
Global Bioenergy Partnership 348–9
Global CCS Institute 230, 231
global energy assessment framework (GEAF) 526,
    527–31
Global Energy Assessment (GEA) 1, 125, 167–8, 180,
    252, 511, 544–5, 551
global governance networks 67
globalization 211
    drivers of 53–62
    and energy systems 4, 52–69, 540–1
    uneven 64–5
    see also trade
greenhouse gas emissions *see* CO_2; carbon; emissions
Greenland 466
Greenpeace 347
gross calorific value 11
Gulf of Mexico 257

heat storage 433
heat system 356–7, 478
heating 559–60
heavy fuel oil (HFO) 220
high-carbon future, policies for 548–9

HVDC (High Voltage Direct Current) 358
hydraulic fracturing 282–5
hydroelectricity *see* hydropower
hydrogen 540, 560–1
    from solar power 356
    industrial feedstock 435
    infrastructure 434–5
    storage 433–4
    see also transport
hydropower 5–6, 16, 394–400, 431
    controversial 395
    costs 398
    in developing countries 397–8
    environmental issues 398–400
    future research 400
    global capacity 394–6
    institutional barriers 398
    large-scale 395
    small 394, 395–6, 397, 397–8, 443
    technology and operation 396–7
    turbine design 396
    see also ocean energy

IAEA 308
Iceland 14
Imperial Oil 274
India 14
    airports 219
    car ownership 191–2
    coal 252, 254
    cook stoves 137
    emissions 543
    hydropower 397
    nuclear power 325
    renewable energy 40
    solar energy 363, 369
    wind power 405, 420
indirect emissions 4, 61, 79, 92–109
    by sector 101–2
    determinants of 102–5
    drivers of 101–5
    see also emissions
indium 460, 463
Indonesia 254, 262, 274, 339
information:
    barriers 171
    energy efficiency 172–3
information technology (IT) 35, 37
infrastructure *see also* roads
innovation:
    definition 34
    and energy systems 34–48
    incremental and radical 35
    inputs 39–43
    models of 35
    oil sands 279–81
    outcomes 44–6
    policy driven 365

systems 35–9
trends 39–46
see also R&D; technology
integrated solar combined–cycle (ISCC) 359
interdependency 20–4
Intergovernmental Panel on Climate Change
        (IPCC) 1, 26–7, 82, 93, 99, 210, 332
    climate change mitigation study 549–50, 551
    Fifth Assessment Report 525
    gas 265
    Kyoto Protocol 96, 108, 544
    renewable energy report 354
    reports 167
    scenarios 507
    Special Report on Renewables 527
    SRES scenarios 214
internal combustion engine (ICE) 195–7, 199
International Air Transport Association
        (IATA) 223
International Civil Aviation Organization (ICAO) 78,
        209, 223
International Electrotechnical Commission Technical
        Committee 379
International Energy Agency (IEA) 3, 5, 11, 20–1,
        39, 210, 217, 229, 248–9, 332,
        378, 551
    Are We Entering a Golden Age for Gas? 291
    energy access definition 542
    Energy Technology Systems Analysis Programme
        (ETSAP) 500
    Mobility Model (MoMo) 189
    model 27–30
    New Policies 255, 262
    scenarios 407, 548
    World Energy Outlook 254–5
International Maritime Organisation (IMO) 209
    Ship Energy Efficiency Management Plan
        (SEEMP) 224
international rate of return 279
interpretivism 164
Iran 256, 260, 262, 543
Iraq 256
Ireland 380, 381
Israel 274
Italy 274, 299
    nuclear power 310, 484
    solar energy 356, 363, 369

Japan:
    coal 253
    emissions 543
    Fukushima 307, 541
    impacts exported 116
    LNG 295–6, 301
    national targets 85
    oil 259
    renewable energy 40
    solar energy 362

transport 215
Jordan 272–3

Kazakhstan 274
Kenya:
    electricity supply 442–3
    network infrastructures 444, 445
    solar power 153
kerogen 269–70
knowledge transfer 104
Korea 215

labour productivity 54
land use 192, 193–4, 332, 340, 399
large combined heat and power (CHP)
        facilities 335–7
Latvia 172–3
learning by doing 35, 45–6
learning curves 45, 409
learning rates 46, 409, 413, 554–6
lifecycle assessment (LCA) 117
Light Emitting Diodes (LEDs) 154
Lighting Africa 151
liquefied natural gas (LNG) 46, 427
    facilities 24–5, 234, 263
    from shale gas 261, 281, 285, 292–5
    Fukushima effect 295–6
    future pricing 300–2
    global market 293–5
    market 46, 262–4
    see also natural gas
liquid petroleum gas (LPG) 542
lithium 468
    cumulative availability curve 464–5
    for electric vehicles (EVs) 459–60
lock–in 37, 39, 83, 139, 504
low–carbon policies 552–3, 561–4
low–carbon transition 65–8, 555–61
    composition effect 66–7
    scale effects 66
    technique effect 66–7
    voluntary 87–8
    see also transitions
Luxembourg 14
Lybia 299

McKelvey box 461
Madagascar 274
MAGICC climate model 502, 507
magnets 459–60
Malaysia 262, 312, 339
Marathon 286
marginal abatement cost curves (MACCs) 165–8
marginal cost pricing 485, 490
marine:
    ecosystem services 529–33
    energy 6
    environment 413

marine: (*cont.*)
  transport 78–9
  see also shipping
Maritime Organisation (IMO) 78
market barriers 167–9
  informational and transaction 171
  and positivism 169–75
market failure 36–7, 204
merit order effects 415–16
metals *see* critical metals
methane 2, 291, 560
  coal bed 237, 249, 260, 269, 286
  from biomass 333–4
  see also natural gas
Mexico 40, 261, 293
micro-banking 153, 155, 156
Millennium Ecosystem Assessment (MEA) 113, 525,
    528, 530
missing money 485, 486, 488, 490
mobile phones 155–6
mobility 190–3
Morocco 154
multi-regional input–output analysis (MRIOA) 115

Namibia 439
National Accounting Matrices including
    Environmental Accounts
(NAMEAs) 93
National Emission Inventories 93
National Renewable Energy Laboratory
    (NREL) 354
natural capital 117, 122
  dividends 112
  of the world 113–14
natural gas:
  compressed 196–7
  market share 15
  production and consumption 262–4
  reserves 259–62
  resources 248
  technically recoverable reserves 260–1
  see also gas; liquefied natural gas (LNG); methane
neodymium 459–60
Netherlands 139, 172, 298, 526
  coal 253, 297
  emission inventories 95
  energy use 132
  National Research Programme 106
  solar PV 518
network infrastructures 6, 35, 426–47
  capacity factor 429
  definition 426
  distribution 428
  economics of 445–6
  high-income countries 437–9
  low-income countries 441–3
  middle-income countries 439–41
  ownership and investment 444–5

  political drivers 443–4
  transmission networks 428, 429
New Zealand 396
Nigeria 262, 274
Nordpool 480
normative rules 139
Norway 526
  CCS 234
  gas 262
  hydropower 394, 396
  wind power 411
NOx 224
Nuclear Non–Proliferation Treaty 312
nuclear power 5, 15–16, 46, 80, 306–26, 483–5
  capacity 308–10
  costs of 324–5
  decommissioning 528
  early development 306–7
  and ecosystem services 529–33
  global supply chains 67
  learning rates 555
  lifecycle stages 528, 529–31
  low–carbon 314–16
  new capacity installation 518
  post–Fukushima trends 308–14
  public opinion 323
  safety 306–7, 310, 316, 319–20
  secure supply 314
  skills gap 321
  subsidies 325
  waste and decommissioning 311, 322, 326
nuclear reactor designs 313, 316–19
  fast breeders 317
  fusion 318
  small modular reactors (SMRs) 318–19
  thorium reactors 318

ocean currents 377
ocean energy 377–93
  cost and performance 387–91
  design consensus 384
  development pathways 384
  feed–in–tariffs 393
  installation costs 390
  policy to support 392–3
  potential 378
  technological development 383–93
  test centres 380
  thermal 378
  see also tides; waves
Ocean Energy Systems Implementing Agreement
    (OES) 378–80
offsetting 108
Offshore Windfarm Egmond aan Zee (OWEZ) 413
oil:
  availability 19
  crisis 9
  cumulative production levels 507–8

future demand for 29
global balance 258
market share 15
peak 19, 256
price 2, 24
production and consumption 257–9
products 18
reserves 244–5, 255–6, 547
storage 432–3
supply cost curves 247–51
tight 268–9
traded 21–2
unconventional 10, 19, 30–1, 239, 249–50, 255–7,
    259, 268, 270, 286–7
oil sands 273–81
environmental footprint 281
global resources 274–6
production 276–8
technological innovation 279–81
oil shales 269–73
ONGC–Videsh 64
OPEC 20, 24, 27–30, 40, 44, 54, 79, 257, 260
Organisation for Economic Co-operation and
    Development (OECD) 11–12, 16–17
'out of the money' 482, 486

Paraguay 396, 431
partial substitution method 11
Patent Co-operation Treaty (PCT) 43–4
patents 43–4, 409
Peak Oil 19, 256
Petrobras 64
PetroChina 64
Petronas 64
photovoltaics (PV) 3, 5–6, 354, 542
cost competitiveness 364–9, 370
demand for 362–4
efficiency 458
feed-in-tariff schemes 362
future of 369–71
indium 460
new capacity installation 518–20
off-grid and grid-connected 361–2, 369–70
supply 356, 364
technologies 360–2, 370
tellurium 460
thin–film 456–9, 463, 467
see also solar power
Physical, Technical, and Economic Model
    (PTEM) 134–5, 164–5, 168–9, 170, 172, 175,
    178, 180–1
pipelines 259, 262, 293, 428, 443
Poland 254, 286
policies 6, 34, 365
on consumption–based carbon
    emissions 97–8
on decarbonization 105–8
high–carbon future 548–9

low–carbon 552–3, 561–4
ocean energy 392–3
pillars of 563–4
trade 107
travel 192–5
trilemma 9, 427, 538, 540–1, 565
wind power 417–19
pollution 1, 436, 533
air 299–300, 302
fossil fuel production 271
local 1, 10, 545–6, 550, 560
marine 37
noise 413
transport 192–5, 197, 199, 202, 530
ports 219–20, 427–8
Portugal 172–3, 278, 296
positivism:
challenges to 175–81
definition 163, 164
and market barriers 169–75
practice theory 136–8
productivity 54, 61
Property Assessed Clean Energy (PACE) 170
provisioning services 528
public opinion 323
public spending 36–7, 47
pumped–storage schemes 396–7, 398
pyrolysis 333

Qatar 260, 262, 263, 299, 301
quality standards 156

railways 195, 215, 427–8
rare earth elements (REE) 67, 460, 466, 468
realist synthesis approach 180
rebound effect 141–2, 203, 225
recycling 99, 344, 526
in CHP 336
metals 461, 466–8, 470, 472–3
regulating services 528–9
regulation 138
electricity markets 477, 478, 493–4
network infrastructure 444, 446
unintended consequences 173–5
Regulatory Assistance Project (RAP) 487, 490
renewable electricity:
global supply chains 67
variable power 483, 488
renewable energy 483–5
intermittent 430
investment 87
spending on 40
subsidized 299
renewable energy technologies (RET)
centralized versus decentralized 151, 154
and free energy 149–50
renewables 16
future demand for 29

renewables (*cont.*)
    learning rates 555
    lifecycles 527–8
    zero–carbon 5–6
Research, Development, and Deployment (RD&D)
    private sector 41–2, 47
    spending on 39–40, 47
research and development (R&D) 30–1, 39–43,
        418–20, 549
    see also innovation; technology
road transport 5, 554, 559
roads 192, 193, 427–8
    deaths 201–2
    user charging 194–5
Romania 274, 300
Rosatom 311
Rosenergoatom 312
Rosneft 64
rural energy service companies (RESCOs) 153–4
Russia 429, 526
    Chernobyl 307, 323, 326
    CO_2 emitter 76
    coal 252
    emissions 543
    fossil fuels 246
    gas 22, 260–1, 262, 263
    national targets 85
    nuclear power 312, 325
    oil 248, 257, 274
    pipelines 443
Russian Federation Classification (RFC)
        system 246
RWE 298

salinity gradient 378
Saudi Arabia 256, 257, 260, 358
security *see* energy security
shale gas 19–20, 22, 25, 56, 56–7, 260–2
    and European gas markets 297–300
    origins and technology 281–6
    revolution 292–3
    wholesale prices 292, 294–5
    see also gas
shale oil 268–9
    see also oil
Shell 27–30, 64, 272–3
shipping 5, 209–25, 427–8, 554
    demand for 210–12
    energy intensity 217–18
    heavy fuel oil (HFO) 220
    size 217, 219
    speed 217, 222
    technology 221, 222
silicon shortage 366
Single European Sky programme 219
social justice 117
social lifecycle assessment (S–LCA) 117
social practice theory 136–8, 215

Society of Petroleum Engineers / Petroleum Resources
        Management
System (SPE/PRMS) 246
socio–economic impacts 117–18, 120
socio–technical systems 38–9
socio–technical theory 209, 218, 223
socio–technical transitions 138–9, 209
solar cooling 357
solar heating:
    active 357
    passive 356–7
solar power 16, 354–71
    cost of 359–60
    CSP 354, 356, 358–60
    hybridization 359
    intermittent 430
    microgeneration 442
    pico–PV 154–5
    and super grids 431, 434
    technical potential 354–5
    see also photovoltaics (PV)
solar technologies 356–62
Sonatrach 233
South Africa 14, 153, 262, 285, 466
South Korea 295–6
Spain 296
    Biscay Marine Energy Platform (BiMEP) 380
    solar energy 356, 358, 364
    transport 215
    variable power 491
    wind power 419
SSE 297
Statoil 64, 233, 234, 295, 298
Stern Review 563, 564
story–telling approach 180
Suncor Energy 274
Sweden:
    bioenergy 335, 336
    emission inventories 95
    nuclear power 326
    rebound effect 203
Switzerland 274, 310, 396
Syria 274
system failures 37, 139
    see also lock–in
System of National Accounts (SNA) 93
systems dynamics modelling 464

Taiwan 296, 312, 543
Tajikistan 274
Talisman Energy 286
Tanzania 466
taxes 107, 561
    Border Tax Adjustment (BTA) 107
    cars 194, 196, 199, 200, 202
    fuel 194, 249, 273
    property 170–1
    see also carbon taxes

techno–economic analysis 35–6, 199–200
techno–economic modelling theory 209, 215
techno–economic and physical technical economic
     models (PTEM) 222
techno–economic potential, obstacles to 167–9
Technological Innovation Systems (TIS)
     framework 38
technology 4
  and climate change regime 79–80
  convergence 64
  downstream 527–8
  and ecosystem services 531–3
  emissions reduction 517–20, 521
  investment in 44–5
  lifecycle emissions 99–101
  ocean energy 383–93
  oil sands 279–81
  policy 563
  shipping and aircraft 221, 222–4
  transfer 104
  vehicle 189–90
  wind power 408–17
  see also innovation; research and development
     (R&D)
tellurium 460, 463, 467
Thailand 312, 339, 543
thermal storage 359
TIAM–UCL global integrated assessment
     model 499–521, 550
tides:
  currents 377, 380–3
  energy converters 386
  range 377, 378, 381
  see also ocean energy; waves
Tonga 274
Total 64, 271, 295
tradable certificate markets 418
trade 211
  comparative advantage 17, 61
  and economic integration 57
  emissions transfer through 61
  policy 107
  see also globalization; indirect emissions
transitions 486
  coal to gas 509–11, 521
  coal to oil 58–9
  gas to coal 297–8
  socio–technical 138–9, 209
  sustainability 38
  to low–carbon 65–8, 555–61
  transport 199–200, 223–4
transnational organizations 62
transport 5, 18, 559
  air 60
  and air quality 202
  carbon intensity 217
  China 441
  declining costs 59, 61

economic growth, welfare and equity 202–3
and energy 53, 57–80
from coal to oil 58–9
infrastructure 427–8
land passenger 189–204
liquid fuels 67
low–carbon 438
marine 59
market failures 204
modal choice 194–5
public 195
road 5, 554, 559
safety 201–2
technologies 199–200
transitions 199–200
and urbanization 545
see also aviation; cars; hydrogen; railways; roads;
     shipping ;
'valley of death' 37, 47
vehicles
travel:
  and GDP 212–14
  intensity 190, 194
  motorized 190–2
  sustainable 189–204
  time budget 190–2
  trends 212–15
  vacation 212
  visiting friends and relatives (VFR) 212
Trinidad and Tobago 274
Tunisia 358
Turkey 420
Turkmenistan 260

UKERC Global and Local Impacts on Ecosystem
     Services project 46, 239, 535
Ukraine 293, 299, 302, 420, 540
United Arab Emirates 232
United Kingdom 11, 172, 298, 300
  aviation 223
  Balancing Mechanism (BM) 480, 481
  behaviour change efforts 180
  BEVs 198
  bioenergy 347
  bioethanol 118–22
  Carbon Abatement Technology Strategy 237
  carbon emission reduction 67
  Carbon Trust 378
  CCS 233, 234, 237–8, 239–40, 241
  Climate Change Act 98, 215
  coal 253
  Commission for Employment and Skills 413
  Committee on Climate Change (CCC) 100, 101
  decent living 141
  Department of Energy and Climate Change 286
  electricity demand 526–7
  Electricity Market Reform (EMR) 238, 490
  electricity networks 432

United Kingdom (*cont.*)
electricity sector 62–3, 67
emission inventories 95–6, 97–8
energy intensity of exports 105
Energy Research Centre 229
Energy Technology Institute 416
energy transition 128–30
Energy Use Energy Demand Centres 139
Engineering and Physical Sciences Research Council
    (EPSRC) Grand
    Challenges 393
gas storage fire 541
gas to coal 297
Green Deal 170–2
indirect emission 101–5
LNG 293–4
MEAD 393
MRCF 393
National Ecosystem Assessment 113
National Grid 480, 493
network infrastructures 445
nuclear power 314, 484
ocean energy 380, 381–3
Offshore Renewable Energy Catapult 419
Offshore Wind Cost Reduction Task Force 413
railways 427
regulation 494
renewable energy 299
Renewable Obligation Certificate (ROC) 393
roads 193
shale gas 283, 286
solar energy 363
Technology Strategy Board 393
Waste and Resources Action Programme
    (WRAP) 106
wind power 411–12, 420
United Nations 11
Conference of the Parties (COP) 74, 75, 84
Copenhagen Accords and Cancun Agreements 74,
    81, 83, 84, 85
Energy Access for All initiative 151
Environmental Performance Index (EPI) index 468
Food and Agriculture Organization (FAO) 340
Framework Classification (UNFC) for Fossil Energy
    and Mineral
Reserves and Resources 246–7
Human Development Index (HDI) 140–1, 466
Sustainable Energy for All 60
United Nations Framework Convention on Climate
    Change (UNFCCC) 4, 26–7, 74–80, 93, 96
Annex 1 and Non–Annex 1 parties 74–7
Cancun Agreements/Conference 27–8, 74, 83–5,
    86, 548
Conference of the Parties (CoP) 26, 543–4
Copenhagen Accord 499, 504–7, 514, 516, 520
Kyoto Protocol 74–80, 84–5, 86
United States 40
45Q 236–7

American Recovery and Investment Act 236
aviation 212, 225
bioenergy 334–5
biofuel 220
California Low Carbon Fuel Standard (LCFS) 235
Canada 380
capital cost financing 170–1
CCS 231–2, 234, 236–7, 239–40
climate change regime 79
coal 252, 253–4, 297
Congressional Budget Office 466
Corporate Average Fuel Economy (CAFE) 201
electricity networks 432
emissions 543
Energy Information Administration (EIA) 27–31,
    257, 260–1, 542
gas 2, 263, 264
Geological Survey (USGS) 470
Home Energy Rating System 175
hydropower 397
impacts exported 116
Kyoto Protocol 75
national targets 85
Next Generation Air Transportation System 219
nuclear power 46, 311
oil 257, 270, 274, 281
political leadership 87
Production Tax Credit 418
rail networks 58
RD&D 42
regulation 493
roads 193
shale gas 19–20, 22, 25, 56–7, 261, 282–5, 292–3
solar energy 358, 363
Three Mile Island 306
wind power 404–5, 414, 420
uranium supplies 320–1
urban design 545
urbanization 544–6
Uzbekistan 274

vehicles
battery 197–8, 459–60
drivetrains 196–9
fuels 196–9
hybrid electric 197
hydrogen fuel cell 198–9
low–carbon drivers 200–1
low–carbon targets 200–1
technologies 189–90
see also cars; transport
Venezuela 256, 274, 396
Vietnam 151, 312

walking 195, 202
want–need spiral 141
water quality 398
waves 377, 378, 380–3

energy converters 385
resources 380, 382–3
see also ocean energy; tides
welfare and transport 202–3
'wicked problem' 538, 565
wind power 3, 5–6, 16, 404–21, 441
advantages of 408
costs 409–10, 412–13, 415–16
deployment policies 417–18
dominant design 408–9
feed–in tariffs 418
and fossil fuel plants 415
frequency response 414–15
and grid stability 414–15
historical development 404–5
industrial development policies 420
new and future innovation 410–13, 416–17
offshore 405, 406–7, 410–13
patents 409
potential 405–8
public opinion 420–1
R&D support 418–20

skills development 413
subsidies 417–18
system inertia 414–15
targets 417
technology 408–17
tradable certificate markets 418
wind turbines:
dominant design 419
and the environment 413
floating 411
high–altitude 416
increased size 416–17
life–expectancy 411
prices 409
vertical–axis 416
wood pellets 337, 347, 348
World Bank 500
World Governance Indicators (WGI) 466
World Energy Council (WEC) 250, 251, 255
World Energy Outlook 308
World Nuclear Association 308
World Trade Organization (WTO) 348, 420